Benedetto Croce

THEORY OF AESTHETIC
HISTORY OF AESTHETIC

〔意〕克罗齐 著

美学理论/美学史

赵美娟 导读 注释

上海译文出版社

World Academic Classics
世界学术经典·英文版

总策划／总主编
庄智象　林骧华

上海时代教育出版研究中心
Shanghai Epoch Education Publishing Research Center
研发

世界学术经典（英文版）
总　序

书之成为经典，乃人类在不同时代的思想、智慧与学术的结晶，优秀文化之积淀，具有不随时代变易的永恒价值。有道是读书须读经典，这是智者的共识。

对于中外经典著作中的思想表述，仅读外文书的中译本或文言著作的白话释文是不够的，尤其是当误译、误释发生的时候，读者容易被误导，或望文生义，或以讹传讹，使原有的文化差异变成更深的文化隔阂。因此，在“世界学术经典（英文版）”的选目中，大部分作品为英文原著；原作是其他语种的经典，则选用相对可靠的英文译本；至于中国古代经典，则采用汉英对照的方式呈现，旨在向西方阐释中国的思想和文化。其中，精选的中国经典是整个系列的重要组成部分。有了这一部分的经典，才真正体现出“世界性”。

以原典和英文方式出版，是为了使读者通过研读，准确理解以英文表达的思想、理论和方法，力求避免舛误，进而通过批判和接受，化为智慧力量。这有利于思想的传播，裨益于新思想的产生，同时亦可提高英语修养。

经典名著的重要性是不言而喻的，但是以下几点意义值得一再重申。

（一）学术经典提供思想源泉

两千六百年来的世界学术经典凝聚了人类思想的精华，世世代代的优秀思想家以他们独特的见识和智慧，留给后人取之不竭、用之不尽的思想源泉。从老子、孔子、柏拉图、亚里士多德以降，天才辈出，思想闪光，精彩纷呈。思想界的大师、名家们在人类思想史传统链条上的每一个环节，都启发后人开拓新的思想领域，探究生命的本质，直抵人性的深层。随着人类思想的不断成熟和完善，各个学科领域的理论从本体论、认识论、方法论、实践论、价值论等维度不断深化。后人继承前人的思想，借经典的滋养保持思想活力，丰富和发展前人的观点，使之形成一波又一波的思想洪流，从而改变人的思想和世界观，改变人类社会的进程。历史已经证明：人类社会的进步，思想的力量大于一切。

（二）学术经典传承精神力量

经典名著中蕴含的人类精神，传承的人类守望的共同价值原则和社会理想，在每一个具体领域里都有诸多丰富的表述，它们从整体上构成了推动人类进步的精神力量。研习和传承人类两千六百多年来的优秀思想，并将它化作求新求变的灵感，是人类文明的要义所在。仅有技术进步，还不足以表明人类的文明程度。若无优秀的思想底蕴，人类存在的意义将大打折扣。

中国思想传统中的基本理念和西方思想传统中的基本理念分别形成了东西方两大具有普遍价值的道德观念和价值系统。值得注意的是：（1）这两大道德系统应该是一个互补、互鉴的整体，两者都不可偏废，因为人类的思想是个多元整体。任何一个民族，缺少其中之一，在精神上都可能是不完整的；(2) 这些基本理念都不是抽象概念，它们都具有很强的实践意义，并且必须由实践来考察，否则就很难体现其价值。

精神传承必然是一种自觉的过程，它靠习得，不靠遗传，因此我们需要研读经典。

（三）学术经典构成文化积淀

“文化”包含三大部分：(1) 思想与精神现象；(2) 制度与习俗；(3) 有形的事物。学术经典是对思想与精神现象的归纳和提炼，对制度与习俗的探究和设计，对有形事物的形而上思考和描述。

每一个学科领域的经典著作中都会提出一些根本性的问题，这些问题直面人的困惑，思考人类社会的疑难，在新思想和新知识中展现人类的智慧。当这些思想成果积淀下来，就构成人类文化的主要组成部分。文化不只是制度或器物的外在形式，更重要的是凝结在其背后的精神与思想。

每一个学科的学术本身都要面对一些形而上的（超越性的、纯理论性的）文化问题。在很多人看来，理性的思考和理论的表述都是很枯燥的，但是热爱真理并且对思想情有独钟的人会从学术经典的理论中发现无限生动的天地，从而产生获得真理的快乐，这才是我们追求的真正文化。

大量阅读经典名著是一种学习、积累文化的根本方法，深度阅读和深刻记忆能使文化积淀在人的身上，并且代代相传。假如这一过程中断了，人世间只剩下花天酒地、歌舞升平，文化也就湮灭了。

（四）学术经典推动社会进步

毫无疑问，凡属学术经典，都必须含有新的学术成果——新思想、新理论、新方法，或者新探索。这样的原创性学术成果越多，人类的思想就越深邃，视野越开阔，理论更全面、完美，方法更先进、有效，社会的进步才能获得新的动力和保障。

人文主义推翻神学，理性主义旨在纠正人的偏激，启蒙精神主张打破思想束缚，多元主义反抗绝对理念。各种新思想层出不穷，带来了学术的进步，启发并推动了更大的社会变革。这些原创思想在历史长河中经过漫长的时间考验，成为经典，在任何一种文明中都是社会进步和发展的动力。

当我们研读完一部学术经典，分析和归纳其原创的思想观点时，可以很清晰地理解和感悟它在同时代的环境里对于社会的变革和进步有着何种意义，以及它对当下有哪些启迪。

相比技术的发展，思想并不浮显在社会的表层，它呈现在书本的字里行间，渗透于人的心智，在人的灵魂中闪光。每当社会需要时，它能让我们看到无形的巨大力量。

温故而知新。今日世界纷乱依旧，但时代已不再朦胧。人类思想史上的各种主张，在实践中都已呈现清晰的面貌。当我们重新梳理各种思想和理论时，自然不会再返回到“全盘接受”或者“全盘否定”的幼稚阶段。二十一世纪人类正确的世界观、人生观、价值观需要优秀思想传统的支撑，并通过批判继承，不断推陈出新，滋衍出磅礴之推力。

我们所选的这些学术经典，成书于不同的时代，代表了不同的思想与理论主张。有些著作带有时代烙印，有其局限性或片面性；有些观点不一定正确，但从另一个方面显示出人类思想的丰富性和复杂性。各门学科建立、各种主张提出之后，都曾经在历代思想的实验场上经受碰撞和检验，被接受或者被批判。我们的学者需要研读这些书，而青年学生们的思想成长更需要读这些书。当然，批评与分析是最有效和最有益的阅读方法。

有鉴于此，我们希望“世界学术经典（英文版）”能够真正做到“开卷有益”，使我们自己在潜移默化中都成长为有思想、有理想、有品位的人。

上海时代教育出版研究中心

2018 年 10 月

目录

导　读

赵美娟

《美学理论 / 美学史》最初以《美学》为书名发表于1902年(以下简称《美学》)，作者是意大利著名唯心主义哲学家、史学家贝奈德托·克罗齐（Benedetto Croce, 1866–1952）。克罗齐在书中阐释了他的美学思想，讨论了美学在整个哲学中的地位，审美活动与其他心灵活动的区别与关系，梳理了不同时期的美学观念、不同学派的美学学说，并分析了美学的近代发展趋势。

一、作者生平

克罗齐1866年出生于意大利中部阿布鲁佐地区贝斯卡赛洛利一个名门望族，少时在那不勒斯的一所天主教公学接受教育。受母亲的影响，他从小热爱读书，尤其爱好文学、历史，但对政治并无兴趣。高中时，克罗齐因宗教信仰动摇而陷入痛苦中，最终他放弃了天主教信仰，成了无神论者。他认为宗教只是一种历史的习俗，人们是通过宗教释放创造性的力量。

1883年夏天克罗齐和家人在卡萨米乔拉度假时，一场地震夺去了他父母、姐姐的生命，被埋在废墟下的他侥幸得救。这场灾难不仅伤害了他的身体，还给他造成了巨大的心理创伤。他自称生活变成了“噩梦”，自此他一直有种孤独感。此后，他和弟弟投奔罗马的堂叔西尔维奥·斯帕芬达（Silvio Spaventa）。西尔维奥是意大利著名的政治家。在西尔维奥家他得到了无微不至的关怀，但家中浓厚的政治氛围让他迷茫，身上的伤痛也使他消沉，甚至萌生自杀的念头。他称那段时间是“最痛苦、灰黯的岁月”（克罗齐，1997，p.43）。那年秋天他在罗马大学修读法律，但对法学毫无兴趣，大部分时间在图书馆自学，直至他开始听安东尼奥·拉布里奥拉（Antonio Labriola）的道德哲学课。那些课程促使他“以理性形式恢复对生活的信心、对生活目的与责任的信仰”（克罗齐，1997，p.43）。他尝试着阅读哲学书籍，对幸福与责任、纯洁与卑污等对立的概念进行哲学探讨。1886 年，克罗齐辍学返回那不勒斯，开始自主学习和独立研究。此后几年，他埋头学习知识并撰写出版了不少叙事作品，例如《1799 年那不勒斯》《那不勒斯的历史与传说》《历史奇闻》等。

随着这些成果的陆续出版，克罗齐认为他人生中的一个时期结束了，他需要脱离城市历史这个狭小的领域，写出更深刻的东西。他大量阅读关于哲学与历史方法论的著作，尝试将历史问题与艺术问题结合起来。1893 年他发表了第一篇学术论文《艺术浓缩了一般意义上的历史》，提出艺术也是历史的论断。1894 年又写了关于文学批评的小册子，这些作品引起了广泛关注和争论。1895 年，他接触到了马克思主义，但经历了最初的兴奋与激情后，他放弃了马克思主义的观点，因为他认为自己“真正的本色是学者和思想家”（克罗齐，1997，p.46）。

1902 年，克罗齐发表了《美学》，其中上篇是“美学理论”，下篇是“美学史”。之后，他又出版了《作为纯粹概念科学的逻辑学》(1904)、《实践活动的哲学》(1908)、《历史学的理论和实际》(1915)。这四部作品被统称为《精神哲学》，标志着他以此建立起了自己独立的哲学体系。

1903 年克罗齐与人合作创办了双月刊《批判》(La Critica)。《批判》是一本历史、文学、哲学杂志，目标是“阐明最近 50 年，即从意大利新国家、新意大利建立以来的意大利精神生活”(克罗齐，1997，p.48)。克罗齐说：“创办《批判》杂志标志着我的生活中一个时代的开端，即成熟期开始，或自身同实在统一的时代的开始……对《批判》的领导与合作是我为意大利文化更为直接的服务。”(克罗齐，1997，p.49) 从创刊到 1944 年《批判》停刊，他在杂志上发表了涉及哲学、历史、文学等领域的许多具有重要价值的文章，形成了自己的新黑格尔哲学体系，成为意大利思想界的代表人物，对意大利知识界和文化界起着先导作用。1910 年克罗齐当选终身参议员。

第一次世界大战爆发后，克罗齐发表了许多有影响的政治见解，后结集成《战争文集》(1919)。1920 年 6 月至 1921 年 6 月，克罗齐出任意大利教育部长，发起了意大利现代史上著名的教育改革运动。他在《拾零集》第二卷中记载了那段时间的活动。

战后，世界经历严重的经济危机，以极权主义和极端民族主义为特征的法西斯主义兴起。起初，克罗齐未加警惕，认为法西斯主义只是战后的暂时现象，但随着法西斯主义的盛行，克罗齐便公开站出来发出反对之声。1925 年他发表了著名的《反法西斯知识分子宣言》，反对墨索里尼的恐怖政策，成为意大利反对法西斯主义的象征。克罗齐也因此遭到了墨索里尼政府的威胁、监视，直至 1943 年法西斯垮塌，方恢复充分自由。在此期间，他闭门著书，写下了大量有关现当代历史的著作。

1943 年克罗齐出任意大利独立党主席，1946 年克罗齐在那不勒斯创建了意大利历史研究所，1947 年他辞去自由党主席职务，完全回归到学术研究中。

克罗齐著作丰硕，作品总共有 80 余部。

二、《美学》的写作背景

高中时期，克罗齐便尝试着学习德国美学，他阅读了弗朗切斯科·德·桑克蒂斯（Francesco De Sanctis）的著作。桑克蒂斯是意大利著名文学批评家，所著的《意大利文学史》至今仍被认为是研究意大利文学的经典著作之一。桑克蒂斯认为文学是社会生活的反映，文学史实质上是文明史；他对意大利文学史上的各个重要文学现象、作家及其重要作品都作了缜密的研究和精辟的分析。受桑克蒂斯的影响，从那时起，克罗齐便断断续续地进行艺术思考。随着哲学在克罗齐的研究中所占比重越来越大，他将这种思考与精神的其他问题联系起来，关于美学的种种构想不断清晰，1898 年秋克罗齐决定撰写一部美学著作。次年夏天他正式着手写作。

动笔之后，他发现要将零散的概念、思想汇集起来困难重重，必须不断厘清原理中暧昧不清的地方，梳理相互的逻辑关系，排除可能的冲突。经过持续的思考推演，经历了 6、7 个月的艰辛笔耕后，他最终将关于美学的基本论点撰写成一篇简明的专题论文——《作为表现的科学和一般语言学的美学的理论》。他说这篇论文“既枯燥又晦涩，但我一旦完成这篇论文，我不仅完全辨明精神问题的方向，甚至以一种清醒、卓越的智慧提出几乎所有主要哲学问题。古典哲学家们都曾艰难地探讨过这些问题；智慧靠单纯阅读他们的著作不能获取，只有在生活的激励下，在自身重复他们的思想脉络时才能获得。”（克罗齐，1997，p.47）

完成了著作的理论部分后，克罗齐稍事休息调整。1900 年 11 月正当他准备续写历史部分时，那不勒斯市政府因一桩诉讼丑闻受到调查并被解散，行政工作由一位特派员负责，克罗齐受邀协助特派员负责市立中小学的行政管理工作，写作不得已被搁置。及至 6 个月后他解脱了那个职务后才得以重拾笔墨，专心写作。1901 年 9 月，《美学》的理论部分和美学史部分全部完稿，次年 4 月出版。

三、作者的主要美学观点

克罗齐的美学思想属于主观唯心主义范畴，是其精神哲学的一个组成部分。克罗齐提出美学就是关于直觉的科学，直觉是其美学思想的核心和出发点。在他看来，知识有两种形式：直觉和逻辑。直觉是由想象而来的，是关于个体的。而逻辑是由理智得来，是关于共相的。直觉产生的是意象，而逻辑产生的是概念。

直觉即表现，艺术即直觉。克罗齐认为，人人都有直觉，直觉来源于情感，情感就是物质、感受、被动或自然，但情感未经心灵综合，即未经直觉时，是无形式的，经过直觉才得到形式，转化为意象。艺术是一种想象活动，它的对象是个体，它所产生的是形象化了的意象。当主体以直觉的形式认识对象时，就等于“表现”了艺术，我们便完成了艺术作品的创造活动。艺术家与一般人相比只是在直觉的量上不同，而非质的区别。

克罗齐认为直觉来源于情感，又表现情感，直觉的成功表现即是美。不成功的表现则是丑，美是一种绝对价值，没有优劣之分，而丑是相对的，有程度之分。

克罗齐指出鉴赏和创造是统一的，创造需要鉴赏力，天才的艺术家都有很高的鉴赏力。鉴赏是用直觉再造艺术家所创造的抒情意象，从而得到和艺术家大致相同的体验和感受，所以鉴赏也是不同程度的创造。

克罗齐还认为语言即艺术。语言在本质上和艺术相同，都是心灵活动的创造，两者同为表现。即使在使用旧词时，我们也会根据客观情境、主观情感的变化赋予它们新的意义。换言之，无论是用新词还是旧词，我们都在不断地凭直觉创造。既然语言与艺术是统一的，语言学与美学也就是统一的，因而《美学》的副标题是“作为表现的科学和一般语言学的美学的理论”。

四、《美学》的内容结构

《美学》分为上下两篇：美学理论、美学史，分别用概念和历史事实来建构克罗齐心中的美学体系。

（一）《美学理论》的内容结构

上篇“美学理论”共 18 章，大致可分为四个部分。

第一至第三章通过探讨直觉与表现、直觉与艺术、艺术与哲学，分析了直觉知识的性质，提出直觉即表现、艺术即直觉的论断，并论证了艺术与哲学、科学之间的关系，指出艺术是一种直觉，而哲学则由概念构成。

第四章至第九章，克罗齐评析了各种美学理论，对其中的错误以及史学与逻辑学中的错误进行批驳，探讨了审美活动与其他心灵的活动的关系。他认为人类心灵的活动分为认识和实践，而认识活动又分为直觉和理性。实践的活动又分为经济的和伦理的。除了直觉、理性、经济、伦理这四种形式外，不存在第五种心理形式，其他的精神活动都是由四种形式复合而成或衍生而来。他进而阐释了艺术的特征，指出艺术不能分类，用各种各样的词语和区分来界定表现是毫无意义的。

在第十章至第十七章中，克罗齐研究了审美感觉以及美与丑的分别，提出美与丑是对立的反价值，美是成功的表现，丑是不成功的表现。克罗齐批评了审美的快乐主义观点，论证移情说属心理学概念，证明物理事实非美学事实，两者不能混为一谈。他指出艺术技巧的意义在于再造。“物理的东西”并不能直接产生艺术的再造，而必须经过回忆那些刺激物原有活动的情况，才能再造。历史学的功能在于建立想象与过去作品的交会，这才是鉴赏力的依据。在审美判断中，所有的印记都带有历史条件。所以，对艺术品的鉴赏不能脱离历史条件。他探讨了文学与艺术的历史方法，指出文艺史也跟其他的历史一样，都要对已发生的事实进行准确记录，只是对象是艺术和文学事实。在编撰文艺史的过程中，还需要超越审美再创造的心理活动，换句话说，就是需要表现，对再创造进行表现。

第十八章是克罗齐美学理论的总结。克罗齐在这章中论证了艺术与语言学的关系。他指出艺术与语言同为表现，语言将声音组织起来构成意义，语言是不断创造的过程。声音与语义不断变化，表现也就不断更新。

（二）《美学史》的内容结构

下篇“美学史”共 19 章，以时间线为主干，从古希腊罗马时期、中世纪、文艺复兴时期、17 世纪、18 世纪到现当代，论述了不同时期的美学观念、思想，各学派之间的渊源、联系、分歧。其中若干章节专门介绍笛卡尔、莱布尼茨、鲍姆加登、维柯、康德、席勒、谢林、佐尔格、黑格尔、叔本华、赫尔巴特、施莱尔马赫、洪堡、施泰因塔尔、德·桑克蒂斯等美学史上各流派代表人物的美学观点。

第一章至第二章讲述了西方美学的起源。克罗齐梳理了古希腊—罗马时代、中世纪和文艺复兴时期的美学意识，介绍了从“摹仿说”到对诗歌、音乐目的的探讨，再到对美、善、智慧的思考。

第三至七章回顾了美学思想的形成与发展。克罗齐分别介绍了中世纪美学概念、笛卡尔学派与莱布尼茨学派的美学思想，鲍姆加登的“美学”观念，评述了维柯的美学思想以及 18 世纪的美学学说。

第八章至第十三章述及德国古典美学及其发展。克罗齐依次介绍了康德的美学思想、以席勒、谢林和黑格尔为代表的德国的唯心主义美学，阐述了叔本华和赫尔巴特、施莱尔马赫等的美学思考，介绍了洪堡的语言哲学，阐释了语言和思维的关系。

第十四、十五章概括了 19 世纪上半叶英国、法国和意大利三国的美学观点和流派，并论述桑克蒂斯的美学主张。

第十六章至第十八章概括了美学的近现代发展。克罗齐介绍了唯心主义之后的后继者美学观点，介绍了美学向经验的、科学的、实证的研究方向发展的思潮，并对现当代的美学趋势进行了论述。

第十九章，克罗齐对下篇“美学史”进行了回顾和总结。从修辞学、文艺史、艺术理论的局限性等方面进行了分类整合。

五、《美学》的学术地位与影响

《美学》是克罗齐的《精神哲学》中的第一部，讨论的不仅是普通的美学问题，更是美学在整个哲学中的地位、审美与其他精神活动的关系。《美学》一书含有他的全部哲学雏形。“克罗齐的直觉说是唯心主义的奠基石”（朱光潜，2012，p.30）。克罗齐在唯心主义形而上学的语境中阐释审美现实，“带来了一个全新的关于审美的概念。”（比厄斯利，2006，p.319）

《美学》探讨了艺术的本质属性，指出艺术不是物理事实，艺术不是功利活动，艺术不是道德活动，艺术不是概念或逻辑活动，开创性地提出了艺术即直觉，排除了人们附加给艺术的许多不纯粹的东西。克罗齐提出评判一件艺术作品，观赏者、批评家必须和艺术家融为一体，对作品进行再创作，他的艺术创造和鉴赏同一的论段对于如何判断艺术具有启发性。

《美学》提出了美学和语言学的统一，将艺术的表现性和语言联系起来，把美学建立在语言学的范式之上，为美学思考提供了一个新方向，开启了西方哲学美学的“语言学转向”。

作为《精神哲学》的第一部，《美学》体现了克罗齐早期的美学思考，阐释观点时所用例证还有些含混之处。但是，瑕不掩瑜，

“克罗齐凭此作让一直被实证主义以及中世纪思想所禁锢的美学松了绑”（韦尔迪基奥，2020，p.39），开启了从以理念为核心的理性主义美学向以直觉为中心的唯心主义美学的转变，引导了 20 世纪西方美学的新趋向，为现代美学做出了开拓性贡献。

参考文献：

[1] 比厄斯利.高建平译，西方美学简史 [M].北京：北京大学出版社，2006.

[2] 克罗齐.田时纲译，自我评论 [J].哲学译丛 1997 第 3 期 P39–49.

[3] 韦尔迪基奥.史菊鸿译，予物以名：克罗齐美学、哲学以及历史思想研究 [M].北京：中国社会科学出版社，2020.

[4] 朱光潜.克罗齐哲学述评 欣慨室逻辑学哲学散论 [M].北京：中华书局，2012.

THEORY OF ÆSTHETIC

AUTHOR'S PREFACE

This volume is composed of a theoretical and of a historical part, which form two independent but complementary books.

The nucleus of the theoretical part is a memoir, bearing the title Fundamental Theses of an Æsthetic as Science of Expression and General Linguistic, which was read at the Accademia Pontaniana of Naples[1] during the sessions of February 18 and May 6, 1900, and printed in vol. xxx. of its Acts. The author has added few substantial variations, but not a few additions and amplifications in rewriting it, also following a somewhat different sequence with a view to rendering the exposition more plain and easy. The first five chapters only of the historical portion were inserted in the Neapolitan review *Flegrea*[2] (April 1901), under the title *Giambattista Vico, First Discoverer of Æsthetic Science*[3], and these also reappear amplified and brought into harmony with the rest.

The author has dwelt, especially in the theoretical part, upon general questions which are side-issues in respect to the theme that he has treated. But this will not seem a digression to those who

---

1. Accademia Pontaniana of Naples：那不勒斯蓬塔尼亚那学会。源于诗人安东尼奥·贝卡德利（Antonio Beccadelli）于 1443 年在那不勒斯创建的非正式学术会议。贝卡德利去世后，作家、诗人乔瓦尼·蓬塔诺（Giovanni Pontano）接任并使其正规化，成为世界第一所现代意义的学会。克罗齐曾在 1917—1923 年任该学会主席。
2. Flegrea：《弗雷格利亚》，那不勒斯当地一本评论专刊。
3. under the title Giambattista Vico, First Discoverer of Æsthetic Science：其标题是“维柯，美学的第一位发现者”。詹巴蒂斯塔·维柯（1668—1744），意大利哲学家、法学家、历史学家和美学家，著有《新科学》等著作，提出了“诗性智慧”的理论。

remember that, strictly speaking, there are no particular philosophical sciences, standing by themselves. Philosophy is unity, and when we treat of Æsthetic or of Logic or of Ethics, we treat always of the whole of philosophy, although illustrating for didactic purposes only one side of that inseparable unity[1]. In like manner, owing to this intimate connexion[2] of all the parts of philosophy, the uncertainty and misunderstanding as to the æsthetic activity, the representative and productive imagination, this firstborn of the spiritual activities, mainstay of the others, generates everywhere else misunderstandings, uncertainties and errors: in Psychology as in Logic, in History as in the Philosophy of Practice. If language is the first spiritual manifestation[3], and if the æsthetic form is language itself, taken in all its true scientific extension, it is hopeless to try to understand clearly the later and more complicated phases of the life of the spirit, when their first and simplest moment is ill known, mutilated and disfigured. From the explanation of the æsthetic activity is also to be expected the correction of several concepts and the solution of certain philosophic problems which generally seem to be almost desperate. Such is precisely the spirit animating the present work. And if the present attempt and the historical illustrations which accompany it may be of use in winning friends to these studies, by leveling obstacles[4] and indicating paths to be followed; if this happen, especially here in Italy, whose æsthetic traditions (as has been demonstrated in its place) are very noble, the author will consider that he has gained his end, and one of his keenest desires will have been satisfied.

NAPLES, *December* 1901

---

1. illustrating for didactic purposes only one side of that inseparable unity：出于教学目的仅阐明不可分割的统一体中的一个方面
2. intimate connexion：紧密联系
3. spiritual manifestation：精神表现形式
4. leveling obstacles：铲平障碍

In addition to a careful literary revision, (in which, as well as in the revision of the notes, I have received valuable help from my friend Fausto Nicolini[1]) I have in this third edition made certain alterations of theory, especially in Chapters X and XI of Part I, suggested by further reflexion and self-criticism.

But I have refrained from introducing corrections or additions of such a kind as to alter the original plan of the book, which was, or was meant to be, a complete but brief æsthetic theory set in the framework of a general sketch of a *Philosophy of the Spirit.*

The reader who desires a complete statement of the general or collateral doctrines[2] or a more particular exposition of the other parts of philosophy (e.g. the lyrical nature of art) is now referred to the volumes on *Logic* and the *Philosophy of Practice*, which together with the present work compose the *Philosophy of the Spirit* which in the author's opinion exhausts the entire field of Philosophy. The three volumes were not conceived and written simultaneously; if they had been, some details would have been differently arranged. When I wrote the first I had no idea of giving it, as I have now done, two such companions; and I therefore designed it to be, as I say, complete in itself. In the second place, the present state of the study of Æsthetic made it desirable to append to the theoretical exposition a somewhat full history of the science, whereas for the other parts of Philosophy I was able to restrict myself to brief historical notes merely designed to show how, from my point of view, such a history would best be composed. Lastly, there are many things which now, after a systematic exposition of the various philosophical sciences, I see in closer connexions and in a clearer, or at least a different, light; a certain hesitation and even some doctrinal errors visible here and there in the Æsthetic, especially

1. Fausto Nicolini：福斯托·尼克里尼（1879—1965），意大利传记作家，文学评论家。
2. general or collateral doctrines：总体学说或旁证学说

where subjects foreign to Æsthetic itself are being treated, would now no longer be justified. For all these reasons the three volumes, in spite of their substantial unity of spirit and of aim, have each its own physiognomy[1], and show marks of the different periods of life at which they were written, so as to group themselves, and to demand interpretation, as a progressive series according to their dates of publication.

With what may be called the minor problems of Æsthetic, and the objections which have been or might be brought against my theory, I have dealt and am continuing to deal in special essays, of which I shall shortly publish a first collection which will form a kind of explanatory and polemical[2] appendix to the present volume.

November 1907

In revising this book once more for a fourth edition, I take the opportunity of announcing that the supplementary volume of essays promised above was published in 1910 under the title *Problems of Æsthetic and Contributions to the History of Æsthetic*[3] in Italy.

B. C.

May 1911

---

1. physiognomy：面貌特征
2. polemical：争辩的，辩论的
3. *Problems of Æsthetic and Contributions to the History of Æsthetic*：《美学问题与美学史稿》

THEORY OF ÆSTHETIC

I INTUITION AND EXPRESSION

Intuitive knowledge.

Knowledge has two forms: it is either intuitive knowledge[1] or logical knowledge; knowledge obtained through the imagination or knowledge obtained through the intellect; knowledge of the individual or knowledge of the universal; of individual things or of the relations between them: it is, in fact, productive either of images or of concepts.

In ordinary life, constant appeal is made to intuitive knowledge. It is said that we cannot give definitions of certain truths; that they are not demonstrable by syllogisms[2]; that they must be learnt intuitively. The politician finds fault with the abstract reasoner[3], who possesses no lively intuition of actual conditions; the educational theorist insists upon the necessity of developing the intuitive faculty in the pupil before everything else; the critic in judging a work of art makes it a point of honour to set aside theory and abstractions, and to judge it by direct intuition; the practical man professes to live rather by intuition than by reason.

But this ample acknowledgment granted to intuitive knowledge in ordinary life, does not correspond to an equal and adequate acknowledgment in the field of theory and of philosophy. There exists

---

1. intuitive knowledge：直觉知识。克罗齐认为直觉知识是初始的、个别的、有创造性的精神活动，可以离开逻辑知识独立存在。
2. syllogisms：三段论，传统逻辑学中的主要推理方式。
3. abstract reasoner：抽象的理性人

a very ancient science of intellectual knowledge[1], admitted by all without discussion, namely, Logic; but a science of intuitive knowledge is timidly and with difficulty asserted by but a few. Logical knowledge has appropriated the lion's share[2]; and if she does not slay and devour her companion outright, yet yields to her but grudgingly the humble place of maid-servant or doorkeeper[3]. — What can intuitive knowledge be without the light of intellectual knowledge? It is a servant without a master; and though a master finds a servant useful, the master is a necessity to the servant, since he enables him to gain his livelihood. Intuition is blind; intellect lends her eyes.

Its independence with respect to intellectual knowledge.

Now, the first point to be firmly fixed in the mind is that intuitive knowledge has no need of a master, nor to lean upon any one; she does not need to borrow the eyes of others, for she has excellent eyes of her own. Doubtless it is possible to find concepts mingled with intuitions. But in many other intuitions there is no trace of such a mixture, which proves that it is not necessary. The impression of a moonlight scene by a painter; the outline of a country drawn by a cartographer[4]; a musical motive, tender or energetic; the words of a sighing lyric, or those with which we ask, command and lament in ordinary life, may well all be intuitive facts without a shadow of intellectual relation. But, think what one may of these instances, and admitting further the contention[5] that the greater part of the intuitions of civilized man

---

1. intellectual knowledge：理性知识
2. has appropriated lion's share：占据最大份额
3. yields to her but grudgingly the humble place of maid-servant or doorkeeper：只是勉强地给予她女仆或看门人那样卑微的地位
4. the outline of a country drawn by a cartographer：地图绘制员绘制的乡村轮廓图
5. contention：论点

are impregnated with concepts[1], there yet remains to be observed something more important and more conclusive . Those concepts which are found mingled and fused with the intuitions are no longer concepts, in so far as they are really mingled and fused, for they have lost all independence and autonomy. They have been concepts, but have now become simple elements of intuition. The philosophical maxims[2] placed in the mouth of a personage[3] of tragedy or of comedy, perform there the function, not of concepts, but of characteristics of such personage; in the same way as the red in a painted face does not there represent the red colour of the physicists, but is a characteristic element of the portrait. The whole is that which determines the quality of the parts. A work of art may be full of philosophical concepts; it may contain them in greater abundance and they may there be even more profound than in a philosophical dissertation[4], which in its turn may be rich to overflowing with descriptions and intuitions. But notwithstanding all these concepts the total effect of the work of art is an intuition; and notwithstanding all those intuitions, the total effect of the philosophical dissertation is a concept. The *Promessi Sposi*[5] contains copious ethical observations and distinctions, but does not for that reason lose as a whole its character of simple story or intuition. In like manner the anecdotes and satirical effusions to be found in the works of a philosopher like Schopenhauer[6] do not deprive those works of their character of intellectual treatises. The difference between a scientific work and a work of art, that is, between an intellectual fact

---

1. impregnated with concepts：包含着概念
2. philosophical maxims：哲学格言
3. personage：角色
4. philosophical dissertation：哲学论著
5. *Promessi Sposi*：《约婚夫妇》，曼佐尼（1785—1873）著，意大利家喻户晓的长篇历史小说。
6. Schopenhauer：叔本华（1788—1860），德国哲学家，唯意志论的代表人物，著有《作为意志和表象的世界》《论意志的自由》等。

and an intuitive fact, lies in the difference of the total effect aimed at by their respective authors. This it is that determines and rules over the several parts of each not these parts separated and considered abstractly in themselves.

Intuition and perception.

But to admit the independence of intuition as regards concept does not suffice to give a true and precise idea of intuition. Another error arises among those who recognize this, or who at any rate do not explicitly make intuition dependent upon the intellect, to obscure and confuse the real nature of intuition. By intuition is frequently understood perception, or the knowledge of actual reality, the apprehension of something as *real*.

Certainly perception is intuition: the perceptions of the room in which I am writing, of the ink-bottle and paper that are before me, of the pen I am using, of the objects that I touch and make use of as instruments of my person, which, if it writes, therefore exists; — these are all intuitions. But the image that is now passing through my brain of a me writing in another room, in another town, with different paper, pen and ink, is also an intuition. This means that the distinction between reality and non-reality is extraneous[1], secondary, to the true nature of intuition. If we imagine a human mind having intuitions for the first time, it would seem that it could have intuitions of actual reality only, that is to say, that it could have perceptions of nothing but the real. But since knowledge of reality is based upon the distinction between real images and unreal images, and since this distinction does not at the first moment exist, these intuitions would in truth not be intuitions either of the real or of the unreal, not

1. extraneous：不相关的

perceptions, but pure intuitions. Where all is real, nothing is real. The child, with its difficulty of distinguishing true from false, history from fable, which are all one to childhood, can furnish us with a sort of very vague and only remotely approximate idea of this ingenuous state[1]. Intuition is the undifferentiated unity of the perception of the real and of the simple image of the possible. In our intuitions we do not oppose ourselves as empirical beings to external reality[2], but we simply objectify[3] our impressions, whatever they be.

Intuition and the concepts of space and time.

Those, therefore, who look upon intuition as sensation formed and arranged simply according to the categories of space and time, would seem to approximate more nearly to the truth. Space and time (they say) are the forms of intuition; to have an intuition is to place it in space and in temporal sequence[4]. Intuitive activity would then consist in this double and concurrent function of spatiality and temporality[5]. But for these two categories must be repeated what was said of intellectual distinctions, when found mingled with intuitions. We have intuitions without space and without time: the colour of a sky, the colour of a feeling, a cry of pain and an effort of will, objectified in consciousness: these are intuitions which we possess, and with their making space and time have nothing to do. In some intuitions, spatiality may be found without temporality, in others, vice versa; and even where both are found, they are perceived by

1. a sort of very vague and only remotely approximate idea of this ingenuous state：对这种率真状态的某种非常模糊的、近似的了解
2. empirical beings to external reality：经验主体与外在实际相对立
3. objectify：客观化
4. temporal sequence：时间顺序
5. double and concurrent function of spatiality and temporality：空间和时间双重的、共同的作用

later reflexion: they can be fused with the intuition in like manner with all its other elements: that is, they are in it *materialiter* and not *formaliter*[1], as ingredients and not as arrangement. Who, without an act of reflexion which for a moment breaks in upon his contemplation, can think of space while looking at a drawing or a view? Who is conscious of temporal sequence while listening to a story or a piece of music without breaking into it with a similar act of reflexion? What intuition reveals in a work of art is not space and time, but character, individual physiognomy. The view here maintained is confirmed in several quarters of modern philosophy. Space and time, far from being simple and primitive functions, are nowadays conceived as intellectual constructions of great complexity. And further, even in some of those who do not altogether deny to space and time the quality of formative principles, categories and functions, one observes an effort to unite them and to regard them in a different manner from that in which these categories are generally conceived. Some limit intuition to the sole category of spatiality, maintaining that even time can only be intuited in terms of space. Others abandon the three dimensions of space as not philosophically necessary, and conceive the function of spatiality as void of all particular spatial determination. But what could such a spatial function be, a simple arrangement that should arrange even time? It represents, surely, all that criticism and refutation have left standing — the bare demand for the affirmation of some intuitive activity in general. And is not this activity truly determined, when one single function is attributed to it, not spatializing nor temporalizing, but characterizing[2]? Or rather, when it is conceived as itself a category or function which gives us knowledge of things in their concreteness and individuality?

---

1. *materialiter* and not *formaliter*：物质的而非形式的（斜体为拉丁文）
2. not spatializing nor temporalizing , but characterizing: 既非空间化，亦非时间化，而是个性化

Intuition and sensation.

Having thus freed intuitive knowledge from any suggestion of intellectualism and from every later and external addition, we must now explain it and determine its limits from another side and defend it from a different kind of invasion and confusion. On the hither side of the lower limit is sensation, formless matter, which the spirit can never apprehend in itself as simple matter. This it can only possess with form and in form, but postulates[1] the notion of it as a mere limit. Matter, in its abstraction, is mechanism, passivity; it is what the spirit of man suffers, but does not produce. Without it no human knowledge or activity is possible; but mere matter produces animality, whatever is brutal and impulsive in man, not the spiritual dominion[2], which is humanity. How often we strive to understand clearly what is passing within us! We do catch a glimpse of something, but this does not appear to the mind as objectified and formed. It is in such moments as these that we best perceive the profound difference between matter and form. These are not two acts of ours, opposed to one another; but the one is outside us and assaults and sweeps us off our feet, while the other inside us tends to absorb and identify itself with that which is outside. Matter, clothed and conquered by form, produces concrete form. It is the matter, the content, which differentiates one of our intuitions from another: the form is constant: it is spiritual activity, while matter is changeable. Without matter spiritual activity would not forsake its abstractness to become concrete and real activity, this or that spiritual content, this or that definite intuition.

It is a curious fact, characteristic of our times, that this very form, this very activity of the spirit, which is essentially ourselves, is so often ignored or denied. Some confound the spiritual activity of man with

1. postulates：假定
2. spiritual dominion：心灵至上

the metaphorical and mythological activity of what is called nature, which is mechanism and has no resemblance to human activity, save when we imagine, with Æsop[1], that "*arbores loquuntur non tantum ferae*[2]." Some affirm that they have never observed in themselves this "miraculous" activity, as though there were no difference, or only one of quantity, between sweating and thinking, feeling cold and the energy of the will. Others, certainly with greater reason, would unify activity and mechanism in a more general concept, though they are specifically distinct. Let us, however, refrain for the moment from examining if such a final unification be possible, and in what sense, but admitting that the attempt may be made, it is clear that to unify two concepts in a third implies to begin with the admission of a difference between the two first. Here it is this difference that concerns us and we set it in relief.

Intuition and association.

Intuition has sometimes been confused with simple sensation. But since this confusion ends by being offensive to common sense, it has more frequently been attenuated or concealed with a phraseology[3] apparently designed at once to confuse and to distinguish them. Thus, it has been asserted that intuition is sensation, but not so much simple sensation as *association* of sensations. Here a double meaning is concealed in the word "association." Association is understood, either as memory, mnemonic association[4], conscious recollection[5], and in that case the claim to unite in memory elements which are

---

1. Æsop：伊索（约公元前 620—前 560），古希腊著名的寓言家。据史料记载，伊索曾经为奴，因智慧超人，最终获得自由并卓有成就。
2. *arbores loquuntur non tantum ferae*：〈拉〉树木也会说话，不仅仅是野兽。
3. attenuated or concealed with a phraseology：用某个措辞弱化或掩饰
4. mnemonic association：记忆间的联想
5. conscious recollection：有意识的回想

not intuited, distinguished, possessed in some way by the spirit and produced by consciousness, seems inconceivable: or it is understood as association of unconscious elements, in which case we remain in the world of sensation and of nature. But if with certain associationists we speak of an association which is neither memory nor flux of sensations, but a *productive* association[1] (formative, constructive, distinguishing[2]); then our contention is admitted and only its name is denied to it. For productive association is no longer association in the sense of the sensationalists , but *synthesis*[3], that is to say, spiritual activity. Synthesis may be called association; but with the concept of productivity is already posited the distinction between passivity and activity, between sensation and intuition.

Intuition and representation.

Other psychologists are disposed to distinguish from sensation something which is sensation no longer, but is not yet intellectual concept: the *representation* or *image*. What is the difference between their representation or image and our intuitive knowledge? Everything and nothing: for "representation" is a very equivocal[4] word. If by representation be understood something cut off and standing out from the psychic basis of the sensations, then representation is intuition. If, on the other hand, it be conceived as complex sensation we are back once more in crude sensation, which does not vary in quality according to its richness or poverty, or according to whether the organism in which it appears is rudimentary[5] or highly developed and full of

---

1. neither memory nor flux of sensations, but a *productive* association：既非记忆亦非感觉的波动，而是创造性的联想
2. formative, constructive, distinguishing：形式上的，建设性的，分辨式的
3. *synthesis*：综合性的
4. equivocal：模棱两可的
5. rudimentary: 原始的

traces of past sensations. Nor is the ambiguity remedied by defining representation as a psychic product[1] of secondary degree in relation to sensation, defined as occupying the first place. What does secondary degree mean here? Does it mean a qualitative, formal difference? If so, representation is an elaboration of sensation[2] and therefore intuition. Or does it mean greater complexity and complication, a quantitative, material difference? In that case intuition is once more confused with simple sensation.

Intuition and expression.

And yet there is a sure method of distinguishing true intuition, true representation, from that which is inferior to it: the spiritual fact from the mechanical, passive, natural fact. Every true intuition or representation is also *expression.* That which does not objectify itself in expression is not intuition or representation, but sensation and mere natural fact. The spirit only intuites in making, forming, expressing. He who separates intuition from expression never succeeds in reuniting them.

Intuitive activity possesses intuitions to the extent that it expresses them. Should this proposition sound paradoxical, that is partly because, as a general rule, a too restricted meaning is given to the word "expression." It is generally restricted to what are called verbal expressions alone. But there exist also non-verbal expressions, such as those of line, colour and sound, and to all of these must be extended our affirmation, which embraces therefore every sort of manifestation of the man, as orator, musician, painter, or anything else. But be it pictorial, or verbal, or musical, or in whatever other form it appear, to no intuition can expression in one of its forms be wanting; it is, in fact,

1. psychic product: 精神产品
2. elaboration of sensation：精致复杂的感觉

an inseparable part of intuition. How can we really possess an intuition of a geometrical figure, unless we possess so accurate an image of it as to be able to trace it immediately upon paper or on the blackboard?

How can we really have an intuition of the contour of a region[1], for example of the island of Sicily, if we are not able to draw it as it is in all its meanderings[2]? Every one can experience the internal illumination which follows upon his success in formulating to himself his impressions and feelings, but only so far as he is able to formulate them. Feelings or impressions, then, pass by means of words from the obscure region of the soul into the clarity of the contemplative spirit[3]. It is impossible to distinguish intuition from expression in this cognitive process. The one appears with the other at the same instant, because they are not two, but one.

Illusion as to their difference.

The principal reason which makes our view appear paradoxical as we maintain it, is the illusion or prejudice that we possess a more complete intuition of reality than we really do. One often hears people say that they have many great thoughts in their minds, but that they are not able to express them. But if they really had them, they would have coined them into just so many beautiful, sounding words, and thus have expressed them. If these thoughts seem to vanish or to become few and meagre[4] in the act of expressing them, the reason is that they did not exist or really were few and meagre. People think that all of us ordinary men imagine and intuite countries, figures and scenes like painters, and bodies like sculptors; save that painters and

---

1. contour of a region：一个地区的地理轮廓
2. meanderings：蜿蜒曲折的道路
3. clarity of the contemplative spirit：沉思心灵的光明境界
4. become few and meagre：日渐消弭

sculptors know how to paint and carve such images, while we bear them unexpressed in our souls. They believe that any one could have imagined a Madonna of Raphæl[1]; but that Raphæl was Raphæl owing to his technical ability in putting the Madonna upon canvas. Nothing can be more false than this view. The world which as a rule we intuite is a small thing. It consists of little expressions, which gradually become greater and wider with the increasing spiritual concentration of certain moments. They are the words we say to ourselves, our silent judgments: "Here is a man, here is a horse, this is heavy, this is sharp, this pleases me," etc. It is a medley of light and colour[2], with no greater pictorial value than would be expressed by a haphazard splash of colours[3], from among which one could barely make out a few special, distinctive traits. This and nothing else is what we possess in our ordinary life; this is the basis of our ordinary action. It is the index of a book. The labels tied to things (it has been said) take the place of the things themselves. This index and these labels (themselves expressions) suffice for small needs and small actions. From time to time we pass from the index to the book, from the label to the thing, or from the slight to the greater intuitions, and from these to the greatest and most lofty. This passage is sometimes far from easy. It has been observed by those who have best studied the psychology of artists that when, after having given a rapid glance at any one, they attempt to obtain a real intuition of him, in order, for example, to paint his portrait, then this ordinary vision, that seemed so precise, so lively, reveals itself as little better than nothing. What remains is found to be at the most some superficial trait[4], which would not even suffice for a caricature[5]. The

1. Madonna of Raphæl：拉斐尔的《圣母像》。拉斐尔（1483—1520），意大利著名画家，文艺复兴三杰之一。
2. medley of light and colour：光与色的混杂
3. haphazard splash of colours：胡乱涂抹的颜料
4. superficial trait：肤浅的特征
5. caricature：讽刺漫画

person to be painted stands before the artist like a world to discover. Michæl-Angelo[1] said, "One paints, not with the hands, but with the brain." Leonardo[2] shocked the prior of the Convent of the Graces[3] by standing for days together gazing at the "Last Supper[4]," without touching it with the brush. He remarked of this attitude: "The minds of men of lofty genius are most active in invention when they are doing the least external work." The painter is a painter, because he sees what others only feel or catch a glimpse of, but do not see. We think we see a smile, but in reality we have only a vague impression of it, we do not perceive all the characteristic traits of which it is the sum, as the painter discovers them after he has worked upon them and is thus able to fix them on the canvas. We do not intuitively possess more even of our intimate friend, who is with us every day and at all hours, than at most certain traits of physiognomy which enable us to distinguish him from others. The illusion is less easy as regards musical expression; because it would seem strange to every one to say that the composer had added or attached notes to a motive which was already in the mind of him who is not the composer; as if Beethoven's Ninth Symphony[5] were not his own intuition and his intuition the Ninth Symphony. Now, just as one who is deluded as to the amount of his material wealth is confuted by arithmetic[6], which states its exact amount, so he who nourishes delusions[7] as to the wealth of his own thoughts and images is brought back to reality, when he is obliged to

---

1. Michæl-Angelo：米开朗琪罗（1475—1564），意大利画家、雕塑家、建筑师和诗人，文艺复兴三杰之一。
2. Leonardo：莱昂纳多·达·芬奇（1452—1519），意大利画家、科学家、博物学家。文艺复兴三杰之一。
3. Convent of the Graces：圣母感恩修道院的主持。达·芬奇的巨幅画作《最后的晚餐》就画在这座修道院餐厅的墙壁上。
4. Last Supper：《最后的晚餐》，达·芬奇的代表画作。
5. Beethoven's Ninth Symphony：贝多芬的《第九交响曲》
6. confuted by arithmetic：被数学纠正
7. nourishes delusions：产生错觉

cross the *Pons Asinorum*[1] of expression. Let us say to the former, count; to the latter, speak; or, here is a pencil, draw, express yourself.

Each of us, as a matter of fact, has in him a little of the poet, of the sculptor, of the musician, of the painter, of the prose writer: but how little, as compared with those who bear those names, just because they possess the most universal dispositions and energies of human nature in so lofty a degree! How little too does a painter possess of the intuitions of a poet! And how little does one painter possess those of another painter! Nevertheless, that little is all our actual patrimony of intuitions or representations[2]. Beyond these are only impressions, sensations, feelings, impulses, emotions, or whatever else one may term what still falls short of the spirit and is not assimilated by man; something postulated for the convenience of exposition, while actually non-existent, since to exist also is a fact of the spirit.

Identity of intuition and expression.

We may thus add this to the various verbal descriptions of intuition, noted at the beginning: intuitive knowledge is expressive knowledge. Independent and autonomous in respect to intellectual function; indifferent to later empirical discriminations[3], to reality and to unreality, to formations and apperceptions of space and time[4], which are also later: intuition or representation is distinguished as *form* from what is felt and suffered, from the flux or wave of sensation, or from psychic matter; and this form, this taking possession, is expression. To intuite is to express; and nothing else (nothing more, but nothing less) than *to express*.

1. *Pons Asinorum*：驴桥定理，即等腰三角形定理，系欧几里得几何中的一个数学定理，指等腰三角形二腰对应的二底角相等。
2. actual patrimony of intuitions or representations：真正具有的直觉或表象
3. later empirical discriminations：随后因经验而形成的辨别力
4. apperceptions of space and time：对时间、空间的感知

II INTUITION AND ART

Corollaries[1] *and explanations.*

Before proceeding further, it may be well to draw certain consequences from what has been established and to add some explanations.

Identity of art and intuitive knowledge.

We have frankly identified intuitive or expressive knowledge with the æsthetic or artistic fact, taking works of art as examples of intuitive knowledge and attributing to them the characteristics of intuition, and *vice versa*. But our identification is combated by[2] a view held even by many philosophers, who consider art to be an intuition of an altogether special sort. "Let us admit" (they say) "that art is intuition; but intuition is not always art: artistic intuition is a distinct species differing from intuition in general by something *more*."

No specific difference.

But no one has ever been able to indicate of what this something more consists. It has sometimes been thought that art is not a simple intuition, but an intuition of an intuition, in the same way as the

1. *corollarie*：推论
2. combated by：与……水火不容

concept of science has been defined, not as the ordinary concept, but as the concept of a concept. Thus man would attain to art by objectifying, not his sensations, as happens with ordinary intuition, but intuition itself. But this process of raising to a second power does not exist; and the comparison of it with the ordinary and scientific concept does not prove what is intended, for the good reason that it is not true that the scientific concept is the concept of a concept. If this comparison proves anything, it proves just the opposite. The ordinary concept, if it be really a concept and not a simple representation, is a perfect concept, however poor and limited. Science substitutes concepts for representations; for those concepts that are poor and limited it substitutes others, larger and more comprehensive; it is ever discovering new relations. But its method does not differ from that by which is formed the smallest universal in the brain of the humblest of men. What is generally called art, collects intuitions that are wider and more complex than those which we generally experience, but these intuitions are always of sensations and impressions.

Art is expression of impressions, not expression of expression.

No difference of intensity.

For the same reason, it cannot be asserted that the intuition, which is generally called artistic, differs from ordinary intuition as intensive intuition[1]. This would be the case if it were to operate differently on the same matter. But since the artistic function is extended to wider fields, yet does not differ in method from ordinary intuition, the difference between them is not intensive but extensive. The intuition of the simplest popular love-song, which says the same thing, or very nearly, as any declaration of love that issues at every moment from

1. intensive intuition：强烈的直觉

the lips of thousands of ordinary men, may be intensively perfect in its poor simplicity, although it be extensively so much more limited than the complex intuition of a love-song by Leopardi[1].

The difference is extensive and empirical.

The whole difference, then, is quantitative, and as such is indifferent to philosophy, *scientia qualitatum*[2]. Certain men have a greater aptitude[3], a more frequent inclination[4] fully to express certain complex states of the soul. These men are known in ordinary language as artists. Some very complicated and difficult expressions are not often achieved, and these are called works of art. The limits of the expression-intuitions that are called art, as opposed to those that are vulgarly called non-art, are empirical and impossible to define. If an epigram[5] be art, why not a simple word? If a story, why not the news-jottings[6] of the journalist? If a landscape, why not a topographical sketch? The teacher of philosophy in Molière's[7] comedy was right: "whenever we speak, we create prose." But there will always be scholars like Monsieur Jourdain[8], astonished at having spoken prose for forty years without knowing it, who will have difficulty in persuading themselves that when they call their servant John to bring their slippers, they have spoken nothing less than — prose.

---

1. Leopardi：贾科莫·莱奥帕尔迪（1798—1837），意大利 19 世纪浪漫主义诗人，开意大利现代自由体抒情诗先河。
2. *scientia qualitatum*：〈拉〉论质的学问
3. greater aptitude：天资聪颖
4. more frequent inclination：更常见的倾向性
5. epigram：隽言
6. news-jottings：新闻报道
7. Molière：莫里哀（1622—1673），法国伟大的喜剧作家、演员、戏剧活动家，代表作有《伪君子》《无病呻吟》等。
8. Jourdain：儒尔丹（1788—1818），法国历史学家、东方学家，古波斯以及亚里士多德著作翻译研究专家。

We must hold firmly to our identification, because among the principal reasons which have prevented Æsthetic, the science of art, from revealing the true nature of art, its real roots in human nature, has been its separation from the general spiritual life, the having made of it a sort of special function or aristocratic club. No one is astonished when he learns from physiology that every cell is an organism and every organism a cell or synthesis of cells. No one is astonished at finding in a lofty mountain the same chemical elements that compose a small stone fragment. There is not one physiology of small animals and one of large animals; nor is there a special chemical theory of stones as distinct from mountains. In the same way, there is not a science of lesser intuition as distinct from a science of greater intuition, nor one of ordinary intuition as distinct from artistic intuition. There is but one Æsthetic, the science of intuitive or expressive knowledge, which is the æsthetic or artistic fact. And this Æsthetic is the true analogue of Logic, which includes, as facts of the same nature, the formation of the smallest and most ordinary concept and the most complicated scientific and philosophical system.

Artistic genius.

Nor can we admit that the word *genius* or artistic genius, as distinct from the non-genius of the ordinary man, possesses more than a quantitative signification[1]. Great artists are said to reveal us to ourselves. But how could this be possible, unless there were identity of nature between their imagination and ours, and unless the difference were only one of quantity? It were better to change *poeta nascitur*[2] into *homo nascitur poeta*[3]: some men are born great poets, some small.

---

1. quantitative signification：量化的意义
2. *poeta nascitur*：〈拉〉诗人是天生的
3. *homo nascitur poeta*：〈拉〉人是天生的诗人

The cult of the genius with all its attendant superstitions[1] has arisen from this quantitative difference having been taken as a difference of quality. It has been forgotten that genius is not something that has fallen from heaven, but humanity itself. The man of genius who poses or is represented as remote from humanity finds his punishment in becoming or appearing somewhat ridiculous. Examples of this are the *genius* of the romantic period[2] and the *superman* of our time.

But it is well to note here, that those who claim unconsciousness as the chief quality of an artistic genius, hurl[3] him from an eminence far above humanity to a position far below it. Intuitive or artistic genius, like every form of human activity, is always conscious; otherwise it would be blind mechanism. The only thing that can be wanting to artistic genius is the *reflective* consciousness[4], the superadded consciousness[5] of the historian or critic , which is not essential to it.

Content and form in Æsthetic.

The relation between matter and form, or between *content* and *form*, as is generally said, is one of the most disputed questions in Æsthetic. Does the æsthetic fact consist of content alone, or of form alone, or of both together? This question has taken on various meanings, which we shall mention, each in its place. But when these words are taken as signifying what we have above defined, and matter is understood as emotionality not æsthetically elaborated, or impressions, and form as

---

1. The cult of the genius with all its attendant superstitions：对天才的膜拜以及随之而来的迷信
2. romantic period：浪漫主义时期。浪漫主义是一种侧重从主观内心世界出发，注重个人情感表达、形式自由奔放的文艺思潮，18 世纪末 19 世纪初出现在德国和英国，19 世纪盛行于法国。
3. hurl：抛下
4. *reflective* consciousness：反省意识
5. superadded consciousness：更进一步的意识

intellectual activity and expression, then our view cannot be in doubt. We must, that is to say, reject both the thesis that makes the æsthetic fact to consist of the content alone (that is, the simple impressions), and the thesis which makes it to consist of a junction between form and content, that is, of impressions plus expressions. In the æsthetic fact, expressive activity is not added to the fact of the impressions, but these latter are formed and elaborated by it. The impressions reappear as it were in expression, like water put into a filter, which reappears the same and yet different on the other side. The æsthetic fact, therefore, is form, and nothing but form.

From this was inferred not that the content is something superfluous (it is, on the contrary, the necessary point of departure for the expressive fact); but that *there is no passage* from the qualities of the content to those of the form. It has sometimes been thought that the content, in order to be æsthetic, that is to say, transformable into form, should possess some determined or determinable qualities. But were that so, then form and content, expression and impression, would be the same thing. It is true that the content is that which is convertible into form, but it has no determinable qualities until this transformation takes place. We know nothing about it. It does not become æsthetic content before, but only after it has been actually transformed. The æsthetic content has also been defined as the *interesting*. That is not an untrue statement; it is merely void of meaning. Interesting to what? To the expressive activity? Certainly the expressive activity would not have raised the content to the dignity of form, had it not been interested in it. Being interested is precisely the raising of the content to the dignity of form. But the word "interesting" has also been employed in another and an illegitimate sense, which we shall explain further on.

Criticism of the imitation of nature and of the artistic illusion.

The proposition that art is *imitation of nature* has also several meanings. Sometimes truths have been expressed or at least shadowed forth in these words, sometimes errors have been promulgated[1]. More frequently, no definite thought has been expressed at all. One of the scientifically legitimate meanings occurs when "imitation" is understood as representation or intuition of nature, a form of knowledge. And when the phrase is used with this intention, and in order to emphasize the spiritual character of the process, another proposition becomes legitimate also: namely, that art is the *idealization* or *idealizing* imitation of nature[2]. But if by imitation of nature be understood that art gives mechanical reproductions, more or less perfect duplicates of natural objects, in the presence of which is renewed the same tumult of impressions as that caused by natural objects, then the proposition is evidently false. The coloured waxen effigies[3] that imitate the life, before which we stand astonished in the museums where such things are shown, do not give æsthetic intuitions. Illusion and hallucination[4] have nothing to do with the calm domain of artistic intuition. But on the other hand if an artist paint the interior of a wax-work museum, or if an actor give a burlesque portrait of a man-statue[5] on the stage, we have work of the spirit and artistic intuition. Finally, if photography have in it anything artistic, it will be to the extent that it transmits the intuition of the photographer, his point of view, the pose and grouping which he has striven to attain. And if photography be not quite an art, that is precisely because the element of nature in it remains more or less unconquered and ineradicable[6].

---

1. promulgated：公布
2. *idealization* or *idealizing* imitation of nature：自然的理想化或者对自然的理想化的模仿
3. waxen effigies：蜡像
4. illusion and hallucination：错觉与幻觉
5. give a burlesque portrait of a man-statue：诙谐地模仿一个蜡像
6. more or less unconquered and ineradicable：或多或少是无法征服、无法消灭的

Do we ever, indeed, feel complete satisfaction before even the best of photographs? Would not an artist vary and touch up much or little, remove or add something to all of them?

Criticism of art conceived as a fact of feeling, not a theoretical fact. Æsthetic appearance and feeling.

The statements repeated so often, that art is not knowledge, that it does not tell the truth, that it does not belong to the world of theory, but to the world of feeling, and so forth, arise from the failure to realize exactly the theoretic character of simple intuition. This simple intuition is quite distinct from intellectual knowledge, as it is distinct from perception of the real; and the statements quoted above arise from the belief that only intellectual cognition[1] is knowledge. We have seen that intuition is knowledge, free from concepts and more simple than the so-called perception of the real. Therefore art is knowledge and form; it does not belong to the world of feeling or to psychic matter. The reason why so many æstheticians have so often insisted that art is *appearance* (*Schein*[2]), is precisely that they have felt the necessity of distinguishing it from the more complex fact of perception, by maintaining its pure intuitiveness[3]. And if for the same reason it has been claimed that art is *feeling,* the reason is the same. For if the concept as content of art, and historical reality as such, be excluded from the sphere of art, there remains no other content than reality apprehended in all its ingenuousness and immediacy in the vital impulse[4], in its *feeling*, that is to say again, pure intuition.

---

1. intellectual cognition：理智的认知
2. *Schein*：〈德〉外表
3. pure intuitiveness：纯粹直觉性
4. ingenuousness and immediacy in the vital impulse: 生命活力的单纯与直接

Criticism of the theory of æsthetic senses.

The theory of the *æsthetic senses* has also arisen from the failure to establish, or from having lost to view, the character of expression as distinct from impression, of form as distinct from matter.

This theory can be reduced to the error just indicated of wishing to find a passage from the qualities of the content to those of the form. To ask, in fact, what the æsthetic senses are, implies asking what sensible impressions are able to enter into æsthetic expressions, and which must of necessity do so. To this we must at once reply, that all impressions can enter into æsthetic expressions or formations, but that none are bound to do so of necessity. Dante[1] raised to the dignity of form not only the "sweet colour of the oriental sapphire[2]" (visual impressions), but also tactual or thermic impressions[3], such as the "dense air" and the "fresh rivulets" which "parch the more" the throat of the thirsty. The belief that a picture yields only visual impressions is a curious illusion. The bloom on a cheek, the warmth of a youthful body, the sweetness and freshness of a fruit, the edge of a sharp knife, are not these, too, impressions obtainable from a picture? Are they visual? What would a picture mean to an imaginary man, lacking all or many of his senses, who should in an instant acquire the organ of sight alone? The picture we are looking at and believe we see only with our eyes would seem to his eyes to be little more than an artist's paint-smeared palette[4].

Some who hold firmly to the æsthetic character of certain groups of impressions (for example, the visual and auditive), and exclude others,

---

1. Dante：但丁（1265—1361），意大利诗人，欧洲文艺复兴运动开拓者之一，代表作为《神曲》。
2. oriental sapphire：东方蓝宝石
3. tactual or thermic impression：触觉印象或热感印象
4. paint-smeared palette：沾着颜料的调色板

are nevertheless ready to admit that if visual and auditive impressions[1] enter *directly* into the æsthetic fact, those of the other senses also enter into it, but only as *associated.* But this distinction is altogether arbitrary. Æsthetic expression is synthesis, in which it is impossible to distinguish direct and indirect. All impressions are placed by it on a level, in so far as they are æstheticized. A man who absorbs the subject of a picture or poem does not have it before him as a series of impressions, some of which have prerogatives and precedence[2] over the others. He knows nothing as to what has happened prior to having absorbed it, just as, on the other hand, distinctions made after reflexion have nothing whatever to do with art as such.

The theory of the æsthetic senses has also been presented in another way; as an attempt to establish what physiological organs are necessary for the æsthetic fact. The physiological organ or apparatus is nothing but a group of cells, constituted and disposed in a particular manner; that is to say, it is a merely physical and natural fact or concept. But expression does not know physiological facts. Expression has its point of departure in the impressions, and the physiological path[3] by which these have found their way to the mind is to it altogether indifferent. One way or another comes to the same thing: it suffices that they should be impressions.

It is true that the want of given organs, that is, of certain groups of cells, prevents the formation of certain impressions (when these are not otherwise obtained through a kind of organic compensation[4]). The man born blind cannot intuite and express light. But the impressions are not conditioned solely by the organ, but also by the stimuli which operate upon the organ. One who has never had the impression of the

1. visual and auditive impressions：视觉印象与听觉印象
2. prerogatives and precedence：优势与优先地位
3. physiological path：生理学途径
4. organic compensation：有机体的补偿

sea will never be able to express it, in the same way as one who has never had the impression of the life of high society or of the political arena will never express either. This, however, does not prove the dependence of the expressive function on the stimulus or on the organ. It merely repeats what we know already: expression presupposes impression, and particular expressions particular impressions. For the rest, every impression excludes other impressions during the moment in which it dominates; and so does every expression.

Unity and indivisibility of the work of art.

Another corollary of the conception of expression as activity is the *indivisibility* of the work of art. Every expression is a single expression. Activity is a fusion of the impressions in an organic whole. A desire to express this has always prompted the affirmation that the work of art should have *unity*, or, what amounts to the same thing, *unity in variety*[1]. Expression is a synthesis of the various, or multiple, in the one.

The fact that we divide a work of art into parts, a poem into scenes,episodes, similes, sentences, or a picture into single figures and objects, background, foreground, etc., may seem opposed to this affirmation. But such division annihilates the work[2], as dividing the organism into heart, brain, nerves, muscles and so on, turns the living being into a corpse. It is true that there exist organisms in which division gives rise to other living beings, but in such a case we must conclude, maintaining the analogy between the organism and the work of art, that in the latter case too there are numerous germs of life each ready to grow, in a moment, into a single complete expression.

It may be said that expression sometimes arises from other expressions. There are simple and there are *compound* expressions.

---

1. *unity in variety*：多样化中的统一性
2. annihilates the work：毁掉作品

One must surely admit some difference between the *eureka*[1], with which Archimedes[2] expressed all his joy at his discovery, and the expressive act (indeed all the five acts) of a regular tragedy. — Not in the least: expression always arises directly from impressions. He who conceives a tragedy puts into a crucible[3] a great quantity, so to say, of impressions: expressions themselves, conceived on other occasions, are fused together with the new in a single mass, in the same way as we can cast into a melting furnace formless pieces of bronze and choicest statuettes. Those choicest statuettes must be melted just like the pieces of bronze, before there can be a new statue. The old expressions must descend again to the level of impressions, in order to be synthesized in a new single expression.

Art as liberator.

By elaborating his impressions, man *frees* himself from them. By objectifying them, he removes them from him and makes himself their superior. The liberating and purifying function of art is another aspect and another formula of its character as activity. Activity is the deliverer, just because it drives away passivity.

This also explains why it is usual to attribute to artists both the maximum of sensibility[4] or *passion*, and the maximum of insensibility[5] or Olympian *serenity*[6]. The two characters are compatible, for they do not refer to the same object. The sensibility or passion relates to

---

1. *eureka*：〈希〉“我发现了！”。相传是阿基米德的根据比重原理测出金子纯度时所说的话。
2. Archimedes：阿基米德（公元前 287—212），古希腊数学家、物理学家、发明家。
3. crucible：熔炉
4. maximum of sensibility：最大限度的敏感性
5. maximum of insensibility：最大限度的无动于衷
6. Olympian *serenity*：奥林匹亚众神的平静安详

the rich material which the artist absorbs into his psychic organism[1]; the insensibility or serenity to the form with which he subdues and dominates the tumult of the sensations and passions.

1. psychic organism：心灵机体

III ART AND PHILOSOPHY

Inseparability of intellectual from intuitive knowledge.

The two forms of knowledge, æsthetic and intellectual or conceptual, are indeed different, but this does not altogether amount to separation and disjunction, as of two forces each pulling in its own direction. If we have shown that the æsthetic form is altogether independent of the intellectual and suffices to itself without external support, we have not said that the intellectual can stand without the æsthetic. To describe the independence as *reciprocal*[1] would not be true.

What is knowledge by concepts? It is knowledge of the relations of things, and things are intuitions. Concepts are not possible without intuitions, just as intuition is itself impossible without the matter of impressions. Intuitions are: this river, this lake, this brook, this rain, this glass of water; the concept is: water, not this or that appearance and particular example of water, but water in general, in whatever time or place it be realized; the material of infinite intuitions, but of one single constant concept.

But the concept, the universal, if it be no longer intuition in one respect, is intuition in another respect, and cannot fail of being intuition. The man who thinks has impressions and emotions, in so far as he thinks. His impression and emotion will be not love or hate, not the passion of the man who is not a philosopher, not hate or love for

1. *reciprocal*：相互的

certain objects and individuals, but *the effort of his thought itself*, with the pain and the joy, the love and the hate joined to it. This effort cannot but assume an intuitive form, in becoming objective to the spirit. To speak is not to think logically; but to *think logically* is also to *speak*.

Criticism of the negations of this thesis.

That thought cannot exist without speech, is a truth generally admitted. The negations of this thesis are all founded on equivocations and errors[1].

The first of the equivocations is that of those who observe that one can likewise think with geometrical figures[2], algebraical numbers[3], ideographic signs[4], without any word, even pronounced silently and almost insensibly within one; that there are languages in which the word, the phonetic sign, expresses nothing, unless the written sign also be examined, and so on. But when we said "speak," we intended to employ a synecdoche[5], by which was to be understood "expression" in general, for we have already remarked that expression is not only so-called verbal expression. It may or may not be true that certain concepts may be thought without phonetic manifestations. But the very examples adduced[6] to show this also prove that those concepts never exist without expressions.

Others point out that animals, or certain animals, think and reason without speaking. Now as to how, whether, and what animals think, whether they be rudimentary men, like savages who refuse to be

1. equivocations and errors：混淆和错误
2. geometrical figures：几何图形
3. algebraical numbers：代数数字
4. ideographic signs：表意符号
5. employ a synecdoche：运用提喻
6. adduced：举例

civilized, rather than physiological machines, as the old spiritualists[1] maintained, are questions that do not concern us here. When the philosopher talks of animal, brutal, impulsive, instinctive nature and the like, he does not base himself on such conjectures as to dogs or cats, lions or ants; but upon observations of what is called animal and brutal in man: of the animal side or basis of what we feel in ourselves. If individual animals, dogs or cats, lions or ants, possess something of the activity of man, so much the better, or so much the worse, for them. This means that in respect to them also we must talk, not of "nature" as a whole, but of its animal basis, as being perhaps larger and stronger in them than the animal basis of man. And if we suppose that animals think and form concepts, what kind of conjecture would justify the assertion that they do so without corresponding expressions? Analogy with man, knowledge of the spirit, human psychology, the instrument of all our conjectures as to animal psychology, would constrain us on the contrary to suppose that if they think in any way, they also somehow speak.

Another objection is derived from human psychology, and indeed literary psychology, to the effect that the concept can exist without the word, for it is certainly true that we all know books *well thought and ill written*: that is to say, a thought which remains *beyond* the expression, or *notwithstanding* faulty expression. But when we talk of books well thought and ill written, we cannot mean anything but that in such books are parts, pages, periods or propositions well thought and well written, and other parts (perhaps the least important) ill thought and ill written, not really thought and so not really expressed. Where Vico's *Scienza nuova*[2] is really ill written, it is also ill thought. If we pass

1. spiritualists：心灵主义者
2. Vico's *Scienza nuova*：维柯的《新科学》。维柯（1668—1744），意大利哲学家，《新科学》是其代表作，全称为《关于各民族共同本性的新科学的原则》，1725 年在那不勒斯出版。

from the consideration of big books to a short sentence, the error or inaccuracy of such a contention will leap to the eyes. How could a single sentence be clearly thought and confusedly written?

All that can be admitted is that sometimes we possess thoughts (concepts) in an intuitive form, which is an abbreviated or rather peculiar expression, sufficient for us, but not sufficient to communicate it easily to any other given person or persons. Hence it is incorrect to say that we have the thought without the expression; whereas we should rather say that we have, indeed, the expression, but in such a form that it is not easy to communicate it to others. This, however, is a very variable, relative fact. There are always those who catch our thought on the wing, prefer it in this abbreviated form, and would be wearied by the greater development of it required by others. In other words, the thought considered abstractly and logically will be the same; but æsthetically we are dealing with two different intuition-expressions, into which different psychological elements enter. The same argument suffices to destroy, that is, to interpret correctly, the altogether empirical distinction[1] between an *internal* and an *external* language.

Art and science.

The most lofty manifestations, the summits of intellectual and of intuitive knowledge shining from afar, are called, as we know, Art and Science. Art and Science, then, are different and yet linked together; they meet on one side, which is the æsthetic side. Every scientific work is also a work of art. The æsthetic side may remain little noticed when our mind is altogether taken up with the effort to understand the thought of the man of science and to examine its truth. But it is no

1. empirical distinction：经验性区分

longer unnoticed when we pass from the activity of understanding to that of contemplation and see that thought either develop itself before us, limpid, exact, well-shaped, without superfluous or insufficient words, with appropriate rhythm and intonation; or confused, broken, embarrassed, tentative. Great thinkers are sometimes called great writers, while other equally great thinkers remain more or less fragmentary writers even if their fragments have the scientific value of harmonious, coherent, and perfect works.

We pardon thinkers and men of science their literary mediocrity[1]. The fragments, the flashes, console us for the whole, because it is far easier to recover the well-arranged composition from the fragmentary work of genius, to liberate the flame latent in the spark, than to achieve the discovery of genius. But how can we pardon mediocre[2] expression in pure artists? "*Mediocribus esse poetis non di, non homines, non concessere columnae*[3]" The poet or painter who lacks form, lacks everything, because he lacks *himself*. Poetical material permeates the souls of all[4]: the expression alone, that is to say, the form, makes the poet. And here appears the truth of the view which denies all content to art, just the intellectual concept being understood as content. In this sense, when we take "content" as equal to "concept" it is most true, not only that art does not consist of content, but also that *it has no content.*

Content and form: another meaning. Prose and poetry.

The distinction between *poetry and prose* also cannot be justified, save as that between art and science. It was seen in antiquity that such

---

1. literary mediocrity：文采平平
2. mediocre：平庸的
3. *Mediocribus esse poetis non di, non homines, non concessere columnae*：〈拉〉诗人若是平庸，无论天、人还是书店的柱子都不能容忍。
4. permeates the souls of all：充斥了所有人的心灵

distinction could not be founded on external elements, such as rhythm and metre, or on rhymed or unrhymed form; that it was, on the contrary, altogether internal. Poetry is the language of feeling, prose of the intellect; but since the intellect is also feeling, in its concreteness and reality, all prose has its poetical side.

The relation of first and second degree.

The relation between intuitive knowledge or expression and intellectual knowledge or concept, between art and science, poetry and prose, cannot be otherwise defined than by saying that it is one of *double degree.* The first degree is the expression, the second the concept: the first can stand without the second, but the second cannot stand without the first. There is poetry without prose, but not prose without poetry. Expression, indeed, is the first affirmation of human activity. Poetry is "the mother tongue of the human race"; the first men "were by nature sublime poets." We assert this in another way, when we observe that the passage from soul to spirit, from animal to human activity, is effected by means of language. And this should be said of intuition or expression in general. But to us it appears somewhat inaccurate to define language or expression as an *intermediate* link between nature and humanity, as though it were a mixture of both. Where humanity appears, the other has already disappeared; the man who expresses himself, certainly emerges from the state of nature, but he really does emerge: he does not stand half within and half without, as the use of the phrase "intermediate link" would imply.

Non-existence of other forms of knowledge.

The cognitive spirit has no form other than these two. Expression and concept exhaust it completely. The whole speculative life of man is spent in passing from one to the other and back again.

Historicity[1]. *Its identity with and difference from art.*

Historicity is incorrectly held to be a third theoretical form. Historicity is not form, but content: as form, it is nothing but intuition or æsthetic fact. History does not seek for laws nor form concepts; it employs neither induction nor deduction[2]; it is directed *ad narrandum, non ad demonstrandum*[3]; it does not construct universals and abstractions, but posits intuitions. The this and here, the *individuum omnimode determinatum*[4], is its domain, as it is the domain of art. History, therefore, is included in the universal concept of art.

As against this doctrine, in view of the impossibility of conceiving a third mode of knowledge, objections have been brought forward which would lead to the affiliation of history to intellectual or scientific knowledge. The greater portion of these objections is animated by the prejudice that in refusing to history the character of conceptual science something of its value and dignity has been taken from it. This really arises from a false idea of art, conceived not as an essential theoretic function, but as an amusement, a superfluity, a frivolity[5]. Without reopening a long debate, which so far as we are concerned is finally closed, we will mention here one sophism[6] which has been and still is widely repeated. Its purpose is to show the logical and scientific nature of history. The sophism consists in admitting that historical knowledge has for its object the individual; but not the representation, it is added, but rather the concept of the individual. From this it is argued that history is also a logical or scientific form of knowledge. History, in fact, is supposed to work out the concept of a personage such as

1. *Historicity*：史实
2. employs neither induction nor deduction：既不用归纳法也不用演绎法
3. *ad narrandum, non ad demonstrandum*：〈拉〉为了讲述，而非为了证明
4. *individuum omnimode determinatum*：〈拉〉各式各样确定的个体
5. a superfluity, a frivolity：多余之物，轻浮之物
6. sophism：诡辩

Charlemagne[1] or Napoleon ; of an epoch, like the Renaissance[2] or the Reformation[3]; of an event, such as the French Revolution[4] and the Unification of Italy[5]. This it is held to do in the same way as Geometry works out the concepts of spatial forms, or Æsthetic that of expression. But all this is untrue. History cannot do otherwise than *represent* Napoleon and Charlemagne, the Renaissance and the Reformation, the French Revolution and the Unification of Italy as individual facts with their individual physiognomy: that is, in the sense in which logicians use the word "represent" when they say that one cannot have a concept of the individual, but only a representation. The so-called concept of the individual is always a universal or general concept, full of characteristics, supremely full, if you like, but however full it be, incapable of attaining to that individuality to which historical knowledge, as æsthetic knowledge, alone attains.

To show how the content of history comes to be distinguished from that of art in the narrow sense, we must recall what has already been observed as to the ideal character of the intuition or first perception, in which all is real and therefore nothing is real. Only at a later stage does the spirit form the concepts of external and internal, of what

---

1. Charlemagne：查理曼大帝（742—814）。他建立了囊括西欧大部分地区的庞大帝国，是法兰克王国加洛林王朝国王，神圣罗马帝国的奠基人，被后世称为"欧洲之父"。
2. Renaissance：文艺复兴（14 世纪—16 世纪），发生在 14 世纪到 16 世纪的一场反映新兴资产阶级要求的欧洲思想文化运动，发起于意大利，后扩展到西欧各国，文艺复兴运动对当时的政治，科学，经济，哲学，神学世界观都产生极大影响，被认为是欧洲中古时代和近代的分界。
3. Reformation：宗教改革（16 世纪—17 世纪），由马丁・路德发起，代表人物有马丁・路德、加尔文等。宗教瓦解了天主教会所主导的政教体系，打破了天主教的精神束缚，为西欧资本主义发展和多元化的现代社会奠定了基础。
4. French Revolution：法国大革命，又称法国资产阶级革命，是 1789 年 7 月 14 日在法国爆发的革命，波旁王朝被推翻，取而代之的是崭新的民主制度。
5. Unification of Italy：意大利统一运动（约 1815—1871），意大利半岛各个国家或政权逐步统一为意大利王国的社会进程。

has happened and what is desired, of object and subject, and the like: only at this later stage, that is, does it distinguish historical from non-historical intuition, the *real* from the *unreal*, real imagination from pure imagination. Even internal facts, what is desired and imagined, castles in the air, and countries of Cockaigne[1], have their reality, and the soul, too, has its history. His illusions form part of the biography of every individual as real facts. But the history of an individual soul is history, because the distinction between the real and the unreal is always active in it, even when the illusions themselves are the real. But these distinctive concepts do not appear in history like the concepts of science, but rather like those that we have seen dissolved and melted in the æsthetic intuitions, although in history they stand out in a manner altogether special to themselves. History does not construct the concepts of the real and unreal, but makes use of them. History, in fact, is not the theory of history. Mere conceptual analysis is of no use in ascertaining whether an event in our lives was real or imaginary. We must mentally reproduce the intuitions in the most complete form, as they were at the moment of production. Historicity is distinguished in the concrete from pure imagination as any one intuition is distinguished from any other: in memory.

Historical criticism.

Where this is not possible, where the delicate and fleeting shades between the real and unreal intuitions are so slight as to mingle the one with the other, we must either renounce for the time being at least the knowledge of what really happened (and this we often do), or we must fall back upon conjecture, verisimilitude[2], probability. The principle of verisimilitude and of probability in fact dominates all

1. Cockaigne：想象中的乐土
2. verisimilitude：近似度

historical criticism. Examination of sources and authorities is devoted to establishing the most credible evidence. And what is the most credible evidence, save that of the best observers, that is, of those who best remember and (be it understood) have not wished to falsify, nor had interest in falsifying the truth of things?

Historical scepticism.

From this it follows that intellectualistic scepticism[1] finds it easy to deny the certainty of any history, for the certainty of history differs from that of science. It is the certainty of memory and of authority, not that of analysis and demonstration. To speak of historical induction or demonstration is to make a metaphorical use of these expressions, which bear a quite different meaning in history to that which they bear in science. The conviction of the historian is the undemonstrable conviction[2] of the juryman, who has heard the witnesses, listened attentively to the case, and prayed Heaven to inspire him. Sometimes, without doubt, he is mistaken, but the mistakes are in a negligible minority compared with the occasions when he grasps the truth. That is why good sense is right against the intellectualists in believing in history, which is not a "fable agreed upon," but what the individual and humanity remember of their past. We strive to enlarge and to render as precise as possible this record, which in some places is dim, in others very clear. We cannot do without it, such as it is, and taken as a whole it is rich in truth. Only in a spirit of paradox can one doubt that there ever was a Greece or a Rome, an Alexander[3] or a Cæsar[4],

---

1. intellectualistic scepticism：唯智论怀疑主义
2. undemonstrable conviction：难以证明的信念
3. Alexander：亚历山大大帝（公元前 356—前 323），马其顿国王，亚历山大帝国皇帝，著名军事家和政治家。
4. Cæsar：恺撒大帝（公元前 102—前 44），罗马共和国末期著名军事统帅和政治家。

a feudal Europe overthrown by a series of revolutions, that on the 1st of November 1517 the theses of Luther[1] were fixed to the door of the church at Wittemberg, or that the Bastile[2] was taken by the people of Paris on the 14th of July 1789.

"What proof hast thou of all this?" asks the sophist, ironically. Humanity replies: "I remember it."

Philosophy as perfect science. The so-called natural sciences, and their limits.

The world of what has happened, of the concrete, of historical fact, is the world called real, natural, including in this definition both the reality called physical and that called spiritual and human. All this world is intuition; historical intuition, if it be shown as it realistically is; imaginary or artistic intuition in the narrow sense, if presented in the aspect of the possible, that is to say, of the imaginable.

Science, true science, which is not intuition but concept, not individuality but universality, cannot be anything but science of the spirit, that is, of what reality has of universal: Philosophy. If natural *sciences* be spoken of, apart from philosophy, we must observe that these are not perfect sciences: they are aggregates of cognitions, arbitrarily abstracted and fixed. The so-called natural sciences indeed themselves recognize that they are surrounded by limitations, and these limitations are nothing but historical and intuitive data. They calculate, measure, establish equalities and uniformities, create classes and types, formulate laws, show in their own way how one fact arises

1. theses of Luther：《路德论纲》，即《九十五条论纲》，是马丁·路德于1517 年 10 月 31 日张贴在德国维滕贝格诸圣堂大门上的《辩论提纲》。马丁·路德（1483—1546），宗教改革发起人，基督教新教路德宗创始人。
2. Bastile：巴士底狱，原名“巴士底要塞”，是根据法国国王查理五世的命令，按照 14 世纪著名军事城堡样式建造的一座非常坚固的要塞。攻占巴士底狱是法国大革命的标志性事件。

out of other facts; but while doing this they are constantly running into facts known intuitively and historically. Even geometry now states that it rests altogether on hypothese, since three dimensional or Euclidean space[1] is but one of the possible spaces, selected for purposes of study because more convenient. What is true in the natural sciences is either philosophy or historical fact. What of properly naturalistic they contain, is abstraction and caprice[2]. When the natural sciences wish to become perfect sciences, they must leave their circle and enter philosophy. They do this when they posit concepts which are anything but naturalistic, such as those of the unextended atom, of ether[3] or vibration, of vital force, of non-intuitional space, and the like. These are true and proper attempts at philosophy, when they are not mere words void of meaning. The concepts of natural science are, without doubt, most useful; but one cannot obtain from them that *system* which belongs only to the spirit.

These historical and intuitive data which cannot be eliminated from the natural sciences furthermore explain not only how, with the advance of knowledge, what was once believed to be true sinks gradually to the level of mythological belief and fantastic illusion, but also how among natural scientists some are to be found who call everything in their sciences upon which reasoning is founded *mythical facts, verbal expedients,* or *conventions*[4]. Natural scientists and mathematicians who approach the study of the energies of the spirit without preparation, are apt to carry thither such mental habits and to speak in philosophy of such and such conventions as "decreed by man[5]."

---

1. Euclidean space：欧几里得空间，也可以称为平直空间，是一个特别的度量空间。
2. abstraction and caprice：东拼西凑
3. ether："以太"，或称"光以太"，亚里士多德所设想的物质之一，是物理学史上一种假想的物质观念，其内涵随物理学发展而演变。
4. *mythical facts, verbal expedients,* or *conventions*：神话细节、口头应急之说或惯例
5. decreed by man：人为规定

They make conventions of truth and morality, and a supreme convention of the Spirit itself! But if there are to be conventions, something must exist which is no convention, but is itself the author of conventions. This is the spiritual activity of man. The limitation of the natural sciences postulates the illimitability of philosophy[1].

The phenomenon and the noumenon.[2]

These explications have firmly established that the pure or fundamental forms of knowledge are two: the intuition and the concept — Art, and Science or Philosophy. With these are to be included History, which is, as it were, the product of intuition placed in contact with the concept, that is, of art receiving in itself philosophic distinctions, while remaining concrete and individual. All other forms (natural sciences and mathematics) are impure, being mingled with extraneous elements[3] of practical origin. Intuition gives us the world, the phenomenon; the concept gives us the noumenon, the Spirit.

---

1. illimitability of philosophy：哲学的无局限性
2. *noumenon*：本体
3. extraneous element：外来因素

IV HISTORICISM AND INTELLECTUALISM IN ÆSTHETIC

These relations between intuitive or æsthetic knowledge and the other fundamental or derivative forms[1] of knowledge having been definitely established, we are now in a position to reveal the errors of a series of theories which have been, or are, presented as theories of Æsthetic.

Criticism of probability and of naturalism.

From the confusion between the demands of art in general and the particular demands of history has resulted the theory (which has lost ground to-day, but was once dominant) of the *probable*[2] as the object of art. As is generally the case with erroneous propositions[3], the meaning of those who employed and employ the concept of probability has no doubt often been much more reasonable than their definition of the word. By probability used really to be meant the artistic *coherence*[4] of the representation, that is to say, its completeness and effectiveness, its actual presence. If "probable" be translated "coherent," a very just meaning will often be found in the discussions, examples, and judgements of the critics who employ this word. An improbable personage, an improbable ending to a comedy, are really badly-drawn personages, badly-arranged endings, happenings without artistic motive. It has been said with reason that even

1. fundamental or derivative forms：基本形式或衍生形式
2. theory of the *probable*：或然说
3. erroneous propositions：错误前提
4. artistic *coherence*：艺术连贯性

fairies and sprites[1] must have probability, that is to say, be really sprites and fairies, coherent artistic intuitions. Sometimes the word "possible" has been used instead of "probable." As we have already remarked in passing, this word possible is synonymous with the imaginable or intuitible[2]. Everything truly, that is to say coherently, imagined, is possible. But also, by a good many critics and theorists, the probable was taken to mean the historically credible, or that historical truth which is not demonstrable but conjecturable, not true but probable. This was the character which these theorists sought to impose upon art. Who does not remember how great a part was played in literary history by criticism based on probability, for example, censure of *Jerusalem Delivered*[3], based upon the history of the Crusades[4], or of the Homeric poems, upon the probable customs of emperors and kings? Sometimes too the æsthetic reproduction of historical reality has been imposed upon art. This is another of the erroneous forms taken by the theory of the *imitation of nature*. Verism[5] and naturalism[6] also have afforded the spectacle of a confusion of the æsthetic fact with the processes of the natural sciences, by aiming at some sort of *experimental* drama or romance.

Criticism of ideas in art, of theses in art and of the typical.

Confusions between the methods of art and those of the philosophic

1. fairies and sprites：神仙鬼怪
2. imaginable or intuitible：可想象的和可直觉的
3. *Jerusalem Delivered*：《解放了的耶路撒冷》，意大利文艺复兴时期的著名诗人托夸多・塔索的叙事长诗。以 11 世纪第一次十字军东征为题材，叙述基督教士历经各种艰难险阻，最终取得对穆斯林的胜利，解放了圣城耶路撒冷的征战历程。
4. Crusades：十字军东征（1095—1291），著名宗教战争，在罗马教皇的准许和支持下，欧洲封建主和骑士向异教徒国家发起攻击，前后共有九次，持续近两百年。
5. Verism：写实主义
6. naturalism：自然主义

sciences have been far more frequent. Thus it has often been held to be the task of art to expound concepts, to unite an intelligible with a sensible, to represent *ideas* or *universals*; putting art in the place of science, that is, confusing the artistic function in general with the particular case in which it becomes æsthetico-logical.

The theory of art as supporting *theses*, of art considered as an individual representation exemplifying scientific laws, can be proved false in like manner. The example, as example, stands for the thing exemplified, and is thus an exposition of the universal, that is to say, a form of science, more or less popular or vulgarizing[1].

The same may be said of the æsthetic theory of the *typical*, when by type is understood, as it frequently is, the abstraction or the concept, and it is affirmed that art should make the *species* shine in the *individual*. If individual be here understood by typical, we have here too a merely verbal variation. To typify would signify, in this case, to characterize; that is, to determine and to represent the individual. Don Quixote[2] is a type; but of what is he a type, save of all Don Quixotes? A type, so to speak, of himself. Certainly he is not a type of abstract concepts, such as the loss of the sense of reality, or of the love of glory. An infinite number of personages can be thought of under these concepts, who are not Don Quixotes. In other words, we find our own impressions fully determined and realized in the expression of a poet (for example in a poetical personage). We call that expression typical, which we might call simply æsthetic. Thus poetical or artistic universals have sometimes been spoken of, only to show that the artistic product is altogether spiritual and ideal.

---

1. more or less popular or vulgarizing：或多或少通俗化了
2. Don Quixote：堂吉诃德。塞万提斯的同名长篇小说主人公，这个没落贵族因迷恋古代骑士小说，幻想自己是个中世纪骑士，自封为堂吉诃德，拉着邻居桑丘·潘沙做自己的仆人，行侠仗义游走天下，作出了种种与时代相悖且令人匪夷所思的举动，结果四处碰壁。

Criticism of the symbol and of the allegory[1].

Continuing to correct these errors, or to clear up misunderstandings, we shall also remark that the *symbol* has sometimes been given as the essence of art. Now, if the symbol be conceived as inseparable from the artistic intuition, it is a synonym for the intuition itself, which always has an ideal character. There is no double bottom to art, but one only; in art all is symbolical, because all is ideal. But if the symbol be conceived as separable — if the symbol can be on one side, and on the other the thing symbolized, we fall back again into the intellectualist error: the so-called symbol is the exposition of an abstract concept, an *allegory*; it is science, or art aping science[2]. But we must also be just toward the allegorical. Sometimes it is altogether harmless. Given the *Gerusalemme liberata*,[3] the allegory was imagined afterwards; given the *Adone* of Marino[4], the poet of the lascivious[5] afterwards insinuated[6] that it was written to show how "immoderate indulgence ends in pain[7]"; given a statue of a beautiful woman, the sculptor can attach a label to the statue saying that it represents *Clemency*[8] or *Goodness*. This allegory that arrives attached to a finished work *post festum*[9] does not change the work of art. What then is it? It is an expression externally *added* to another expression. A little page of prose is added to the *Gerusalemme*, expressing another thought of

1. *allegory*：寓言
2. art aping science：模仿艺术的科学
3. *Gerusalemme liberata*：〈意〉《解放了的耶路撒冷》
4. *Adone* of Marino：马里诺的《安东尼斯》。马里诺（1569—1526），意大利诗人。《安东尼斯》运用比喻，典故和象征手法，叙述了维纳斯与安东尼的爱情故事。
5. the poet of the lascivious：轻浮的诗人
6. insinuated：暗示
7. immoderate indulgence ends in pain：荒淫无度必有恶报
8. *Clemency*：仁慈
9. *post festum*：〈拉〉事后

the poet; a verse or a strophe[1] is added to the *Adone*, expressing what the poet would like to make a part of his public believe; to the statue nothing but the single word: *Clemency* or *Goodness*.

Criticism of the theory of artistic and literary kinds.

But the greatest triumph of the intellectualist error lies in the theory of artistic and literary kinds, which still has vogue in literary treatises and disturbs the critics and the historians of art. Let us observe its genesis.

The human mind can pass from the æsthetic to the logical, just because the former is a first step in respect to the latter. It can destroy expression, that is, the thought of the individual, by thinking of the universal. It can gather up expressive facts into logical relations. We have already shown that this operation becomes in its turn concrete in an expression, but this does not mean that the first expressions have not been destroyed. They have yielded their place to the new æsthetico-logical expressions. When we are on the second step, we have left the first.

One who enters a picture-gallery, or who reads a series of poems, having looked and read, may go further: he may seek out the nature and the relations of the things there expressed. Thus those pictures and compositions, each of which is an individual inexpressible in logical terms, are gradually resolved into universals and abstractions, such as *costumes, landscapes, portraits, domestic life, battles, animals, flowers, fruit, seascapes, lakes, deserts; tragic, comic, pathetic, cruel, lyrical, epic, dramatic, chivalrous, idyllic facts,* and the like. They are often also resolved into merely quantitative categories, such as *miniature, picture,*

1. a verse or a strophe：一句或一章

statuette, group, madrigal[1], *ballad, sonnet, sonnet-sequence*[2], *poetry, poem, story, romance,* and the like.

When we think the concept *domestic life,* or *chivalry*[3], or *idyll*[4], or *cruelty*, or one of the quantitative concepts mentioned above, the individual expressive fact from which we started has been abandoned. From æsthetes that we were, we have changed into logicians; from contemplators of expression, into reasoners. Certainly no objection can be made to such a process. In what other way could science arise, which, if it have æsthetic expressions presupposed in it, must yet go beyond them in order to fulfil its function? The logical or scientific form, as such, excludes the æsthetic form. He who begins to think scientifically has already ceased to contemplate æsthetically; although his thought assumes of necessity in its turn an æsthetic form, as has already been said, and as it would be superfluous to repeat.

Error begins when we try to deduce the expression from the concept, and to find in what takes its place the laws of the thing whose place is taken; when the difference between the second and the first step has not been observed, and when, in consequence, we declare that we are standing on the first step, when we are really standing on the second. This error is known as the *theory of artistic and literary kinds.*

"What is the *æsthetic* form of domestic life, of chivalry, of the idyll, of cruelty, and so forth? How should these contents be *represented*?" Such is the absurd problem implied in the theory of artistic and literary classes, when it has been shorn of excrescences[5] and reduced to a simple formula. It is in this that consists all search after laws or rules

1. madrigal：情歌
2. *sonnet-sequence*：十四行诗
3. *chivalry*：骑士风度
4. *idyll*：田园生活
5. shorn of excrescences：化繁为简

of classes. Domestic life, chivalry, idyll, cruelty and the like, are not impressions, but concepts. They are not contents, but logical-æsthetic forms. You cannot express the form, for it is already itself expression. For what are the words cruelty, idyll, chivalry, domestic life, and so on, but the expression of those concepts?

Even the most refined of such distinctions, which possess the most philosophic appearance, do not resist criticism; as when works of art are divided into subjective and objective kinds, into lyric and epic, into works of feeling and decorative works. In æsthetic analysis it is impossible to separate subjective from objective, lyric from epic, the image of feeling from that of things.

Errors derived from this theory in judgements on art.

From the theory of artistic and literary kinds derive those erroneous modes of judgement and of criticism, thanks to which, instead of asking before a work of art if it be expressive and what it expresses, whether it speak or stammer or is altogether silent, they ask if it obey the *laws* of epic or of tragedy, of historical painting or of landscape. While making a verbal pretence of agreeing, or yielding a feigned obedience, artists have, however, really always disregarded these *laws of the kinds*. Every true work of art has violated some established kind and upset the ideas of the critics, who have thus been obliged to broaden the kinds, until finally even the broadened kind has proved too narrow, owing to the appearance of new works of art, naturally followed by new scandals, new upsettings and — new broadenings.

To the same theory are due the prejudices, owing to which at one time (is it really passed?) people used to lament[1] that Italy had no

1. lament：惋惜

tragedy (until one arose who bestowed such a wreath, which alone of adornments was wanting to her glorious locks), nor France the epic poem (until the *Henriade*[1], which slaked the thirsty throats of the critics[2]). Eulogies[3] accorded to the inventors of new kinds are connected with these prejudices, so much so, that in the seventeenth century the invention of the *mock-heroic* poem[4] seemed an important event, and the honour of it was disputed, as though it were the discovery of America. But the works adorned with this name (the *Secchia rapita*[5] and the *Scherno degli Dei*[6]) were still-born, because their authors (a slight drawback) had nothing new or original to say. Mediocrities racked their brains[7] to invent new kinds artificially. The *piscatorial* eclogue[8] was added to the *pastoral*[9], and finally the *military* eclogue[10]. The *Aminta*[11] was dipped and became the *Alceo*. Finally, there have been historians of art and literature, so much fascinated with these ideas of kinds, that they claimed to write the history, not of individual and real literary and artistic works, but of those empty phantoms, their kinds. They have claimed to portray, not the evolution of the *artistic spirit*, but the *evolution of kinds*.

The philosophical condemnation of artistic and literary kinds is found in the formulation and demonstration of what artistic activity

---

1. *Henriade*：《亨利亚德》，伏尔泰创作的长诗。
2. slaked the thirsty throats of the critics：滋润了一下批评家们的干渴的嗓子
3. Eulogies：颂歌
4. *mock-heroic* poem：仿英雄体诗
5. *Secchia rapita*：〈意〉《被绑架的水桶》，意大利诗人亚历山大·塔索尼（Alessandro Tassoni，1565—1635）的讽刺诗。
6. *Scherno degli Dei*：〈意〉《封神的蔑视》，弗朗切斯科·布拉乔利尼（Francesco Bracciolini）著。
7. Mediocrities racked their brains：平庸之辈们绞尽脑汁
8. *piscatorial* eclogue：渔歌
9. *pastoral*：田园诗
10. *military* eclogue：军旅诗
11. *Aminta*：《阿明塔》，意大利诗人塔索 (Tasso) 创作的田园剧，塑造了一位对爱情忠贞不渝的痴情者形象。

has always done and good taste always recognized. What are we to do if good taste and the real fact, when reduced to formulas, sometimes assume the air of paradoxes?

Empirical sense of the divisions of kinds.

It is not scientifically incorrect to talk of tragedies, comedies, dramas, romances, pictures of everyday life, battle-pieces, landscapes, seascapes, poems, versicles[1], lyrics, and the like, if it be only with a view to be understood, and to draw attention to certain groups of works, in general and approximately, to which, for one reason or another, it is desired to draw attention. To employ *words* and *phrases* is not to establish *laws* and *definitions*. The mistake only arises when the weight of a scientific definition is given to a word, when we ingenuously let ourselves be caught in the meshes of that phraseology. Pray permit me a comparison. The books in a library must be arranged in one way or another. This used generally to be done by a rough classification of subjects (among which the categories of miscellaneous and eccentric were not wanting); they are now generally arranged by sizes or by publishers. Who can deny the necessity and the utility of such arrangements? But what should we say if some one began seriously to seek out the literary laws of miscellanies and of eccentricities, of the Aldines[2] or Bodonis[3], of shelf A or shelf B, that is to say, of those altogether arbitrary groupings whose sole object was their practical utility. Yet should any one attempt such an undertaking,

---

1. versicles：小诗，短诗
2. Aldines：阿尔丁版。印刷世家出身的阿尔杜斯・马努蒂乌斯（Aldus Manutius）1494 年在威尼斯创立阿尔丁出版社。该出版社因在印刷中引入斜体字而闻名于世。
3. Bodonis：波多尼版。由意大利字体设计师波多尼（Bodoni 1740—1813）创设。波多尼字体体现了现代主义风格，后被广泛用于出版印刷。

he would be doing neither more nor less than those do who seek out the *æsthetic laws* which must in their belief control literary and artistic kinds.

V ANALOGOUS ERRORS IN THE THEORY OF HISTORY AND IN LOGIC

The better to confirm these criticisms, it will be useful to cast a rapid glance over analogous and opposite errors, due to ignorance as to the true nature of art and its relation to history and to science. These errors have injured alike the theory of history and that of science, Historic (or Historiology) and Logic.

Criticism of the philosophy of history.

Historical intellectualism[1] has opened the way to the many attempts, made especially during the last two centuries and continued to-day, to discover *a philosophy of history,* an *ideal history,* a *sociology*, a *historical psychology*, or whatever else a science may be called, whose object is to extract from history concepts and universal laws. What must these laws, these universals be? Historical laws and historical concepts? In that case, an elementary acquaintance with the theory of knowledge suffices to make clear the absurdity of the attempt. When such expressions as a *historical law,* a *historical concept* are not simply metaphors colloquially employed, they are truly contradictory terms: the adjective is as unsuitable to the substantive as in the expressions "qualitative quantity[2]" or "pluralistic monism[3]." History implies concreteness and individuality, law and concept mean abstractness

1. Historical intellectualism：历史唯智主义
2. qualitative quantity: 论质的量
3. pluralistic monism：多元化一元论

and universality. But if the attempt to extract *historical* laws and concepts from history be abandoned, and it be merely desired to draw from it laws and concepts, the attempt is certainly not frivolous[1]; but the science thus obtained will be, not a philosophy of history, but rather, according to circumstances, either philosophy in its various forms of Ethics, Logic, etc., or empirical science with its infinite divisions and subdivisions. The search is in fact either for those philosophical concepts which, as already remarked, are the basis of every historical construction and differentiate perception from intuition, historical intuition from pure intuition, history from art; or already formed historical intuitions are collected and arranged in types and classes, which is exactly the method of the natural sciences. Great thinkers have sometimes donned the ill-fitting cloak of the philosophy of history[2], and notwithstanding the covering, they have attained philosophical truths of the greatest magnitude. The cloak discarded, the truth has remained. Modern sociologists are rather to be blamed, not so much for the illusion in which they are involved when they talk of an impossible science of sociology, as for the infecundity[3] which almost always accompanies their illusion. It matters little that Æsthetic should be called "sociological Æsthetic," or Logic, "sociological Logic." The grave evil is that such Æsthetic is an old-fashioned expression of sensationalism, such Logic verbal and incoherent. The philosophical movement to which we have referred has however borne two good fruits in relation to history. First of all, a keener desire has arisen for a theory of history, that is, a theory of the nature and the limits of history, a theory which, in conformity with the analysis made above, cannot obtain satisfaction save in a

---

1. frivolous: 愚蠢的；轻率的
2. donned the ill-fitting cloak of the philosophy of history: 披上不合身的历史哲学外衣
3. infecundity：收效甚微

general science of intuition, in an Æsthetic, in which the theory of history would form a special chapter, distinguished by the insertion of universal functions. Furthermore, concrete truths relating to historical events have often been expressed beneath the false and presumptuous cloak[1] of a philosophy of history; rules and warnings have been formulated, empirical no doubt, yet by no means useless to students and critics. It does not seem possible to deny this utility even to the most recent of philosophies of history, known as historical materialism, which has thrown a very vivid light upon many sides of social life formerly neglected or ill understood.

Æsthetic intrusions into Logic.

The principle of authority, of the *ipse dixit*[2], is an intrusion by historicity into the domains of science and philosophy which has dominated the schools and substitutes for introspection and philosophical analysis this or that evidence, document, or authoritative statement, with which history certainly cannot dispense. But Logic, the science of thought and of intellectual knowledge, has suffered the most grave and destructive of all disturbances and errors through an imperfect understanding of the æsthetic fact. How could it be otherwise, if logical activity come after and contain in itself æsthetic activity? An inexact Æsthetic must of necessity drag after it an inexact Logic.

Whoever opens a logical treatise, from the *Organon* of Aristotle[3] to the modern works on the subject, must agree that all contain a haphazard mixture of verbal facts and facts of thought, of grammatical forms and of conceptual forms, of Æsthetic and of Logic. Not that

1. false and presumptuous cloak：虚伪且肆无忌惮的伪装
2. *ipse dixit*：〈拉〉武断的言辞
3. *Organon* of Aristotle：亚里士多德的《工具论》

attempts have been wanting to escape from verbal expression and to seize thought in its true nature. Aristotelian logic itself did not become mere syllogistic and verbalism[1] without some hesitation and indecision. The problem proper to logic was often touched upon in their disputes by the nominalists, realists and conceptualists[2] of the Middle Ages . With Galileo[3] and with Bacon[4], the natural sciences gave an honourable place to induction. Vico combated formalist and mathematical logic in favour of inventive methods. Kant[5] called attention to the *a priori*[6] synthesis. Absolute idealism[7] despised the Aristotelian Logic. The followers of Herbart[8], though still loyal to Aristotle, emphasized those judgements which they called narrative and which have a character altogether differing from that of other logical judgements. Finally, the linguists insisted upon the irrationality of the word, in relation to the concept. But a conscious, sure and radical movement of reform can find no basis or point of departure, save in the science of Æsthetic.

Logic in its essence.

In a Logic suitably reformed on this basis, this truth must first and foremost be proclaimed, and all its consequences deduced: the logical

1. syllogistic and verbalism：三段论和无意义的套话
2. nominalists, realists and conceptualists：唯名论者，唯实论者和概念论者
3. Galileo：伽利略（1564—1642），意大利物理学家、天文学家、数学家和哲学家，近代实验科学先驱。
4. Bacon：培根（1561—1626），英国近代唯物主义哲学家、思想家和科学家，著有《新工具》《培根散文集》等。
5. Kant：康德（1724—1804），哲学家，德国古典美学奠基者，著有《纯粹理性批判》《实践理性批判》《判断力批判》等。
6. *a priori*：〈拉〉先验
7. Absolute idealism：绝对唯心主义
8. Herbart：即约翰·弗里德里希·赫尔巴特（1776—1841），德国哲学家、心理学家，教育学创立者，著有《普通教育学》《形而上学的要点》《一般实践哲学》等。

fact, *the only logical fact*, is *the concept*, the universal, the spirit that forms, and in so far as it forms, the universal. And if by induction be understood, as sometimes it has been, the formation of universals, and by deduction their verbal development, then it is clear that true Logic can be nothing but inductive Logic. But since by the word "deduction" has been more frequently understood the special processes of mathematics, and the word "induction" those of the natural sciences, it will be best to avoid both words and say that true Logic is Logic of the concept. The Logic of the concept, while employing a method which is both induction and deduction, will employ neither exclusively, that is, it will employ the speculative method which is intrinsic to it.

The concept, the universal, considered abstractly in itself, is *inexpressible*. No word is proper to it. So true is this, that the logical concept remains always the same, notwithstanding the variation of verbal forms. In respect to the concept, expression is a simple *sign* or *indication*. There must be an expression, it cannot be absent; but what it is to be, this or that, is determined by the historical and psychological conditions of the individual who is speaking. The quality of the expression is not deducible from the nature of the concept. There does not exist a true (logical) sense of words. The true sense of words is that which is conferred upon them on each occasion by the person forming a concept.

Distinction between logical and non-logical judgements.

This being so, the only truly logical (that is, æsthetico-logical) propositions, the only rigorously logical judgements, must be those whose proper and sole content is the determination of a concept. These propositions or judgements are *definitions*. Science itself is nothing but a collection of definitions, unified in a supreme definition; a system of concepts, or highest concept.

It is therefore necessary (at least as a preliminary) to exclude from

Logic all those propositions which do not affirm universals. Narrative judgements, not less than those termed non-enunciative[1] by Aristotle, such as the expression of desires, are not properly logical judgements. They are either purely æsthetic propositions[2] or historical propositions. "Peter is passing; it is raining to-day; I am sleepy; I want to read": these and an infinity of propositions of the same kind are nothing but either a mere enclosing in words the impression of the fact that Peter is passing, of the falling rain, of my organism inclining to sleep, and of my will directed to reading, or an existential affirmation concerning those facts. They are expressions of the real or of the unreal, historical-imaginative or pure-imaginative; they are certainly not definitions of universals.

Syllogistic.

This exclusion cannot meet with great difficulties. It is already almost an accomplished fact, and the only thing required is to render it explicit, decisive and coherent. But what is to be done with all that part of human thought called *syllogistic*, consisting of judgements and reasonings based upon concepts? What is syllogistic? Is it to be looked down upon with contempt, as something useless, as has so often been done by the humanists in their reaction against scholasticism, by absolute idealism, by the enthusiastic admiration of our times for the methods of observation and experiment of the natural sciences? — Syllogistic, reasonings *forma*[3], is not the discovery of truth; it is the art of expounding, debating, disputing with oneself and others. Proceeding from concepts already formed, from facts already observed, and appealing to the persistence of the true or of thought

1. non-enunciative：未阐明的判断
2. æsthetic proposition：审美命题
3. reasonings *forma*：公式化推理

(such is the meaning of the laws of identity and contradiction[1]), it infers consequences from those data, that is, it re-states what has already been discovered. Therefore, if it be an *idem per idem*[2] from the point of view of invention, it is most efficacious[3] in teaching and in exposition. To reduce affirmations to a syllogistic form is a way of controlling one's own thought and of criticizing the thought of others. It is easy to laugh at syllogizers, but, if syllogistic has been born and persists, it must have good reasons of its own. Satire on it can concern only its abuses, such as the attempt to prove syllogistically questions of fact, observation and intuition, or the neglect of profound meditation and unprejudiced investigation of problems, in favour of syllogistic externality. And if so-called *mathematical Logic* can sometimes aid us in our attempt to remember with ease, rapidly to control the results of our own thought, let us welcome this form of syllogistic also, anticipated by Leibnitz[4] among others and again attempted by some in our own days.

But precisely because syllogistic is the art of exposition and debate, its theory cannot hold the first place in a philosophical Logic, thus usurping[5] that belonging to the doctrine of the concept, which is the central and dominating doctrine, to which everything logical in syllogistic is reducible, without leaving a residuum (relations of concepts, subordination, co-ordination, identification and so on). Nor must it ever be forgotten that concept and (logical) judgement and syllogism are not in the same line. The first alone is the logical fact, the second and third are the forms in which the first manifests itself. These, in so far as they are forms, can only be examined æsthetically

1. laws of identity and contradiction：同一律和矛盾律
2. *idem per idem*：〈拉〉the same for the same
3. efficacious: 灵验的
4. Leibnitz：莱布尼茨（1646—1716），德国哲学家、数学家和物理学家，著有《抽象运动的理论》《新物理学假说》等。
5. usurping：篡夺

(grammatically), and in so far as they possess logical content, only by ignoring the forms themselves and passing to the doctrine of the concept.

Logical falsehood and æsthetic truth.

This confirms the truth of the ordinary remark to the effect that he who reasons ill, also speaks and writes ill, that exact logical analysis is the basis of good expression. This truth is a tautology[1], for to reason well is in fact to express oneself well, because the expression is the intuitive possession of one's own logical thought. The principle of contradiction itself is at bottom nothing but the æsthetic principle of coherence. It may be maintained that it is possible to write and to speak exceedingly well, as it is also possible to reason well though starting from erroneous concepts; that some, though lacking the acuteness[2] that makes a great discoverer, are nevertheless exceedingly lucid writers; because to write well depends upon having a clear intuition of one's own thought, even if it be erroneous; not of its scientific, but of its æsthetic truth, which indeed is the same thing as writing well. A philosopher like Schopenhauer can imagine that art is a representation of the Platonic ideas. This doctrine is scientifically false, yet he may develop this false knowledge in excellent prose, æsthetically most true. But we have already replied to these objections, when observing that at that precise point where a speaker or a writer enunciates[3] an ill-thought concept, he is at the same time a bad speaker and a bad writer, although he may afterwards recover himself in the many other parts of his thought which contain true propositions not connected with the preceding error, and therefore lucid

1. tautology：同义反复
2. lacking the acuteness：缺乏敏锐性
3. enunciates：阐述

expressions[1] following upon confused expressions.

Reformed logic.

All researches as to the forms of judgements and of syllogisms, their conversions and their various relations, which still encumber[2] treatises on Logic, are therefore destined to diminish, to be transformed, to be converted into something else. The doctrine of the concept and of the organism of concepts, of definition, of system, of philosophy and the various sciences, and the like, will occupy the field and alone will constitute true and proper Logic.

Those who first had some suspicion of the intimate connexion between Æsthetic and Logic and conceived Æsthetic as a *Logic of sensible knowledge* were peculiarly addicted to applying logical categories to the new knowledge, talking of *æsthetic concepts, æsthetic judgements, æsthetic syllogisms,* and so on. We who are less superstitious as regards the permanence of the traditional Logic of the schools, and better informed as to the nature of Æsthetic, do not recommend the application of Logic to Æsthetic, but the liberation of Logic from æsthetic forms. These have given rise to non-existent forms or categories of Logic, due to the adoption of altogether arbitrary and ill-considered distinctions.

Logic thus reformed will still be *formal* Logic; it will study the true form or activity of thought, the concept, excluding individual and particular concepts. The old Logic is ill called formal; it would be better to call it *verbal* or *formalistic*. Formal Logic will drive out formalistic Logic. To attain this object, it will not be necessary to have recourse, as some have done, to a real or material Logic, which is no longer a science of thought, but thought itself in action; not only a

---

1. lucid expressions：清晰的表达
2. encumber：塞满

Logic, but the whole of Philosophy, in which Logic is also included. The science of thought (Logic) is that of the concept, as that of imagination (Æsthetic) is that of expression. The well-being of both sciences lies in exactly carrying out in every particular the distinction between the two domains.

VI THE THEORETIC ACTIVITY AND THE PRACTICAL ACTIVITY

The intuitive and intellectual forms contain between them, as we have said, the whole theoretic domain of the spirit. But it is not possible to know them thoroughly, nor to criticize another series of erroneous æsthetic theories, without first establishing clearly the relations of the theoretic spirit with the *practical* spirit.

The will.

The practical form or activity is the *will*. We do not here employ this word in the sense of some philosophical systems, where the will is the foundation of the universe, the ground of things and the true reality. Nor do we employ it in the wide sense of other systems, which understand by will the energy of the spirit, spirit or activity in general, making of every act of the human spirit an act of will. Neither such metaphysical nor such metaphorical[1] meaning is ours. For us, the will is, as generally understood, that activity of the spirit which differs from the merely theoretical contemplation of things, and is productive, not of knowledge, but of actions. Action is really action, in so far as it is voluntary. It is not necessary to remark that in the will to do, we include, in the scientific sense, also what is usually called not-doing: the will to resist, to reject, the will of a Prometheus[2], which also is action.

1. Neither such metaphysical nor such metaphorical：既非形而上学，也非玄学
2. Prometheus：普罗米修斯，古希腊神话中盗火的英雄。

The will as an ulterior stage in respect to knowledge.

Man understands things with the theoretical form, with the practical form he changes them; with the one he appropriates the universe, with the other he creates it. But the first form is the basis of the second; and the relation of *double degree,* which we have already found existing between æsthetic and logical activity, is repeated between these two on a larger scale. A knowing independent of the will is thinkable, at least in a certain sense; will independent of knowing is unthinkable. Blind will is not will; true will has eyes.

How can we will, without having before us historical intuitions (perceptions) of objects, and knowledge of (logical) relations, which enlightens us as to the nature of those objects? How can we really will, if we do not know the world which surrounds us or how to change things by acting upon them?

Objections and explanations.

It has been objected that men of action, practical men *par excellence*[1], are the least disposed to contemplate and to theorize: their energy is not delayed in contemplation, it rushes at once into will. And conversely, that contemplative men, philosophers, are often very mediocre in practical matters, weak willed, and therefore neglected and thrust aside in the tumult of life. It is easy to see that these distinctions are merely empirical and quantitative. Certainly, the practical man has no need of a philosophical system in order to act, but in the spheres where he does act, he starts from intuitions and concepts which are perfectly clear to him. Otherwise the most ordinary actions could not be willed. It would not be possible to will

---

1. *par excellence*：〈拉〉出类拔萃的

to feed oneself, for instance, without knowledge of the food, and of the link of cause and effect between certain movements and certain satisfactions. Rising gradually to the more complex forms of action, for example to the political, how could we will anything politically good or bad without knowing the real conditions of society, and consequently the means and expedients to be adopted? When the practical man feels himself in the dark about one or more of these points, or when he is seized with doubt, action either does not begin or stops. It is then that the theoretical moment, which in the rapid succession[1] of human actions is hardly noticed and rapidly forgotten, becomes important and occupies consciousness for a longer time. And if this moment be prolonged, then the practical man may become a Hamlet[2], divided between desire for action and his deficient theoretical clarity as regards the situation and the means to be employed. And if he develop a taste for contemplation and discovery, and leave willing and acting, to a greater or less extent, to others, there is formed in him the calm disposition of the artist, of the man of science, or of the philosopher, who in practice are sometimes incompetent or downright immoral[3]. These observations are all obvious. Their exactitude cannot be denied. Let us, however, repeat that they are founded on quantitative distinctions and do not disprove but confirm the fact that an action, however slight it be, cannot really be an action, that is, an action that is willed, unless it be preceded by the cognitive activity.

Criticism of practical judgements or judgements of value.

Some psychologists, on the other hand, place before practical action

1. rapid succession：快速更替
2. Hamlet：哈姆雷特，莎士比亚笔下的经典悲剧人物。
3. incompetent or downright immoral：无能甚至彻头彻尾的不道德

an altogether special class of judgements, which they call *practical* judgements or *judgements of value.* They say that in order to resolve on performing an action there must have been a judgement to the effect: "this action is useful, this action is good." And at first sight this seems to have the testimony of consciousness on its side. But closer observation and analysis of greater subtlety reveal that such judgements follow instead of preceding the affirmation of the will, and are nothing but the expression of the volition[1] already exercised. A good or useful action is an action willed. It will always be impossible to distil a single drop of usefulness or goodness from the objective study of things. We do not desire things because we know them to be good or useful; but we know them to be good and useful, because we desire them. Here too, the rapidity with which the facts of consciousness follow one another has given rise to an illusion. Practical action is preceded by knowledge, but not by practical knowledge, or rather, knowledge of the practical: to obtain this, we must first have practical action. The third moment, therefore, of practical judgements, or judgements of value, is altogether imaginary. It does not come between the two moments or degrees of theory and practice. For the rest, normative sciences[2] in general, which regulate or command, discover and indicate values to the practical activity, do not exist; indeed none exist for any sort of activity, since every science presupposes that activity to be already realized and developed, which it afterwards takes as its object.

Exclusion of the practical from the æsthetic.

These distinctions established, we must condemn as erroneous every theory which annexes the æsthetic activity to the practical, or introduces the laws of the second into the first. That science is theory

1. expression of the volition：意志的表达
2. normative sciences：规范科学

and art practice has been many times affirmed. Those who make his statement, and look upon the æsthetic fact as a practical fact, do not do so capriciously[1] or because they are groping in the void; but because they have their eye on something which is really practical. But the practical which they aim is not Æsthetic, nor within Æsthetic; it is *outside and beside it*; and although often found united, they are not united necessarily or by the bond of identity of nature.

The æsthetic fact is altogether completed in the expressive elaboration of impressions. When we have achieved the word within us, conceived definitely and vividly a figure or a statue, or found a musical motive, expression is born and is complete; there is no need for anything else. If after this we should open our mouths — *will* to open them to speak, or our throats to sing, that is to say, utter by word of mouth and audible melody what we have completely said or sung to ourselves; or if we should stretch out *–will* to stretch out our hands to touch the notes of the piano, or to take up the brush and chisel, thus making on a large scale movements which we have already made in little and rapidly, in a material in which we leave more or less durable traces; this is all an addition, a fact which obeys quite different laws from the former, with which we are not concerned for the moment, although we recognize henceforth that this second movement is a production of things, a *practical* fact, or fact of *will*. It is usual to distinguish the internal from the external work of art: the terminology seems to us infelicitous[2], for the work of art (the æsthetic work) is always *internal*; and what is called *external* is no longer a work of art. Others distinguish between *æsthetic* fact and *artistic* fact, meaning by the second the external or practical stage, which may follow and generally does follow the first. But in this case, it is simply a question of a linguistic usage, doubtless permissible, though perhaps not

1. capriciously：任性地
2. infelicitous：不恰当的

advisable.

Criticism of the theory of the end of art and of the choice of content.

For the same reasons the search for the *end of art* is ridiculous, when it is understood of art as art. And since to fix an end is to choose, the theory that the content of art must be *selected* is another form of the same error. A selection among impressions and sensations implies that these are already expressions, otherwise how could a selection be made among the continuous and indistinct? To choose is to will: to will this and not to will that: and this and that must be before us, expressed. Practice follows, it does not precede theory; expression is free inspiration.

The true artist, in fact, finds himself big with his theme, he knows not how; he feels the moment of birth drawing near, but he cannot will it or not will it. If he were to wish to act in opposition to his inspiration, to make an arbitrary choice, if, born Anacreon[1], he should wish to sing of Atreus[2] and of Alcides[3], his lyre would warn him of his mistake, sounding only of Venus[4] and of Love, notwithstanding his efforts to the contrary.

Practical innocence of art.

The theme or content cannot, therefore, be practically or morally charged with epithets of praise or blame[5]. When critics of art remark

1. Anacreon：阿那克里翁（约公元前 570—约前 480），古希腊抒情诗人，他的诗作现仅存片段。
2. Atreus：阿特柔斯，古希腊神话人物，伯罗奔尼撒半岛西北部伊利斯国国王。
3. Alcides：阿尔喀得斯，本名赫拉克拉斯，古希腊神话中的大力神。
4. Venus：维纳斯，古罗马神话中的女神，是美的化身，古罗马十二主神之一。
5. epithets of praise or blame：溢美之词或指责之论

that a theme is *badly selected,* in cases where that observation has a just foundation, it is a question of blaming, not the selection of the theme (which would be absurd), but the manner in which the artist has treated it, the failure of the expression due to the contradictions which it contains. And when the same critics object to the theme or content of works which they proclaim to be artistically perfect as being unworthy of art and blameworthy; if these expressions really are perfect, there is nothing to be done but to advise the critics to leave the artists in peace, for they can only derive inspiration from what has moved their soul. They should rather direct their attention towards effecting changes in surrounding nature and society, that such impressions and states of soul should not recur. If ugliness were to vanish from the world, if universal virtue and felicity[1] were established there, perhaps artists would no longer represent perverse or pessimistic feelings[2], but calm, innocent and joyous feelings, Arcadians of a real Arcady[3]. But so long as ugliness and turpitude[4] exist in nature and impose themselves upon the artist, to prevent the expression of these things also is impossible; and when it has arisen, *factum infectum fieri nequit*[5]. We speak thus entirely from the æsthetic point of view, and of pure criticism of art.

We are not concerned to estimate the damage which the criticism of "choice" does to artistic production, with the prejudices which it produces or maintains among the artists themselves, and with the conflict to which it gives rise between artistic impulse and critical demands. It is true that sometimes it seems also to do some good, by aiding artists to discover themselves, that is, their own impressions

1. felicity：福祉
2. perverse or pessimistic feelings：异样或悲观的情绪
3. Arcady：阿卡迪亚，古希腊神话传说中的世外桃源。
4. ugliness and turpitude：丑恶与堕落
5. *factum infectum fieri nequit*：〈拉〉木已成舟

and their own inspiration, and to acquire consciousness of the task which is, as it were, imposed upon them by the historical moment in which they live, and by their individual temperament. In these cases, criticism of "choice," while believing that it generates, merely recognizes and aids the expressions which are already being formed. It believes itself to be the mother, where, at most, it is only the midwife.

The independence of art.

The impossibility of choice of content completes the theorem of the *independence of art,* and is also the only legitimate meaning of the expression: *art for art's sake.* Art is independent both of science and of the useful and the moral. There should be no fear lest frivolous or cold art should thus be justified, since what is truly frivolous or cold is so because it has not been raised to expression; or in other words, frivolity and frigidity come always from the form of the æsthetic treatment, from failure to grasp a content, not from the material qualities of the content itself.

Criticism of the saying: the style is the man.

The saying: *the style is the man*, can also not be completely criticized, save by starting from the distinction between the theoretic and the practical, and from the theoretic character of the æsthetic activity. Man is not simply knowledge and contemplation: he is will, which contains the cognitive moment in itself. Hence the saying is either altogether void, as when it is taken to mean that the style is the man *qua* style[1] — is the man, that is, but only so far as he is expressive activity; or it is erroneous, as when the attempt is made to deduce

---

1. the style is the man *qua* style：风格即人格

what a man has done and willed from what he has seen and expressed, thereby asserting that there is a logical connexion between knowing and willing. Many legends in the biographies of artists have sprung from this erroneous identification, since it seemed impossible that a man who gives expression to generous feelings should not be a noble and generous man in practical life; or that the dramatist whose plays are full of stabbing[1], should not himself have done a little stabbing in real life. Artists protest vainly: "*Lasciva est nobis pagina, vita proba.*[2]" They are merely taxed in addition with lying and hypocrisy[3]. How far more prudent you were, poor women of Verona, when you founded your belief that Dante had really descended to hell upon his blackened countenance[4]! Yours was at any rate a historical conjecture.

Criticism of the concept of sincerity in art.

Finally, *sincerity* imposed as a duty upon the artist (a law of ethics also said to be a law of æsthetic) rests upon another double meaning. For by sincerity may be meant, in the first place, the moral duty not to deceive one's neighbour; and in that case it is foreign to the artist. For indeed he deceives no one, since he gives form to what is already in his soul. He would only deceive if he were to betray his duty as an artist by failing to execute his task in its essential nature. If lies and deceit are in his soul, then the form which he gives to these things cannot be deceit or lies, precisely because it is æsthetic. If the artist be a charlatan, a liar, or a miscreant[5], he purifies his other self by reflecting it in art. If by sincerity be meant, in the second place, fulness and truth

1. stabbing：讽刺
2. *Lasciva est nobis pagina, vita proba.*：〈拉〉诗词淫佚而行为端正
3. taxed in addition with lying and hypocrisy：受到指责，背负撒谎和虚伪的骂名
4. blackened countenance：黝黑的脸庞
5. miscreant：异端者

of expression, it is clear that this second sense has no relation to the ethical concept. The law, called both ethical and æsthetic, reveals itself here as nothing but a word used both by Ethics and Æsthetic.

VII ANALOGY BETWEEN THE THEORETIC AND THE PRACTICAL

The two forms of the practical activity.

The double degree of the theoretical activity, æsthetic and logical, has an important parallel in the practical activity, which has not yet been placed in due relief[1]. The practical activity is also divided into a first and second degree, the second implying the first. The first practical degree is the simply *useful* or *economical* activity; the second the *moral* activity.

Economy is, as it were, the Æsthetic of practical life; Morality its Logic.

The economically useful.

If this has not been clearly seen by philosophers; if the correct place in the system of the spirit has not been given to the economic activity, if it has been left to wander about in the prolegomena[2] to treatises on political economy, often vague and but little developed, this is due, among other reasons, to the fact that the useful or economic has been confused, sometimes with the concept of the *technical*, sometimes with that of the *egoistical*[3].

1. in due relief：给予应有的强调
2. prolegomena：序论
3. *egoistical*：利己主义的

Distinction between the useful and the technical.

Technique is certainly not a special activity of the spirit. Technique is knowledge; or rather, it is knowledge itself in general which takes this name when it serves as basis, as we have seen it does, for practical action. Knowledge which is not followed, or is supposed not to be easily followed by practical action, is called "pure": the same knowledge, if effectively followed by action, is called "applied"; if it is supposed that it can be easily followed by a particular action, it is called "applicable" or "technical." This word, then, indicates a *situation* in which knowledge is, or may easily be, not a special form of knowledge. So true is this, that it would be altogether impossible to establish whether a given order of knowledge were, intrinsically, pure or applied. All knowledge, however abstract and philosophical it may be believed to be, may be a guide to practical acts; a theoretical error in the ultimate principles of morality may be reflected and always in some way is reflected in practical life. One can only speak roughly and unscientifically of certain truths as pure and of others as applied.

The same knowledge that is called technical may also be called *useful*. But the word "*useful*" in conformity with the criticism of judgements of value made above, is to be understood as used here in a verbal or metaphorical sense. When we say that water is useful for putting out fire, the word "useful" is used in a non-scientific sense. Water thrown on the fire is the cause of its going out: this is the knowledge that serves for basis to the action, let us say, of firemen. There is a link, not of nature, but of simple succession[1], between the useful action of the person who extinguishes the conflagration[2] and that knowledge. The technique of the effects of the water is the theoretical activity which precedes; the only useful thing is the *action* of

1. simple succession：先后顺序
2. conflagration：大火

the man who extinguishes the fire.

Distinction of the useful from the egoistic.

Some economists identify utility, that is to say, merely economic action or will, with the *egoistic*, that is to say, with what is profitable to the individual, in so far as individual, without regard to and indeed in complete opposition to the moral law. The egoistic is the immoral. In this case Economics would be a very strange science, standing not beside but opposite Ethics, like the devil facing God, or at least like the *advocatus diaboli*[1] in the processes of canonization[2]. Such a conception is altogether inadmissible: the science of immorality is implied in that of morality, as the science of the false is implied in Logic, science of the true, and a science of unsuccessful expression in Æsthetic, science of successful expression. If, then, Economics were the scientific treatment of egoism, it would be a chapter of Ethics, or Ethics itself; because every moral determination implies, at the same time, a negation of its contrary.

Further, conscience tells us that to conduct oneself economically is not to conduct oneself egoistically; that even the most morally scrupulous man[3] must conduct himself usefully (economically), if he does not wish to act at hazard and consequently in a manner quite the reverse of moral. If utility were egoism, how could it be the duty of the altruist[4] to behave like an egoist?

Economic will and moral will.

If we are not mistaken, the difficulty is solved in a manner perfectly

---

1. *advocatus diaboli*：〈拉〉魔鬼的辩护士
2. processes of canonization：教会封圣仪式
3. scrupulous man：谨言慎行者
4. duty of the altruist：无私者的责任

analogous to that in which is solved the problem of the relations between expression and concept, Æsthetic and Logic.

To will economically is to *will an end;* to will morally is to *will the rational end.* But whoever wills and acts morally, cannot but will and act usefully (economically). How could he will the *rational* end, unless he also willed it *as his particular end*?

Pure economicity.

The converse is not true; as it is not true in æsthetic science that the expressive fact must of necessity be linked with the logical fact. It is possible to will economically without willing morally; and it is possible to conduct oneself with perfect economic coherence, while pursuing an end which is objectively irrational (immoral), or, rather, an end which would be held to be so at a higher grade of consciousness.

Examples of the economic, without the moral character, are Machiavelli's hero Cæsar Borgia[1], or the Iago[2] of Shakespeare. Who can help admiring their strength of will, although their activity is only economic, and is developed in opposition to what we hold moral? Who can help admiring the Ser Ciappelletto[3] of Boccaccio[4], who pursues and realizes his ideal of the perfect rascal[5] even on his death-bed, making the petty and timid little thieves who are present at his

1. Machiavelli's hero Cæsar Borgia：马基雅维利《君主论》的主人公恺撒·博尔吉亚。马基雅维利（1469—1527），意大利政治思想家、历史学家。他以恺撒·博尔吉亚为原型撰写了《君主论》。恺撒·博尔吉亚（1475 或 1476—1507），曾担任过瓦伦西亚大主教和枢机主教，以残忍无情、善弄权术著称，马基雅维利却认为他是一个完美的统治者。
2. Iago：埃古，莎士比亚剧作《奥赛罗》中的反面人物。
3. Ser Ciappelletto：即西亚佩雷多，薄伽丘笔下的人物。
4. Boccaccio：薄伽丘（1313—1375），意大利文艺复兴运动时期作家、诗人，著有《十日谈》，与但丁、彼特拉克并称“文学三杰”。
5. his ideal of the perfect rascal：他心目中地道的流氓模样

burlesque confession[1] exclaim: "What manner of man is this, whose perversity neither age, nor infirmity, nor the fear of death which he sees at hand, nor the fear of God before whose judgement-seat he must stand in a little while, have been able to remove, nor to make him wish to die otherwise than as he has lived?"

The economic side of morality.

The moral man unites with the pertinacity[2] and fearlessness of a Cæsar Borgia, of an Iago, or of a Ser Ciappelletto, the good will of the saint or of the hero. Or, rather, good will would not be will, and consequently not good, if it did not possess, in addition to the side which makes it *good*, also that which makes it *will*. So a logical thought which does not succeed in expressing itself is not thought, but at the most a confused presentiment[3] of a thought beyond yet to come.

It is not correct, then, to conceive of the amoral man as also anti-economical, or to make of morality an element of coherence in the acts of life, and therefore of economicity. Nothing prevents us from conceiving (an hypothesis which is verified at least during certain periods and moments, if not during whole lifetimes) a man altogether without moral conscience. In a man thus organized, what for us is immorality is not so for him, because it is not felt as such. The consciousness of the contradiction between what is desired as a rational end and what is pursued egoistically cannot arise in him. This contradiction is anti-economicity. Immoral conduct becomes also anti-economical only in the man who possesses moral conscience. The moral remorse[4] which is the indication of this, is also economical

1. burlesque confession：装模作样的忏悔
2. pertinacity：行侠仗义
3. confused presentiment：（令人）困惑的预感
4. moral remorse：道德上的内疚

remorse; that is to say, sorrow at not having known how to will completely and to attain that moral ideal which was willed at first, instead of allowing himself to be led astray by the passions. *Video meliora proboque, deteriora sequor*[1]. The *video* and the *probo* are here an initial volo immediately contradicted and overthrown. In the man without moral sense, we must admit a remorse that is *merely economic*; like that of a thief or of an assassin who, when on the point of robbing or of assassinating should abstain from doing so, not owing to a conversion of his being, but to nervousness and bewilderment[2], or even to a momentary awakening of moral consciousness. When he has come back to himself, such a thief or assassin will regret and be ashamed of his incoherence; his remorse will not be due to having done wrong, but to *not* having done wrong; it is therefore economic, not moral, since the latter is excluded by hypothesis. But since a lively moral consciousness is generally found among the majority of men and its total absence is a rare and perhaps non-existent monstrosity[3], it may be admitted that morality, in general, coincides with economicity in the conduct of life.

The merely economic and the error of the morally indifferent.

There need be no fear lest the parallelism that we support should introduce afresh into science the category of the *morally indifferent,* of that which is in truth action and volition, but is neither moral nor immoral; the category in short of the *licit*[4] and of the *permissible*, which has always been the cause or reflexion of ethical corruption, as was the case with Jesuitical[5] morality, which it dominated. It remains quite

1. *Video meliora proboque, deteriora sequor.*：〈拉〉看到更好的方法并认可它，但却反其道而行之。
2. nervousness and bewilderment：张皇失措
3. non-existent monstrosity：不存在的怪胎
4. *licit*：合法的
5. Jesuitical：耶稣会的

certain that indifferent moral actions do not exist, because moral activity pervades and must pervade every least volitional movement of man. But far from upsetting the established parallelism, this confirms it. Are there by any chance intuitions which science and the intellect do not pervade and analyse, resolving them into universal concepts, or changing them into historical affirmations? We have already seen that true science, philosophy, knows no external limits which bar its way, as happens with the so-called natural sciences. Science and morality entirely dominate, the one the æsthetic intuitions, the other the economic volitions of man, although neither of them can appear in the concrete, save the one in the intuitive, the other in the economic form.

Criticism of utilitarianism[1] *and the reform of Ethics and of Economics.*

This combined identity and difference of the useful and the moral, of the economic and the ethical, explains the success at the present time and formerly of the utilitarian theory of Ethics. Indeed it is easy to discover and to illustrate a utilitarian side in every moral action; as it is easy to reveal the æsthetic side in every logical proposition. The criticism of ethical utilitarianism cannot begin by denying this truth and seeking out absurd and non-existent examples of *useless* moral actions. It must admit the utilitarian side and explain it as the concrete form of morality, which consists in this, that it is *inside* this form. Utilitarians do not see this inside. This is not the place for the fuller development that such ideas deserve. Ethics and Economics cannot however fail to be gainers (as we have said of Logic and Æsthetic) by a more exact determination of the relations that exist between them. Economic science is now rising to the activistical concept[2] of the useful, as it attempts to surpass the mathematical phase in which

1. *utilitarianism*：功利主义
2. activistical concept：活动主义概念

it is still entangled; a phase which was in its turn a progress when it superseded historicism, or the confusion of the theoretical with the historical, and destroyed a number of capricious distinctions and false economic theories. With this conception, it will be easy on the one hand to absorb and to verify the semi-philosophical theories of so-called pure economics, and on the other, by the introduction of successive complications and additions, to effect a transition from the philosophical to the empirical or naturalistic method[1] and thus to embrace the particular theories expounded in the so-called political or national economy of the schools.

Phenomenon and noumenon in practical activity.

As æsthetic intuition knows the phenomenon or nature, and the philosophic concept the noumenon or spirit; so the economic activity wills the phenomenon or nature, and the moral activity the noumenon or spirit. *The spirit which wills itself,* its true self, the universal which is in the empirical and finite spirit: that is the formula which perhaps defines the essence of morality with the least impropriety[2]. This will for the true self is *absolute freedom.*

1. empirical or naturalistic method：经验主义方法或自然主义方法
2. least impropriety：最恰当的

VIII EXCLUSION OF OTHER SPIRITUAL FORMS

The system of the spirit.

In this summary sketch that we have given of the entire philosophy of the spirit in its fundamental moments, the spirit is thus conceived as consisting of four moments or degrees, disposed in such a way that the theoretical activity is to the practical as the first theoretical degree is to the second theoretical, and the first practical degree to the second practical. The four moments imply one another regressively[1] by their concreteness. The concept cannot exist without expression, the useful without both and morality without the three preceding degrees. If the æsthetic fact is in a certain sense alone independent while the others are more or less dependent, then the logical is the least dependent and the moral will the most. Moral intention acts on given theoretic bases, with which it cannot dispense, unless we are willing to accept that absurd procedure known to the Jesuits as *direction of intention,* in which people pretend to themselves not to know what they know only too well.

The forms of genius.
The system of the spirit.

If the forms of human activity are four, four also are the forms

1. regressively：递减地

of *genius*. Men endowed with genius in art, in science, and in moral will or heroes, have always been recognized. But the genius of pure economicity has met with repugnance[1]. It is not altogether without reason that a category of bad geniuses or of *geniuses of evil* has been created. The practical, merely economic genius, which is not directed to a rational end, cannot but excite an admiration mingled with alarm[2]. To dispute as to whether the word "genius" should be applied only to creators of æsthetic expression or also to men of scientific research and of action would be a mere question of words. To observe, on the other hand, that "genius," of whatever kind it be, is always a quantitative conception and an empirical distinction, would be to repeat what has already been explained as regards artistic genius.

Non-existence of a fifth form of activity. Law; sociability.

A fifth form of spiritual activity does not exist. It would be easy to show how all the other forms either do not possess the character of activity, or are verbal variants of the activities already examined, or are complex and derivative facts, in which the various activities are mingled, and are filled with particular and contingent contents[3].

The *juridical* fact, for example, considered as what is called objective law, is derived both from the economic and from the logical activities. Law is a rule, a formula (whether oral or written matters little here) in which is fixed an economic relation willed by an individual or by a community, and this economic side at once unites it with and distinguishes it from moral activity. Take another example. Sociology (among the many meanings the word bears in our times) is sometimes conceived as the study of an original element, which is called

1. met with repugnance: 遭人嫌弃
2. admiration mingled with alarm：羡慕，伴随着警觉
3. particular and contingent contents：个别的、偶然的内容

sociability. Now what is it that distinguishes sociability, or the relations which are developed in a meeting of men, and not in a meeting of sub-human beings, if it be not just the various spiritual activities which exist among the former and which are supposed not to exist, or to exist only in a rudimentary degree[1], among the latter? Sociability, then, far from being an original, simple, irreducible conception, is very complex and complicated. A proof of this would be the impossibility, generally recognized, of enunciating a single law which could be described as purely sociological. Those that are improperly so called are shown to be either empirical historical observations, or spiritual laws, that is to say judgements into which the conceptions of the spiritual activities are translated, when they are not simply empty and indeterminate generalities[2], like the so-called law of evolution. Sometimes, too, nothing more is understood by "sociability" than "social rule," and so law; thus confounding sociology with the science or theory of law itself. Law, sociability, and similar concepts, are to be dealt with in a mode analogous to that employed by us in the consideration and analysis of historicity and technique.

Religion.

It may seem that *religious* activity should be judged otherwise. But religion is nothing but knowledge, and does not differ from its other forms and sub-forms. For it is in turn either the expression of practical aspirations and ideals (religious ideals), or historical narrative (legend), or conceptual science (dogma).

It can therefore be maintained with equal truth either that religion is destroyed by the progress of human knowledge, or that it is always present there. Their religion was the whole intellectual patrimony

---

1. only in a rudimentary degree：仅具雏形的
2. empty and indeterminate generalities：空洞游离的泛泛之谈

of primitive peoples: our intellectual patrimony is our religion. The content has been changed, bettered, refined, and it will change and become better and more refined in the future also; but its form is always the same. We do not know what use could be made of religion by those who wish to preserve it side by side with the theoretic activity of man, with his art, with his criticism and with his philosophy. It is impossible to preserve an imperfect and inferior kind of knowledge, such as religion, side by side with what has surpassed and disproved it. Catholicism[1], which is always consistent, will not tolerate a Science, a History, an Ethics, in contradiction to its views and doctrines. The rationalists are less coherent: they are disposed to allow a little space in their souls for a religion in contradiction with their whole theoretic world.

The religious affectations and weaknesses prevalent among the rationalists of our time have their origin in the superstitious worship so recklessly lavished upon the natural sciences[2]. We know ourselves and their chief representatives admit that these sciences are all surrounded by *limits*. Science having been wrongly identified with the so-called natural sciences, it could be foreseen that the remainder would be sought in religion; that remainder with which the human spirit cannot dispense. We are therefore indebted to materialism, to positivism[3], to naturalism for this unhealthy and often disingenuous recrudescence of religious exaltation, which belongs to the hospital, when it does not belong to the politician.

Metaphysic.

Philosophy removes from religion all reason for existing, because

1. Catholicism：天主教教义
2. recklessly lavished upon the natural sciences: 草率地滥加于自然科学
3. positivism：实证主义

it substitutes itself for religion. As the science of the spirit, it looks upon religion as a phenomenon, a transitory historical fact, a psychic condition that can be surpassed. Philosophy shares the domain of knowledge with the natural sciences, with history and with art. To the first it leaves enumeration, measurement and classification[1]; to the second, the chronicling[2] of what has individually happened; to the third, the individually possible. There is nothing left to allot to religion. For the same reason, philosophy, as the science of the spirit, cannot be philosophy of the intuitive datum[3]; nor, as has been seen, *philosophy of history*, nor *philosophy of nature*; and therefore there cannot be a philosophical science of what is not form and universal, but material and particular. This amounts to affirming the impossibility of *Metaphysic*.

The methodology or logic of history has supplanted[4] the philosophy of history; an epistemology[5] of the concepts employed in the natural sciences succeeded the Philosophy of Nature. What philosophy can study of history is its mode of construction (intuition, perception, document, probability, etc.); of the natural sciences the forms of the concepts which constitute them (space, time, motion, number, types, classes, etc.). Philosophy as metaphysic in the sense above described would, on the other hand, claim to compete with history and with the natural sciences, which alone are legitimate and effective in their field. Such a challenge could do nothing but reveal the incompetence of those who made it. In this sense we are *anti-metaphysicans*[6], while declaring ourselves to be *ultra-metaphysicians*[7], when the word is used to

1. enumeration, measurement and classification：列举，测量和分类
2. chronicling：记事；叙述
3. intuitive datum：直觉数据
4. supplanted：替代
5. epistemology：认识论
6. *anti-metaphysicans*：反形而上学者
7. *ultra-metaphysicians*：极端形而上学者

claim and to affirm the office of philosophy as self-consciousness of the spirit, distinguished from the merely empirical and classificatory office of the natural sciences.

Mental imagination and the intuitive intellect.

Metaphysic has been obliged to assert the existence of a specific spiritual activity producing it, in order to maintain itself side by side with the sciences of the spirit. This activity, called in antiquity *mental or superior imagination*, and more often in modern times *intuitive intellect or intellectual intuition*[1], was held to unite the characters of imagination and intellect in an altogether special form. It was supposed to provide the means of passing by deduction or dialectic from the infinite to the finite, from form to matter, from the concept to the intuition, from science to history, acting by a method which was held to penetrate both the universal and the particular, the abstract and the concrete, intuition and intellect. A faculty marvellous indeed and most valuable to possess; but we, who do not possess it, have no means of establishing its existence.

Mystical Æsthetic.

Intellectual intuition has sometimes been considered to be the true æsthetic activity. At others a no less marvellous æsthetic activity has been placed beside, below, or above it, a faculty altogether different from simple intuition. The glories of this faculty have been celebrated, and the production of art attributed to it, or at least of certain groups of artistic production, arbitrarily chosen. Art, religion and philosophy have seemed in turn to be one only, or three distinct faculties of the

---

1. *intuitive intellect or intellectual intuition*：直觉理智或理智直觉

spirit, sometimes one, sometimes another of them being supreme in the dignity shared by all.

It is impossible to enumerate all the various attitudes assumed or capable of being assumed by this conception of Æsthetic, which we will call *mystical.* We are here in the kingdom, not of the science of imagination, but of imagination itself, which creates its world out of varying elements drawn from impressions and feelings. Suffice it to mention that this mysterious faculty has been conceived, sometimes as practical, sometimes as a mean between the theoretic and the practical, at others again as a theoretic form side by side with philosophy and religion[1].

Mortality and immortality of art.

The immortality of art has sometimes been deduced from this last conception, as belonging with its sisters to the sphere of absolute spirit. At other times, on the other hand, when religion has been looked upon as mortal and as dissolved in philosophy, then has been proclaimed the mortality, even the death, actual or at least imminent[2], of art. This question has no meaning for us, because, seeing that the function of art is a necessary degree of the spirit, to ask if art can be eliminated is the same as to ask if sensation or intelligence can be eliminated. But Metaphysic, in the above sense, transplanting itself into an arbitrary world, is not to be criticized in its particulars, any more than we can criticize the botany of the garden of Alcina[3] or the navigation of the voyage of Astolfo[4]. Criticism can only exist when we refuse to

1. side by side with philosophy and religion：与哲学宗教并驾齐驱
2. actual or at least imminent：已死的或濒死的
3. Alcina：阿尔辛娜，意大利诗人阿里奥斯托作品《疯狂的奥兰多》中的女巫，以魔法迷惑男性，在她的花园里将情人石化。
4. Astolfo：阿斯托尔福，《疯狂的奥兰多》中的人物，在月球寻找奥兰多的灵魂。

join in the game; that is to say, when we reject the very possibility of Metaphysic, always in the sense above indicated.

There is therefore no intellectual intuition in philosophy, as there is no surrogate or equivalent of it in art, or any other mode by which this imaginary function may be called and represented. There does not exist (if we may repeat ourselves) a fifth degree, a fifth or supreme faculty, theoretic or practical-theoretic, imaginative-intellectual, or intellectual-imaginative, or however otherwise it may be attempted to conceive such a faculty.

IX INDIVISIBILITY OF EXPRESSION INTO MODES OR DEGREES AND CRITICISM OF RHETORIC

The characters of art.

It is customary to give long catalogues of the *characters* of art. Having reached this point of the treatise, after having studied art as spiritual activity, as theoretic activity, and as special theoretic activity (intuitive), we are able to discover that those varied and numerous determinations of characters, where they refer to anything real, do nothing but represent what we have already met with as genera[1], species and individuality of the æsthetic form. To the generic are reducible, as we have already observed, the characters, or rather, the verbal variants of *unity*, and of *unity* in *variety*, of *simplicity*, or *originality*, and so on; to the specific, the characters of *truth*, of *sincerity*, and the like; to the individual, the characters of *life*, of *vivacity*[2], of *animation*, of *concreteness*, of *individuality*, of *characteristicality*. The words may change again, but they will not contribute anything scientifically new. The analysis of expression as such is completely effected in the results expounded above.

Non-existence of modes of expression.

It might, on the other hand, be asked at this point if there be *modes*

1. genera：性质；类别
2. *vivacity*：生命活力

or *degrees* of expression; if, having distinguished two degrees of activity of the spirit, each of which is subdivided into two other degrees, one of these, the intuitive-expressive, is not in its turn subdivided into two or more intuitive modes, into a first, second or third degree of expression. But this further division is impossible; a classification of intuition-expressions is certainly permissible, but is not philosophical: individual expressive facts are so many individuals, not one of which is interchangeable with another, save in its common quality of expression. To employ the language of the schools: expression is a species which cannot function in its turn as a genus[1]. Impressions or contents vary; every content differs from every other content, because nothing repeats itself in life; and the irreducible variety of the forms of expression corresponds to the continual variation of the contents, the æsthetic synthesis of impressions.

Impossibility of translations.

A corollary of this is the impossibility of *translations*, in so far as they pretend to effect the re-moulding of one expression into another, like a liquid poured from a vase of a certain shape into a vase of another shape. We can elaborate logically what we have already elaborated in æsthetic form only; but we cannot reduce what has already possessed its æsthetic form to another form also æsthetic. Indeed, every translation either diminishes and spoils, or it creates a new expression, by putting the former back into the crucible and mingling it with the personal impressions of the so-called translator. In the former case, the expression always remains one, that of the original, the translation being more or less deficient, that is to say, not properly expression: in the other case, there would certainly be two expressions, but with two

---

1. genus：门类

different contents. "Faithful ugliness or faithless beauty" is a proverb that well expresses the dilemma with which every translator is faced. Un-æsthetic translations, such as those that are word for word, or paraphrastic[1], are to be looked upon as simple commentaries upon the original.

Criticism of the rhetorical categories.

The illegitimate division of expressions into various grades is known in literature by the name of doctrine of *ornament* or of *rhetorical categories*[2]. But similar attempts at distinctions in other artistic groups are not wanting: suffice it to recall the *realistic* and *symbolic* forms, so often mentioned in relation to painting and sculpture.

Realistic and *symbolic*, *objective* and *subjective*, *classical* and *romantic*, *simple* and *ornate*[3], *proper* and *metaphorical*, the fourteen forms of metaphor, the figures of *word* and *sentence*, *pleonasm*, *ellipse*, *inversion*, *repetition*, *synonyms* and *homonyms*[4], these and all other determinations of modes or degrees of expression reveal their philosophical nullity when the attempt is made to develop them in precise definitions, because they either grasp the void or fall into the absurd. A typical example of this is the very common definition of metaphor as of *another word used in place of the proper word*. Now why give oneself this trouble? Why substitute the improper for the proper word? Why take the worse and longer road when you know the shorter and better road? Perhaps, as is commonly said, because the proper word is in certain cases not so *expressive* as the so-called improper word or metaphor? But if this be so

---

1. paraphrastic：意译的
2. *rhetorical categories*：修辞种类论
3. *ornate*：雕琢
4. *pleonasm, ellipse, inversion, repetition, synonyms and homonyms*：重复格，省略格，倒装格，重叠格，同义格和同形异义格

the metaphor is exactly the proper word in that case, and the so-called "proper" word, if it were used, would be *inexpressive* and therefore most improper. Similar observations of elementary good sense can be made regarding the other categories, as, for example, the general one of the *ornate*. Here for instance it may be asked how an ornament can be joined to expression. Externally? In that case it is always separated from the expression. Internally? In that case, either it does not assist the expression and mars it; or it does form part of it and is not an ornament, but a constituent element of the expression, indivisible and indistinguishable in its unity.

It is needless to say how much harm has been done by rhetorical distinctions. Rhetoric has often been declaimed against, but although there has been rebellion against its consequences, its principles have, at the same time, been carefully preserved (perhaps in order to show proof of philosophic consistency). In literature the rhetorical categories have contributed, if not to make dominant, at least to justify theoretically, that particular kind of *bad writing* which is called *fine writing* or writing according to rhetoric.

Use of these categories as synonyms of the æsthetic fact.

The terms above mentioned would never have gone beyond the schools, where we all of us learned them (only we never found an opportunity of using them in strictly æsthetic discussions, or at most of doing so jocosely[1] and with a comic intention), were it not that they can sometimes be employed in one of the following significations: as *verbal variants* of the æsthetic concept; as indications of the *anti-æsthetic*, or, finally (and this is their most important use), no longer in the service of art and æsthetic, but of *science* and *logic*.

---

1. jocosely：幽默地

Empirical sense of the rhetorical categories.

First. Expressions considered directly or positively are not divisible into classes, but some are successful, others half-successful, others failures. There are perfect and imperfect, successful and unsuccessful expressions. The words recorded, and others of the same sort, may therefore sometimes indicate the successful expression, and the various forms of the failures. But they do this in the most inconstant and capricious manner, so much so that the same word serves sometimes to proclaim the perfect, sometimes to condemn the imperfect.

For example, some will say of two pictures — one without inspiration, in which the author has copied natural objects without intelligence; the other inspired, but without close relation to existing objects — that the first is *realistic*, the second *symbolic*. Others, on the contrary, utter the word *realistic* before a picture strongly felt representing a scene of ordinary life, while they apply that of *symbolic* to another picture that is but a cold allegory[1]. It is evident that in the first case symbolic means artistic and realistic inartistic, while in the second, realistic is synonymous with artistic and symbolic with inartistic. What wonder, then, that some hotly maintain the true art form is the symbolic, and that the realistic is inartistic; others, that the realistic is artistic and the symbolic inartistic? We cannot but grant that both are right, since each uses the same words in such a different sense.

The great disputes about *classicism* and *romanticism* were frequently based upon such equivocations. Sometimes the former was understood as the artistically perfect, and the second as lacking balance and imperfect; at others "classic" meant cold and artificial, "romantic" pure, warm, powerful, truly expressive. Thus it was always possible

1. cold allegory：枯燥乏味的寓言

reasonably to take the side of the classic against the romantic, or of the romantic against the classic.

The same thing happens as regards the word *style*. Sometimes it is said that every writer must have style. Here style is synonymous with form of expression. At others the form of a code of laws or of a mathematical work is said to be without style. Here the error is again committed of admitting diverse modes of expression, an ornate and a naked form, because, if style is form, the code and the mathematical treatise must also be asserted, strictly speaking, to have each its style. At other times, one hears the critics blaming some one for "having too much style" or for "writing a style." Here it is clear that style signifies, not the form, nor a mode of it, but improper and pretentious expression, a form of the inartistic.

Their use to indicate various æsthetic imperfections.

Second. The second not altogether meaningless use of these words and distinctions is to be found when we hear in the examination of a literal composition such remarks as these: here is a pleonasm, here an ellipse, there a metaphor, here again a synonym or an ambiguity. The meaning is: Here is an error consisting of using a larger number of words than necessary (pleonasm); here, on the other hand, the error arises from too few having been used (ellipse), here from the use of an unsuitable word (metaphor), here of two words which seem to say two different things, but really say the same thing (synonym); here, on the contrary, of one word which seems to express the same thing, whereas it says two different things (ambiguity). This depreciatory and pathological use[1] of the terms is, however, less common than the preceding.

1. depreciatory and pathological use：贬义的、病态的用法

Their use in a sense transcending æsthetic, in the service of science.

Thirdly and finally, when rhetorical terminology possesses no æsthetic signification similar or analogous to those passed in review, and yet one feels that it is not void of meaning and designates something that deserves to be noted, this means that it is used in the service of logic and of science. Granted that a concept used by a writer in a scientific sense is designated by a definite term, it is natural that other terms found in use by that writer on which he incidentally employs himself to signify the same thought, become *in respect to* the vocabulary fixed upon by him as true, metaphors, synecdoches, synonyms, elliptical forms and the like. We ourselves in the course of this treatise have several times made use of, and intend again to make use of such language, in order to make clear the sense of the words we employ, or may find employed. But this proceeding, which is of value in discussions pertaining to the criticism of science and philosophy, has none whatever in literary and artistic criticism. There are words and metaphors proper to science: the same concept may be psychologically formed in various circumstances and therefore differ in its intuitional expression. When the scientific terminology[1] of a given writer has been established and one of these modes fixed as correct, then all other uses of it become improper or tropical. But in the æsthetic fact there are none but proper words: the same intuition can be expressed in one way only, precisely because it is intuition and not concept.

Rhetoric in the schools.

Some, while admitting the æsthetic non-existence of the rhetorical

1. scientific terminology：科学术语

categories, yet make a reservation as to their utility and the service they are supposed to render, especially in schools of literature. We confess that we fail to understand how error and confusion can educate the mind to logical distinction, or aid the teaching of a science which they disturb and obscure. Perhaps what is meant is that such distinctions, as empirical classes, can aid memory and learning, as was admitted above for literary and artistic kinds. To this there is no objection. There is certainly another purpose for which the rhetorical categories should continue to appear in schools: to be criticized there. The errors of the past must not be forgotten and no more said, and truths cannot be kept alive save by making them combat errors. Unless an account of the rhetorical categories be given, accompanied by a criticism of them, there is a risk of their springing up again, and it may be said that they are already springing up among certain philologists[1] as the latest *psychological* discoveries.

The resemblances of expressions.

It might seem that we thus wished to deny all bond of resemblance between different expressions and works of art. Resemblances exist, and by means of them, works of art can be arranged in this or that group. But they are likenesses such as are observed among individuals, and can never be rendered with abstract determinations. That is to say, it would be incorrect to apply identification, subordination, co-ordination and the other relations of concepts to these resemblances, which consist wholly of what is called a *family likeness*, derived from the historical conditions in which the various works have appeared and from relationship of soul among the artists.

---

1. philologists：语文学家

The relative possibility of translations.

It is in these resemblances that lies the *relative* possibility of translations; not as reproductions of the same original expressions (which it would be vain to attempt), but as productions of *similar* expressions more or less nearly resembling the originals. The translation called good is an approximation which has original value as a work of art and can stand by itself.

X ÆSTHETIC FEELINGS AND THE DISTINCTION BETWEEN THE BEAUTIFUL AND THE UGLY

Various significations of the word feeling.

Passing to the study of more complex concepts, where the æsthetic activity is to be considered in conjunction with other orders of facts, and showing the mode of their union or complication, we find ourselves first face to face with the concept of *feeling* and with those feelings that are called *æsthetic.*

The word "feeling" is one of the richest in meanings in philosophic terminology. We have already had occasion to meet with it once, among those used to designate the spirit in its passivity, the matter or content of art, and so as synonym of *impressions.* Once again (and then the meaning was altogether different), we have met with it as designating the *non-logical* and *non-historical* character of the æsthetic fact, that is to say, pure intuition, a form of truth which defines no concept and affirms no fact.

Feeling as activity.

But here it is not regarded in either of these two meanings, nor in the others which have also been conferred upon[1] it to designate other

---

1. conferred upon：赋予

cognitive forms of the spirit, but only in that where feeling is understood as a special activity, of non-cognitive nature, having its two poles, positive and negative[1], in *pleasure* and *pain*.

This activity has always greatly embarrassed philosophers, who have therefore attempted either to deny it as activity, or to attribute it to *nature*, excluding it from the spirit. But both these solutions bristle with[2] difficulties of such a kind as to prove them finally unacceptable to any one who examines them with care. For what could a non-spiritual activity ever be, an *activity of nature*, when we have no other knowledge of activity save as spirituality, nor of spirituality save as activity? Nature is in this case, by definition, the merely passive, inert, mechanical, material. On the other hand, the negation of the character of activity to feeling is energetically disproved by those very poles of pleasure and of pain which appear in it and manifest activity in its concreteness, or, so to say, quivering.

Identification of feeling with economic activity.

This critical conclusion should place us especially in the greatest embarrassment, for in the sketch of the system of the spirit given above we have left no room for the new activity of which we are now obliged to recognize the existence. But the activity of feeling, if it is activity, is not new. It has already had its place assigned to it in the system that we have sketched, where, however, it has been given another name, *economic* activity. What is called the activity of feeling is nothing but that more elementary and fundamental practical activity which we have distinguished from the ethical activity and made to consist of the appetition and volition for some individual end, apart from any moral determination.

---

1. two poles, positive and negative: 正负两极
2. bristle with：充满着

If feeling has been sometimes considered to be an organic or natural activity, this has happened just because it does not coincide either with logical, æsthetic or ethical activity. Looked at from the standpoint of those three (which were the only ones admitted), it has seemed to lie *outside* the true and real spirit, spirit in its aristocracy, and to be almost a determination of nature, or of the soul in so far as it is nature. From this two results the truth of another thesis, often maintained, that the æsthetic activity, like the ethical and intellectual activities, is not feeling. This thesis is inexpugnable[1], when feeling has already been understood implicitly and unconsciously as economic volition.

Criticism of hedonism[2].

The view refuted in this thesis is known as *hedonism*. This consists in reducing all the various forms of the spirit to one, which thus also loses its own distinctive character and becomes something obscure and mysterious, like "the night in which all cows are black." Having brought about this reduction and mutilation[3], the hedonists naturally do not succeed in seeing anything else in any activity but pleasure and pain. They find no substantial difference between the pleasure of art and that of easy digestion, between the pleasure of a good action and that of breathing the fresh air with wide-expanded lungs.

Feeling as a concomitant[4] *of every form of activity.*

But if the activity of feeling in the sense here defined must not be substituted for all the other forms of spiritual activity, we have not

---

1. inexpugnable：不容反驳的
2. *hedonism*：享乐主义
3. reduction and mutilation：简化和剥离
4. *concomitants*：伴生物

said that it cannot *accompany* them. Indeed it accompanies them of necessity, because they are all in close relation both with one another and with the elementary volitional form. Therefore each of them has for concomitants individual volitions and volitional pleasures and pains[1], known as feeling. But we must not confound a concomitant with the principal fact, and substitute the one for the other. The discovery of a truth, or the fulfilment of a moral duty, produces in us a joy which makes vibrate our whole being, which, by attaining the aim of those forms of spiritual activity, attains at the same time that to which it was *practically* tending, as its end. Nevertheless, *economic* or *hedonistic* satisfaction, *ethical* satisfaction, *æsthetic* satisfaction, *intellectual* satisfaction, though thus united, remain always distinct.

A question often asked is thus answered at the same time, one which has correctly seemed to be a matter of life or death[2] for æsthetic science, namely, whether feeling and pleasure precede or follow, are cause or effect of the æsthetic fact. We must widen this question to include the relation between the various spiritual forms, and answer it by maintaining that one cannot talk of cause and effect and of a chronological before and after in the unity of the spirit.

And once the relation above expounded is established, all necessity for inquiry as to the nature of æsthetic, moral, intellectual and even what was sometimes called economic feelings, must disappear. In this last case, it is clear that it is a question, not of two terms, but of one, and inquiry as to economic feeling must be the same as that relating to economic activity. But in the other cases also, we must attend, not to the substantive, but to the adjective[3]: the æsthetic, moral and logical character will explain the colouring of the feelings as æsthetic, moral

1. volitional pleasures and pains：意志上的快乐与痛苦
2. a matter of life or death：生死攸关的问题
3. we must attend, not to the substantive, but to the adjective：我们必须关注的不是名词，而是形容词。

and intellectual, whereas feeling, studied alone, will never explain those refractions and colorations[1].

Meaning of certain ordinary distinctions of feelings.

A further consequence is, that we no longer need retain the well-known distinctions between values or feelings *of value*, and feelings that are merely hedonistic and *without value*; *disinterested* and *interested* feelings, *objective* feelings and feelings not *objective* but simply *subjective* feelings of *approbation*[2] and of *mere pleasure* (cf. the distinction of *Gefallen*[3] and *Vergnügen*[4] in German). Those distinctions were used to save the three spiritual forms, which were recognized as the triad of the *True*, the *Good* and the *Beautiful*[5], from confusion with the fourth form, still unknown, and therefore insidious in its indeterminateness[6] and mother of scandals. For us this triad has completed its task, because we are capable of reaching the distinction far more directly, by receiving also the selfish, subjective, merely pleasurable feelings among the respectable forms of the spirit; and where formerly antitheses were conceived (by ourselves and others), between value and feelings, as between spirituality and naturality, henceforth we see nothing but differences between value and value.

Value and disvalue: the contraries and their union.

As has already been said, feeling or the economic activity presents itself as divided into two poles, positive and negative, pleasure and

---

1. refractions and colorations：折射和着色
2. approbation：赞赏
3. *Gefallen*：〈德〉满意
4. *Vergnügen*：〈德〉愉快
5. the triad of the *True*, the *Good* and the *Beautiful*：真、善、美三元素
6. insidious in its indeterminateness：无从知晓，游移不定

pain, which we can now translate into useful and disuseful (or hurtful). This bipartition[1] has already been noted above, as a mark of the activistic character of feeling, and one which is to be found in all forms of activity. If each of these is *value*, each has opposed to it *antivalue* or *disvalue*[2]. Absence of value is not sufficient to cause disvalue, but activity and passivity must be struggling between themselves, without the one getting the better of the other; hence the contradiction and disvalue of the activity that is embarrassed, impeded, or interrupted. Value is activity that unfolds itself freely: disvalue is its contrary.

We will content ourselves with this definition of the two terms, without entering into the problem of the relation between value and disvalue, that is, the problem of contraries (that is to say, whether they are to be thought of dualistically, as two beings or two orders of beings, like Ormuzd and Ahriman[3], angels and devils, enemies to one another; or as a unity, which is also contrariety). This definition of the two terms will be sufficient for our purpose, which is to make clear the nature of æsthetic activity, and at this particular point one of the most obscure and disputed concepts of Æsthetic: the concept of the *Beautiful.*

The Beautiful as the value of expression, or expression without qualification.

Æsthetic, intellectual, economic and ethical values and disvalues are variously denominated in current speech: *beautiful, true, good, useful, expedient, just, right* and so on — thus designating the free development of spiritual activity, action, scientific research, artistic production,

---

1. bipartition：二分法
2. *antivalue* or *disvalue*：反价值和反面价值
3. Ormuzd and Ahriman：奥玛兹德和阿里曼。波斯摩尼教相信两神交替治理世界，其一是奥玛兹德，光明之神，善的化身；另一是阿利曼，黑暗之神，恶的化身。

when they are successful; *ugly, false, bad, useless, inexpedient, unjust,* wrong designating embarrassed activity, the product that is a failure. In linguistic usage, these denominations are being continually shifted from one order of facts to another. *Beautiful*, for instance, is said not only of a successful expression, but also of a scientific truth, of an action successfully achieved, and of a moral action: thus we talk of an *intellectual beauty*, of a *beautiful action*, of a *moral beauty*. The attempt to keep up with these infinitely varying usages leads into a trackless labyrinth of verbalism[1] in which many philosophers and students of art have lost their way. For this reason we have thought it best studiously to avoid the use of the word "beautiful" to indicate successful expression in its positive value. But after all the explanations that we have given, all danger of misunderstanding being now dissipated, and since on the other hand we cannot fail to recognize that the prevailing tendency, both in current speech and in philosophy, is to limit the meaning of the word "beautiful" precisely to the æsthetic value, it seems now both permissible and advisable to define beauty as *successful expression,*or rather, as *expression* and nothing more, because expression when it is not successful is not expression.

The ugly, and the elements of beauty which compose it.

Consequently, the ugly is unsuccessful expression. The paradox is true, for works of art that are failures, that the beautiful presents itself as *unity*, the ugly as *multiplicity*[2]. Hence we hear of *merits* in relation to works of art that are more or less failures, that is to say, of *those parts of them that are beautiful*, which is not the case with perfect works. It is in fact impossible to enumerate the merits or to point out what parts of

1. labyrinth of verbalism：文字游戏的迷宫
2. the beautiful presents itself as *unity*, the ugly as *multiplicity*：美以统一形式呈现，丑以多重形式出现

the latter are beautiful, because being a complete fusion they have but one value. Life circulates in the whole organism: it is not withdrawn into the several parts.

Illusion that there exist expressions neither beautiful nor ugly.

Unsuccessful works may have merit in various degrees, even the greatest. The beautiful does not possess degrees, for there is no conceiving a more beautiful, that is, an expressive that is more expressive, an adequate that is more than adequate. Ugliness, on the other hand, does possess degrees, from the rather ugly (or almost beautiful) to the extremely ugly. But if the ugly were *complete*, that is to say, without any element of beauty, it would for that very reason cease to be ugly, because it would be without the contradiction in which is the reason of its existence. The disvalue would become non-value; activity would give place to passivity, with which it is not at war, save when activity is really present to oppose it.

And because the distinctive consciousness of the beautiful and of the ugly is based on the conflicts and contradictions in which æsthetic activity is developed, it is evident that this consciousness becomes attenuated to the point of disappearing altogether, as we descend from the more complicated to the more simple and to the simplest instances of expression. Hence the illusion that there are expressions neither beautiful nor ugly, those which are obtained without sensible effort and appear easy and natural being considered such.

True æsthetic feelings and concomitant or accidental feelings.

The whole mystery of the *beautiful* and the *ugly* is reduced to these henceforth most easy definitions. Should any one object that there exist perfect æsthetic expressions before which no pleasure is felt, and others, perhaps even failures, which give him the greatest pleasure, we must recommend him to concentrate his attention in the æsthetic fact, upon that which is truly æsthetic pleasure. Æsthetic pleasure is sometimes reinforced or rather complicated by pleasures arising from extraneous facts, which are only accidentally found united with it. The poet or any other artist affords an instance of purely æsthetic pleasure at the moment when he sees (or intuites) his work for the first time; that is to say, when his impressions take form and his countenance is irradiated with the divine joy of the creator. On the other hand, a mixed pleasure is experienced by one who goes to the theatre, after a day's work, to witness a comedy: when the pleasure of rest and amusement, or that of laughingly snatching a nail from his coffin, accompanies the moment of true æsthetic pleasure in the art of the dramatist and actors. The same may be said of the artist who looks upon his labour with pleasure when it is finished, experiencing, in addition to the æsthetic pleasure, that very different one which arises from the thought of self-complacency[1] satisfied, or even of the economic gain which will come to him from his work. Instances could be multiplied.

Criticism of apparent feelings.

A category of *apparent* æsthetic feelings has been formed in modern Æsthetic, not arising from the form, that is to say, from the works of art as such, but from their content. It has been remarked that artistic representations arouse pleasure and pain in their infinite shades of variety. We tremble with anxiety, we rejoice, we fear, we laugh, we weep, we desire, with the personages of a drama or of a romance, with the figures in a picture and with the melody of music. But these feelings are not such as would be aroused by the real fact outside

1. self-complacency：自鸣得意

art; or rather, they are the same in quality, but are quantitatively an attenuation of real things. Æsthetic and *apparent* pleasure and pain show themselves to be light, shallow, mobile. We have no need to treat here of these *apparent feelings*, for the good reason that we have already amply discussed them; indeed, we have hitherto treated of nothing but them. What are these apparent or manifested feelings, but feelings objectified, intuited, expressed? And it is natural that they do not trouble and afflict us as passionately as those of real life, because those were matter, these are form and activity; those true and proper feelings, these intuitions and expressions. The formula of *apparent feelings* is therefore for us nothing but a tautology, through which we can run the pen without scruple[1].

1. without scruple：毫无顾忌地

XI CRITICISM OF ÆSTHETIC HEDONISM

As we are opposed to hedonism in general, that is to say, to the theory based upon the pleasure and pain intrinsic to the economic activity and accompanying every other form of activity, which, confounding container and content[1], fails to recognize any process but the hedonistic; so we are opposed to æsthetic hedonism in particular, which looks at any rate upon the æsthetic, if not also upon all other activities, as a simple fact of feeling, and confounds the pleasurable expression, which is the beautiful, with the simply pleasurable and all its other species.

Criticism of the beautiful as that which pleases the higher senses.

The æsthetic-hedonistic point of view has been presented in several forms. One of the most ancient conceives the beautiful as that which pleases sight and hearing, that is to say, the so-called *higher senses*. When analysis of æsthetic facts first began, it was, indeed, difficult to avoid the false belief that a picture and a piece of music are impressions of sight or hearing and correctly to interpret the obvious remark that the blind man does not enjoy the picture, nor the deaf man the music. To show, as we have shown, that the æsthetic fact does not depend upon the nature of the impressions, but that all sensible impressions can be raised to æsthetic expression and that none need of necessity be so raised, is an idea which presents itself only when all other doctrinal

1. confounding container and content：将外在、内在混为一谈

constructions of this problem have been tried. Any one who holds that the æsthetic fact is something pleasing to the eyes or to the hearing, has no line of defence against him who consistently proceeds to identify the beautiful with the pleasurable in general, and includes in Æsthetic cooking, or (as some positivists have called it) the viscerally beautiful[1].

Criticism of the theory of play.

The theory of *play* is another form of æsthetic hedonism. The concept of play has sometimes helped towards the realization of the activistic character of the expressive fact: man (it has been said) is not really man, save when he begins to play (that is to say, when he frees himself from natural and mechanical causality[2] and works spiritually); and his first game is art. But since the word "play" also means that pleasure which arises from the expenditure of the exuberant energy[3] of the organism (which is a practical fact), the consequence of this theory has been that every game has been called an æsthetic fact, or that the æsthetic function has been called a game, because like science and everything else, it may form part of a game. Morality alone cannot ever be caused by the will to play (for it will never consent to such an origin), but on the contrary itself dominates and regulates the act itself of playing.

Criticism of the theories of sexuality and of triumph.

Finally, some have tried to deduce the pleasure of art from the echo of that of the sexual organs. And some of the most recent æstheticians

---

1. viscerally beautiful：本能的美
2. natural and mechanical causality：自然的、机械的因果论
3. exuberant energy：旺盛精力

confidently find the genesis of the æsthetic fact in the pleasure of *conquering* and in that of *triumphing*, or, as others add, in the wish of the male to conquer the female. This theory is seasoned with much anecdotal erudition[1], heaven knows of what degree of credibility, as to the customs of savage peoples. But there was really no need for such assistance, since in ordinary life one often meets poets who adorn themselves with their poetry, like cocks raising their crests, or turkeys spreading out their tails. But any one who does this, in so far as he does it, is not a poet but a poor fool, in fact, a poor fool of a cock or turkey, and the desire for the victorious conquest of women has nothing to do with the fact of art. It would be just as correct to look upon poetry as *economic*, because there once were court poets and salaried poets, and there are poets now who find in the sale of their verses an aid to life if not a complete living. This deduction and definition has not failed to attract some zealous neophytes[2] in historical materialism.

Criticism of the Æsthetic of the sympathetic. Meaning in it of content and form.

Another less vulgar current of thought considers Æsthetic as the science of the *sympathetic*, as that with which we sympathize, which attracts, rejoices, arouses pleasure and admiration. But the sympathetic is nothing but the image or representation of what pleases. And as such it is a complex fact, resulting from a constant element, the æsthetic element of representation, and a variable element, the pleasing in its infinite forms, arising from all the various classes of values.

In ordinary language, there is sometimes a feeling of repugnance at calling an expression "beautiful," unless it is an expression of the

1. anecdotal erudition：野史逸闻
2. zealous neophytes：狂热的新信徒

sympathetic. Hence the continual conflicts between the point of view of the æsthetician or art critic and that of the ordinary person, who cannot succeed in persuading himself that the image of pain and baseness[1] can be beautiful or at least that it has as much right to be beautiful as the pleasing and the good.

The conflict could be put an end to by distinguishing two different sciences, one of expression and the other of the sympathetic, if the latter could be the object of a special science; that is to say, if it were not, as has been shown, a complex and equivocal concept. If predominance be given to the expressive fact, it enters Æsthetic as science of expression; if to the pleasurable content, we fall back to the study of facts essentially hedonistic (utilitarian), however complicated they may appear. The particular origin of the doctrine which conceives the relation between form and content as the sum of two values is also to be sought in the doctrine of the sympathetic.

Æsthetic hedonism and moralism.

In all the doctrines just now discussed, art is considered as a merely hedonistic thing. But æsthetic hedonism cannot be maintained, save by uniting it with a general philosophical hedonism, which does not admit any other form of value. Hardly has this hedonistic conception of art been received by philosophers who admit one or more spiritual values, truth or morality, when the following question must necessarily be asked: What must be done with art? To what use should it be put? Should a free course be allowed to the pleasures it procures? And if so, to what extent? The question of the *end of art*, which in the Æsthetic of expression is inconceivable, has a clear significance in the Æsthetic of the Sympathetic and demands a solution.

1. image of pain and baseness：痛苦卑贱的形象

The rigoristic negation, and the pedagogic justification[1] *of art.*

Now it is evident that such solution can have but two forms, one altogether negative, the other of a restrictive nature. The first, which we shall call *rigoristic* or *ascetic*[2], appears several times, although not frequently, in the history of ideas. It looks upon art as an inebriation of the senses[3] and therefore as not only useless but harmful. According to this theory, then, we must exert all our strength to liberate the human soul from its disturbing influence. The other solution, which we shall call *pedagogic* or *moralistic-utilitarian*, admits art, but only in so far as it co-operates with the end of morality; in so far as it assists with innocent pleasure the work of him who points the way to the true and the good; in so far as it anoints[4] the edge of the cup of wisdom and morality with sweet honey.

It is well to observe that it would be an error to divide this second view into intellectualistic and moralistic-utilitarian, according as to whether be assigned to art the end of leading to the true or to what is practically good. The educational task which is imposed upon it, precisely because it is an end which is sought after and advised, is no longer merely a theoretical fact, but a theoretical fact already become the ground for practical action; it is not, therefore, intellectualism, but pedagogism and practicism. Nor would it be more exact to subdivide the pedagogic view into pure utilitarian and moralistic-utilitarian; because those who admit only the satisfaction of the individual (the desire of the individual), precisely because they are absolute hedonists, have no motive for seeking an ulterior justification[5] for art.

---

1. pedagogic justification：学究辩护
2. *ascetic*：苦修主义
3. inebriation of the senses：感官麻醉剂
4. anoints：涂油
5. ulterior justification：终极辩护

But to enunciate these theories at the point to which we have attained is to confute them. We prefer to restrict ourselves to observing that in the pedagogic theory of art is to be found another of the reasons why the claim has erroneously been made that the content of art should be *chosen* with a view to certain practical effects.

Criticism of pure beauty.

The thesis that art consists of *pure beauty* has often been brought forward against hedonistic and pedagogic Æsthetic, and eagerly taken up by artists: "Heaven places all our joy in *pure beauty,* and the Verse is everything." If by this be understood that art is not to be confounded with sensual pleasure (utilitarian practicism), nor with the exercise of morality, then our Æsthetic also must be permitted to adorn itself with the title of *Æsthetic of pure beauty*. But if (as is often the case) something mystical and transcendent be meant by this, something unknown to our poor human world, or something spiritual and beatific[1], but not expressive, we must reply that while applauding the conception of a beauty *free from all that is not the spiritual form of expression*, we are unable to conceive a beauty superior to this and still less that it should be *purified of expression*, or severed from itself[2].

---

1. beatific：赐福的
2. severed from itself：脱离自身

XII THE ÆSTHETIC OF THE SYMPATHETIC AND PSEUDO-ÆSTHETIC CONCEPTS

Pseudo-æsthetic concepts, and the æsthetic of the sympathetic.

The doctrine of the sympathetic (very often animated and seconded in this by the capricious metaphysical and mystical Æsthetic, and by that blind traditionalism which assumes an intimate connection between things fortuitously treated together[1] by the same authors in the same books), has introduced and rendered familiar in systems of Æsthetic a series of concepts a rapid mention of which suffices to justify our resolute expulsion[2] of them from our own treatise.

Their catalogue is long, not to say interminable: *tragic, comic, sublime, pathetic, moving, sad, ridiculous, melancholy, tragi-comic, humorous, majestic, dignified, serious, grave, imposing, noble, decorous*[3], *graceful, attractive, piquant*[4], *coquettish*[5], *idyllic, elegiac*[6], *cheerful, violent, ingenuous, cruel, base, horrible, disgusting, dreadful, nauseating*; the fist can be increased at will.

Since that doctrine took the sympathetic as its special object, it was naturally unable to neglect any of the varieties of the sympathetic, any of the mixtures or gradations by means of which, starting from the

1. fortuitously treated together：随意堆砌在一起的
2. resolute expulsion：坚决驳斥
3. *decorous*：端庄的
4. *piquant*：刺激的
5. *coquettish*：妖媚的
6. *elegiac*：哀婉的

sympathetic in its loftiest and most intense manifestation, its contrary, the antipathetic and repugnant, is finally reached. And since the sympathetic content was held to be the *beautiful* and the antipathetic the *ugly*, the varieties (tragic, comic, sublime, pathetic, etc.) constituted for that conception of Æsthetic the shades and gradations intervening between the beautiful and the ugly.

Criticism of the theory of the ugly in art and of the overcoming of it.

Having enumerated and defined as well as it could, the chief of these varieties, the Æsthetic of the sympathetic set itself the problem of the place to be assigned to the *ugly in art*. This problem is without meaning for us, who do not recognize any ugliness save the anti-æsthetic or inexpressive, which can never form *part* of the æsthetic fact, being, on the contrary, its *antithesis*[1]. But in the doctrine which we are here criticizing the positing and discussion of that problem meant neither more nor less than the necessity of reconciling in some way the false and defective idea of art from which it started — art reduced to the representation of the pleasurable — with real art, which occupies a far wider field. Hence the artificial attempt to settle what examples of the *ugly* (antipathetic) could be admitted in artistic representation, and for what reasons, and in what ways.

The answer was: that the ugly is admissible, only when it can be *overcome*; an unconquerable ugliness, such as the *disgusting* or the *nauseating*, being altogether excluded. Further, that the duty of the ugly, when admitted in art, is to contribute towards heightening the effect of the beautiful (sympathetic), by producing a series of contrasts, from which the pleasurable may issue more efficacious and joy-giving. It is, indeed, a common observation that pleasure is more vividly felt

1. *antithesis*：对立；对照

when preceded by abstinence[1] and suffering. Thus the ugly in art was looked upon as adapted for the service of the beautiful, a stimulant and condiment[2] of æsthetic pleasure.

That special refinement of hedonistic theory which used to be pompously called the doctrine of the *overcoming of the ugly* falls with the Æsthetic of the sympathetic, and with it the enumeration and definition of the concepts mentioned above, which show themselves to be completely foreign to Æsthetic. For Æsthetic does not recognize the sympathetic or the antipathetic or their varieties, but only the spiritual activity of representation.

Pseudo-æsthetic concepts belong to Psychology.

Nevertheless, the important place which, as we have said, those concepts have hitherto occupied in æsthetic treatises makes it advisable to supply a rather more complete explanation as to their nature. What shall be their lot? Excluded from Æsthetic, in what other part of Philosophy will they be received?

In truth, nowhere; for all those concepts are without philosophical value. They are nothing but a series of classes, which can be fashioned in the most various ways and multiplied at pleasure, to which it is sought to reduce the infinite complications and shadings of the values and disvalues of life. Of these classes, some have an especially positive significance, like the beautiful, the sublime, the majestic, the solemn, the serious, the weighty, the noble, the elevated; others a significance chiefly negative, like the ugly, the painful, the horrible, the dreadful, the tremendous, the monstrous, the insipid[3], the extravagant; finally in others a mixed significance prevails, such as the comic, the tender,

1. abstinence：节制；禁欲
2. condiment：调味品
3. insipid：枯燥的

the melancholy, the humorous, the tragi-comic. The complications are infinite, because the individuations[1] are infinite; hence it is not possible to construct the concepts, save in the arbitrary and approximate manner proper to the natural sciences, satisfied with making the best classification they can of that reality which they can neither exhaust by enumeration, nor understand and conquer speculatively. And since *Psychology* is the naturalistic science which undertakes to construct types and schemes of the spiritual life of man (a science whose merely empirical and descriptive character becomes more evident day by day), these concepts do not belong to Æsthetic, nor to Philosophy in general, but must simply be handed over to Psychology.

Impossibility of rigorous definitions of them.

The case of those concepts is that of all other psychological constructions: no rigorous definitions of them are possible; and consequently they cannot be deduced from one another nor be connected in a system, though this has often been attempted, with great waste of time and without obtaining thereby any useful results. Nor can it be claimed as possible to obtain empirical definitions, universally acceptable as precise and true in the place of those philosophical definitions recognized as impossible. For no single definition of a single fact can be given, but there are innumerable definitions of it, according to the cases and the purposes for which they are made; and it is clear that if there were only one which had the value of truth it would no longer be an empirical, but a rigorous and philosophical definition. And as a matter of fact whenever one of the terms to which we have referred has been employed (or indeed

1. individuations：个性化

any other belonging to the same class), a new definition of it has been given at the same time, expressed or understood. Each one of those definitions differed somehow from the others, in some particular, however minute, and in its implied reference to some individual fact or other, which thus became a special object of attention and was raised to the position of a general type. Thus it is that not one of such definitions satisfies either the hearer or the constructor of it. For a moment later he finds himself before a new instance to which he recognizes that his definition is more or less insufficient, ill-adapted, and in need of retouching. So we must leave writers and speakers free to define the sublime or the comic, the tragic or the humorous, on every occasion as they please and as may suit the end they have in view. And if an empirical definition of universal validity be demanded, we can but submit this one: — The sublime (or comic, tragic, humorous, etc.) is *everything* that is or shall be so *called* by those who have employed or shall employ these *words*.

Examples: definitions of the sublime, the comic, the humorous.

What is the sublime? The unexpected affirmation of an overwhelming moral force[1]: that is one definition. But the other definition is equally good, which recognizes the sublime also where the force which affirms itself is certainly overwhelming, but immoral and destructive. Both remain vague and lack precision, until applied to a concrete case, to an example which makes clear what is meant by "overwhelming," and what by unexpected. They are quantitative concepts, but falsely quantitative, since there is no way of measuring them; they are at bottom metaphors, emphatic phrases, or logical tautologies. The humorous will be laughter amid tears, bitter laughter, the sudden

1. overwhelming moral force：无法抗拒的道德力量

spring from the comic to the tragic and from the tragic to the comic, the romantic comic, the opposite of the sublime, war declared against every attempt at insincerity, compassion ashamed to weep, a laugh, not at the fact, but at the ideal itself; and what you will beside, according as it is wished to get a view of the physiognomy of this or that poet, of this or that poem, which, in its uniqueness, is its own definition, and though momentary and circumscribed, is alone adequate. The comic has been defined as the displeasure arising from the perception of a deformity immediately followed by a greater pleasure arising from the relaxation of our psychical forces, strained in expectation of a perception looked upon as important. While listening to a narrative, which might, for example, be a description of the magnificently heroic purpose of some individual, we anticipate in imagination the occurrence of a magnificent and heroic action, and we prepare for its reception by concentrating our psychic forces. All of a sudden, however, instead of the magnificent and heroic action, which the preliminaries and the tone of the narrative had led us to expect, there is an unexpected change to a small, mean, foolish action, which does not satisfy to our expectation. We have been deceived, and the recognition of the deceit brings with it an instant of displeasure. But this instant is as it were conquered by that which immediately follows: we are able to relax our strained attention, to free ourselves from the provision of accumulated psychic energy henceforth superfluous, to feel ourselves light and well. This is the pleasure of the comic, with its physiological equivalent of laughter. If the unpleasant fact that has appeared should painfully affect our interests, there would not be pleasure, laughter would be at once suffocated, the psychic energy would be strained and overstrained by other more weighty perceptions. If on the other hand such more weighty perceptions do not appear, if the whole loss be limited to a slight deception of our foresight, then the feeling of our psychic wealth that ensues affords ample compensation for this very slight disappointment. Such, expressed in a few words, is one of the most accurate modern definitions of the comic. It boasts of containing in itself, justified or corrected and verified, the manifold attempts to define the comic, from Hellenic antiquity to our own day, from Plato's definition in the *Philebus*[1], and from Aristotle's, which is more explicit, and looks upon the comic as an *ugliness without pain*, to that of Hobbes[2], who replaced it in the feeling of *individual superiority*; of Kant, who saw in it the *relaxation of a tension*; or from the other proposals of those for whom it was *the conflict between great and small, between the finite and the infinite* and so on. But on close observation, the analysis and definition above given, although in appearance most elaborate and precise, yet enunciates characteristics which are applicable, not only to the comic, but to every spiritual process; such as the succession of painful and pleasing moments and the satisfaction arising from the consciousness of strength and of its free expansion. The differentiation is here given by quantitative determinations whose limits cannot be laid down. They therefore remain vague words, possessing some degree of meaning from their reference to this or that particular comic fact, and from the psychic disposition of qualities of the speaker. If such definitions be taken too seriously, there happens to them what Jean Paul Richte said of all the definitions of the comic: namely, that their sole merit is *to be themselves comic* and to produce in reality the fact which they vainly try to fix logically. And who will ever logically determine the dividing line between the comic and the non-comic, between laughter and smiles, between smiling and gravity, or cut the ever varying continuum into which life melts into clearly divided parts?

Relation between these concepts and æsthetic concepts.

The facts, classified as far as possible in these psychological concepts,

---

1. *Philebus*：《斐力布斯篇》，柏拉图著。
2. Hobbes：霍布斯（1588—1679），英国政治家、哲学家，创立了机械唯物主义的完整体系，著有《利维坦》《论公民》《论人》等。

bear no relation to the artistic fact, beyond the general one, that all of them, in so far as they constitute the material of life, can become the object of artistic representation; and the other, an accidental relation, that æsthetic facts also may sometimes enter the processes described, such as the impression of the sublime aroused by the work of a Titanic artist[1], such as Dante or Shakespeare, and of the comic produced by the attempts of a dauber or scribbler[2].

But here too the process is external to the æsthetic fact, to which is linked only the feeling of æsthetic value and disvalue, of the beautiful and of the ugly. Dante's Farinata[3] is æsthetically beautiful and nothing but beautiful: if the force of will of that personage seem also sublime, or the expression that Dante gives him seem, by reason of his great genius, sublime in comparison with that of a less energetic poet, these are things altogether outside æsthetic consideration. We repeat again that this last pays attention always and only to the adequateness of the expression, that is to say, to beauty.

---

1. Titanic artist：艺术巨匠
2. dauber or scribbler：拙劣的画家或乱涂乱画的人
3. Farinata：法里纳塔，但丁《神曲·地狱篇》中的人物。

XIII THE "PHYSICALLY BEAUTIFUL" IN NATURE AND IN ART

Æsthetic activity and physical concepts.

Æsthetic activity, distinct from the practical activity, is always accompanied by it in its manifestations. Hence its utilitarian or hedonistic side, and the pleasure and pain which are, as it were, the practical echo of æsthetic value and disvalue, of the beautiful and of the ugly. But this practical side of the æsthetic activity has in its turn a *physical* or *psycho-physical* accompaniment, which consists of sounds, tones, movements, combinations of lines and colours, and so on.

Does it *really* possess this side, or does it only seem to possess it, through the construction which we put on it in physical science, and the useful and arbitrary methods which we have already several times set in relief as proper to the empirical and abstract sciences? Our reply cannot be doubtful, that is, it must affirm to the second of the two hypotheses.

However, it will be better to leave this point in suspense, since it is not at present necessary to press this line of inquiry further. The mere mention suffices to secure our speaking (for reasons of simplicity and adhesion to ordinary language) of the physical element as something objective and existing, against leading to hasty conclusions as to the concepts of spirit and nature and their relation.

Expression in the æsthetic sense, and expression in the naturalistic sense.

It is important, on the other hand, to make clear that as the existence of the hedonistic side in every spiritual activity has given rise to the confusion between the æsthetic activity and the useful or pleasurable, so the existence of, or rather the possibility of constructing, this physical side, has caused the confusion between *æsthetic* expression and expression *in a naturalistic sense*; that is to say, between a spiritual fact and a mechanical and passive fact (not to say, between a concrete reality and an abstraction or fiction). In common speech, sometimes it is the words of the poet that are called *expressions*, the notes of the musician, or the figures of the painter; sometimes the blush which generally accompanies the feeling of shame, the pallor[1] often due to fear, the grinding of the teeth[2] proper to violent anger, the shining of the eyes and certain movements of the muscles of the mouth, which manifest cheerfulness. We also say that a certain degree of heat is the *expression* of fever, that the falling of the barometer is the *expression* of rain, and even that the height of the exchange *expresses* the depreciation of the paper currency of a State, or social discontent the approach of a revolution. One can well imagine what sort of scientific results would be attained by allowing oneself to be governed by verbal usage and classing together facts so widely different. But there is, in fact, an abyss[3] between a man who is the prey of anger with all its natural manifestations and another man who expresses it æsthetically; between the appearance, the cries and contortions of some one grieving at the loss of a dear one and the words or song with which the same individual portrays his suffering at another time; between the grimace of emotion and the gesture of the actor. Darwin's[4] book on the expression of the emotions in man and animals does not

---

1. pallor：面色苍白
2. grinding of the teeth：咬牙切齿
3. abyss：深渊
4. Darwin：达尔文（1809—1882），英国生物学家、地质学家、自然主义者，著有《物种起源》《人与动物的表情表达》等。

belong to Æsthetic; because there is nothing in common between the science of spiritual expression and a Semiotic[1], whether it be medical, meteorological, political, physiognomic, or chiromantic[2].

Expression in the naturalistic sense simply lacks *expression in the spiritual sense*, that is to say, the very character of activity and of spirituality, and therefore the bipartition into the poles of beauty and of ugliness. It is nothing but a relation between cause and effect, fixed by the abstract intellect. The complete process of æsthetic production can be symbolized in four stages, which are: *a,* impressions; *b,* expression or spiritual æsthetic synthesis; *c,* hedonistic accompaniment, or pleasure of the beautiful (æsthetic pleasure); *d,* translation of the æsthetic fact into physical phenomena (sounds, tones, movements, combinations of lines and colours, etc.). Any one can see that the capital point, the only one that is properly speaking æsthetic and truly real, is in *b,* which is lacking to the merely naturalistic manifestation or construction also metaphorically called expression.

The expressive process is exhausted when these four stages have been passed through. It begins again with new impressions, a new æsthetic synthesis, and the accompaniments that belong to it.

Representations and memory.

Expressions or representations follow one another, the one drives out the other. Certainly, this passing away, this being driven out, is not a perishing, it is not total elimination: nothing that is born dies with that complete death which would be identical with never having been born. If all things pass away, nothing can die. Even the representations that we have forgotten persist somehow in our spirit, for without this we could not explain acquired habits and capacities. Indeed the strength

---

1. Semiotic：符号学
2. chiromantic：手相的

of life lies in this apparent forgetting: one forgets what has been absorbed and what life has superseded.

But other representations are also powerful elements in the present processes of our spirit; and it is incumbent upon[1] us not to forget them, or to be capable of recalling them when they are wanted. The will is always vigilant[2] in this work of preservation, which aims at preserving (we may say) the greater, the more fundamental part of all our riches. But its vigilance does not always suffice. Memory, as we say, abandons or betrays us in different ways. For this very reason, the human spirit devises expedients[3] which succour the weakness[4] of memory and are its *aids*.

The production of aids to memory.

How these aids are possible we have been informed from what has been said. Expressions or representations are *also* practical facts, which are also called physical in so far as physics classifies and reduces them to types. Now it is clear that if we can succeed in making those practical or physical facts somehow permanent, it will always be possible (all other conditions remaining equal) on perceiving them to reproduce in ourselves the already produced expression or intuition.

If that be called the object or physical stimulus in which the practical concomitant acts, or (to use physical terms) in which the movements have been isolated and made in some sort permanent, and if that object or stimulus be designated by the letter *e*; the process of reproduction will take place in the following order: *e*, the physical stimulus; *d-b*, perception of physical facts (sounds, tones, mimetic,

---

1. incumbent upon：对……来说义不容辞
2. vigilant：警醒的
3. expedients：补救措施
4. succour the weakness：弥补缺陷

combinations of lines and colours, etc.), which is together the æsthetic synthesis, already produced; *c*, the hedonistic accompaniment, which is also reproduced.

And what else are those combinations of words called poetry, prose, poems, novels, romances, tragedies or comedies, but *physical stimulants of reproduction* (the stage *e*); what else are those combinations of sound called operas, symphonies, sonatas; or those combinations of lines and colours called pictures, statues, architecture? The spiritual energy of memory, with the assistance of the physical facts above mentioned, makes possible the preservation and the reproduction of the intuitions produced by man. The physiological organism and with it the memory become weakened; the monuments of art are destroyed, and lo, all that æsthetic wealth, the fruit of the labours of many generations, diminishes and rapidly disappears.

Physical beauty.

Monuments of art, the stimulants of æsthetic reproduction, are called *beautiful things* or *physical beauty*. This combination of words constitutes a verbal paradox, for the beautiful is not a physical fact; it does not belong to things, but to the activity of man, to spiritual energy. But it is now clear through what transferences and associations, physical things and facts which are simply aids to the reproduction of the beautiful are finally called elliptically beautiful things and physical beauty. And now that we have explained this elliptical usage, we shall ourselves employ it without hesitation.

Content and form: another meaning.

The intervention of "physical beauty" serves to explain another meaning of the words "*content*" and "*form*," as used by æstheticians. Some call "content" the internal fact or expression (for us, on the other hand, form), and "form" the marble, the colours, the rhythm, the sounds (for us the antithesis of form); thus looking upon the physical fact as the form, which may or may not be joined to the content. It also serves to explain another aspect of what is called æsthetic "ugliness." Somebody who has nothing definite to express may try to conceal his internal emptiness in a flood of words, in sounding verse, in deafening polyphony[1], in painting that dazzles the eye, or by heaping together great architectural masses which arrest and astonish us without conveying anything whatever. Ugliness, then, is the capricious, the charlatanesque[2]; and, in reality, if practical caprice did not intervene in the theoretic function, there might be absence of beauty, but never the real presence of something deserving the adjective "ugly."

Natural and artificial beauty.

Physical beauty is usually divided into *natural* and *artificial* beauty. Thus we reach one of the facts which have given the greatest trouble to thinkers: *natural beauty*. These words often designate facts of merely practical pleasure. Any one who calls a landscape beautiful where the eye rests upon verdure, where the body moves briskly and the warm sun envelops and caresses the limbs, does not speak of anything æsthetic. But it is nevertheless indubitable that on other occasions the adjective "beautiful," applied to objects and scenes existing in nature, has a completely æsthetic signification.

It has been observed that in order to enjoy natural objects æsthetically, we must abstract from their external and historical reality, and separate their simple semblance or appearance from existence; that if we contemplate a landscape with our head between our legs, so as to cancel our wonted relations with it, the landscape appears to us to be an

---

1. deafening polyphony：震耳欲聋的音节
2. charlatanesque：江湖骗子

ideal spectacle; that nature is beautiful only for him who contemplates her *with the eye of the artist*; that zoologists and botanists do not recognize *beautiful* animals and flowers; that natural beauty is *discovered* (and examples of discovery are the points of view, pointed out by men of taste and imagination, to which more or less æsthetic travellers and excursionists afterwards have recourse in pilgrimage, whence a kind of collective *suggestion*); that, without the *aid of the imagination*, no part of nature is beautiful, and that with such aid the same natural object or fact is, according to the disposition of the soul, now expressive, now insignificant, now expressive of one definite thing, now of another, sad or glad, sublime or ridiculous, sweet or laughable; finally, that a *natural beauty* which an artist would not *to some extent correct, does not exist.*

All these observations are just, and fully confirm the fact that natural beauty is simply a *stimulus* to æsthetic reproduction, which presupposes previous production. Without the previous æsthetic intuitions of the imagination, nature cannot awaken any at all. As regards natural beauty, man is like the mythical Narcissus[1] at the fountain. Leopardi[2] said that natural beauty is "rare, scattered, and fugitive[3]": it is imperfect, equivocal, variable. Each refers the natural fact to the expression in his mind. One artist is thrown into transports by a smiling landscape, another by a rag-shop[4], another by the pretty face of a young girl, another by the squalid countenance of an old rascal[5]. Perhaps the first will say that the rag-shop and the ugly face of the old rascal are *repulsive*[6]; the second, that the smiling landscape and the face of the young girl are

1. Narcissus：那喀索斯，希腊神话中的美男子，因迷恋自己水中的倒影抑郁而死，化身为水仙花，寓意自恋情结。
2. Leopardi：莱奥帕尔迪（1798—1837），意大利十九世纪著名浪漫主义诗人。他的诗，语言洗练朴素，格律自由多变，开意大利现代自由体抒情诗的先河。
3. rare, scattered, fugitive：稀有、零星、转瞬即逝的
4. rag-shop：杂货店
5. squalid countenance of an old rascal：老流氓的肮脏面目
6. *repulsive*：令人厌恶的

insipid. They may dispute for ever; but they will never agree, save when they are supplied with a sufficient dose of æsthetic knowledge to enable them to recognize that both are right. *Artificial* beauty, created by man, supplies an aid that is far more ductile and efficacious[1].

Mixed beauty.

In addition to these two classes, æstheticians also sometimes talk in their treatises of a *mixed* beauty. A mixture of what? Precisely of natural and artificial. Whoever fixes and externalizes, operates with natural data which he does not create but combines and transforms. In this sense, every artificial product is a mixture of nature and artifice; and there would be no occasion to speak of a mixed beauty, as of a special category. But it sometimes happens that combinations already given in nature can be used a great deal more than in others; as, for instance, when we design a beautiful garden and include in our design groups of trees or ponds already in place. On other occasions externalization is limited by the impossibility of producing certain effects artificially. Thus we can mix colouring matters, but we cannot create a powerful voice or a face and figure appropriate to this or that character in a play. We must therefore seek them among already existing things, and make use of them when found. When, therefore, we employ a great number of combinations already existing in nature, such as we should not be able to produce artificially if they did not exist, the resulting fact is called *mixed* beauty.

Writings.

We must distinguish from artificial beauty those instruments

---

1. far more ductile and efficacious：更灵活有效的

of reproduction called *writings*, such as alphabets, musical notes, hieroglyphics[1], and all pseudolanguages[2], from the language of flowers and flags to the language of patches (so much in vogue in the society of the eighteenth century). Writings are not physical facts which arouse directly impressions answering to æsthetic expressions; they are simple *indications* of what must be done in order to produce such physical facts. A series of graphic signs serves to remind us of the movements which we must execute with our vocal apparatus in order to emit certain definite sounds. If, through practice, we become able to hear the words without opening our mouths and (what is much more difficult) to hear the sounds by running the eye along the stave, all this does not alter in any way the nature of the writings, which are altogether different from direct physical beauty. No one calls the book which contains the *Divine Comedy*[3], or the score which contains *Don Giovanni*[4], beautiful in the same sense in which the block of marble which contains Michæl-Angelo's *Moses*[5], or the piece of coloured wood which contains the Transfiguration[6], is metaphorically called beautiful. Both serve the reproduction of the beautiful, but the former by a far longer and more indirect route than the latter.

Free and non-free beauty.

Another division of the beautiful, still found in treatises, is that into *free and not free*. By not-free beauties have been understood those

---

1. hieroglyphics：象形文字
2. pseudolanguages：虚拟语言
3. *Divine Comedy*：《神曲》，但丁著。
4. *Don Giovanni*：歌剧《唐璜》，奥地利音乐家莫扎特作品。
5. *Moses*：《摩西》，意大利艺术家米开朗琪罗雕塑作品。
6. Transfiguration：主显圣容。据《新约・圣经》，耶稣与门徒彼得、约翰、詹姆斯登山祈祷，耶稣开始发出耀眼的光，接着先知摩西与伊利亚出现在耶稣身旁。拉斐尔据此题材创作了《主显圣容》。

objects which have to serve a double purpose, extra-æsthetic and æsthetic (stimulants of intuitions); and since it seems that the first purpose sets limits and barriers in the way of the second, the resulting beautiful object has been considered as not-free beauty.

Architectural works are especially cited; and just for this reason, architecture has often been excluded from the number of what are called the fine arts. A temple must above all things be for the use of a cult; a house must contain all the rooms needed for the convenience of life, and they must be arranged with a view to this convenience; a fortress must be a construction capable of resisting the attacks of given armies and the blows of given instruments of war. It is therefore concluded that the architect's field is restricted: he may *embellish*[1] to some extent the temple, the house, the fortress; but he is bound by the *object* of those edifices, and he can only manifest that part of his vision of beauty which does not impair their extra-æsthetic but fundamental objects.

Other examples are taken from what is called art applied to industry. Plates, glasses, knives, guns and combs can be made beautiful; but it is held that their beauty must not be pushed so far as to prevent our eating from the plate, drinking from the glass, cutting with the knife, firing off the gun, or combing one's hair with the comb. The same is said of the art of typography: a book should be beautiful, but not to the extent of being difficult or impossible to read.

Criticism of non-free beauty.

In respect of all this we must observe in the first place that the extrinsic purpose is not necessarily, precisely because it is such, a limit or impediment to the other purpose of being a stimulus to æsthetic

1. *embellish*：美化

reproduction. It is therefore quite false to maintain that architecture, for example, is by its nature imperfect and not free, since it must also obey other practical purposes; in fact, the mere presence of fine works of architecture is enough to dispel any such illusion.

In the second place, not only are the two purposes not necessarily contradictory, but we must add that the artist always has the means of preventing this contradiction from arising. How? by simply making the *destination* of the object which serves a practical end enter as material into his æsthetic intuition and externalization. He will not need to add anything to the object, in order to make it the instrument of æsthetic intuitions: it will be so, if perfectly adapted to its practical purpose. Rustic dwellings and palaces, churches and barracks, swords and ploughs, are beautiful, not in so far as they are embellished and adorned, but in so far as they express their end. A garment is only beautiful because it is exactly suitable to a given person in given conditions. The sword bound to the side of the warrior Rinaldo by the amorous Armida[1] was not beautiful: "so adorned that it may seem a useless ornament, not the free instrument of war," or it was beautiful, if you will, but to the eyes and imagination of the sorceress[2], who liked to see her lover equipped in that effeminate[3] way. The æsthetic activity can always agree with the practical, because expression is truth.

It cannot however be denied that æsthetic contemplation sometimes hinders practical usage. For instance, it is a quite common experience to find certain new objects seem so well adapted to their

---

1. The sword bound to the side of the warrior Rinaldo by the amorous Armida：含情脉脉的阿米达放在勇士里纳尔多身边的剑。里纳尔多是十字军的一员战将。阿尔米达是伊斯兰教魔女，发誓要以自己的魔法和美色战胜十字军。传说中，阿尔米达引诱里纳尔多登上 — 座小岛，施魔法把他迷醉。本想杀死沉睡中的里纳尔多，但是，爱乱点鸳鸯谱的丘比特突然闯入，令她由恨转爱，堕入了情网。
2. sorceress：女魔法师，指阿米达
3. effeminate：女性化的

purpose, and therefore so beautiful, that people occasionally feel scruples in maltreating them[1] by passing from their contemplation to their use. It was for this reason that King Frederick William of Prussia[2] showed such repugnance to sending his magnificent grenadiers[3], so well adapted to war, into the mud and fire of battle, while his less æsthetic son, Frederick the Great, obtained from them excellent service.

Stimulants of production.

It might be objected to the explanation of the physically beautiful as a simple aid to the reproduction of the internally beautiful, or expressions, that the artist creates his expressions by painting or by sculpturing, by writing or by composing, and that therefore the physically beautiful, instead of following, sometimes precedes the æsthetically beautiful. This would be a somewhat superficial mode of understanding the procedure of the artist, who never in reality makes a stroke with his brush without having previously seen it with his imagination; and if he has not yet seen it, he will make the stroke, not in order to externalize his expression (which does not yet exist), but as a kind of experiment and in order to have a point of departure for further meditation and internal concentration. The physical point of departure is not the physically beautiful instrument of reproduction, but a means that may be called *pedagogic*, like retiring into solitude, or the many other expedients frequently very strange, adopted by artists and scientists, who vary in these according to their various

1. feel scruples in maltreating them：不忍暴殄天物
2. King Frederick William of Prussia：普鲁士国王威廉一世（1797—1888），1871 年加冕为德意志帝国皇帝，通过三次王朝战争，统一德意志，并建立德意志帝国。
3. magnificent grenadiers: 英姿飒爽的炮兵部队

idiosyncrasies[1]. The old æsthetician Baumgarten[2] advised poets seeking inspiration to ride on horseback, to drink wine in moderation, and (provided they were chaste) to look at beautiful women.

1. idiosyncrasies：个人癖好
2. Baumgarten：鲍姆加登（1714—1762），德国哲学家，莱布尼茨学说追随者。“美学”一词由他首次提出。

XIV ERRORS ARISING FROM THE CONFUSION BETWEEN PHYSICS AND ÆSTHETIC

We must mention a series of fallacious scientific doctrines[1] which have arisen from the failure to understand the purely external relation between the æsthetic fact or artistic vision and the physical fact or instrument which aids in its reproduction, together with brief criticisms of them deduced from what has already been said.

Criticism of æsthetic associationism.

That form of associationism[2] which identifies the æsthetic fact with the *association* of two images finds support in such lack of apprehension. By what path has it been possible to arrive at such an error, so repugnant to our æsthetic consciousness, which is a consciousness of perfect unity, never of duality? Precisely because the physical and æsthetic facts have been considered separately, as two distinct images, which enter the spirit, the one drawn in by the other, first one and then the other. A picture has been divided into the image of the *picture* and the image of the *meaning* of the picture; a poem, into the image of the *words* and the image of the *meaning* of the words. But this dualism of images[3] is non-existent: the physical fact does not enter the spirit as an image, but causes the reproduction of the image (the

1. fallacious scientific doctrines：错误的科学学说
2. associationism：联想主义
3. dualism of images：形象的二元性

only image, which is the æsthetic fact), in so far as it blindly stimulates the psychic organism and produces the impression which answers to the æsthetic expression already produced.

The efforts of the associationists (the usurpers of to-day in the field of Æsthetic) to emerge from the difficulty, and to reaffirm in some way the unity which has been destroyed by their principle of association, are highly instructive. Some maintain that the image recalled is unconscious; others, leaving unconsciousness alone, hold that, on the contrary, it is vague, vaporous, confused, thus reducing the *force* of the æsthetic fact to the *weakness* of bad memory. But the dilemma is inexorable[1]: either keep association and give up unity, or keep unity and give up association. No third way out of the difficulty exists.

Criticism of æsthetic physics.

From the failure to analyse so-called natural beauty thoroughly and to recognize that it is simply an incident of æsthetic reproduction, and from having looked upon it, on the contrary, as given in nature, is derived all that portion of treatises upon Æsthetic entitled *Beauty of Nature* or *Æsthetic Physics*; sometimes even subdivided, save the mark, into æsthetic Mineralogy, Botany and Zoology. We do not wish to deny that such treatises contain many just observations, and are sometimes themselves works of art, in so far as they represent beautifully the imaginings and fancies or impressions of their authors. But we must affirm it to be scientifically false to ask oneself if the dog be beautiful and the ornithorhynchus[2] ugly, the lily beautiful and the artichoke[3] ugly. Indeed, the error is here double. On the one hand, æsthetic Physics falls back into the equivocation of the theory of artistic and

1. inexorable：无可回避的
2. ornithorhynchus：鸭嘴兽
3. artichoke：朝鲜蓟

literary kinds, of attempting to attach æsthetic determinations to the abstractions of our intellect; on the other, it fails to recognize, as we said, the true formation of so-called natural beauty, a formation which excludes even the possibility of the question as to whether some given individual animal, flower or man be beautiful or ugly. What is not produced by the æsthetic spirit, or cannot be referred to it, is neither beautiful nor ugly. The æsthetic process arises from the ideal connexions in which natural objects are placed.

Criticism of the theory of the beauty of the human body.

The double error can be exemplified by the question as to the *Beauty of the human body*, upon which whole volumes have been written. Here we must before everything turn those who discuss this subject from the abstract toward the concrete, by asking: "What do you mean by the human body, that of the male, the female, or the hermaphrodite[1]?" Let us assume that they reply by dividing the inquiry into two distinct inquiries, as to male and female beauty (there really are writers who seriously discuss whether man or woman is the more beautiful); and let us continue: "Masculine or feminine beauty[2]; but of what race of men — the white, the yellow or the black, or any others that may exist, according to the division you prefer?" Let us assume that they limit themselves to the white race, and drive home the argument: "To what sub-species of the white race?" And when we have restricted them gradually to one corner of the white world, going, let us say, from the Italian to the Tuscan[3], the Siennese[4], the Porta Camollia[5] quarter, we

---

1. hermaphrodite：雌雄同体
2. Masculine or feminine beauty: 阳刚美或阴柔美
3. Tuscan：托斯卡纳人，古意大利人。
4. Siennese：锡耶纳人。锡耶纳位于意大利中西部佛罗伦萨南部。
5. Porta Camollia：卡莫利亚城门，锡耶纳老城区。

will proceed: "Very good; but at what age of the human body, and in what condition and stage — that of the newborn babe, of the child, of the boy, of the adolescent, of the man of middle age, and so on? and of him who is at rest or of him who is at work, or of him who is occupied like Paul Potter's[1] bull, or the Ganymede of Rembrand[2]".

Having thus arrived, by successive reductions, at the individual *omnimode determinatum*, or rather at "this man here," pointed out with the finger, it will be easy to expose the other error, by recalling what we have said about the natural fact, which is now beautiful, now ugly, according to the point of view and to what is passing in the soul of the artist. If even the Gulf of Naples[3] have its detractors[4], and if there be artists who declare it inexpressive, preferring the "gloomy firs," the "clouds and perpetual north winds," of northern seas; is it really possible that such relativity does not exist for the human body, source of the most varied suggestions?

Criticism of the beauty of geometrical figures.

The question of the *beauty of geometrical figures* is connected with æsthetic Physics. But if by geometrical figures be understood the concepts of geometry (the concepts of the triangle, the square, the cone), these are neither beautiful nor ugly, just because they are concepts. If, on the other hand, by such figures be understood bodies which possess definite geometrical forms, they will be beautiful or ugly, like every natural fact,

---

1. Paul Potter：保罗·波特（1625—1654），荷兰画家，擅长画动物画与风景画。
2. Ganymede of Rembrand：伦勃朗画作《甘尼米德》。伦勃朗（1606—1669），荷兰画家，欧洲 17 世纪最伟大的画家之一。甘尼米德是古希腊神话中的美少年。
3. Gulf of Naples：《那不勒斯湾》，法国印象派画家雷诺阿（1841—1919）的作品。
4. detractors：批评者

according to the ideal connexions in which they are placed. Some hold that those geometrical figures are beautiful which point upwards, since they give the suggestion of firmness and of power. We do not deny that this may be so. But it must not be denied on the other hand that those also may possess beauty which give the impression of instability and weakness, where they represent just the insecure and the feeble; and that in these last cases the firmness of the straight fine and the lightness of the cone or of the equilateral triangle would seem to be on the contrary elements of ugliness.

Certainly, such questions as to the beauty of nature and the beauty of geometry, like others analogous as to the historically beautiful and human beauty, seem less absurd in the Æsthetic of the sympathetic, which really means by the words "æsthetic beauty" the representation of the pleasing. But the claim to determine scientifically what are sympathetic contents and what are irremediably antipathetic[1] is none the less erroneous, even in the sphere of that doctrine and after laying down those premises. One can only answer such questions by repeating with an infinitely long postscript the *Sunt quos*[2] of the first ode of the first book of Horace[3], and the *Havvi chi* of Leopardi's letter to Carlo Pepoli[4]. To each man his beautiful (= sympathetic), as to each man his fair one. Philography[5] is not science.

Criticism of another aspect of the imitation of nature.

The artist sometimes has naturally existing facts before him, in

---

1. irremediably antipathetic：无可救药地令人反感的
2. *Sunt quos*：〈拉〉有这样一种人……=there is (of) which
3. first ode of the first book of Horace ：贺拉斯的第一部书的第一章。贺拉斯（前 65 年—前 8 年），古罗马时期著名诗人、批评家、翻译家，古罗马三大诗人之一，代表作有《诗艺》。
4. Carlo Pepoli：卡罗・佩波利，19 世纪意大利剧作家、三幕正歌剧《清教徒》的编剧。
5. Philography：文字学

producing the artificial instrument, or physically beautiful. These are called his *models*: bodies, stuffs, flowers and so on. Let us run over the sketches, studies and notes of artists: Leonardo noted down in his pocket-book, when he was working on the Last Supper : "Giovannina, weird face, is at St. Catherine's, at the Hospital; Cristofano di Castiglione is at the Pietà, he has a fine head; Christ, Giovan Conte, of Cardinal Mortaro's suite." And so on. From this comes the illusion that the artist *imitates nature*, when it would perhaps be more exact to say that nature imitates the artist, and obeys him. The illusion that *art imitates nature* has sometimes found ground and support in this illusion, as also in its variant, more easily maintained, which makes of art the *idealizer of nature*. This last theory presents the process out of its true order, which indeed is not merely upset but actually inverted; for the artist does not proceed from external reality, in order to modify it by approximating it to the ideal; he goes from the impression of external nature to expression, that is to say, his ideal, and from this passes to the natural fact, which he employs as instrument of reproduction of the ideal fact.

Criticism of the theory of the elementary forms of the beautiful.

Another consequence of the confusion between the æsthetic fact and the physical fact is the theory of the *elementary forms of the beautiful*. If expression, if the beautiful, be indivisible, the physical fact on the contrary, in which it externalizes itself, can easily be divided and subdivided: for example, a painted surface, into lines and colours, groups and curves of lines, kinds of colours, and so on; a poem, into strophes, verses, feet, syllables; a piece of prose, into chapters, paragraphs, headings, periods, phrases, words and so on. The parts thus obtained are not æsthetic facts, but smaller physical facts, arbitrarily divided. If this path were followed and the confusion persisted in, we should end by concluding that the true elementary forms of the beautiful are *atoms*.

The æsthetic law, several times promulgated[1], that beauty must have *bulk*, could be invoked against the atoms. It cannot be the imperceptibility of the too small[2], or the inapprehensibility of the too large[3]. But a greatness determined by perceptibility, not by measurement, implies a concept widely different from the mathematical. Indeed, what is called imperceptible and inapprehensible does not produce an impression, because it is not a real fact, but a concept: the demand for bulk in the beautiful is thus reduced to the actual presence of the physical fact, which serves for the reproduction of the beautiful.

Criticism of the search for the objective conditions of the beautiful.

Continuing the search for the *physical laws* or for the *objective conditions of the beautiful*, it has been asked: To what physical facts does the beautiful correspond? To what the ugly? To what unions of tones, colours, sizes, mathematically determinable? Such inquiries are as if in Political Economy one were to seek for the laws of exchange in the physical nature of the objects exchanged. The persistent fruitlessness of the attempt should have given rise before long to some suspicion of its vanity. In our times, especially, necessity for an *inductive* Æsthetic has been often proclaimed, of an Æsthetic starting *from below*, proceeding like natural science and not jumping to its conclusions. Inductive? But Æsthetic has always been both inductive and deductive, like every philosophical science; induction and deduction cannot be separated, nor can they separately avail to characterize a true science. But the word "induction" was not pronounced here by chance. The

---

1. several times promulgated：一再重申
2. imperceptibility of the too small：小到无法察觉
3. inapprehensibility of the too large：大到无法分辨

intention was to imply that the æsthetic fact is really nothing but a physical fact, to be studied by the methods proper to the physical and natural sciences.

With such a presupposition and in such a faith did inductive Æsthetic or Æsthetic *from below* (what pride in this modesty!) begin its labours. It conscientiously began by making a collection of *beautiful things*, for example, a great number of envelopes of various shapes and sizes, and asked which of these give the impression of beauty and which of ugliness. As was to be expected, the inductive æstheticians speedily found themselves in a difficulty, for the same objects that appeared ugly in one aspect appeared beautiful in another. A coarse yellow envelope, which would be extremely ugly for the purpose of enclosing a love-letter, is just what is wanted for a writ served by process[1] on stamped paper, which in its turn would look very bad, or seem at any rate an irony, enclosed in a square envelope of English paper. Such considerations of simple common sense should have sufficed to convince inductive æstheticians that the beautiful has no physical existence, and cause them to desist from their vain and ridiculous quest. But no: they had recourse to an expedient, as to which we should hardly like to say how far it belongs to the strict method of natural science. They sent their envelopes round and opened a *referendum*[2] trying to settle in what beauty or ugliness consists by the votes of the majority.

The astrology of Æsthetic.

We will not waste time over this subject, lest we should seem to be turning ourselves into tellers of comic tales rather than expositors of æsthetic science and of its problems. It is a matter of fact that the

1. writ served by process：按诉讼程序发出的令状
2. *referendum*：全民公决

inductive æstheticians have not yet discovered *one single law.*

He who despairs of doctors is apt to abandon himself to charlatans. This has befallen those who have believed in the naturalistic laws of the beautiful. Artists sometimes adopt empirical canons, such as that of the proportions of the human body, or of the golden section[1], that is to say, of a line divided into two parts in such a manner that the less is to the greater as is the greater to the whole line ($bc : ac = ac : ab$). Such canons easily become their superstitions, and they attribute to them the success of their works. Thus Michæl Angelo left as a precept[2] to his disciple Marco del Pino da Siena[3] that "he should always make a pyramidal serpentine figure[4] multiplied by one two and three," a precept which did not enable Marco da Siena to emerge from that mediocrity which we can yet observe in many of his paintings that exist here in Naples. Others took Michæl Angelo's words as authority for the precept that serpentine undulating lines[5] were the true *lines of beauty.* Whole volumes have been composed on these laws of beauty, on the golden section and on the undulating and serpentine lines. These should in our opinion be looked upon as the *astrology of Æsthetic.*

---

1. golden section：黄金分割
2. precept：秘方
3. Marco del Pino da Siena：锡耶纳人马克·德·皮诺，米开朗琪罗弟子。
4. pyramidal serpentine figure：金字塔之光影蛇形
5. serpentine undulating lines：蛇形曲线

XV THE ACTIVITY OF EXTERNALIZATION. TECHNIQUE AND THE THEORY OF THE ARTS

The practical activity of externalization.

The fact of the production of physical beauty implies, as has already been remarked, a vigilant will, which persists in not allowing certain visions, intuitions or representations to be lost. Such a will must be able to act with the utmost rapidity and as it were instinctively, and may also need long and laborious deliberations[1]. In any case, thus and thus only does the practical activity enter into relations with the æsthetic, that is to say, no longer as its simple accompaniment, but as a really distinct moment of it. We cannot will or not will our æsthetic vision: we can however will or not will to externalize it, or rather, to preserve and communicate to others, or not, the externalization produced.

The technique of externalization.

This volitional fact of externalization is preceded by a complex of various kinds of knowledge. These are known as *technique*, like all knowledge which precedes a practical activity. Thus we talk of an *artistic technique* in the same metaphorical and elliptic manner that

---

1. deliberations：深思熟虑

we talk of the physically beautiful, that is to say (in more precise language), *knowledge at the service of the practical activity directed to producing stimuli to æsthetic reproduction*. In place of employing so lengthy a phrase, we shall here avail ourselves of ordinary terminology, whose meaning we now understand.

The possibility of this technical knowledge, at the service of artistic reproduction, is what has led minds astray to imagine the existence of an æsthetic technique of internal expression, which is tantamount to[1] saying, a doctrine of the *means of internal expression*, a thing that is altogether inconceivable. And we know well the reason of its inconceivability; expression, considered in itself, is a primary theoretic activity, and as such precedes practice and intellectual knowledge which illumines practice and is independent alike of both. It aids for its part to illumine practice, but is not illuminated by it. Expression does not possess *means*, because it has not an *end*; it has intuitions of things, but it does not will and is therefore unanalysable into the abstract components of volition, means and end. Sometimes a certain writer is said to have invented a new technique of fiction or of drama, or a painter is said to have discovered a new technique of distributing light. The word is used here at hazard; because the so-called *new technique* is really *that romance itself, or that new picture* itself and nothing else. The distribution of light belongs to the vision of the picture itself; as the technique of a dramatist is his dramatic conception itself. On other occasions, the word "technique" is used to designate certain merits or defects in a work that is a failure; and it is euphemistically said that the conception is bad but the technique good, or that the conception is good but the technique bad.

On the other hand, when we talk of the different ways of painting

1. tantamount to：相当于

in oils, or of etching[1], or of sculpturing in alabaster[2], then the word "technique" is in its place; but in such a case the adjective "artistic" is used metaphorically. And if a dramatic technique in the æsthetic sense be impossible, a theatrical technique of processes of externalization of certain particular æsthetic works is not impossible. When, for instance, women were introduced on the stage in Italy in the second half of the sixteenth century, in place of men dressed as women, this was a true and real discovery in theatrical technique; such too was the perfecting in the following century of machines for the rapid changing of scenery by the impresarios[3] of Venice.

The collection of technical knowledge at the service of artists desirous of externalizing their expressions, can be divided into groups, which may be entitled *theories of the arts*. Thus arises a theory of Architecture, comprising mechanical laws, information relating to the weight or resistance of the materials of construction or of fortification[4], manuals relating to the method of mixing lime or stucco[5]; a theory of Sculpture, containing advice as to the instruments to be used for sculpturing the various sorts of stone, for obtaining a successful mixture of bronze, for working with the chisel, for the accurate casting of the clay or plaster model, for keeping clay damp[6]; a theory of Painting, on the various techniques of tempera[7], of oil-painting, of water-colour, of pastel, on the proportions of the human body, on the laws of perspective[8]; a theory of Oratory[9], with precepts

1. etching：蚀刻
2. sculpturing in alabaster：石膏雕刻
3. impresarios：（剧院）经理
4. weight or resistance of the materials of construction or of fortification: 建筑材料或防御设施材料的重量及阻力
5. mixing lime or stucco：混合石灰水泥
6. keeping clay damp：保持黏土湿度
7. techniques of tempera：蛋彩画绘制技巧
8. laws of perspective：透视规律
9. theory of Oratory：演说技巧

as to the method of producing, of exercising and of strengthening the voice, of attitude in impersonation and gesture; a theory of Music, on the combinations and fusions of tones and sounds; and so on. Such collections of precepts abound in all literatures. And since it is impossible to say what is useful and what useless to know, books of this sort become very often a sort of *encyclopædias* or *catalogues* of *desiderata*[1]. Vitruvius[2], in his treatise on Architecture, claims for the architect a knowledge of letters, of drawing, of geometry, of arithmetic, of optic, of history, of natural and moral philosophy, of jurisprudence, of medicine, of astrology, of music, and so on. Everything is worth knowing: learn the art and have done with it.

Technical theories of the different arts.

It should be evident that such empirical collections are not reducible to science. They are composed of notions, taken from various sciences and disciplines, and their philosophical and scientific principles are to be found in the latter. To propose to construct a scientific theory of the different arts would be to wish to reduce to the single and homogeneous what is by nature multiple and heterogeneous; to wish to destroy the existence as a collection of what was put together precisely to form a collection. Were we to try to give scientific form to the manuals of the architect, the painter, or the musician, it is clear that nothing would remain in our hands but the general principles of Mechanics, Optics, or Acoustics. And if we were to extract and isolate what may be scattered among them of properly artistic observations,

1. *encyclopædias* or *catalogues* of *desiderata*：百科全书或亟须之物目录（斜体为拉丁文）
2. Vitruvius：维特鲁威（约前 80—前 15），古罗马作家、建筑师与工程师，著有《建筑十书》，最早提出了建筑的三要素“实用、坚固、美观”，并总结出了人体结构的比例规律，提出将人体的自然比例应用到建筑的丈量上。

to make of them a scientific system, then the sphere of the individual art would be abandoned and that of Æsthetic entered, for Æsthetic is always general Æsthetic, or rather it cannot be divided into general and special. This last case (that is, the attempt to furnish a technique which ends in composing an Æsthetic) arises when men possessing strong scientific instincts and a natural tendency to philosophy set themselves to work to produce such theories and technical manuals.

Criticism of æsthetic theories of particular arts.

But the confusion between Physics and Æsthetic has attained to its highest degree, when æsthetic theories of particular arts are imagined, to answer such questions as: What are the *limits* of each art? What can be represented with colours, and what with sounds? What with simple monochromatic lines[1] and what with touches of various colours? What with tones, and what with metres and rhythms? What are the limits between the figurative and the auditive arts[2], between painting and sculpture, poetry and music?

This, translated into scientific language, is tantamount to asking: What is the connexion between Acoustics and æsthetic expression? What between the latter and Optics? — and the like. Now, if *there is no passage* from the physical fact to the æsthetic, how could there be from the æsthetic to particular groups of physical facts, such as the phenomena of Optics or of Acoustics?

Criticism of the classification of the arts.

The so-called *arts* have no æsthetic limits, because, in order to have them, they would need to have also æsthetic existence in their

1. monochromatic lines：单色线条
2. figurative and auditive arts：具象艺术与听觉艺术

particularity; and we have demonstrated the altogether empirical genesis of those partitions. Consequently, any attempt at an æsthetic classification of the arts is absurd. If they be without limits, they are not exactly determinable, and consequently cannot be philosophically classified. All the books dealing with classifications and systems of the arts could be burned without any loss whatever. (We say this with the utmost respect to the writers who have expended their labours upon them.)

The impossibility of such systematizations finds something like a proof in the strange attempts made to carry it out. The first and most common partition is that into arts of *hearing, sight*, and *imagination*; as if eyes, ears, and imagination were on the same level and could be deduced from the same logical variable[1] as *fundamentum divisionis*[2]. Others have proposed the division into arts of *space* and arts of *time*, arts of *rest*; and *movement*; as if the concepts of space, time, rest and motion could determine special æsthetic forms and possess anything in common with art as such. Finally, others have amused themselves by dividing them into *classic* and *romantic*, or into *oriental*, *classic*, and *romantic*, thereby conferring the value of scientific concepts upon simple historical denominations[3], or falling into those rhetorical partitions of expressive forms, already criticized above; or into arts *that can only be seen from one side*, like painting, and arts *that can be seen from all sides*, like sculpture — and similar extravagances, which hold good neither in heaven nor on earth.

The theory of the limits of the arts was perhaps at the time when it was put forward a beneficial critical reaction against those who believed in the possibility of remodelling one expression into another,

---

1. logical variable：逻辑变量
2. *fundamentum divisionis*：〈拉〉划分依据
3. simple historical denominations：单纯的历史名称

as the *Iliad*[1] or *Paradise Lost*[2] into a series of paintings, and indeed held a poem to be of greater or lesser value according as it could or could not be translated into pictures by a painter. But if the rebellion were reasonable and resulted in victory, this does not mean that the arguments employed and the systems constructed for the purpose were sound.

Criticism of the theory of the union of the arts.

Another theory which is a corollary to that of the arts and their limits, falls with them; that of the *union of the arts*. Given particular arts, distinct and limited, it was asked: Which is the most *powerful*? Do we not obtain *more powerful* effects by *uniting* several? We know nothing of this: we know only that in each particular case certain given artistic intuitions have need of definite physical means for their reproduction and other artistic intuitions of other means. We can obtain the effect of certain plays by simply reading them; others need declamation and scenic display: there are some artistic intuitions which need for their full externalization words, song, musical instruments, colours, statuary, architecture, actors; while others are quite complete in a slight outline made with the pen, or a few strokes of the pencil. But it is false to suppose that declamation and scenic effects and all the other things together that we have mentioned are *more powerful* than a simple reading or a simple outline of pen or pencil; because each of those facts or groups of facts has, so to say, a different purpose, and the power of the means cannot be compared when the purposes are

---

1. *Iliad*：《伊利亚特》，古希腊荷马创作的史诗，讲述特洛伊战争第十年中51天内发生的故事。
2. *Paradise Lost*：《失乐园》，英国诗人弥尔顿（1608—1674）的史诗作品，与《复乐园》构成上下篇，主题是亚当、夏娃受撒旦诱惑并被上帝驱逐出伊甸园。

different.

Relation of the activity of externalization to utility and morality.

Finally, it is only from the point of view of a clear and rigorous distinction between the true and proper æsthetic activity and the practical activity of externalization that we can solve the complicated and confused questions as to the relations between *art and utility* and *art and morality.*

We have demonstrated above that art as art is independent both of utility and of morality, as also of all practical value. Without this independence, it would not be possible to speak of an intrinsic value of art, nor indeed to conceive an æsthetic science, which demands the autonomy of the æsthetic fact as its necessary condition.

But it would be erroneous to maintain that this independence of the vision or intuition or *internal expression* of the artist should be simply extended to the practical activity of externalization and communication which may or may not follow the æsthetic fact. If by art be understood the externalization of art, then utility and morality have a perfect right to enter into it; that is to say, the right to be master in one's own house.

Indeed we do not externalize and fix all the many expressions and intuitions which we form in our spirit; we do not declare our every thought in a loud voice, or write it down, or print, or draw, or paint, or expose it to the public. We *select* from the crowd of intuitions which are formed or at least sketched within us; and the selection is ruled by the criteria of the economic disposition of life and of its moral direction. Therefore, when we have fixed an intuition, we have still to decide whether or no we should communicate it to others, and to whom, and when, and how; all which deliberations come equally under the utilitarian and ethical criterion.

Thus we find the concepts of *selection*, of the *interesting*, of *morality*, of an *educational end*, of *popularity*, etc., to some extent justified, although these can in no way be justified when imposed upon art as art, and we have ourselves rejected them in pure Æsthetic. Error always contains an element of truth. He who formulated those erroneous æsthetic propositions in reality had his eye on practical facts, which attach themselves externally to the æsthetic fact and belong to economic and moral life.

It is well to advocate yet greater freedom in making known the means of æsthetic reproduction; we are of the same opinion, and leave projects for legislation and for legal action against immoral art, to hypocrites, to the ingenuous and to wasters of time. But the proclamation of this freedom, and the fixing of its limits, how wide soever they be, is always the task of morality. And it would in any case be out of place to invoke that highest principle, that *fundamentum æsthetices*[1], which is the independence of art, to deduce from it the guiltlessness of the artist who calculates like an immoral speculator[2] upon the unhealthy tastes of his readers in the externalization of his imaginings, or the freedom of hawkers to sell obscene statuettes[3] in the public squares. This last case is the affair of the police, as the first must be brought before the tribunal of the moral consciousness. The æsthetic judgement on the work of art has nothing to do with the morality of the artist as a practical man, or with the provisions to be taken that the things of art may not be diverted to evil ends alien to her nature, which is pure theoretic contemplation.

---

1. *fundamentum æsthetices*：〈拉〉美学原理
2. immoral speculator：无良投机客
3. obscene statuettes：不雅小雕像

XVI TASTE AND THE REPRODUCTION OF ART

Æsthetic judgement. Its identity with æsthetic reproduction.

When the entire æsthetic and externalizing process has been completed, when a beautiful expression has been produced and it has been fixed in a definite physical material, what is meant by *judging ill to reproduce it in oneself*, answer the critics of art, almost with one voice. Very good. Let us try thoroughly to understand this fact, and with that object in view, let us represent it schematically[1].

The individual A is seeking the expression of an impression which he feels or anticipates, but has not yet expressed. See him trying various words and phrases which may give the sought-for expression[2], that expression which must exist, but which he does not possess. He tries the combination *m*, but rejects it as unsuitable, inexpressive, incomplete, ugly: he tries the combination *n*, with a like result. *He does not see at all, or does not see clearly.* The expression still eludes him. After other vain attempts, during which he sometimes approaches, sometimes retreats from the mark at which he aims, all of a sudden (almost as though formed spontaneously of itself) he forms the sought-for expression, and *lux facta est*[3]. He enjoys for an instant æsthetic

---

1. represent it schematically：用图表说明
2. sought-for expression：所寻求的表达
3. *lux facta est*：〈拉〉便有了光。完整的句子为“dixitque Deus fiat lux et facta est lux”，意为：上帝说要有光，于是便有了光。

pleasure or the pleasure of the beautiful. The ugly, with its correlative displeasure, was the æsthetic activity which had not succeeded in conquering the obstacle; the beautiful is the expressive activity which now displays itself triumphant.

We have taken this example from the domain of speech, as being nearer and more accessible, and because we all talk, though we do not all draw or paint. Now if another individual, whom we shall call B, is to judge that expression and decide whether it be beautiful or ugly, he *must of necessity place himself at A's point of view*, and go through the whole process again, with the help of the physical sign supplied to him by A. If A has seen clearly, then B (who has placed himself at A's point of view) will also see clearly and will see this expression as beautiful. If A has not seen clearly, then B also will not see clearly, and will find the expression more or less ugly, *just as A did.*

Impossibility of divergences.

It may be observed that we have not taken into consideration two other cases: that of A having a clear and B an obscure vision; and that of A having an obscure and B a clear vision. Strictly speaking, these two cases are *impossible.*

Expressive activity, just because it is activity, is not caprice, but spiritual necessity; it cannot solve a definite æsthetic problem save in one way, which is the right way. It will be objected to this plain statement that works which seem beautiful to the artists are afterwards found to be ugly by the critics; while other works with which the artists were discontented and held to be imperfect or failures are, on the contrary, held to be beautiful and perfect by the critics. But in this case, one of the two is wrong: either the critics or the artists, sometimes the artists, at other times the critics. Indeed, the producer of an expression does not always fully realize what is happening in his soul. Haste, vanity, want of reflexion, theoretic prejudices, make people say, and others sometimes almost believe, that works of ours are beautiful, which, if we really looked into ourselves, we should see to be ugly, as they are in reality. Thus poor Don Quixote, when he had reattached to his helmet as well as he could the vizor of cardboard[1] — the vizor that had showed itself to possess but the feeblest force of resistance at the first encounter, — took good care not to test it again with a well-delivered sword-thrust, but simply declared and maintained it to be (says the author) *por celada finisima de encaxe*[2]. And in other cases, the same reasons, or opposite but analogous ones, trouble the consciousness of the artist, and cause him to value badly what he has successfully produced, or to strive to undo! and do again for the worse what he has done well in artistic spontaneity[3]. An instance of this is Tasso and his passage from the *Gerusalemme liberata* to the *Gerusalemme conquistata*[4]. In the same way, haste, laziness, want of reflexion, theoretic prejudices, personal sympathies or animosities[5], and other motives of a similar sort, sometimes cause the critics to proclaim ugly what is beautiful, and beautiful what is ugly. Were they to eliminate such disturbing elements, they would feel the work of art as it really is, and would not leave it to posterity[6], that more diligent and more dispassionate judge[7], to award the palm, or to do that justice which they have refused.

Identity of taste and genius.

It is clear from the preceding theorem that the activity of judgement

1. vizor of cardboard：纸板做的面甲
2. *por celada finisima de encaxe*：〈西〉就当它是个完美的头盔
3. artistic spontaneity：艺术自发性
4. *Gerusalemme conquistata*：〈意〉《征服耶路撒冷》，塔索以十字军东征为主题的史诗作品。
5. animosities：憎恶
6. posterity：后人
7. more diligent and more dispassionate judge：更勤勉冷静的裁决者

which criticizes and recognizes the beautiful is identical with what produces it. The only difference lies in the diversity of circumstances, since in the one case it is a question of æsthetic production, in the other of reproduction. The activity which judges is called *taste*; the productive activity is called *genius*: genius and taste are therefore substantially *identical*.

The common remark that the critic should possess something of the genius of the artist and that the artist should possess taste, gives a glimpse of this identity; or the remark that there exists an active (productive) and a passive (reproductive) taste. But it is also negated in other equally common remarks, as when people speak of taste without genius, or of genius without taste. These last observations are meaningless, unless they allude to quantitative or psychological differences, those being called geniuses without taste who produce works of art, inspired in their chief parts and neglected or defective in their secondary parts, and men of taste without genius, those who, while they succeed in obtaining certain isolated or secondary merits, do not possess sufficient power for a great artistic synthesis. Analogous explanations can easily be given of other similar expressions. But to posit a substantial difference between genius and taste, between artistic production and reproduction, would render both communication and judgement alike inconceivable. How could we judge what remained external to us? How could that which is produced by a given activity be judged by a *different* activity? The critic may be a small genius, the artist a great one; the former may have the strength of ten, the latter of a hundred; the former, in order to reach a certain height, will have need of the assistance of the other; but the nature of both must remain the same. To judge Dante, we must raise ourselves to his level: let it be well understood that empirically we are not Dante, nor Dante we; but in that moment of contemplation and judgement, our spirit is one with that of the poet, and in that moment we and he are one thing. In this identity alone resides the possibility that our little souls can echo great souls, and grow great with them in the universality of the spirit.

Analogy with other activities.

Let us remark in passing that what has been said of the æsthetic judgement holds good equally for every other activity and for every other judgement; and that scientific, economic, and ethical criticism is effected in a like manner. To limit ourselves to this last, only if we place ourselves ideally in the same conditions in which he found himself who took a given resolution, can we form a judgement as to whether his decision were moral or immoral. An action would otherwise remain incomprehensible and therefore impossible to judge. A homicide[1] may be a rascal or a hero: if this be, within limits, indifferent as regards the defence of society, which condemns both to the same punishment, it is not indifferent to one who wishes to distinguish and judge from the moral point of view, and we therefore cannot dispense with reconstructing the individual psychology of the homicide, in order to determine the true nature of his deed, not merely in its legal, but also in its moral aspect. In Ethics, a moral taste or tact is sometimes mentioned, answering to what is generally called the moral consciousness, that is to say, to the activity of the good will itself.

Criticism of æsthetic absolutism (intellectualism) and relativism.

The explanation above given of æsthetic judgement or reproduction both agrees with and condemns the absolutists and relativists, those who affirm and those who deny the absoluteness of taste.

In affirming that the beautiful can be judged, the absolutists are

1. homicide：杀人者

right; but the theory on which they found their affirmation is not tenable, because they conceive of the beautiful, that is, æsthetic value, as something placed outside the æsthetic activity, as a concept or a model which an artist realizes in his work, and of which the critic avails himself afterwards in judging the work itself. These concepts and models have no existence in art, for when proclaiming that every art can be judged only in itself and that it has its model in itself, they implicitly denied the existence of objective models of beauty, whether these are intellectual concepts, or ideas suspended in a metaphysical heaven.

In proclaiming this, their adversaries, the relativists, are perfectly right, and effect an advance upon them. However, the initial rationality of their thesis in its turn becomes converted into a false theory. Repeating the ancient adage that there is no accounting for tastes[1], they believe that æsthetic expression is of the same nature as the pleasant and the unpleasant, which every one feels in his own way, and about which there is no dispute. But we know that the pleasant and the unpleasant are utilitarian, practical facts. Thus the relativists deny the specific character of the æsthetic fact, and again confound expression with impression, the theoretic with the practical.

The true solution lies in rejecting alike relativism or psychologism and false absolutism; and in recognizing that the criterion of taste is absolute, but absolute in a different way from that of the intellect, which expresses itself in ratiocination[2]. The criterion of taste is absolute, with the intuitive absoluteness of the imagination. Thus any act of expressive activity, which is so really, is to be recognized as beautiful, and any fact as ugly in which expressive activity and passivity are found engaged with one another in an unfinished

1. ancient adage that there is no accounting for tastes：老话说萝卜青菜各有所爱
2. ratiocination：推理

struggle.

Criticism of relative relativism.

Between absolutists and relativists is a third class, which may be called that of the relative relativists. These affirm the existence of absolute values in other fields, such as Logic and Ethic, but deny it in the field of Æsthetic. To dispute about science or morals seems to them to be rational and justifiable, because science depends upon the universal, common to all men, and morality upon duty, which is also a law of human nature; but how dispute about art, which depends upon imagination? Not only, however, is the imaginative activity universal and no less inherent in human nature than the logical concept and practical duty; but there is a preliminary objection to the thesis in question. If the absoluteness of the imagination be denied, we must also deny intellectual or conceptual truth and implicitly morality. Does not morality presuppose logical distinctions? How could these be known, otherwise than in expressions and words, that is to say, in imaginative form? If the absoluteness of the imagination were removed, the life of the spirit would tremble to its foundations. One individual would no longer understand another, nor indeed his own self of a moment before, which is already another individual considered a moment after.

Objection founded on the variation of the stimulus and of psychic disposition.

Nevertheless, variety of judgements is an indubitable fact. Men disagree as to logical, ethical, and economical valuations; and they disagree equally or even more as to the æsthetic. If certain reasons recorded by us above, such as haste, prejudices, passions, etc., may lessen the importance of this disagreement, they do not on that account annul it. When speaking of the stimuli of reproduction we have added a caution, for we said that reproduction takes place, *if all the other conditions remain equal.* Do they remain equal? Does the hypothesis correspond to reality?

It would appear not. In order to reproduce an impression several times by means of a suitable physical stimulus it is necessary that this stimulus be not changed, and that the organism remain in the same psychical conditions as those in which was experienced the impression that it is desired to reproduce. Now it is a fact that the physical stimulus is continually changing, and in like manner the psychological conditions.

Oil-paintings grow dark, frescoes fade[1], statues lose noses, hands and legs, architecture becomes totally or partially a ruin, the tradition of the execution of a piece of music is lost, the text of a poem is corrupted by bad copyists or bad printing. These are obvious instances of the changes which daily occur in objects or physical stimuli. As regards psychological conditions, we will not dwell upon the cases of deafness or blindness, that is to say, upon the loss of entire orders of psychical impressions; these cases are secondary and of less importance compared with the fundamental, daily, inevitable and perpetual changes of the society around us and of the internal conditions of our individual life. The phonetic manifestations or words and verses of Dante's *Commedia* must produce a very different impression on an Italian citizen engaged in the politics of the third Rome[2], from that experienced by a well-informed and intimate contemporary of the poet. The Madonna of Cimabue[3] is still in the

---

1. frescoes fade：壁画褪色
2. third Rome：第三罗马。一些欧洲城市或国家用此词自称是第一、第二罗马的继承者。
3. Madonna of Cimabue：契马布埃画的圣母。契马布埃（约 1240—1302），意大利画家，意大利拜占庭传统的最后一位画家，开写实主义画风之先河。

Church of Santa Maria Novella[1]; but does she speak to the visitor of to-day as to the Florentines of the thirteenth century? Even though she were not also darkened by time, must we not suppose that the impression which she now produces is altogether different from that of former times? And even in the case of the same individual poet, will a poem composed by him in youth make the same impression upon him when he re-reads it in his old age, with psychic conditions altogether changed?

Criticism of the distinction of signs into natural and conventional.

It is true that certain æstheticians have attempted a distinction between stimuli and stimuli, between *natural* and *conventional* signs. The former are held to have a constant effect upon all; the latter only upon a limited circle. In their belief, signs employed in painting are natural, those used in poetry conventional. But the difference between them is at the most only one of degree. It has often been said that painting is a language understood by all, while with poetry it is otherwise. Here, for example, Leonardo found one of the prerogatives[2] of his art, "which hath not need of interpreters of different tongues as have letters," and it pleases man and beast. He relates the anecdote of that portrait of the father of a family "which the little grandchildren were wont to caress while they were still in swaddling-clothes[3], and the dogs and cats of the house in like manner." But other anecdotes, such as those of the savages who took the portrait of a soldier for a boat, or considered the portrait of a man on horseback to be furnished with only one leg, are apt to shake one's faith in the understanding

---

1. Church of Santa Maria Novella：新圣母玛利亚教堂，位于佛罗伦萨最大的火车站对面，是佛罗伦萨第一座宗座教堂。
2. prerogatives：特性
3. still in swaddling-clothes：襁褓之中

of painting by sucklings[1], dogs and cats. Fortunately, no arduous researches are necessary to convince oneself that pictures, poetry and all works of art only produce effects upon souls prepared to receive them. Natural signs do not exist; because all are equally conventional, or, to speak with greater exactness, *historically conditioned*[2].

The surmounting of variety.

Granting this, how are we to succeed in causing the expression to be reproduced by means of the physical object? How obtain the same effect, when the conditions are no longer the same? Would it not, rather, seem necessary to conclude that expressions cannot be reproduced, despite the physical instruments made for the purpose, and that what is called reproduction consists in ever new expressions? Such would indeed be the conclusion if the varieties of physical and psychical conditions were intrinsically insurmountable[3]. But since the insuperability[4] has none of the characteristics of necessity we must on the contrary conclude that reproduction always occurs when we can replace ourselves in the conditions in which the stimulus (physical beauty) was produced.

Not only can we replace ourselves in these conditions as an abstract possibility, but as a matter of fact we do so continually. Individual life, which is communion with ourselves (with our past), and social life, which is communion with our like, would not otherwise be possible.

Restorations and historical interpretation.

As regards the physical object, palæographers[5] and philologists, who

1. sucklings：乳儿
2. *historically conditioned*：受历史条件约束
3. intrinsically insurmountable：本质上不可超越
4. insuperability：不可超越性
5. palæographers：古文字学家

restore to texts their original physiognomy, *restorers* of pictures and of statues and other industrious toilers[1] strive precisely to preserve or to restore to the physical object all its primitive energy. These efforts are certainly not always successful, or are not completely successful, for it is never or hardly ever possible to obtain a restoration complete in its smallest details. But the insurmountable is here only present accidentally and must not lead us to overlook the successes which actually are achieved.

Historical interpretation labours for its part to reintegrate in us the psychological conditions which have changed in the course of history. It revives the dead, completes the fragmentary, and enables us to see a work of art (a physical object) as its author saw it in the moment of production.

A condition of this historical labour is tradition, with the help of which it is possible to collect the scattered rays and concentrate them in one focus. With the help of memory we surround the physical stimulus with all the facts among which it arose; and thus we enable it to act upon us as it acted upon him who produced it.

Where the tradition is broken, interpretation is arrested; in this case, the products of the past remain silent for us. Thus the expressions contained in the Etruscan or Mexican inscriptions[2] are unattainable; thus we still hear discussions among ethnographers[3] as to whether certain products of the art of savages are pictures or writings; thus archæologists and prehistorians[4] are not always able to establish with certainty whether the figures found on the pottery of a certain region,

1. industrious toilers：辛劳工作者
2. Etruscan or Mexican inscriptions：埃特鲁里亚或墨西哥碑文。埃特鲁里亚是古代城邦国家，地处现代意大利中部。该文明的全盛时期为公元前 6 世纪。
3. ethnographers：人种志学家
4. prehistorians：史前史学家

and on other instruments employed, are of a religious or profane[1] nature. But the arrest of interpretation, as that of restoration, is never a definitely insurmountable barrier; and the daily discoveries of new historical sources and of new methods of better exploiting the old, which we may hope to see ever improving, link up again broken traditions.

We do not wish to deny that erroneous historical interpretation sometimes produces what may be called palimpsests[2], new expressions imposed upon the ancient, artistic fancies instead of historical reproductions. The so-called "fascination of the past" depends in part upon these expressions of ours, which we weave upon the historical. Thus has been discovered in Greek plastic art[3] the calm and serene intuition of life of those peoples, who nevertheless felt the universal sorrow so poignantly[4]; thus "the terror of the year 1000" has recently been discerned on the faces of the Byzantine saints, a terror which is a misunderstanding, or an artificial legend invented later by men of learning. But *historical criticism* tends precisely to circumscribe fancies and to establish exactly the point of view from which we must look.

By means of the above process we live in communication with other men of the present and of the past; and we must not conclude because we sometimes, and indeed often, meet with an unknown or an ill-known, that therefore, when we believe we are engaged in a dialogue, we are always speaking a monologue; or that we are unable even to repeat the monologue which we formerly held with ourselves.

---

1. profane：世俗的
2. palimpsests：隐迹稿本；重写羊皮书卷（全部或部分原有文字被刮去，在上面另行书写）
3. Greek plastic art：希腊造型艺术
4. poignantly：深刻地

XVII THE HISTORY OF LITERATURE AND ART

This brief exposition of the method by which is obtained the reintegration of the original conditions in which the work of art was produced, and consequently reproduction and judgement are made possible, shows how important is the function fulfilled by historical research in relation to artistic and literary works which is what is usually called *historical criticism* or method in literature and art.

Historical criticism in literature and art. Its importance.

Without tradition and historical criticism the enjoyment of all or nearly all the works of art produced by humanity would be irrevocably lost[1]: we should be little more than animals, immersed in the present alone, or in the most recent past. It is fatuous[2] to despise and laugh at one who reconstitutes an authentic text, explains the sense of forgotten words and customs, investigates the conditions in which an artist lived, and accomplishes all those labours which revive the qualities and the original colouring of works of art.

Sometimes a depreciatory or negative judgement[3] is passed upon historical research because of the presumed or proved inability of such researches, in many cases, to give us a true understanding of works of art. But it must be observed, in the first place, that historical research

---

1. irrevocably lost：无可挽回地失去
2. fatuous：愚昧
3. depreciatory or negative judgement：鄙视的或否定的判断

does not only fulfil the task of helping to reproduce and judge artistic works: the biography of a writer or of an artist, for example, and the study of the customs of a period, have an interest of their own, that is to say, extraneous to the history of art[1], but not to other forms of historiography[2]. If allusion be made to those researches which do not appear to have interest of any kind, nor to fulfil any purpose, it must be replied that the historical student must often reconcile himself to the useful but inglorious function[3] of a collector of facts. These facts remain for the time being formless, incoherent and meaningless, but they are preserves or mines for the historian of the future and for whosoever may afterwards want them for any purpose. In the same way in a library, books which nobody asks for are placed on the shelves and catalogued, because they may be asked for at some time or other. Certainly, just as an intelligent librarian gives the preference to the acquisition and cataloguing of those books which he foresees may be of more or better service, so intelligent students possess an instinct as to what is or may more probably be of use among the material of facts which they are examining; while others less well endowed[4], less intelligent or more hasty in producing, accumulate useless rubbish, refuse and sweepings, and lose themselves in details and petty discussions. But this appertains to the economy of research[5], and does not concern us. It concerns at most the master who selects the subjects, the publisher who pays for the printing, and the critic who is called upon to praise or to blame the research workers.

On the other hand, it is clear that historical research directed to illuminate a work of art does not alone suffice to bring it to birth

1. extraneous to the history of art：对艺术史是题外话
2. historiography：史学
3. inglorious function：无甚光彩的作用
4. less well endowed：天资平平的
5. appertains to the economy of research：属于研究的管理方面的问题

in our spirit and place us in a position to judge it, but presupposes taste, that is to say, an alert and cultivated imagination. The greatest historical erudition[1] may accompany a gross or otherwise defective taste, a slow imagination, or, as they say, a cold hard heart closed to art. Which is the lesser evil, great erudition with defective taste, or natural taste and much ignorance? The question has often been asked, and perhaps it will be best to deny that it has any meaning, because one cannot tell which of two evils is the less, or what exactly that means. The merely learned man never succeeds in entering into direct communion with great spirits; he keeps wandering for ever about the outer courts, the staircases and antechambers[2] of their palaces; but the gifted ignoramus[3] either passes by masterpieces to him inaccessible, or instead of understanding works of art as they really are, invents others with his fancy. Now, the labour of the former may at least serve to enlighten others; but the genius of the latter remains altogether sterile in relation to knowledge. How then can we in a certain respect fail to prefer the conscientious learned man to the inconclusive though gifted man[4], who is not really gifted, if he resign himself and in so far as he resigns himself, to his inconclusiveness?

Literary and artistic history. Its distinction from historical criticism and from the æsthetic judgement.

We must accurately distinguish *the history of art and literature* from those historical labours where works of art are used, but for extraneous purposes (such as biography, civil, religious and political history, etc.), and also from historical erudition directed to the preparation of the

---

1. historical erudition：渊博的历史知识
2. antechambers：前厅
3. gifted ignoramus：有天赋而无知之人
4. inconclusive though gifted man：虽有天赋却无法令人信服之人

æsthetic synthesis of reproduction.

The difference of the first two is obvious. The history of art and literature has the works of art themselves as its principal subject; those other labours invoke and interrogate works of art, but only as witnesses from whom to discover the truth of facts which are not æsthetic. The second difference to which we have referred may seem less profound. It is, however, very great. Erudition directed to illuminate the understanding of works of art aims simply at calling into existence a certain internal fact, an æsthetic reproduction. Artistic and literary history, on the other hand, does not appear until after such reproduction has been obtained. It implies, therefore, a further stage of labour.

Like all other history, its object is to record precisely such facts as have really taken place, in this case artistic and literary facts. A man who, after having acquired the requisite historical erudition, reproduces in himself and tastes a work of art, may remain simply a man of taste, or at the most express his own feeling with an exclamation of praise or condemnation. This does not suffice for the making of a historian of literature and art. Something else is needed, namely, that a new mental operation succeed in him the simple reproduction. This new operation is in its turn an expression: the expression of the reproduction; the historical description, exposition or representation. There is this difference, then, between the man of taste and the historian: the first merely reproduces in his spirit the work of art; the second, after having reproduced it, represents it historically, or applies those categories by which, as we know, history is differentiated from pure art. Artistic and literary history is therefore *a historical work of art founded upon one or more works of art.*

The name "artistic" or "literary" critic is used in various senses: sometimes it is applied to the scholar who devotes his services to literature; sometimes to the historian who reveals the works of art of the past in their reality; more often to both. By critic is sometimes understood in a more restricted sense he who judges and describes contemporary literary works, and by historian, he who treats of those less recent. These are linguistic uses and empirical distinctions, which may be neglected; because the true difference lies between *the scholar, the man of taste* and *the historian of art*. These words designate three successive stages of work, each one independent relatively to the one that follows, but not to that which precedes. As we have seen, a man may be a mere scholar, and possess little capacity for understanding works of art; he may even both be learned and possess taste, yet be unable to portray them by writing a page of artistic and literary history. But the true and complete historian, while containing in himself both the scholar and the man of taste as necessary pre-requisites, must add to their qualities the gift of historical comprehension and representation.

The method of artistic and literary history.

The theory of artistic and literary historical method presents problems and difficulties, some common to the theory of historical method in general, others peculiar to it, because derived from the concept of art itself.

Criticism of the problem of the origin of art.

History is commonly divided into human history, natural history, and the mixture of both. Without examining here the question of the solidity of this distinction, it is clear that artistic and literary history belongs in any case to the first, since it concerns a spiritual activity, that is to say, an activity proper to man. And since this activity is its subject, the absurdity of propounding the historical problem of the *origin* of *art* becomes at once evident. We should note that by this formula many different things have in turn been included on many different occasions. *Origin* has often meant *nature* or *character* of the artistic fact, in which case an attempt was made to deal with a real scientific or philosophic problem, the very problem in fact which our treatise has attempted to solve. At other times, by origin has been understood the *ideal genesis*[1], the search for the reason of art, the deduction of the artistic fact from a first principle containing in itself both spirit and nature. This is also a philosophical problem, complementary to the preceding, coinciding indeed with it, although it has sometimes been strangely interpreted and solved by means of an arbitrary and semi-imaginary metaphysic. But when the object was to discover further exactly in what way the artistic function was *historically formed*, the result has been the absurdity which we have mentioned. If expression be the first form of consciousness, how can we look for the historical origin of what is not a product of nature and is presupposed by human history? How can we assign a historical genesis to a thing which is a category by means of which all historical processes and facts are understood? The absurdity has arisen from the comparison with human institutions, which have been formed in the course of history, and have disappeared or may disappear in its course. Between the æsthetic fact and a human institution (such as monogamic marriage[2] or the fief[3]) there exists a difference comparable with that between simple and compound bodies in chemistry. It is impossible to indicate the formation of the former, otherwise they would not be simple, and if this be discovered, they cease to be simple and become compound.

The problem of the origin of art, historically understood, is only justified when it is proposed to investigate, not the formation of the artistic category, but where and when art has appeared for the first time (appeared, that is to say, in a striking manner), at what point or

---

1. *ideal genesis*：理想起源
2. monogamic marriage：一夫一妻制
3. fief：采邑制

in what region of the globe and at what point or epoch of its history; when, that is to say, not the origin of art, but its earliest or primitive history is the object of research. This problem forms one with that of the appearance of human civilization on the earth. Data for its solution are certainly wanting, but there yet remains the abstract possibility of a solution, and certainly tentative and hypothetical solutions[1] abound.

The criterion of progress and history.

Every representation of human history has the concept of *progress* as foundation. But by progress must not be understood the imaginary *law of progress* which is supposed to lead the generations of man with irresistible force to some unknown destiny, according to a providential plan which we can divine and then understand logically. A supposed law of this sort is the negation of history itself, of that accidentality, that empiricity, that contingency, which distinguish concrete fact from abstraction. And for the same reason, progress has nothing to do with the so-called law of *evolution*, which, if it mean that reality evolves (and it is only reality in so far as it evolves or becomes), cannot be called a law, and if it be given as a law, becomes identical with the law of progress in the sense just described. The progress of which we speak here is nothing but *the very concept of human activity*, which, working upon the material supplied to it by nature, conquers its obstacles and bends it to its own ends.

Such conception of progress, that is to say, of human activity applied to a given material, is the *point of view* of the historian of humanity. No one but a mere collector of unrelated facts, a mere antiquary[2] or inconsequent annalist[3], can put together the smallest narrative of

1. tentative and hypothetical solutions：试探性和假设性解决方案
2. antiquary：古文物研究者
3. annalist：编年史作者

human doings unless he have a determined point of view, that is to say, a personal conviction of his own regarding the facts whose history he has undertaken to relate. No one can start from the confused and discordant mass of crude facts and arrive at the historical work of art save by means of this apperception[1], which makes it possible to carve a definite representation in that rough and formless mass. The historian of a practical action should know what is economy and what is morality; the historian of mathematics, what is mathematics; the historian of botany, what is botany; the historian of philosophy, what is philosophy. If he does not really know these things, he must at least have the illusion of knowing them; otherwise he will not even be able to delude himself into believing that he is writing history.

We cannot here expand the demonstration of the necessity and inevitability of this subjective criterion in every narrative of human affairs (which is compatible with the utmost objectivity, impartiality and scrupulousness[2] in dealing with data of fact and indeed forms a constitutive element in these virtues), in every narrative of human doings and happenings. It suffices to read any book of history to discover at once the point of view of the author, if he be a historian worthy of the name and know his own business. There are liberal and reactionary, rationalist and catholic historians, who deal with political or social history; for the history of philosophy there are metaphysical, empirical, sceptical, idealist and spiritualist historians. Purely historical historians do not and cannot exist. Were Thucydides[3] and Polybius[4], Livy[5]

---

1. apperception：透彻的理解
2. utmost objectivity, impartiality and scrupulousness：最高客观性、公正性与严谨性
3. Thucydides：修昔底德（公元前 471—前 400），古希腊历史学家，著有《伯罗奔尼撒战争史》。
4. Polybius：波利比乌斯（公元前 203—前 120），古希腊历史学家，著有《历史》。
5. Livy：李维（公元前 59 年—17 年），古罗马史学家，著有《罗马史》。

and Tacitus[1], Machiavelli and Guicciardini[2], Giannone[3] and Voltaire[4], wholly without moral and political views; and, in our time, was Guizot[5] or Thiers[6], Macaulay[7] or Balbo[8], Ranke[9] or Mommsen[10]? And in the history of philosophy, from Hegel[11], who was the first to raise it to a great height, to Ritter[12], Zeller[13], Cousin[14], Lewes[15] and our Spaventa[16], was there one who did not possess his conception of progress and his criterion of judgement? Is there one single work of any value on the history of Æsthetic which has not been written from this or that

1. Tacitus：塔西佗（55—120），古罗马历史学家、演讲家、律师，著有《历史》与《编年史》。
2. Guicciardini：圭恰迪尼（1483—1540），意大利历史学家、政治家，马基雅维利的好友。著有《佛罗伦萨史》和《意大利史》。
3. Giannone：詹农（1676—1748），意大利历史学家、哲学家、法学家，著有《那不勒斯王国民史》。
4. Voltaire：伏尔泰（1694—1748），法国启蒙主义作家、散文家、哲学家，著有历史著作《路易十四时代》。
5. Guizot：基佐（1787—1874），法国历史学家、政治家，著有《英国革命史》《欧洲文明史》《法国文明史》等。
6. Thiers：梯也尔（1797—1877），法国政治家、历史学家、奥尔良党人，1871 年至 1873 年之间担任法兰西第三共和国总统。
7. Macaulay：麦考莱（1800—1859），英国诗人、历史学家、作家，著有《英国史》《古罗马抒情短诗》《论马基雅维利》。
8. Balbo：巴尔博（1789—1853），意大利作家与政治家，著有《但丁传》《意大利的希望》等。
9. Ranke：兰克（1795—1886），德国历史学家，著有《世界史》和《罗马教皇史》。
10. Mommsen：蒙森（1817—1903），德国历史学家、法学家、政治家，考古学家，著有《罗马史》。1902 年获诺贝尔文学奖。
11. Hegel：黑格尔（1770—1834），德国哲学家，德国唯心主义代表人物，著有《精神现象学》《逻辑学》《哲学全书》等。
12. Ritter：芮特（1791—1869），德国哲学史学家。著有《哲学史》等。
13. Zeller：策勒（1814—1908），德国哲学家、新教神学家、历史学家，著有《希腊哲学史纲要》。
14. Cousin：库辛（1792—1867），法国哲学家，著有《哲学片段》。
15. Lewes：即亨利 · 路易斯（1817—1878），英国作家，著有《哲学的传记史》《歌德传》。
16. Spaventa：斯巴文陀（1817—1883），意大利哲学家，著有《逻辑学与形而上学》《康德哲学及其与意大利哲学的关系》等。

point of view, with this or that bias (Hegelian or Herbartian[1]), from a sensationalist[2] or from an eclectic[3] or some other point of view? If the historian is to escape from the inevitable necessity of taking a side, he must become a political or scientific eunuch; and history is not an occupation for eunuchs. Such would at most be of use in compiling those great tomes of not useless erudition, *elumbis atque fracta*[4], which are called, not without reason, monkish[5].

If, then, a concept of progress, a point of view, a criterion, be inevitable, the best to be done is not to try and escape from it, but to obtain the best possible. Every one tends to this end when he forms his own convictions, seriously and laboriously. Historians who profess to wish to interrogate the facts without adding anything of their own to them are not to be trusted. This is at best the result of ingenuousness and illusion on their part: they will always add something of their own, if they be truly historians, even without knowing it, or they will only believe that they have avoided doing so because they have conveyed it only by hints, which is the most insinuating, penetrative and effective of methods[6].

Non-existence of a single line of progress in artistic and literary history.

Artistic and literary history cannot dispense with the criterion of progress any more easily than other history. We cannot show what a given work of art is, save by proceeding from a conception of art, in

1. Hegelian or Herbartian：黑格尔派或赫尔巴特派。前者主张从唯心主义哲学出发建构美学，后者针锋相对，主张就艺术论艺术。
2. sensationalist：感觉主义者
3. eclectic：折中主义者
4. *elumbis atque fracta*：〈拉〉软弱无力又支离破碎
5. monkish：僧侣气的
6. most insinuating, penetrative and effective of methods：最委婉、最透彻也是最有效的办法

order to fix the artistic problem which the author of such work of art had to solve, and by determining whether or no he has solved it, or by how much and in what way he has failed to do so. But it is important to note that the criterion of progress assumes a different form in artistic and literary history to that which it assumes (or is believed to assume) in the history of science.

It is customary to represent the whole history of knowledge by one single line of progress and regress[1]. Science is the universal, and its problems are arranged in one single vast system or comprehensive problem. All thinkers labour upon the same problem as to the nature of reality and of knowledge: contemplative Indians and Greek philosophers, Christians and Mohammedans, bare heads and turbaned heads, wigged heads and college-capped heads[2] (as Heine[3] said); and future generations will weary themselves with it, as ours has done. It would take too long to inquire here if this be true or not of science. But it is certainly not true of art; art is intuition, and intuition is individuality, and individuality does not repeat itself. To conceive of the history of the artistic production of the human race as developed along a single line of progress and regress would therefore be altogether erroneous.

At the most, and working to some extent with generalizations and abstractions, it may be asserted that the history of æsthetic productions shows progressive cycles, but each cycle with its own problem and each progressive only in respect to that problem. When many are at work in a general way upon the same subject, without succeeding in giving to it the suitable form, yet drawing always more near to it,

1. progress and regress：进步与倒退
2. turbaned heads, wigged heads and college-capped heads：包头巾者，戴假发者或者戴学院帽者
3. Heine：海涅（1797—1856），德国著名诗人、散文家、文艺评论家，著有《哈尔茨游记》《德国，一个冬天的童话》《论浪漫主义》等。

there is said to be progress, and when appears the man who gives it definite form, the cycle is said to be complete, and progress is ended. A typical example of this would here be the progress in the elaboration of the mode of using the subject-matter of chivalry, during the Italian Renaissance, from Pulci[1] to Ariosto[2] (using this as an example and excusing excessive simplification). Nothing but repetition and imitation, diminution or exaggeration, a spoiling of what had already been done, in short decadence could be the result of employing that same material after Ariosto. The epigoni[3] of Ariosto prove this. Progress begins with the beginning of a new cycle. Cervantes, with his more open and conscious irony, is an instance of this. In what did the general decadence of Italian literature at the end of the sixteenth century consist? Simply in having nothing more to say and in repeating and exaggerating motives already discovered. If the Italians of this period had even been able to express their own decadence, they would not have been altogether failures, but would have anticipated the literary movement of the Risorgimento[4]. Where the matter is not the same, a progressive cycle does not exist. Shakespeare does not represent an advance on Dante, nor Goethe upon Shakespeare. Dante, however, represents an advance on the visionaries of the Middle Ages, Shakespeare on the Elizabethan dramatists, Goethe, with *Werther*[5] and the first part of *Faust*[6], on the writers of the *Sturm und Drang*[7] period. This mode of presenting the history of poetry and art contains,

---

1. Pulci：浦尔契（1432—1484），意大利诗人，著有史诗《摩尔干提》。
2. Ariosto：阿里奥斯托（1474—1533），意大利诗人，著有浪漫史诗《疯狂的奥兰多》。
3. epigoni：追随者
4. Risorgimento：意大利复兴时期，指意大利 19 世纪中叶的统一运动。
5. *Werther*：《维特》，即《少年维特之烦恼》，歌德书信体小说。
6. *Faust*：《浮士德》，歌德根据 16 世纪的关于浮士德博士的民间传说而创作的诗剧。
7. Sturm und Drang：〈德〉“狂飙突进”，德国 18 世纪 60 年代至 80 年代之间文学与音乐领域浪漫主义运动，领导者为歌德、席勒和赫尔德。

however, as we have remarked, something of the abstract, of the merely practical, and is without strict philosophical value. Not only is the art of savages not inferior, as art, to that of civilized peoples, if it be correlative to the impressions of the savage; but every individual, indeed every moment of the spiritual life of an individual, has its artistic world; none of these worlds can be compared with any other in respect of artistic value.

Errors committed against this law.

Many have sinned and continue to sin against this special form of the criterion of progress in artistic and literary history. Some, for instance, talk of the infancy of Italian art in Giotto[1], and of its maturity in Raphæl[2] or in Titian[3]; as though Giotto were not complete and absolutely perfect, granted the material of feeling with which his mind was furnished. He was certainly incapable of drawing a figure like Raphæl, or of colouring it like Titian; but was Raphæl or Titian capable of creating the *Marriage of Saint Francis with Poverty*[4] or the *Death of Saint Francis*[5]? The spirit of Giotto had not felt the attraction of the body beautiful, which the Renaissance studied and raised to a place of honour; the spirits of Raphæl and of Titian were no longer interested in certain movements of ardour and of tenderness with which the man of the fourteenth century was in love. How, then, can a

1. Giotto：乔托（1266—1337），意大利画家、雕刻家与建筑师，意大利文艺复兴时期的开创者，被誉为“欧洲绘画之父”，作品包括《基督在水上行走》《最后的审判》《荣耀圣母》等。
2. Raphæl：拉斐尔（1483—1520），意大利文艺复兴时期画家、建筑师。代表作品有《西斯廷圣母》《雅典学派》等。
3. Titian：提香（1490—1576），意大利文艺复兴时期画家，威尼斯画派重要成员之一。作品有《圣母升天》《乌比诺的维纳斯》等。
4. *Marriage of Saint Francis with Poverty*：《圣弗朗西斯与贫穷结婚》，意大利画家安德烈·萨基（Andrea Sacchi）（1599—1661）的油画作品。
5. *Death of Saint Francis*：《圣弗朗西斯之死》，为乔托 1325 年所作。

comparison be made, where there is no comparative term?

The celebrated divisions of the history of art into an oriental period, representing a lack of equilibrium between idea and form, the latter dominating, a classical representing an equilibrium between idea and form, a romantic representing a new lack of equilibrium between idea and form, the former dominating, suffer from the same defect. The same is true of the division into oriental art, representing imperfection of form; classical, perfection of form; romantic or modern, perfection of content and of form. Thus classic and romantic have also received, among their many other meanings, that of progressive or regressive periods, in respect to the realization of some alleged artistic ideal of all humanity.

Other meanings of the word "progress" in respect to Æsthetic.

There is no such thing, then, as an *æsthetic* progress of humanity. However, by æsthetic progress is sometimes meant, not what the two words coupled together really signify, but the ever-increasing accumulation of our historical knowledge, which makes us able to sympathize with all the artistic products of all peoples and of all times, or, as they say, makes our taste more catholic. The difference appears very great if the eighteenth century, so incapable of escaping from itself, be compared with our own time, which enjoys alike Greek and Roman art, now better understood, Byzantine, mediæval, Arabic and Renaissance art, the art of the Cinquecento[1], baroque art[2], and the art of the eighteenth century. Egyptian, Babylonian, Etruscan, and even prehistoric art are more profoundly studied every day. Certainly, the difference between the savage and civilized man does not lie in

1. Cinquecento：〈意〉16 世纪，指 16 世纪意大利艺术与文学。
2. baroque art：巴洛克艺术，1600 年至 1750 年间在欧洲盛行的一种重视装饰效果的艺术风格。

the human faculties. The savage has speech, intellect, religion and morality in common with civilized man, and is a complete man. The only difference lies in this, that civilized man penetrates and dominates a larger portion of the universe with his theoretic and practical activity. We cannot claim to be more spiritually alert than, for example, the contemporaries of Pericles[1]; but no one can deny that we are richer than they — rich with their riches and with those of how many other peoples and generations besides our own?

By æsthetic progress is also meant, in another sense, which is also improper, the greater abundance of artistic intuitions and the smaller number of imperfect or inferior works which one epoch produces in respect to another. Thus it may be said that there was æsthetic progress, an artistic awakening in Italy, at the end of the thirteenth or of the fifteenth century.

Finally, æsthetic progress is talked of in a third sense, with an eye to the refinement and complications of soul-states exhibited in the works of art of the most civilized peoples, as compared with those of less civilized peoples, barbarians and savages. But in this case the progress is of the comprehensive psycho-social conditions, not of the artistic activity, to which the material is indifferent.

These are the most important points to note concerning the method of artistic and literary history.

1. Pericles：伯利克里（约前 495—前 429），雅典执政官，古希腊民主政治的杰出代表。

XVIII CONCLUSION: IDENTITY OF LINGUISTIC AND ÆSTHETIC

Summary of the study.

A glance over the path traversed will show that we have completed the entire programme of our treatise. We have studied the nature of intuitive or expressive knowledge, which is the æsthetic or artistic fact (I. and II.), and described the other form of knowledge, the intellectual, and the successive complications of these forms (III.); it thus became possible for us to criticize all erroneous æsthetic theories arising from the confusion between the various forms and from the illicit transference[1] of the characteristics of one form to another (IV.), noting at the same time the opposite errors to be found in the theory of intellectual knowledge and of historiography (V.). Passing on to examine the relations between the æsthetic activity and the other activities of the spirit, no longer theoretic but practical, we indicated the true character of the practical activity and the place which it occupies in respect to the theoretic activity: hence the criticism of the intrusion into æsthetic theory of practical concepts (VI.); we have distinguished the two forms of the practical activity, as economic and ethical (VII.), reaching the conclusion that there are no other forms of the spirit beyond the four which we have analyzed; hence (VIII.) the criticism of every mystical or imaginative Æsthetic. And since

1. illicit transference：不正当的转移

there are no other spiritual forms co-ordinate with these, so there are no original subdivisions of the four established, and in particular of Æsthetic. From this arises the impossibility of classes of expressions and the criticism of Rhetoric, that is, of ornate expression distinct from simple expression, and of other similar distinctions and subdistinctions (IX.) But by the law of the unity of the spirit, the æsthetic fact is also a practical fact, and as such, occasions pleasure and pain. This led us to study the feelings of value in general, and those of æsthetic value or of the beautiful in particular (X.), to criticize æsthetic hedonism in all its various manifestations and complications (XI.), and to expel from the system of Æsthetic the long series of psychological concepts which had been introduced into it (XII.). Proceeding from æsthetic production to the facts of reproduction, we began by investigating the external fixing of the æsthetic expression, for the purpose of reproduction. This is called the physically beautiful, whether natural or artificial (XIII.). We derived from this distinction the criticism of the errors which arise from confounding the physical with the æsthetic side of facts (XIV.). We determined the meaning of artistic technique, or that technique which is at the service of reproduction, thus criticizing the divisions, limits and classifications of the individual arts, and establishing the relations of art, economy and morality (XV.). Since the existence of physical objects does not suffice to stimulate æsthetic reproduction to the full, and since, in order to obtain it, we must recall the conditions in which the stimulus first operated, we have also studied the function of historical erudition, directed toward re-establishing the communication between the imagination and the works of the past, and to serve as the basis of the æsthetic judgement (XVI.). We have concluded our treatise by showing how the reproduction thus obtained is afterwards elaborated by the categories of thought, that is to say, by an examination of the method of literary and artistic history (XVII.).

The æsthetic fact has in short been considered both in itself and in its relations with the other spiritual activities, with the feelings of pleasure and pain, with what are called physical facts, with memory and with historical treatment. It has passed before us as *subject* until it became *object*, that is to say, from the moment of *its birth* until it becomes gradually changed for the spirit into *subject-matter of history*.

Our treatise may appear to be somewhat meagre when externally compared with the great volumes usually dedicated to Æsthetic. But it will not seem so when we perceive that those volumes are nine-tenths full of matter that is not pertinent[1], such as definitions, psychological or metaphysical, of pseudo-æsthetic concepts (the sublime, the comic, the tragic, the humorous, etc.), or of the exposition of the supposed Zoology, Botany and Mineralogy of Æsthetic, and of universal history æsthetically judged; that the whole history of concrete art and literature has also been dragged into those Æsthetics and generally mangled, and that they contain judgements upon Homer and Dante, Ariosto and Shakespeare, Beethoven and Rossini[2], Michæl Angelo and Raphæl. When all this has been deducted from them, we flatter ourselves that our treatise will no longer be held to be too meagre, but, on the contrary, far richer than ordinary treatises, which either omit altogether, or hardly touch at all, the greater part of the difficult problems proper to Æsthetic which we have felt it to be our duty to study.

Identity of linguistic and Æsthetic.

But although Æsthetic as science of expression has been studied by us in its every aspect, it remains to justify the sub-title which we have added to the title of our book, *General Linguistic*, to state and make clear the thesis that the science of art and that of language, Æsthetic

1. not pertinent：互不相干的
2. Rossini：罗西尼（1792—1868），意大利著名歌剧作曲家，作品有《塞维尔的理发师》《威廉·泰尔》等。

and Linguistic, conceived as true sciences, are not two distinct things, but one thing only. Not that there is a special Linguistic; but the much-sought-for science of language[1], general Linguistic, *in so far as what it contains is reducible to philosophy*, is nothing but Æsthetic. Whoever studies general Linguistic, that is to say, philosophical Linguistic, studies æsthetic problems, and *vice versa. Philosophy of language and philosophy of art are the same thing.*

Were Linguistic really a *different* science from Æsthetic it would not have for its object expression, which is the essentially æsthetic fact; that is to say, we must deny that language is expression. But an emission of sounds which expresses nothing is not language. Language is sound articulated, circumscribed and organized for the purposes of expression. If, on the other hand, linguistic were a *special* science in respect to Æsthetic, it would necessarily have for its object a *special class* of expressions. But the non-existence of classes of expression is a point which we have already demonstrated.

Æsthetic formulation of linguistic problems. Nature of language.

The problems which Linguistic tries to solve, and the errors in which Linguistic has been and is involved, are the same that respectively occupy and complicate Æsthetic. If it be not always easy, it is on the other hand always possible to reduce the philosophic questions of Linguistic to their æsthetic formula.

The disputes themselves as to the nature of the one find their parallel in those as to the nature of the other. Thus it has been disputed whether Linguistic be a historical or a scientific discipline, and, the scientific having been distinguished from the historical, it has been asked whether it belong to the order of the natural

1. much-sought-for science of language：备受推崇的语言科学

or of the psychological sciences, understanding by these latter empirical Psychology as well as the Sciences of the spirit. The same has happened with Æsthetic, which some have looked upon as a natural science (confusing the æsthetic and the physical sense of the word expression). Others have looked upon it as a psychological science (confusing expression in its universality with the empirical classification of expressions). Others again, denying the very possibility of a science of such a subject, change it into a simple collection of historical facts; not one of these attaining to the consciousness of Æsthetic as a science of activity or of value, a science of the spirit.

Linguistic expression, or speech, has often seemed to be a fact of interjection, which belongs to the so-called physical expressions of the feelings, common alike to men and animals. But it was soon perceived that an abyss yawns between the "Ah!" which is a physical reflex of pain and a word; as also between that "Ah!" of pain and the "Ah!" employed as a word. The theory of the interjection being abandoned (jocosely termed the "Ah! Ah!" theory by German linguists), the theory of *association* or *convention* appeared. This is liable to the same objection which destroyed æsthetic associationism in general: speech is unity, not multiplicity of images, and multiplicity does not explain, but indeed presupposes the expression to be explained. A variant of linguistic associationism is the imitative, that is to say, the theory of *onomatopœia*[1], which the same philologists deride under the name of the "bow-wow" theory, from the imitation of the dog's bark, which, according to the onomatopœists, must have given its name to the dog.

The most usual theory of our times as regards language (apart from mere crass naturalism) consists of a sort of eclecticism or mixture of the various theories to which we have referred. It is assumed that language is in part the product of interjections and in part of

1. *onomatopœia*：拟声

onomatopœia and convention. This doctrine is altogether worthy of the philosophical decadence of the second half of the nineteenth century.

Origin of language and its development.

We must here note an error into which have fallen those very philologists who have best discerned the activistic nature of language, when they maintain that although language was *originally a spiritual creation*, yet that it afterwards increased by *association*. But the distinction does not hold, for origin in this case cannot mean anything but nature or character; and if language be spiritual creation, it must always be creation; if it be association, it must have been so from the beginning. The error has arisen from having failed to grasp the general principle of Æsthetic, known to us: that expressions already produced must descend to the rank of impressions before they can give rise to new impressions. When we utter new words we generally transform the old ones, varying or enlarging their meaning; but this process is not associative, it is *creative*, although the creation has for material the impressions, not of the hypothetical primitive man, but of man who has lived long ages in society, and who has, so to say, stored so many things in his psychic organism, and among them so much language.

Relation between Grammar and Logic.

The question of the distinction between the æsthetic and the intellectual fact appears in Linguistic as that of the relations between Grammar and Logic. This problem has been solved in two partially true ways: the *inseparability* and the *separability* of Logic and Grammar. But the complete solution is this: if the logical form be inseparable from the grammatical (æsthetic), the grammatical is separable from the logical.

Grammatical kinds or parts of speech.

If we look at a picture which for instance portrays a man walking on a country road we may say: "This picture represents a fact of *movement*, which, if conceived as voluntary, is called *action*; and since every movement implies a *material object*, and every action a *being* that acts, this picture also represents a *material object* or *being*. But this movement takes place in a definite place, which is a piece of a definite heavenly body (the Earth), and precisely of a piece of it which is called terra-firma[1], and more precisely of a part of it that is wooded and covered with grass, which is called *country*, cut naturally or artificially into a form called *road*. Now, there is only one example of that star, which is called Earth: the earth is an *individual*. But *terra-firma, country, road* are genera or *universals*, because there are other terra-firmas, other countries, other roads." And it would be possible to continue for a while with similar considerations. By substituting a phrase for the picture that we have imagined, for example one to this effect: "Peter is walking on a country road," and by making the same remarks, we obtain the concepts of *verb* (motion or action), of *noun* (material object or agent), of *proper noun*, of *common noun*; and so on.

What have we done in both cases? Neither more nor less than submit to logical elaboration what first presented itself only æsthetically; that is to say, we have destroyed the æsthetic for the logical. But since in general Æsthetic error begins when we wish to return from the logical to the æsthetic and ask what is the *expression* of motion, action, matter, being, of the general, of the individual, etc.; so in the case of language, error begins when motion or action are called *verb*, being or matter,

---

1. *terra-firma*：〈拉〉陆地

noun or substantive[1], and when linguistic categories, or *parts of speech*, are made of all these, noun and verb and so on. The theory of the parts of speech is really identical with that of artistic and literary kinds, already criticized in our Æsthetic.

It is false to say that the verb or noun is expressed in definite words, truly distinguishable from others. Expression is an indivisible whole. Noun and verb do not exist in it, but are abstractions made by us, destroying the sole linguistic reality, which is the *sentence*. This last is to be understood, not in the way common to grammars, but as an organism expressive of a complete meaning, which includes alike the simplest exclamation and a great poem. This sounds paradoxical, but is nevertheless the simplest truth.

And since in Æsthetic the artistic productions of certain peoples have been looked upon as imperfect, owing to the error above mentioned, because the supposed kinds have seemed not yet to have been discriminated, or to be in part wanting; so in Linguistic, the theory of the parts of speech has caused the analogous error of judging languages as *formed* and *unformed*, according to whether there appear in them or not some of those supposed parts of speech; for example, the verb.

The individuality of speech and the classification of languages.

Linguistic also discovered the irreducible individuality of the æsthetic fact, when it affirmed that the word is what is really spoken, and that two truly identical words do not exist. Thus were synonyms and homonyms destroyed, and thus was shown the impossibility of really translating one word into another, from so-called dialect into so-called language, or from the so-called mother-tongue into the so-called

1. substantive：实词

foreign tongue.

But the attempt to classify languages ill agrees with this just view. Languages have no reality beyond the propositions and complexes of propositions really written and pronounced by given peoples at definite periods; that is to say, they have no existence outside the works of art (whether little or great, oral or written, soon forgotten or long remembered, does not matter) in which they exist concretely. And what is the art of a given people but the whole of its artistic products? What is the character of an art (for example of Greek art or Provençal literature[1]) but the whole physiognomy of those products? And how can such a question be answered, save by narrating in its particulars the history of the literature, that is to say, of the language in its actuality?

It may be thought that this argument, although possessing validity as against many of the usual classifications of languages, yet is without any as regards that queen of classifications, the historico-genealogical[2], that glory of comparative philology. And this it certainly is; but why? Precisely because that historico-genealogical method is not a mere classification. He who writes history does not classify, and the philologists themselves have hastened to say that languages which can be arranged in historical series (those whose series have hitherto been traced) are not distinct and separate species but a single whole of facts in the various phases of its development.

Impossibility of a normative grammar.

Language has sometimes been regarded as a voluntary or arbitrary act. But at others the impossibility of creating language artificially,

---

1. Provençal literature：普罗旺斯文学。11 世纪至 14 世纪流行于法国东南部普罗旺斯及邻近地区，以抒情诗的形式，颂扬骑士和宫廷爱情。
2. historico-genealogical：历史谱系

by an act of will, has been clearly seen. "*Tu, Caesar, civitatem dare potes homini, verbo non potes*[1]" was once said to a Roman Emperor. And the æsthetic (and therefore theoretic as opposed to practical) nature of expression supplies the method of discovering the scientific error which lies in the conception of a (normative) *Grammar*, establishing the rules of correct speech. Good sense has always rebelled against this error. An example of such rebellion is the "So much the worse for grammar" attributed to Monsieur de Voltaire. But the impossibility of a normative grammar[2] is also recognized by those who teach it, when they confess that to write well cannot be learned by rules, that there are no rules without exceptions, and that the study of Grammar should be conducted practically, by reading and examples, which should form the literary taste. The scientific reason of this impossibility lies in the principle that we have demonstrated: that a technique of the theoretical amounts to a contradiction in terms. And what could a (normative) grammar be, but precisely a technique of linguistic expression, that is to say of a theoretic fact?

Didactic organisms[3].

The case in which Grammar is understood merely as an empirical discipline, that is to say, as a collection of schemes useful for learning languages, without any claim whatever to philosophic truth, is quite different. Even the abstractions of the parts of speech are in this case both admissible and useful. And we must tolerate as merely didactic many books entitled "Treatises of Linguistic," where we generally find a little of everything, from the description of the vocal

---

1. *Tu, Caesar, civitatem dare potes homini, verbo non potes*：〈拉〉恺撒，你可以向民众颁布法令，却无法向他们颁布文法。
2. normative grammar：规范语法
3. *Didactic organisms*：说教式著作

apparatus[1] and of the artificial machines (phonographs) which can imitate it, to summaries of the most important results obtained by Indo-European, Semitic[2], Coptic[3], Chinese, or other philologies; from philosophical generalizations as to the origin or nature of language, to advice on format, calligraphy and the arrangement of notes relating to philological work. But this mass of notions, here administered in a fragmentary and incomplete manner about language in its essence, about language as expression, resolves itself into notions of Æsthetic. Nothing exists outside *Æsthetic*, which gives knowledge of the nature of language, and *empirical Grammar*, which is a pedagogic expedient, save the *History of languages* in their living reality, that is to say, the history of concrete literary productions, which is substantially identical with the *History of literature.*

Elementary linguistic facts or roots.

The same error of taking the physical for the æsthetic, from which the search for the *elementary forms* of the beautiful originates, is made by those who go in search of *elementary linguistic facts*, decorating with that name the divisions of the longer series of physical sounds into shorter series. Syllables, vowels and consonants, and the series of syllables called words, all these elements of speech, which give no definite sense when taken alone, must be called not *facts of language*, but mere sounds, or rather sounds abstracted and classified physically.

Another error of the same sort is that of *roots*[4], to which the most distinguished philologists now accord but small value. Having confused physical with linguistic or expressive facts, and considering

---

1. vocal apparatus：发音器官
2. Semitic：闪米特语系
3. Coptic：科普特语系，古埃及语言。
4. *roots*：词根

that the simple precedes the complex in the order of ideas, they necessarily ended by thinking that the smallest physical facts indicated the simplest linguistic facts. Hence the imaginary necessity that the most ancient primitive languages had a monosyllabic character[1], and that historical research must always lead to the discovery of monosyllabic roots. But (to follow up the imaginary hypothesis) the first expression that the first man conceived may have had not a phonetic but a mimetic physical reflex; may have been externalized not in a sound but in a gesture. And assuming that it was externalized in a sound, there is no reason to suppose that sound to have been monosyllabic rather than polysyllabic[2]. Philologists readily blame their own ignorance and impotence[3], when they do not always succeed in reducing polysyllabism to monosyllabism, and rely upon the future to accomplish the reduction. But their faith is without foundation, and their blame of themselves is an act of humility arising from an erroneous presumption.

For the rest, the limits of syllables, as those of words, are altogether arbitrary, and distinguished somehow or other by empirical use. Primitive speech, or the speech of uneducated man, is a *continuum*, unaccompanied by any consciousness of divisions of the discourse into words or syllables, imaginary beings created by schools. No true law of Linguistic can be founded on such divisions. Proof of this is to be found in the confession of linguists, that there are no truly phonetic laws of the hiatus[4], of cacophony[5], of diæresis or synæresis[6], but merely laws of taste and convenience; that is to say, *æsthetic* laws. And what are

1. monosyllabic character：单音节字符
2. polysyllabic：多音节的
3. readily blame their own ignorance and impotence：往往怪自己无知又无能
4. cacophony：杂音
5. hiatus：元音连读
6. diæresis or synæresis：二合元音对立或元音省略

laws of *words* which are not at the same time laws of *style*?

Æsthetic judgement and the model language.

Finally, the search for a *model language*, or for a method of reducing linguistic usage to *unity*, arises from the superstition of a rationalistic measure of the beautiful, from that concept which we have called false æsthetic absoluteness. In Italy we call this the question of the *unity of the language.*

Language is perpetual creation. What has been linguistically expressed is not repeated, save by reproduction of what has already been produced. The ever-new impressions give rise to continuous changes of sound and meaning, that is, to ever-new expressions. To seek the model language, then, is to seek the immobility of motion. Everyone speaks and should speak according to the echoes which things arouse in his soul, that is, according to his impressions. It is not without reason that the most convinced supporter of any one of the solutions of the problem of the unity of language (whether by adopting a standard Italian approximating to Latin, or to fourteenth-century usage, or to the Florentine dialect) feels repugnance in applying his theory, when he is speaking to communicate his thoughts and to make himself understood. The reason is that he feels that in substituting the Latin, fourteenth-century Italian, or Florentine word for that of different origin, but which answers to his natural impressions, he would be falsifying the genuine form of truth. He would become a vain listener to himself instead of a speaker, a pedant in place of a serious man, an actor instead of a sincere person. To write according to a theory is not really to write: at the most, it is making *literature.*

The question of the unity of language is always reappearing, because, stated as it is, it is insoluble, being based upon a false conception of what language is. Language is not an arsenal of arms already made, and it is not a *vocabulary*, a collection of abstractions, or a cemetery of corpses more or less well embalmed.

Our dismissal of the question of the model language, or of the unity of the language, may seem somewhat abrupt, and yet we would not wish to appear otherwise than respectful towards the long line of literary men who have debated this question in Italy for centuries. But those ardent debates were fundamentally concerned with debates of æstheticity, not of æsthetic science, of literature rather than of literary theory, of effective speaking and writing, not of linguistic science. Their error consisted in transforming the manifestation of a need into a scientific thesis, the desirability, for example, of easier mutual understanding among a people divided by dialects into the philosophic demand for a single, ideal language. Such a search was as absurd as that other search for a *universal language*, a language possessing the immobility of the concept and of abstraction. The social need for a better understanding of one another cannot be satisfied save by the spread of education becoming general, by the increase of communications, and by the interchange of thought among men.

Conclusion.

These scattered observations must suffice to show that all the scientific problems of Linguistic are the same as those of Æsthetic, and that the truths and errors of the one are the truths and errors of the other. If Linguistic and Æsthetic appear to be two different sciences, this arises from the fact that people think of the former as grammar, or as a mixture between philosophy and grammar, that is, an arbitrary mnemonic schematism[1] or a pedagogic medley[2], and not of a rational science and a pure philosophy of speaking. Grammar, or something not unconnected with grammar, also introduces into the mind the

1. arbitrary mnemonic schematism：任意的图标式助记表述
2. pedagogic medley：学究式的大杂烩

prejudice that the reality of language lies in isolated and combinable words, not in living discourse, in the expressive organisms, rationally indivisible.

Those linguists or philologists, philosophically endowed, who have penetrated deepest into the problems of language, find themselves (to employ a trite but effective simile[1]) like workmen piercing a tunnel: at a certain point they must hear the voices of their companions, the philosophers of Æsthetic, who have been at work on the other side. At a certain stage of scientific elaboration, Linguistic, in so far as it is philosophy, must merge itself in Æsthetic: and this indeed it does without leaving a residue.

---

1. trite but effective simile：老套的但很有效的比喻

HISTORY OF ÆSTHETIC

I ÆSTHETIC IDEAS IN GRÆCO-ROMAN ANTIQUITY

Point of view of this History of Æsthetic.

The question whether Æsthetic is to be considered as an ancient or a modern science has on several occasions been a matter of controversy; whether, that is to say, it arose for the first time in the eighteenth century, or had previously arisen in the Græco-Roman world. This is a question, not only of facts, but of criteria, as is easily to be understood: whether one answers it in this way or that depends upon one's idea of that science, an idea afterwards adopted as a standard or criterion. [1]

Our view is that Æsthetic is the *science of the expressive* (representative or imaginative) *activity*. In our opinion, therefore, it does not appear until a precise concept is formulated of imagination, representation or expression, or in whatever other manner we prefer to name that attitude of the spirit, which is theoretical but not intellectual, a producer of knowledge, but of the individual, not of the universal. Outside this point of view, we for our part are not able to discover anything but deviations and errors.

These deviations can lead in various directions. Following the distinctions and terminology of an eminent Italian philosopher [2] in an analogous case[1], we shall be inclined to say that they arise either

---

1. distinctions and terminology of an eminent Italian philosopher in an analogous case：相似情形下，意大利某大哲学家所使用的分类法和术语。指罗斯米尼和他的《理念起源新说》。

from *excess* or from *defect*. The deviation from defect would be that which denies the existence of a special æsthetic and imaginative activity, or, which amounts to the same thing, denies its autonomy, and thus mutilates the reality of the spirit[1]. Deviation by excess is that which substitutes for it or imposes upon it another activity, altogether undiscoverable in the experience of the interior life, a mysterious activity which does not really exist. Both these deviations, as can be deduced from the theoretical part of this work, take various forms. The first, that due to defect, may be: (*a*) *purely hedonistic*, in so far as it considers and accepts art as a simple fact of sensuous pleasure; (*b*) *rigoristic-hedonistic*, in so far as, looking upon it in the same way, it declares it to be irreconcilable with the highest life of man; (*c*) *hedonistic-moralistic* or *pedagogic*, in so far as it consents to a compromise, and while still considering art to be a fact of sense, declares that it need not be harmful, indeed that it may render some service to morality, provided always that it is submissive and obedient. [3] The forms of the second deviation (which we shall call "mystical") are not determinable *a priori*, for they belong to feeling and imagination in their infinite variety and shades of meaning. [4]

Mistaken tendencies, and attempts towards an Æsthetic, in Græco-Roman antiquity.

The Græco-Roman world presents all these fundamental forms of deviation: pure hedonism, moralism or pedagogism, mysticism, and together with them the most solemn and celebrated rigoristic negation of art which has ever been made. It also exhibits attempts at the theory of expression or pure imagination; but nothing more than approaches and attempts. Hence, since we must now take sides in the controversy

---

1. mutilates the reality of the spirit: 割裂精神实在

as to whether Æsthetic is an ancient or modern science, we cannot but place ourselves upon the side of those who affirm its modernity.

A rapid glance at the theories of antiquity will suffice to justify what we have said. We say rapid, because to enter into minute particulars, collecting all the scattered observations of ancient writers upon art, would be to do again what has been done many times and sometimes very well. Further, those ideas, propositions and theories have passed into the common patrimony of knowledge[1], together with what else remains of the classical world. It is therefore more advisable here than in any other part of this history merely to indicate the general lines of development.

Origin of the æsthetic problem in Greece.

Art, the artistic faculty, only became a philosophical problem in Greece after the sophistical movement[2] and as a consequence of the Socratic dialectic[3]. The historians of literature generally point to the origins of Greek Æsthetic in the first appearance of criticism and reflection upon poetical works, painting and sculpture; in the judgements pronounced on the occasion of poetical competitions, in the observations that were made as to the methods of the different artists, in the analogies between painting and poetry as expressed in the sayings attributed to Simonides[4] and Sophocles[5]; or, finally, in the appearance of that word which served to group together the various arts and to indicate in a certain way their relationship — the word mimesis or mimetic (μίμησις)[6] —

---

1. common patrimony of knowledge：共同的知识财富
2. sophistical movement：诡辩运动
3. Socratic dialectic：苏格拉底辩证法
4. Simonides：西蒙尼戴斯（公元前556—前469），古希腊抒情诗人。
5. Sophocles：索福克勒斯（公元前5世纪初 — 前406），希腊雅典剧作家、政治家，希腊三大悲剧作家之一。
6. mimesis or mimetic (μίμησις)：模拟或模仿。μίμησις：〈希〉模仿。

which oscillates between the meaning of "imitation" and that of "representation." Others make the origin of Æsthetic go back to the polemics[1] which were conducted by the first naturalistic and moralistic philosophers against the tales, fantasies and morals of poets, and to the interpretations of the hidden meaning (υπόνοια[2]), or, as the moderns call it, allegory, employed to defend the good name of Homer and of the other poets; finally, to the *ancient quarrel* between philosophy and poetry, as Plato was afterwards to call it. [5] But, to tell the truth, none of these reflections, observations and arguments implied a true and proper philosophical discussion of the nature of art. Nor was the sophistical movement favourable to its appearance. For although attention was at that time certainly given to internal psychical facts[3], yet these were conceived as mere phenomena of opinion and feeling, of pleasure and pain, of illusion, whim or caprice. And where there is no true and no false, no good and no evil, there can be no question of beautiful and ugly, nor of a difference between the true and the beautiful or between the beautiful and the good. The most one has in that case is the general problem of the irrational and the rational, but not that of the nature of art, which assumes the difference between rational and irrational, material and spiritual, mere fact and value, to have been already stated and grasped. If, then, the sophistical period was the necessary antecedent to the discoveries of Socrates, the æsthetic problem could only arise after Socrates. And it did indeed arise with Plato, author of the first, or indeed of the only really great negation of art of which there remains documentary proof in the history of ideas.

Plato's rigoristic negation.

---

1. polemics：辩论术
2. υπόνοια：〈希〉隐喻
3. internal psychical facts：内在心理事实

Is art, mimesis, a rational or an irrational fact? Does it belong to the noble region of the soul, where philosophy and virtue are found, or does it dwell in that base lower sphere, with sensuality and crude passionality[1]? This is the question asked by Plato, [6] who thus states the problem of Æsthetic for the first time. The sophist Gorgias[2] was able to note, with his sceptical acuteness, that tragic representation is a deception, which (strangely enough) turns out to the honour both of him who deceives and of him who is deceived, in which it is shameful not to know how to deceive oneself and not to let oneself be deceived. [7] With that remark he could rest content. That was for him a fact like another. But Plato, the philosopher, was bound to solve the problem: if it were a deception, then down with tragedy and the rest of mimetic productions: down with them among the other things to be despised, among the animal qualities of man. But if it were not deception, what was it? What place did art occupy among the lofty activities of philosophy and of good action?

The answer that he gave is well known. Mimetic does not realize the ideas, that is to say the truth of things, but reproduces natural or artificial things, which are pale shadows of them; it is a diminution of a diminution[3], a third-hand work. Art, then, does not belong to the lofty and rational region of the soul (του λογιστικοϋ ἐν ψυχή) but to the sensual; it is not a strengthening but a corruption of the mind (λώβη τής διάνοιας); it can serve only sensual pleasure, which troubles and obscures. For this reason, mimetic, poetry and poets, must be excluded from the perfect Republic.

Plato is the most consistent example of those who do not succeed in discovering any other form of knowledge but the intellectual. It

1. crude passionality：兽欲
2. The sophist Gorgias：诡辩家高尔吉亚。高尔吉亚（公元前 483—前 375），古希腊哲学家、修辞学家。著有《论虚无》等。
3. a diminution of a diminution：一降再降

was correctly observed by him that imitation stops at natural things, at the image (το φάντασμα), and does not reach the concept, logical truth (ἀλήθεια), of which poets and painters are altogether ignorant. But his error consisted in believing that there is no other form of truth below the intellectual; that there is nothing but sensuality and passionality outside or prior to the intellect, that which discovers the ideas. Certainly, the fine æsthetic sense of Plato did not echo that depreciatory[1] judgement of art; he himself declared that he would have been very glad to have been shown how to justify art and to place it among the forms of the spirit. But since none was able to give him this assistance, and since art with its *appearance* that yet lacks *reality* was repugnant to his ethical consciousness , and reason compelled him (ὁ λόγος ᾕρει) to banish it and place it with its peers, he resolutely obeyed his conscience and his reason. [8]

Æsthetic hedonism and moralism.

Others were not troubled with these scruples, and although art was always looked upon as a mere thing of pleasure among the later hedonistic schools of various sorts, among rhetoricians and worldly people the duty of combating or of abolishing it was not felt. Nevertheless, this opposite extreme was also not calculated to meet with the endorsement of public opinion, for the latter, if tender towards art, is no less tender towards rationality and morality. For this reason both rationalists and moralists, compelled to recognize the force of such a condemnation as Plato's, sought for a compromise, a half measure. Away with the sensual and with art: certainly. But can we expel the sensual and the pleasurable without more ado[2]? Can fragile human nature nourish itself exclusively with the strong food of

1. depreciatory：贬低的
2. without more ado：干脆地

philosophy and morality? Can we obtain observance of the true and of the good from the young and from the people, without allowing them at the same time some amusement? And has not man himself always something of the child, has he not always something of the people in him, is he not to be treated with the same precautions? Is there not a risk that the over-bent bow will break? — These considerations prepared the way for the justification of art, for they showed that if it were not rational in itself, it could on the other hand serve a rational end. Hence the search for the *external end* of art, which takes the place of the search for the essence or *internal end*. When art had been lowered to the level of a simple pleasurable illusion, an inebriation of the senses[1], it was necessary to subordinate the practical action of producing such an illusion and inebriation, like any other action, to the moral end. Art, being deprived of any dignity of its own, as obliged to assume a reflected or secondhand dignity. Thus the moralistic and pedagogic theory was constructed upon a hedonistic basis. The artist, who, for the pure hedonist, was comparable to a *hetaira*[2], became for the moralist a *pedagogue*[3]. Hetaira and pedagogue, these are the symbols of the two conceptions of art that were disseminated in antiquity, and the second was grafted upon the first.

Even before Plato's peremptory negation[4] had directed thought to this way of issue, the literary criticism of Aristophanes[5] was already full of the pedagogic idea: "What schoolmasters are to children, poets are to young men" (τοῖς ἡβῶσιν δὲ ποιηταί)[6], he says in a

---

1. inebriation of the senses：感官陶醉
2. *hetaira*：古希腊的高级妓女
3. *pedagogue*：教师
4. peremptory negation：断然否定
5. Aristophanes：阿里斯托芬（约公元前 446 年—前 385 年）古希腊早期喜剧作家，著有《云》《蛙》等。
6. τοῖς ἡβῶσιν δὲ ποιηταί：〈希〉年轻人（young men），下同，括号内希腊文与前面英文意思一致，不一一注释。

celebrated verse [9]. But we can find traces of it in Plato himself (in the dialogues in which he seems to withdraw from the too rigid conclusions of the *Republic*) and in Aristotle, both in the *Politics*[1], where he determines the use of music in education, and perhaps in the *Poetics*[2], where he speaks obscurely of a tragical *catharsis*[3]; although as regards this latter, it is not to be altogether denied that he may have had a sort of glimpse of the modern idea of the liberating power of art. [10] Later on, the pedagogic theory takes a form that was much affected by the Stoics[4]. Strabo[5] develops and defends this at great length, in the introduction to his geographical work, where he combats Eratosthenes[6], who has made poetry consist in mere pleasure without any notion of teaching. Strabo, on the contrary, maintained the opinion of the ancients, that it was "a first philosophy" (φιλοσοφίαν τινα πρωτήν), which educated young men for life, and created customs, affections and actions, by means of pleasure." Therefore, he said, poetry has always been a part of education; one cannot be a good poet unless one is a good man (άνδρα άγαθόν). Legislators and founders of cities were the first to employ fables to admonish and to terrify: then this duty, which must be performed for women and children and even for adults, passed to the poets. We caress and dominate the multitude with fiction and with falsehood. [11] "The poets tell many lies" (πολλά ψεύδονται άοιδοί) is a hemistich[7]

---

1. *Politics*：《政治学》，亚里士多德根据他和他的学生对希腊 158 个城邦政治法律制度的调查结果写就的政治学著作。
2. *Poetics*：《诗学》，亚里士多德的美学著作。
3. tragical *catharsis*：悲剧的净化功能
4. Stoics：斯多葛学派，主张苦修主义。
5. Strabo：斯特拉博（约公元前 64—24），希腊地理学家、哲学家、历史学家，著有《地理》《历史手稿》。
6. Eratosthenes：埃拉托色尼（公元前 276—前 194），古希腊数学家、地理学家和天文学家，设计了世界地图，并推算了地球周长以及地球到月球和太阳之间的距离，被西方地理学家推崇为"地理学之父"。
7. hemistich：半句诗

recorded by Plutarch[1], who describes minutely in one of his lesser works how the poets should be read to youths. [12] For him too poetry is a preparation for philosophy; it is a disguised philosophy, and therefore delights us in the same way as do fish and meat at feasts, so prepared as not to seem to be fish and meat; it is philosophy softened with fables, like the vine that grows close to the mandragora[2], and produces a wine that is the giver of sweet slumbers. It is not possible to pass from dense darkness to sunlight; one should first accustom the eyes to moderate light. Philosophers, in order to exhort and instruct, take their examples from true things; poets aim at a like result, when they create fictions and fables. [13] Lucretius[3], in Roman literature, gives us the well-known comparison of the boys for whom the doctors "*prius or as pocula circum Contingunt mellis dulci flavoque liquore*[4]," in order to administer the bitter wormwood. [14] Horace, in certain verses of the Epistle to the Pisones[5] which have become proverbial (perhaps his source for them was the Greek of Neoptolemus of Paros[6]?), offers both views (that of art as courtesan and of art as pedagogue) in his "*Aut prodesse volunt aut deledare poetae ... omne tulit punctum qui miscuit utile dulci.*[7]" [15]

Thus looked at, the office of the poet was confounded with that of the orator, for he too was a practical man aiming at practical

1. Plutarch：普鲁塔克（约公元 46—120），古罗马时代的希腊历史学家、传记作家和散文家，著有《希腊罗马名人列传》。
2. mandragora：曼陀罗草
3. Lucretius：卢克莱修（约公元前 99—前 55），古罗马哲学家与诗人，著有《物性论》。
4. *prius or as pocula circum Contingunt mellis dulci flavoque liquore*：〈拉〉先在装满了苦药的杯子的杯口涂满了蜂蜜和甜酒。
5. Epistle to the Pisones：写给皮索父子的诗体信简，指贺拉斯的《诗艺》。
6. Neoptolemus of Paros：帕罗斯岛的尼奥普拉托列墨斯，希腊神话人物。
7. *Aut prodesse volunt aut deledare poetae ... omne tulit punctum qui miscuit utile dulci.*：〈拉〉诗人既讲求娱乐也讲求教育。他把两者融为一体，大有裨益。

effects; hence there arose discussions as to whether Virgil[1] was to be considered as a poet or as an orator ("*Virgilius poeta an orator*?[2]"). To both was assigned the triple end of *delectare, movere, docere*[3]; in any case this tripartition[4] was very empirical, for we clearly perceive that the *delectare* is here a means and the *docere* a simple part of the *movere*[5]: to move in the direction of the good, and therefore, among other goods, towards that of instruction. In like manner, it was said of the orator and poet (recording the meretricious[6] basis of their task, and with a metaphor significant in its *naïveté*[7]) that they were bound to avail themselves of the *allurements*[8] (*lenocinium*) of form.

Mystical æsthetic in antiquity.

The mystical view, which considers art as a special mode of self-beatification, of entering into relation with the Absolute, with the Summum Bonum[9], with the ultimate root of things, appeared only in late antiquity, almost at the entrance to the Middle Ages. Its representative is the founder of the neo-Platonic school, Plotinus[10].

It is strange that Plato should be usually selected as the founder and head of this æsthetic tendency, and that for this very reason to him should be attributed the honour of being the father of Æsthetic.

---

1. Virgil：维吉尔（公元前70—前19），罗马诗人，著有史诗《埃涅阿斯》。
2. *Virgilius poeta an orator*?〈拉〉维吉尔到底是诗人还是演说家？
3. triple end of *delectare, movere, docere*：〈拉〉娱乐、感化、教育三大目的
4. tripartition：三部分
5. the *delectare* is here a means — and the *docere* a simple part of the *movere*：娱乐是手段，而教育是感化的一个简单部分。
6. meretricious：华而不实的
7. naïveté：纯真
8. allurement：诱惑
9. *Summum Bonum*：〈拉〉至善
10. Plotinus：普罗提诺（205—270），新柏拉图学派奠基人，古罗马帝国时代哲学家。

But how could he, who had expounded with such great limpidity and clearness the reasons for which he was not able to accord to art a high place among the activities of the spirit, be credited with having accorded to it one of the highest places, equal, if not superior, to philosophy itself? This misunderstanding has evidently arisen out of the enthusiastic effusions about the Beautiful that we read in the *Gorgias*, the *Philebus*, the *Phædrus*, the *Symposium*[1], and other Platonic dialogues. It is well to dissipate it by declaring that the *Beauty* of which Plato discourses has nothing to do with art or with *artistic beauty*.

Investigations as to the Beautiful.

The search for the meaning and scientific content of the word "beautiful" could not but early attract the attention of the subtle and elegant Greek dialecticians. Indeed, we find Socrates engaged in discussing this question in one of the discourses that have been preserved for us by Xenophon[2]; and we find him disposed to stop for the moment at the conclusion that the beautiful is *that which is convenient and which answers to the end desired*, or at the other conclusion that it is *that which one loves* [16] . Plato too examines this sort of problem and proposes various sorts of solutions or attempts at solutions of it. He sometimes speaks of a beauty that dwells not only in bodies, but also in laws, in actions, in the sciences; sometimes he seems to conjoin and almost to identify it with the true, the good and the divine; now he returns to the view of Socrates and confuses it with the useful; now he distinguishes between a beautiful in itself (καλά καθ' αυτά) and a relatively beautiful (πρός τι καλά); or he makes true

1. the *Gorgias*, the *Philebus*, the *Phædrus*, the *Symposium*：《高尔吉斯篇》《菲力布斯篇》《费德罗篇》《会饮篇》，柏拉图对话体著作。
2. Xenophon：色诺芬（约公元前 427—前 355），古希腊作家、军人、苏格拉底门徒，著有《远征记》。

beauty consist in pure pleasure (ἡδονη καθαρά), free from all shadow of pain; or he places it in measure and proportion (μετριότης καί ξνμμετρία); or talks of colours and sounds as possessing a beauty in themselves. It was impossible to find an independent dominion for the beautiful, if the artistic or mimetic activity were deserted. This explains his wandering among so many different conceptions, among which it is just possible to say that the identification of the Beautiful with the Good prevails. Nothing better describes this uncertainty than the dialogue of the *Hippias maior*[1] (which, if it be not Plato's, is Platonic). He here wishes to find out not what things are beautiful things, but what the beautiful is; that is to say, what it is that makes beautiful, not only a beautiful virgin, but also a beautiful mare, a beautiful lyre, a beautiful pot with two graceful ears of clay. Hippias and Socrates himself propose in turn the most various solutions; but the latter ends by confuting them all. "That which makes things beautiful is the gold that is added to them by way of ornament." No: gold only embellishes where it is *fitting* (πρέπων): for instance, a pot should have a wooden rather than a golden handle. "That is beautiful which cannot seem ugly to any one." But it is not a question of *seeming*: the question is to define what the beautiful is, whether it seems so or not. It is the *fitting* which makes things seem to be beautiful. But in that case, the fitting (which makes them *appear*, not *be*) is one thing, and the beautiful another. "The beautiful is what leads to the end, that is to say, the *useful* (χρήσιμον)." But if that were so, then evil would also be beautiful, because the useful leads also to the evil. "The beautiful is the *helpful*, that which leads to the good (ωφέλιμον)." But in this case, the good would not be beautiful nor the beautiful good; for the cause is not the effect, and the effect is not the cause. "The beautiful is that which delights the sight and hearing." But this fails to persuade for

1. *Hippias maior*：《大希琵阿斯篇》，柏拉图著。以诡辩学者希琵阿斯与苏格拉底对话形式展开，探讨美的本质问题。

three reasons: firstly, because beautiful studies and laws are beautiful, which have nothing to do with the eye or with the ear; secondly, because we cannot discover a reason for limiting the beautiful to those senses, while excluding the pleasure of eating and smelling, and the extremely vivid pleasures of sex; thirdly, because, if the foundation of the beautiful were *visibility* , it would not be *audibility* , and if it were audibility it would not be visibility; hence that which constitutes the beautiful cannot dwell in either of the two qualities. And the question which has been repeated so insistently in the course of the dialogue: *what is the beautiful?* (τί εστι το καλόν;) remains unanswered. [17]

Later writers also conducted inquiries into the beautiful, and we possess the titles of several treatises upon the theme, which have been lost. Aristotle shows himself changeable and uncertain upon the point. In the scanty references which he makes to it, he at one time confounds the beautiful with the good, defining it as that which is both good and pleasing; [18] at another he notes that the good consists of action (εν πράξει) and the beautiful also in things that are immoveable (εν τοῖς άκινήτοις), drawing from this the argument that mathematics should be studied in order to determine its characters, order, symmetry and limit; [19] sometimes he places it in bigness and in order (εν μεγεθει καί τάξει); [20] at others he was led to look upon it as something apparently indefinable. [21] Antiquity also established canons of beautiful things, such as that attributed to Polycletus[1] on the proportions of the human body. And Cicero said of the beauty of bodies that they were "*quaedam apta figura membrorum cum coloris quadam suavitate*[2]." [22] All these affirmations, even when they are not mere empirical observations, or verbal glosses and substitutions, meet with unsurmountable obstacles.

---

1. Polycletus：波利克里托斯（公元前 450—前 415）。古希腊著名雕塑家、艺术家。著有论述人体比例的《法则》（*Canon*）。
2. *quaedam apta figura membrorum cum coloris quadam suavitate*：〈拉〉任何肢体语言都和悦目的色彩有关

Distinction between the theory of Art and the theory of the Beautiful.

In any case, not only is the conception of the beautiful, taken as a whole, identified with art in none of them; but sometimes art and beauty, mimesis and pleasing or displeasing material of mimesis, are clearly distinguished. Aristotle notes in his *Poetics* that it pleases us to see the most faithful images of things that are repugnant to us in reality, such, for instance, as the most contemptible forms of animals, or corpses (τάς εικόνας τάς μάλιστα ήκριβωμενας χαίρομεν θεωρουντες). [23] Plutarch demonstrates at length that works of art please us not as beautiful but as *resembling* (ούχ ως καλόν, άλλ, ώς ομοιον); he affirms that if the artist beautified things that are ugly in nature he would be offending against fitness and resemblance (το πρεπον και το είκός); and he proclaims the principle that *the beautiful is one thing and beautiful imitation another* (ού γαρ εστι ταυτό, το καλον και καλως τι μιμεισθαι). Paintings of horrible events are pleasing, such as *Medea slaying her sons* by Timomachus[1], *Orestes the matricide* by Theon[2], and the *Pretended madness of Ulysses* by Parrhasius[3]; and if the grunting of a pig, the grating of a machine, the noise of the winds and the tumult of the sea are unpleasing, they pleased on the contrary in the case of Parmenon[4], who imitated the pig perfectly, and in Theodorus[5], who was not less expert in rendering the grating of machines. [24] If the ancients had really wanted to place the beautiful and art in relation, a

---

1. Medea slaying her sons by Timomachus：提莫马修斯的《美狄亚杀子》。提莫马修斯系公元前 1 世纪希腊画家。《美狄亚杀子》主题出自欧里庇得斯的悲剧《美狄亚》。
2. *Orestes the matricide* by Theon：希恩的《俄瑞斯忒斯弑母》。希恩系亚历山大大帝时期画家。
3. Pretended madness of Ulysses by Parrhasius：巴赫西斯的《尤利西斯装疯》。巴赫西斯，古希腊画家。
4. Parmenon：帕耳梅农，是普鲁塔克（Ploutarkhos）在《道德论丛》(Moralia) 中提到的一位演员，此人学猪叫几可乱真。
5. Theodorus：提奥多鲁斯，公元前 6 世纪希腊雕塑家、建筑师。

secondary and partial connexion of the two conceptions was to hand in the shape of the category of the *relatively* as distinguished from the *absolutely* beautiful. But where the word *καλόν* or *pulchrum*[1] is applied to artistic productions in the writings of literary critics, it does not seem to be more than a linguistic usage, as we find, for instance, in the case of Plutarch's *beautiful* imitation, or also in the terminology of the rhetoricians, who sometimes called elegance and adornment of discourse *beauty* of elocution[2] (το τής φράσεως κάλλος).

Fusion of the two by Plotinus.

It is only with Plotinus that the two divided territories are united and *the beautiful and art are fused into a single concept*, not by means of a beneficial absorption of the *equivocal*[3] Platonic conception of beauty into the *unequivocal* conception of art, but by absorption of the clear into the confused, of *imitative art* in the so-called *beautiful*. And thus we reach an altogether new view: the beautiful and art are now both alike melted into a mystical passion and elevation of the spirit.

Beauty, observes Plotinus, resides chiefly in things visible; but it is also to be found in things audible, such as verbal and musical compositions, and it is not lacking in things supersensible[4], such as works, offices, actions, habits, sciences and virtues. What is it that makes beautiful sensible and supersensible things alike? Not, he answers, the symmetry of their parts among themselves, and with the whole (συμμετρία των μερών προς αλληλα και προς το ολον) and their colour (ενχροια), according to one of the definitions most in vogue, which we have quoted above in the words of Cicero; because

1. *καλόν* or *pulchrum*：美（斜体分别为希腊语、拉丁语表示的“美”）
2. elocution：演说术，辩论术
3. *equivocal*：模棱两可的
4. supersensible：超感觉的

there are proportions in things ugly, and there are things that are simply beautiful without any relation of proportion: beauty, then, is one thing and symmetry another. [25] The beautiful is what we welcome as akin to our own nature; the ugly is what repels us as our opposite, and the affinity of beautiful things with our souls that perceive them has its origin in the Idea, which produces both. That is beautiful which is *formed*; the ugly is what is *unformed*, that is to say, something which is capable of receiving form, but does not receive it or is not entirely dominated by it. A beautiful body is such, because of its communion (κοινωνία) with the Divine; beauty is the Divine, the Idea, shining through; and matter is beautiful, not in itself, but only when it is illuminated by the Idea. Light and fire, which are nearest to this state, shed beauty upon visible things, as the most spiritual among bodies. But the soul must purify itself, in order to perceive the beautiful, and make the power of the Idea that lies in it efficacious. Moderation, strength, prudence, and every other virtue, what else are they, according to the oracle, but *purification*? Thus there opens another eye in the soul, beside that of sensible beauty, which permits it to contemplate divine Beauty coincident with the Good, which is the supreme condition of beatitude. [26] Art enters into such contemplation, because beauty, in things made by man, comes from the mind. Compare two blocks of stone, the one placed beside the other: one rough and crude, the other reduced to the statue of a god or of a man, for example of a Grace or of a Muse, or of a human being of such a shape, as art has collected from many particular beauties. The beauty of a block of this shape does not consist in its being of stone, but in the form that art has been able to give to it (παρά του ειδους ο ενηκεν η τέχνη); and when the form is fully impressed upon it, the thing of art is more beautiful than any other natural thing. Hence he who despised the arts (Plato), because they imitated nature, was wrong; whereas the truth is, in the first place, that nature itself imitates the idea, and then that the arts do not simply limit themselves to imitating what the eyes see, but go back to those reasons or ideas from which nature itself is derived (ώς ούχ απλώς το όρώμενον μεμούνται, αλλ'άνατρέχουσιν επι τούς λόγους έξ ων η φύσις). Art therefore does not belong to nature, but adds beauty where it is wanting in nature: Phidias[1] did not represent Jove[2] because he had seen him, but such as he would appear if he wished to reveal himself to mortal eyes. [27] The beauty of natural things is the archetype[3] existing in the soul, the sole source of natural beauty. [28]

The scientific tendency. Aristotle.

This affirmation of Plotinus and of neo-Platonism is the first true and proper affirmation of mystical Æsthetic, destined to such high fortunes in modern times, especially in the first half of the nineteenth century. But the attempts at a true Æsthetic, excluding certain luminous but incidental observations to be found even in Plato: for instance, that the poet should weave fables, not arguments (μύθους άλλ'ού λόγους), [29] go back to Aristotle and are altogether independent of his few and feeble speculations as to the beautiful. Aristotle by no means agreed with the Platonic condemnation; he felt (as indeed Plato himself had suspected) that such a result could not be altogether true, and that some aspect of the problem must have been neglected. When in his turn he attempted to find a solution, he found himself in more advantageous conditions than his great predecessor[4], since he had already overcome the obstacle that arose from the Platonic doctrine of ideas, a hypostasis of concepts and abstractions[5]. The ideas were for him simply concepts, and reality

---

1. Phidias：菲狄亚斯（约公元前 490—前 430），古希腊雕塑家。
2. Jove：朱威，即朱庇特（Jupiter），罗马神话中的主神，对应希腊神话中的宙斯（Zeus）。
3. archetype：原型
4. predecessor：前辈，指柏拉图
5. hypostasis of concepts and abstractions：理念的或抽象的本体

presented itself in a far more lively manner, not as a diminution of ideas, but as a synthesis of matter and form, it was thus much more easy for him to recognize the rationality of mimesis in his general philosophical doctrine and to assign to it its right place; and indeed it seems generally clear to Aristotle that mimesis, being proper to man by nature, is contemplation or theoretic activity; although he sometimes seems to forget this (as when he confuses imitation with the case of boys, who acquire their first knowledge by following an example [30]), and although his system, which admits practical sciences and poietic activities (distinguished from the practical as leaving a material object behind them), disturbed the firm and constant consideration of artistic mimesis and poetry as a theoretical activity. But if it is a theoretical activity, by what characteristic is poetry distinguished both from *scientific* knowledge and from *historical* knowledge? This is the way Aristotle states the problem concerning the nature of art, and this is the true and only way of stating it. Even we moderns ask ourselves in what way art is distinguished from history and from science, and what this artistic form can be, which has the ideality of science and the concreteness and individuality of history. Poetry, answers Aristotle, differs from history, because, while the latter draws things that have happened (τα γενόμενα), poetry draws things that may possibly happen (οια αν γένοιτο), and differs from science, because, although it regards the universal and not the particular (τα καθ'εκαστον) like history, it does not regard it in the same way as science, but in a certain measure, which the philosopher indicates by the word *rather* (μαλλον τα καθόλου). The point then is to establish the precise meaning of the *possible*, the *rather* and the *historical particular*. But no sooner does Aristotle attempt to determine the meaning of these words, than he falls into contradictions and fallacies[1]. That *universal* of poetry, which is the *possible*, seems to identify

1. fallacies：谬论

itself for him with the probable or the necessary (τα *κατά το είκος η το ἀναγκαῖον*), and the particular of history is not explained at all, except by giving instances: "that which Alcibiades[1] did and what happened to him." [31] Aristotle, in fact, after having made so good a beginning in the discovery of the purely imaginative, proper to poetry, remains half-way, perplexed and uncertain. Thus he sometimes makes the truth of imitation consist in a certain learning and syllogizing that takes place when we look at imitations, by which we recognize that "this is that," that a copy answers to the original; [32] or, worse, he loses the grains of truth that he has found and forgets that poetry has for its content the possible, admitting, not only that it may also depict the *impossible* (το αδύνατον), and even the *absurd* (το ἄτοπον), seeing that both are *credible* and that they do not injure the end of art, but even that we must prefer impossible probabilities to incredible possibilities. [33] Art, since it has to do even with the impossible and absurd, will not therefore have in it anything of the rational, but in accordance with the Platonic theory it will be an imitation of the appearance in which empty sense indulges itself; that is to say, a thing of pleasure. Aristotle does not attain to this result, because he does not attain to any clear and precise result in this part of the subject, but it is one of the results that can be deduced from what he has said, or that, at any rate he is not able to exclude. This means that he did not fulfil his tacitly assumed task, and that although he re-examined the problem with marvellous acuteness[2] after Plato, he failed truly to rid himself of the Platonic definition, by substituting a firmly-established one of his own.

The concepts of imitation and of imagination after Aristotle. Philostratus.

But the field of investigation toward which Aristotle had turned was

1. Alcibiades：亚西比德（约公元前 440—前 404），雅典政治家，将军。
2. marvellous acuteness：令人赞赏的智慧

generally neglected in antiquity: the very *Poetics* of Aristotle does not seem to have been widely known or influential. Ancient psychology knew fancy or imagination as a faculty midway between sense and intellect[1], but always as conservative and reproductive of sensuous impressions or conveying conceptions to the senses, never properly as a productive autonomous activity. That faculty was rarely and with little result placed in relation with the problem of art. Several historians of Æsthetic attach singular importance to certain passages in the *Life of Apollonius of Tyana* by the elder Philostratus[2], in which they believe that they discover a correction of the theory of *mimesis* and the first affirmation in history of the conception of *imaginative creation*. Phidias and Praxiteles[3] (says the extract in question) did not need to go to heaven to see the gods, in order to be able to depict them in their works, as would have been necessary according to the theory of imitation. Imagination, without any need of models, made them able to do what they did: imagination, which is a wiser agent than simple imitation (φαντασία ... σοφωτόρα μιμήσεως δημιουργός), and gives form, like the other, not only to what has been seen, but also to what has never been seen, imagining it on the basis of existing things and in that way creating Jupiters and Minervas[4]. [34] However, the imagination of which Philostratus speaks here is not something different from the Aristotelian mimesis, which, as has been noted, was concerned not only with real things but also and chiefly with possible things. And had not Socrates observed (in the dialogue with the painter Parrhasius,

---

1. faculty midway between sense and intellect：感性与知性之间的才能
2. *Life of Apollonius of Tyana* by the elder Philostratus：大斐罗斯特拉图所著的《狄阿那的阿波洛尼阿斯传》。大斐罗斯屈拉特（约 170—245），古希腊传记作家、批评家。他在《狄阿那的阿波洛尼阿斯传》中对古希腊以来的艺术摹仿论进行了改造，对想象进行了初步论证。
3. Praxiteles：普拉克西特利斯（公元前 4 世纪），古希腊古典后期雕塑艺术的代表人物。
4. Minervas：密涅瓦，罗马神话智慧女神，相等于希腊神话中的雅典娜。

preserved for us by Xenophon) that painters work by collecting what they need to form their figures from several bodies (εκ πολλων συνάγοντες τα εξ εκάστου καλλιστα)? [35] And was not the anecdote of Zeuxis[1], who was supposed to have taken the best of five Crotonian maidens in order to paint his Helen, and other anecdotes of a like sort, sufficiently widespread in antiquity? And had not Cicero eloquently explained, some years before Philostratus, how Phidias, when he was carving Jupiter, did not copy anything real, but kept his looks fixed upon "*species pulchritudinis eximia quaedam*[2]," which he had in his soul and which directed his art and his hand? [36] Nor can it be said that Philostratus opened the way to Plotinus, for whom the superior or intellectual imagination (νοητή), or eye of supersensible beauty, when it is not a new designation for beautiful imitation, is mystical intuition[3].

The vagueness of the concept of mimesis reached its apex[4] in those writers who gave it as a general title to any sort of work that had nature for its object, employing the Aristotelian phrase to affirm that "*omnis ars naturae imitatio est*[5]," [37] or saying, like the painter Eupompus[6] when he blamed his servile imitators, that "*natura est imitanda, non artifex*[7]." [38] And those who wished to escape this vagueness did not know how to do so, save by conceiving the activity of imitation as the practical producer of duplicates of natural objects,

---

1. the anecdote of Zeuxis：宙西斯的轶事。宙西斯，公元前 5 世纪希腊画家，在绘画中首次使用明暗法。相传，克罗东的居民请宙西斯绘制特洛伊的海伦画像，宙西斯集中了全城五个最美丽的女孩，将每个人最美丽的特质综合在一起，从而绘制出无可挑剔的女神。
2. *species pulchritudinis eximia quaedam*：〈拉〉美的形式
3. mystical intuition：神秘直觉
4. apex：顶峰
5. *omnis ars naturae imitatio est*：〈拉〉任何艺术都是对自然的模仿。
6. Eupompus：欧珀姆浦斯，公元前 4 世纪的古希腊画家，创建西锡安画派。
7. *natura est imitanda, non artifex*：〈拉〉自然是被模仿的，不是被抄袭的。

a prejudice born in the bosom of the pictorial and plastic arts, against which Philostratus perhaps intended to argue, in common with the other advocates of imagination.

Speculations on language.

The speculations upon language had a close connexion with those upon the nature of art begun by the sophists, for whom it became a matter for wonder that sounds could signify colours or things inaudible; that is to say, *speech* presented itself as a *problem*. [39] It was then discussed whether language was by nature (φύσει[1]) or by convention (νόμω[2]). By nature was sometimes understood mental necessity, and by convention what we should call a merely natural fact, psychological mechanism or sensationalism[3]. In that sense of the terms, language would have been better called φύσει than νόμω. But at other times the distinction led to the question whether language answers to objective or logical truth and to the real relations between things (όρθότης των ονομάτων); and in this case, those would seem to be nearer the truth who proclaimed it to be conventional or arbitrary in respect to logical truth: νόμω or θέσει[4], and not φύσει. Two different questions were consequently being treated together, and both were confusedly and equivocally discussed. They find their monument in the obscure *Cratylus*[5] of Plato, which seems to fluctuate between different solutions. Nor did the later affirmation that the word is a sign (σημείον) of the thought solve anything, for it still remained to be shown in what way the sign was to be understood, whether φύσει

1. φύσει：〈希〉自然的
2. νόμω：〈希〉约定俗成
3. psychological mechanism or sensationalism：心理机械论或感知论
4. θέσει：〈希〉人为的
5. Cratylus:《克拉底鲁篇》，柏拉图著，记录了苏格拉底与赫谟根尼对于约定论的讨论。

or νόμω. Aristotle, who looked upon words as imitations (μιμηματα), in the same way as poetry, [40] made an observation of first-rate importance: in addition to the *enunciative*[1] propositions, which express the (logically) true or false, there are others which do not express either the (logically) true or false, as for example the expressions of aspirations and of desires (εὐχή), which therefore belong, not to logical exposition, but to poetical and rhetorical exposition[2]. [41] And in another place we find him affirming in opposition to Bryson[3] (who had said that a base thing remained such with whatever word it were designated) that base things can be expressed both with words that place them beneath the eye in all their crudity, and with other words which surround them with a veil. [42] All this might have led to the separation of the linguistic faculty from the properly logical, and to its consideration in union with the poetical and artistic faculty; but here too the attempt stopped half-way. The Aristotelian logic assumed a verbal and formalistic character, which became more and more accentuated as time went on and formed an obstacle to the distinction between the two theoretical forms. Nevertheless, Epicurus[4] asserted that the diversity of names designating the same thing with various peoples was due, not to convention and caprice, but to the fact that the impressions produced by things were different in each one of them. [43] And the Stoics, although they connected language with thought (διάνοια) and not with imagination, seem to have had a suspicion of the non-logical nature of language, for they interposed between thought and sound a *certain something* which was indicated in Greek by the word λεκτόν, and by the words *effatum* or *dicibile*[5] in Latin. But we are not

1. enunciative：明确的
2. poetical and rhetorical exposition：诗性阐述和修辞型阐述
3. Bryson：布莱森（公元前 450—前 390），古希腊数学家。
4. Epicurus：伊壁鸠鲁（公元前 341—前 271），古希腊哲学家，创建了主张快乐主义的伊壁鸠鲁学派。
5. λεκτόν ... *effatum* or *dicibile*：可以说出的（斜体分别为希腊语和拉丁语）。

sure what they really meant, and whether that vague concept were intended by them to distinguish the linguistic representation from the abstract concept (which would bring them into touch with the modern view), or the meaning of sound in general. [44]

We cannot collect any other germ of truth from the ancient writers. A philosophical Grammar, like a philosophical Poetics, remained unattainable in antiquity.

[1] See above, pp. 171–175. Quotations which give only the name of the author, or are otherwise abbreviated, refer to historical or critical works of which the complete title is given in the Bibliographical Appendix.

[2] Rosmini, *Nuovo saggio sull' origine delle idee*, sections iii. and iv., where theories of knowledge are classified.

[3] See above, pp. 115–116.

[4] See above, p. 93.

[5] *Republic*, x. 607.

[6] *Republic*, x. 607.

[7] Plutarch, *De audiendis poetis*, ch. i.

[8] *Republic* x.

[9] *Frogs*, 1, 1055.

[10] Plato, *Laws*, bk. ii.; Aristotle, *Poet.* ch. 14; *Polit*, bk. viii.

[11] Strabo, *Geographica*, i. ch. 2, § § 3–9.

[12] Texts collected in E. Müller, *Gesch. d. Th. d. K.* i. pp. 57–85.

[13] Plutarch, *De aud. poetis,* chs. 1–4, 14.

[14] *De rerum natura*, i. 935–947.

[15] *Ad Pisones*, 333–334.

[16] *Memorab.* iii. ch. 8; iv. ch. 6.

[17] *Hippias maior, passim.*

[18] *Rhet.* i. ch. 9.

[19] *Metaphys.* xii. ch. 3.

[20] *Poet.* ch. 7.

[21] Diog. Lært. v. ch. i, § 20.

[22] *Tuscul. quæst.* bk. iv. § 13.

[23] *Poet.* ch. iv. 3.

[24] *De aud. poetis*, ch. 3.

[25] *Enneads*, I. bk. vi. ch. i.

[26] *Enneads, loc. cit.* chs. 2–9.

[27] *Enneads*, V. bk. viii. ch. i.

[28] *Enneads, loc. cit.* chs. 2–3.

[29] *Phædrus*, ch. 4.

[30] *Poet.* ch. 4, § 2.

[31] *Poet.* ch. 9, §§ 1–4.

[32] *Poet.* ch. 4, §§ 4–5.

[33] *Poet.* chs. 24–25.

[34] *Apoll. vita*, vi. ch. io.

[35] *Memorab.* iii. ch. io.

[36] *Orator ad Brutum*, ch. 2.

[37] For example, Seneca, *Epist.* 65.

[38] Pliny, *Nat. Hist.* xxxiv. ch. 19.

[39] Gorgias in *De Xenoph., Zen. et Gorg.* (in Aristot., ed. Didot), chs. 5–6.

[40] *Rhet.* bk. iii. ch. 1.

[41] *Rhet.* bk. iii. ch. 2.

[42] *De interp.* ch. 4.

[43] Diog. Lært. bk. x. § 75.

[44] Steinthal, *Gesch. d. Sprachw.*, 2nd ed., i. pp. 288, 293, 296–297.

II ÆSTHETIC IDEAS IN THE MIDDLE AGES AND RENAISSANCE

Middle Ages. Mysticism: Ideas on the Beautiful.

Almost all the developments of ancient Æsthetic were continued by tradition or reappeared by spontaneous generation in the course of the Middle Ages. Neo-Platonic mysticism[1] continued, entrusted to the care of the pseudo-Dionysius the Areopagite[2] (*De cœlesti hierarchia*[3], *De ecclesiastica hierarchia*[4], *De divinis nominibus*[5], etc.), to the translations of these works made by John Scotus Eriugena[6], and to the divulgations of the Spanish Jews (Avicebron[7]). The Christian God took the place of the Summum Bonum or Idea: God, wisdom, goodness, supreme beauty, source of beautiful things in nature, which are a ladder to the contemplation of the Creator. But these speculations continued to recede further and further from the consideration of art, with

1. Neo-Platonic mysticism：新柏拉图学派神秘主义
2. the pseudo-Dionysius the Areopagite：亚略巴古的伪狄奥尼修斯。狄奥尼修斯，生活在约公元 1 世纪，雅典法官，基督教会推崇的主保圣人。中世纪欧洲盛行的神秘主义著作都假托他的名字。
3. *De cœlesti hierarchia*：〈拉〉《天国等级》
4. *De ecclesiastica hierarchia*：〈拉〉《宗教等级》
5. *De divinis nominibus*：〈拉〉《神的名字》
6. John Scotus Eriugena：约翰·斯卡图斯·埃里金纳（810—877），爱尔兰神学家、新柏拉图派哲学家、诗人，翻译并评论了假托狄奥尼修斯的著作。
7. Avicebron：阿维塞布隆，11 世纪西班牙安达卢西亚犹太学者，新柏拉图主义哲学家。

which Plotinus had connected them; and the empty definitions of the beautiful by Cicero and other ancient writers were often repeated. Saint Augustine[1] defined beauty in general as unity (*omnis pulchritudinis forma unitas est*[2],) and that of the body as *congruentia partium cum quadam coloris suavitate*[3], and the old distinction between something that is beautiful in itself and relative beauty reappeared in a book of his, which has been lost, entitled *De pulchro et apto*[4]; the very name shows that he reasserted the old distinction between the beautiful in itself and the relatively beautiful, *quoniam apte accommodaretur alicui*[5]. Elsewhere he notes that an image is called beautiful *si perfecte implei illud cujus imago est, et coaequatur ei*[6]. [1]

Thomas Aquinas[7] varied but little from him in positing three requisites for beauty: integrity or perfection, due proportion, and clearness[8]; following Aristotle, he distinguished the beautiful from the good, defining the first as that which pleases in the mere contemplation of it (*pulcrum ... id cujus ipsa apprehensio placet*[9]); he referred to the beauty that even base things possess if well imitated, and applied the doctrine of imitation to the beauty of the Second Person of the Trinity[10]

1. Saint Augustine：圣奥古斯丁（354—430），古罗马帝国时期天主教神学家，欧洲中世纪基督教神学、教父哲学的重要代表人物。著有《忏悔录》与《上帝之城》。
2. *omnis pulchritudinis forma unitas est*：〈拉〉任何美的表现都是统一的
3. *congruentia partium cum quadam coloris suavitate*：〈拉〉各部分的黄金比例加上悦目的颜色
4. *De pulchro et apto*：《论美与适合》，圣奥古斯丁著作。
5. *quoniam apte accommodaretur alicui*：〈拉〉适用于任何事物的
6. *si perfecte implei illud cujus imago est, et coaequatur ei*：〈拉〉假如自身美铭刻在物象的事物上，并与之融为一体
7. Thomas Aquinas：托马斯·阿奎那（约 1225—1274），中世纪神学家、经院哲学家，自然神学最早的提倡者之一，著有《神学大全》。
8. integrity or perfection due proportion, and clearness：完整或完美、适当比例和鲜明
9. *pulcrum ... id cujus ipsa apprehensio placet*：〈拉〉美，只需静观就足以让人愉悦。
10. the Second Person of the Trinity：三位一体之第二人格，即圣子耶稣。

(*in quantum est imago expressa Patris*[1]). [2] If it were wished to discover references to the hedonistic conception of art, it would be possible to do this, with a little goodwill, in some of the sayings of jongleurs and troubadours[2]. Æsthetic rigorism[3], the total negation of art for religion or for divine and human science, shows itself in Tertullian[4] and among certain Fathers of the Church, at the entrance to the Middle Ages; at their conclusion, in a certain crude scholastic spirit, for example in Cecco d'Ascoli[5], who proclaimed against Dante: "I leave trifles behind me and return to the *true*; fables are always unpleasing to me," and later, in the reactionary Savonarola[6]. But the narcotic[7] theory of pedagogic or moralistic art prevailed over every other. It had contributed to send to sleep the æsthetic doubts and inquiries of the ancients, and was well suited to a period of relative decadence of culture. This was all the more the case, seeing that it accorded well with the moral and religious ideas of the Middle Ages, and afforded a justification not only for the new art of Christian inspiration, but also for the surviving works of classical and pagan art[8].

The pedagogic theory of art in the Middle Ages.

The allegorical interpretation was again a means of salvation for

---

1. *in quantum est imago expressa Patris*：〈拉〉因为它是天父的真实形象。
2. jongleurs and troubadours：行吟诗人与说唱艺人
3. Æsthetic rigorism：美学严格主义
4. Tertullian：德尔图良（155—230），迦太基神学家，首先提出 trinity（三位一体）一词，完整地陈述了三位一体的教义。
5. Cecco d'Ascoli：切科·达斯科利（1257—1327），意大利著名百科全书学家、物理学家、诗人，著有长诗《阿采尔巴》。
6. Savonarola：萨佛纳罗拉（1452—1498），意大利狂热的政治和宗教改革家，以清教主义反对文艺复兴，因批评教皇亚历山大六世被逐出教会并处死。
7. narcotic：麻醉的
8. classical and pagan art：古典艺术与异教艺术

these last. The *De continentia Virgiliana* of Fulgentius[1] (sixth century) is a curious monument to this fact. This work made Virgil compatible with the Middle Ages and opened his way to that great reputation which he was destined to attain, as the "gentle sage who knew all things." Even John of Salisbury[2] says of the Roman poet, that "*sub imagine fabularum totius philosophiae exprimit veritatem.*[3]" [3] The process of interpretation became fixed in the doctrine of the *four meanings*, literal, allegorical, moral and anagogic[4], which Dante afterwards transferred to vernacular poetry[5]. It would be easy to accumulate quotations from mediæval writers, repeating in all keys the theory that art inculcates[6] the truths of morality and of faith and constrains hearts to Christian piety, beginning with those well-known verses of Theodulf[7]: "*In quorum dictis* (that is to say, in the utterances of the poets) *quamquam sint frivola multa, Plurima sub falso tegmine vera latent*[8]," and so on, until we reach the doctrines and opinions of our own great men, Dante and Boccaccio. For Dante, poetry "*nihil aliud est quam fictio rhethorica in musicaque posita.*[9]" [4] The poet should have a "reasoning" in his verses "under a cloak of figure or of rhetorical colour"; and it would be a shameful thing for him, if, "when asked, he were not able to divest his

---

1. Fulgentius：傅箴修（5 世纪末—6 世纪初），拉丁文语法学家、作家，著有《神话学》《论世界的时代与人的时代》。
2. John of Salisbury：索尔兹伯里的约翰（约 1120—1180），自称小约翰，英国作家、哲学家、教育家、外交家。
3. *sub imagine fabularum totius philosophiae exprimit veritatem.*：〈拉〉通过寓言中的形象表达哲学真理。
4. the doctrine of the *four meanings*, literal, allegorical, moral and anagogic：四义说，即字面义、寓意义、道德义、神秘义
5. vernacular poetry：俗语诗
6. inculcates：教诲
7. Theodulf：狄奥多尔夫（750—821），法国奥尔良教区主教、诗人。
8. *In quorum dictis quamquam sint frivola multa, Plurima sub falso tegmine vera latent*：〈拉〉虽然诗人的语言苍白无力，但在虚饰的外表下却蕴含着真理。
9. *nihil aliud est quam fictio rhethorica in musicaque posita.*：〈拉〉诗无非是按照乐理组织起来的修辞虚构。

words of such a garment, in such a way as to show that they possessed a true meaning." [5] Readers sometimes stop at the external vesture alone, and this indeed suffices for those who, like the vulgar, do not succeed in penetrating the hidden meaning. Poetry will say to the vulgar, which does not understand "its argument," what a song of Dante's says at its conclusion, "At least behold how *beautiful* I am": if you are not able to obtain instruction from me, at least enjoy me as a pleasing thing. Many, indeed, "their beauty more than their goodness will delight," in poems, unless they are assisted by commentaries in the nature of the *Convivio*[1], "a light which will allow every shade of meaning to reach them." [6] Poetry was the "gay science," "*un fingimiento*" (as the Spanish poet the Marquis of Santillana[2] wrote) "*de cosas utiles, cubiertas ó veladas con muy fermosa cobertura, compuestas, distinguidas é scandidas, por cierto cuento, pessoé medida.*[3]" [7]

It would not then be correct to say that the Middle Ages simply identified art with theology and with philosophy. Indeed it sharply distinguished the one from the other, defining art and poetry, like Dante, with the words *fictio rhethorica*[4], "figure" and "rhetorical colour," "cloak," "beauty," or like Santillana with those of *fingimiento* or *fermosa cobertura*[5]. This pleasing falsity was justified from the practical point of view, very much in the same way as sexual union and love were justified and sanctified in matrimony. This did not exclude, indeed it implied, that the perfect state was certainly celibacy[6] — that

---

1. *Convivio*：《飨宴篇》，但丁未完成作品。
2. Marquis of Santillana：桑蒂利亚纳侯爵（1398—1458），西班牙政治家、诗人。
3. *un fingimiento de cosas utiles, cubiertas ó veladas con muy fermosa cobertura, compuestas, distinguidas é scandidas, por cierto cuento, pessoé medida.*：〈西〉（诗）是被漂亮的纱幔遮掩住的有益东西的虚构，它由节奏明快的发音写就，也就是说诗由韵律和格调组成。
4. *fictio rhethorica*：〈拉〉修辞虚构
5. *fermosa cobertura*：〈西〉美丽的遮掩物
6. celibacy：独身主义

is to say, pure science, free from admixture of art.

Hints of an Æsthetic in scholastic philosophy.

The only tendency that had no true and proper representatives was the sound scientific tendency. The *Poetics* of Aristotle itself was hardly known or rather it was ill-known, from the Latin translation that a German of the name of Hermann made, not earlier than 1256, of the paraphrase or commentary of Averroes[1]. Perhaps the best of the mediæval investigations into language is that supplied by Dante's *De vulgari eloquentia*[2], where the word is, however, still looked upon as a sign ("*rationale signum et sensuale ... natura sensuale quidem, in quantum sonus est, rationale vero in quantum aliquid significare videtur ad piacitum*[3]"). [8] The study of the expressive, æsthetic, linguistic faculty would, however, have found an appropriate occasion and a point of departure in the secular debate between nominalism and realism, which could not avoid touching to some extent the relations between the word and the flesh, thought and language. Duns Scotus[4] wrote a treatise *De modis significandi seu* (the addition is due perhaps to the editors) *grammatica speculativa*[5]. [9] Abelard[6] had defined sensation as *confusa*

---

1. Averroes：阿威罗伊（1126—1198），阿拉伯哲学家、思想家，亚里士多德思想拥趸，著有《亚里士多德注疏集》《论理智》《论三段论》《伊斯兰百科全书》等。
2. *De vulgari eloquentia*：〈拉〉《论俗语》，但丁著。
3. *rationale signum et sensuale ... natura sensuale quidem, in quantum sonus est, rationale vero in quantum aliquid significare videtur ad piacitum*：〈拉〉感性符号与理性符号，所谓感性，因为它代表着一种声音。而所谓理性，因为它代表着对富有理性意义的真理的偏好。
4. Duns Scotus：邓斯·司各脱（1266—1308），苏格兰神学家、哲学家与逻辑学家，著有《论第一原理》。
5. *De modis significandi seu grammatica speculativa*：〈拉〉《论表意方法或思辨语法》，司各脱著。
6. Abelard：阿伯拉尔（1079—1142），法国神学家、经院主义哲学家，著有《论理智》《伦理学或自我认知》等。

conceptio[1], and *imaginatio*[2] as a faculty that preserved sensations; the intellect renders discursive what is intuitive in the preceding stage, and we have finally the perfection of knowledge in the intuitive knowledge of the discursive. We find the same importance attached to intuitive knowledge, perception, of the individual or *species specialissima*[3], in Duns Scotus, together with the progressive denominations of the different sorts of knowledge as *confusæ*, *indistinctæ* and *distinctæ*[4]. We shall see this terminology reappear, big with consequences, at the very commencement of modern Æsthetic. [10]

Renaissance. Philography and philosophical and empirical inquiries concerning the Beautiful.

It may be said that the literary and artistic doctrines and opinions of the Middle Ages have, with few exceptions, a value rather for the history of culture than for the general history of science. The like observation holds good of the Renaissance, for here, too, the circle of the ideas of antiquity was not overstepped. Culture increases; original sources are studied; the ancient writers are translated and commented upon; many treatises are written and henceforth printed upon poetry and the arts, grammars, rhetorics, dialogues, and dissertations upon the beautiful: the proportions have increased, the world has become bigger; but truly original ideas do not yet show themselves in the domain of æsthetic science. The mystical tradition is refreshed and strengthened by the renewed cult of Plato: Marsilio Ficino[5], Pico

1. *confusa conceptio*：〈拉〉混乱的认知
2. *imaginatio*：〈拉〉想象力
3. *species specialissima*：〈拉〉特定种类
4. knowledge as *confusæ*, *indistinctæ* and *distinctæ*：混乱的知识，不明确的知识以及明确的知识
5. Marsilio Ficino：马尔西利奥·费奇诺（1433—1499），意大利思想家，新柏拉图主义的捍卫者。

della Mirandola[1], Cattani[2], Leon Battista Alberti[3], in the fifteenth century, and Pietro Bembo[4], Mario Equicola[5], Castiglione[6], Nobili[7], Betussi[8], and very many others in the following century, wrote upon the Beautiful and upon Love. Among the most noteworthy productions of the sort, a crossing of the mediæval and classical currents, is the book of the *Dialogues of Love*[9] (1535), composed in Italian by the Spanish Jew Leo, and translated into all the cultured languages of the time. [11] The three parts into which it is divided treat of the nature and essence, of the universality, and of the origin of love; and it is demonstrated that every beautiful thing is good, but not every good thing is beautiful; that beauty is a grace which dilates the soul and moves it to love, and that knowledge of lesser beauties leads to that of higher spiritual beauties. The author gave the name of "Philography" to these and similar affirmations and effusions of which the book is composed. Equicola's [12] work is also interesting, because it contains historical accounts of those who wrote upon the subject before he did so himself. The same intuition was versified and sighed forth by the

1. Pico della Mirandola：皮克·德拉·米兰多拉（1463—1494），意大利文艺复兴时期人文主义哲学家。
2. Cattani：卡塔尼（1466—1522），意大利文艺复兴时期新柏拉图主义哲学家，著有《论美》。
3. Leon Battista Alberti：阿尔伯蒂（1404—1472），意大利文艺复兴时期画家、诗人、语言学家、哲学家、音乐家、建筑家以及密码学家，著有《论画》《论雕塑》等。
4. Pietro Bembo：皮埃特罗·本博（1470—1547），意大利红衣主教、诗人、文学理论家，对意大利语的发展有举足轻重的影响。
5. Mario Equicola：马里奥·艾奎克拉（1470—1525），意大利文艺复兴时期人文主义者、新拉丁文作家、藏书家，著有《论爱的本质》。
6. Castiglione：卡斯蒂廖内（1478—1529），意大利文艺复兴时期外交家、作家，著有《朝臣之书》。
7. Nobili：诺比利（1518—1553），意大利文艺复兴时期作家。
8. Betussi：贝图西，意大利 16 世纪作家。
9. *Dialogues of Love*：《爱的对话》，西班牙人莱奥内·埃布雷奥的意大利语作品，1535 年首次在罗马出版，是文艺复兴时期的重要著作。

Petrarchists[1] in their sonnets and ballads, while others, rebellious and mocking, derided it in comedies, verses in *terza rima*[2] and parodies[3] of all sorts. Some mathematicians, reincarnations of Pythagoras[4], set to work to determine beauty by exact relations: for instance Leonardo's friend, Luca Paciolo[5], in the *De divina proportione* (1509), in which he laid down the pretended æsthetic law of the golden section. [13] And side by side with these new Pythagoreans were those who revived the canon of Polycletus as to the beauty of the human body, especially of the female body, such as Firenzuola, Franco, Luigini, and Dolce. Michæl-Angelo fixed an empirical canon for painting in general, when he stated that the means of giving movement and grace to figures [14] consisted in the observance of a certain arithmetical relation. Others, such as Fulvio Pellegrino Morato[6], investigated the symbolism or meaning of colours. The Platonists generally placed beauty in the soul, the Aristotelians rather in the physical qualities. The Averroist, Agostino Nifo[7], amid much chatter and many inconclusive remarks, demonstrated the existence of the beautiful in nature by describing the supremely beautiful body of Joan of Aragon, Princess of Tagliacozzo, to whom the book is dedicated. [15] Torquato Tasso, in the "Minturno[8]," [16]

---

1. Petrarchists：彼特拉克主义者。弗兰齐斯科 · 彼特拉克 (1304—1374)，意大利学者，诗人，被誉为“文艺复兴之父”。
2. *terza rima*：〈拉〉十四行诗
3. parodies：戏仿
4. Pythagoras：毕达哥拉斯（公元前 582—前 496），古希腊哲学家、数学家，勾股定理发现者，被誉为“数学之父”。
5. Luca Paciolo：卢卡 · 帕奇奥利，达 · 芬奇好友，著有《论黄金比例》(De Divina Proportione)。
6. Fulvio Pellegrino Morato：即莫拉图（Polycletus the elder），16 世纪意大利哲学家，著有《论颜色的意义》。
7. Agostino Nifo：阿戈斯蒂诺 · 尼佛（1473—1538），意大利哲学家、评论家，著有《理智之书》《论不朽生命》等。
8. Minturno：明图尔诺，意大利拉希奥南部城市，塔索以此城市为书名写过书。

imitated the uncertainties of the *Hippias*[1] of Plato, not without making a free use of the speculations of Plotinus. A chapter of the *Poetica* of Campanella[2] possesses greater importance, where he describes the good as *signum boni*[3] and the ugly as *signum mali*[4], understanding by good the three prime forces of Power, Wisdom and Love. Although Campanella was still tied to the Platonic idea of the beautiful, the conception of a sign or symbol, here introduced by him, represents progress. By this means he succeeded in perceiving that material things or external facts are neither beautiful nor ugly in themselves. "Mandricard called the wounds in the bodies of his friends the Moors beautiful, for they were large and gave evidence of the great strength of Roland who dealt them; Saint Augustine called the gashes and the dislocations in the body of Saint Vincent[5] beautiful, because they were evidence of his endurance, but they were on the other hand ugly in so far as they were signs of the cruelty of the tyrant Dacianus and of his executioners. It is beautiful to die fighting, said Virgil, for it is the sign of a strong soul. The pet dog of his mistress will seem beautiful to the lover, and doctors call even urine and fæces beautiful, when they indicate health. Everything is both beautiful and ugly" (*quapropter nihil est quod non sit pulcrum simul et turpe*[6]). [17] In such observations as these we have not a mere state of mystical exaltation, but to some extent a movement in the direction of analysis.

The pedagogic theory of art and the Poetics of Aristotle.

---

1. *Hippias*：《希琵阿斯前篇》，柏拉图著，记录了希琵阿斯与苏格拉底关于什么是美的讨论。希琵阿斯是公元前 5 世纪的著名智术师。
2. Campanella：坎帕内拉（1568—1639），意大利哲学家、神学家、星象学家、诗人，著有《诗学》《太阳之城》等。
3. *signum boni*：〈拉〉善的符号
4. *signum mali*：〈拉〉恶的符号
5. the gashes and the dislocations in the body of Saint Vincent：圣文森特身上的刀伤和脱臼处。圣文森特（？—304），基督徒殉道者。
6. *quapropter nihil est quod non sit pulcrum simul et turpe*：〈拉〉任何事物都兼具美和丑。

Nothing better serves to demonstrate that the Renaissance did not pass beyond the confines of ancient æsthetic thought than the fact that notwithstanding the renewed acquaintance with the thought of Aristotle, the pedagogic theory of art not only persisted and triumphed, but was transplanted bodily into the text of Aristotle, where its interpreters read it with a certainty that we have to make efforts to achieve. Certainly, a Robortelli[1] (1548) or a Castelvetro[2] (1570) stopped short at the simple, purely hedonistic solution, giving simple pleasure as the end of art: poetry, says Castelvetro, "was discovered solely for the purpose of delighting and of recreating ... the souls of the rude multitude and of the common people." [18] And here and there some were able to free themselves from both the pleasure theory and that of the didactic end; but the majority, such as Segni[3], Maggi[4], Vettori[5], [19] were for the *docere delectando*[6]. Scaliger[7](1561) declared that mimesis or imitation was "*finis medius ad illum ultimum qui est docendi cum delectatione*[8]," and believing himself to be altogether in agreement with Aristotle as to this, he continued, "*docet affectus poeta per actiones, ut bonos amplectamur atque imitemur ad agendum, malos aspernemur ad abstinendum.*[9]" [20]

---

1. Robortelli: 罗伯特里（1516—1567），文艺复兴时期人文主义者，绰号"语法狗"，著有《论历史学方法》《论修辞手法》。
2. Castelvetro：卡斯特尔维特罗（1505—1571），剧作家，新古典主义发展过程中的重要人物，著有《平民语言中的亚里士多德诗学》。
3. Segni：塞尼（1498—1561），1530年任威尼斯圣马可教堂的首席管风琴师，译著有《亚里士多德修辞学与诗学》《亚里士多德伦理学》。
4. Maggi：麦吉（1523—1572），意大利法学家、诗人、军事工程师、建筑师、城市规划师、语言学家、考古学家、数学家和自然主义者，著有《烈火吞噬世界与审判日》《论城防》。
5. Vettori：韦托里（1474—1539），意大利外交家、政治家与作家。
6. *docere delectando*：〈拉〉寓教于乐
7. Scaliger：斯卡利格（1484—1558），意大利生理学家。
8. *finis medius ad illum ultimum qui est docendi cum delectatione*：〈拉〉（模仿）只是手段，最终目的还是寓教于乐。
9. *docet affectus poeta per actiones, ut bonos amplectamur atque imitemur ad agendum, malos aspernemur ad abstinendum.*：〈拉〉诗人用行动激发情感，通过模仿并选择善行教导人们鄙弃恶行。

Piccolomini[1] (1575) observed that "It must not be thought that so many excellent poets and artists, ancient and modern, would have devoted such care and diligence to this most noble study, had they not known and believed that in so doing they were aiding human life," and if "they had not thought that we were to be instructed, directed, and well established by it." [21] The "truth preserved in soft verses, which attracts and persuades the most reluctant" (Tasso), [22] with the comparison from Lucretius attached, is the conception that even Campanella repeats. Poetry is for him "*Rhetorica quaedam figurata, quasi magica, quae exempla ministrat ad suadendum bonum et dissuadendum malum delectabiliter iis qui simplici verum et bonum audire nolunt, aut non possunt aut nesciunt.*[2]" [23] Thus returned the comparison of poetry with oratory[3]; according to Segni they only differ because the first occupies a more lofty situation: "for since imitation representing itself in act by means of poetry, in mighty, chosen words, in metaphors, images, and indeed the whole of figured speech, which is to be found more in poetry than in the art of oratory, the metrical qualities that are also required in verse, the subjects of which it treats, which have something of the great and delightful, make it appear most beautiful and worthy of being held all the greater marvel." [24] "Three most noble arts" (wrote Tassoni in 1620, and he repeated common opinion), "History, Poetics, and Oratory, come under the heading of Politics and depend upon it; the first of these has reference to the instruction of princes and gentlemen, the second of the people, the third of those who give counsel in public trials or defend private ones that come up for judgment." [25]

1. Piccolomini：皮克罗米尼（1508—1579），意大利天文学家，哲学家。
2. *Rhetorica quaedam figurata, quasi magica, quae exempla ministrat ad suadendum bonum et dissuadendum malum delectabiliter iis qui simplici verum et bonum audire nolunt, aut non possunt aut nesciunt.*：〈拉〉（诗）是形象化的，魔幻的修辞学。它用愉悦的方式提供典范，提醒人们去恶向善。因为普通人无意，无力更不知听从真与善的劝导。
3. comparison of poetry with oratory：诗与雄辩术的对比

According to these views, the tragical catharsis was regarded as designed in general to demonstrate the instability of fortune, or to terrify by example, or to proclaim the triumph of justice, or to render the spectators insensible to the strokes of fortune, owing to their familiarity with suffering. The pedagogic theory, thus renewed and sustained by the authority of the ancients, was popularized in France, Spain, England and Germany, together with all the Italian poetic doctrines of the Renaissance. The French writers of the period of Louis XIV are altogether penetrated with it. "*Cette science agréable qui mêle la gravité des préceptes avec la douceur du langage*[1]," is what La Ménardière[2] calls poetry (1640), in the same way as Le Bossu[3] (1675), for whom "*le premier but du poète est d'instruire*[4]," [26] as Homer taught, when he wrote two interesting didactic manuals relating to military and political events: the *Iliad* and the *Odyssey*[5].

The "Poetics of the Renaissance."

This pedagogic theory has therefore been reasonably described by all the modern critics in concert, as if by antonomasia[6], as the *Poetics of the Renaissance.* It must, however, always be understood that it did not appear for the first time in the fifteenth or sixteenth century, but that it was prevalent and generally accepted at that time. It may even be remarked, as has already been acutely done, [27] that the Renaissance naturally did not distinguish the didactic kind of poetry[7]

---

1. *Cette science agréable qui mêle la gravité des préceptes avec la douceur du langage*：〈法〉这门快乐的学问将教育的意义和语言的甜美结合在一起。
2. La Ménardière：米纳尔迪尔（1610—1663），法国生理学家、文学家、戏剧家。
3. Le Bossu：包素（1631—1680），法国评论家，著有《论史诗》。
4. *le premier but du poète est d'instruire*：〈法〉诗的首要任务是教育。
5. the *Iliad* and the *Odyssey*：史诗《伊利亚特》与《奥德赛》
6. antonomasia：换称；代类名
7. didactic kind of poetry：教育类诗歌

from the other kinds, since for it every kind of poetry was didactic. But the Renaissance was not a real Renaissance, save when and where it continued the interrupted spiritual work of antiquity, and in this sense it would perhaps be more just to describe as its Poetics, or rather, as the important element in its Poetics, not the repetition of the pedagogic theory of antiquity and of the Middle Ages, but the resumption, which also took place, of the discussions upon the possible, the probable (*verisimile*, εικός) of Aristotle, on the reasons of Plato's condemnation and on the procedure of the artist who creates by imagining.

Dispute concerning the universal and the probable in art.

It is in such discussions that is to be found the true contribution of that epoch, not to learning, but to the formation of the science of Æsthetic. The ground was prepared and enriched through the work of the interpreters and commentators of Aristotle and of the new writers on Poetics, especially the Italians, and it was also enriched with some seed that was destined to sprout and to become a vigorous plant in the future. The study of Plato also contributed not a little to call attention to the function of the idea, or of the universal, in poetry. What meaning was to be attached to the statement that poetry should aim at the universal and history at the particular? What was the meaning of the proposition that poetry should proceed according to *probability*? What could that *certain idea* consist of, which Raphæl said that he followed in his painting?

G. Fracastoro.

Girolamo Fracastoro[1] was among the first to ask himself this

1. Girolamo Fracastoro：吉罗拉莫·弗莱卡斯特罗（约 1476—1553），意大利生理学家、数学家、地理学家、天文学家、诗人，著有《论酒温》《论诗艺》等。

question seriously, in the dialogue *Naugerius, sive De poetica*[1] (1555). He disdainfully rejected the thesis that the end of poetry is pleasure: far be from us, he exclaimed, so bad an opinion of the poets, who the ancients said were the inventors of all the good arts. Nor did the end of instruction seem to him to be acceptable, which is the task, not of poetry, but of other faculties, such as geography, history, agronomy[2], philosophy. The poet's task is to represent or to imitate, and he differs from the historian, not in the matter, but in the manner of representation. The others imitate the particular, the poet the universal: the others are like the painters of portraits, the poet produces things as he contemplates the universal and most beautiful idea of them: the others say only what they need to say for their purposes, the poet that he may say everything beautifully and fully.

But the beauty of a poem must always be understood as relative to the class of subject of which it treats; it is the most beautiful in this class, not the supremely beautiful: one must be careful to guard against the equivocal or double meaning of this word "beauty" (*æquivocatio illius verbi*[3]). A poet never utters what is false or expresses what does not exist, for his words inevitably harmonize in appearance or signification either with the opinions of men or with the universal. Nor can we accept the Platonic axiom that the poet has no knowledge of the things of which he treats; he does know them, but in his own poet's manner. [28]

L. Castelvetro.

While Fracastoro strives to elaborate the important passage in

1. *Naugerius, sive De poetica*：〈意〉《瑙格吕斯论诗艺》
2. agronomy：农学
3. *æquivocatio illius verbi*：〈拉〉一个词的双重含义

Aristotle touching the universal of poetry, and though somewhat vague in his treatment, keeps fairly close to the mark; Castelvetro, on the contrary, judges the Aristotelian fragment with the freedom and superior knowledge of the true critic. He recognizes that the *Poetics* is merely a notebook recording certain principles and methods of compiling the art, not the art fully compiled. He remarks, moreover, not without logical acumen[1], that Aristotle having adopted the criterion of probability or of that "which presents an appearance of historic truth," should have applied his theory in the first case to history, not to poetry; for history being a "narrative according to truth of memorable human actions," and poetry a narrative according to probability of events which might possibly occur, the second cannot receive "all its radiance" from the first. Nor does it escape him that Aristotle describes two different things by the one word "imitation": (*a*) "following the example of another," which is "acting in exactly the same way as another without knowing the reason of such action": and (*b*) the imitation "demanded by poetry," which "does things in a manner totally different from that in which they have been done hitherto and proposes a new example for imitation." Nevertheless Castelvetro cannot extricate himself from the confusion between the imaginary and the historical; for he himself says "the realm of the former is generally that of certainty," but "the field of certainty is often crossed with bars of uncertainty just as the field of uncertainty is often crossed with bars of certainty." Also what can be said of this curious interpretation of the Aristotelian theory of pleasure experienced in the imitation of ugly models, that such pleasure is based on the fact that since an imitation is always imperfect, it is incapable of exciting the disgust and fear which would arise from the contemplation of real ugliness? And what of his remark that the characteristics of painting

1. logical acumen：逻辑敏感度

and poetry are so diverse as to be in opposition one to the other; imitation of objects giving rise to great pleasure in the former art and as great displeasure in the latter? And so on in numberless cases of bold but scarcely felicitous subtleties.[1] [29]

Piccolomini and Pinciano.

In opposition to Robortelli, who asserted the identity of the probable and the false, Piccolomini held that the probable (*verisimile*) is inherently neither false nor true, only by accident becoming one or other. [30] Of the same mind is the Spaniard Alonso Lopez Pinciano[2] (1596), who says the scope of poetry "*no es la mentira, que seria coincider con la sophística, ni la historia que seria tomar la material al histórico; y no siendo historia porque toca fabúlas ni mentira porque toca historia, tiene por objeto el verisimil, que todo lo abraza. De aqui resulta que es un arte superior á la metaphysica, porqué comprende mucho mas, y se extiende a lo que es y á lo que no es.*[3]" [31] What may lie behind this notion of probability is still indefinite and impenetrable.

Fr. Patrizzi (Patricius).

Moved by a wish to place poetry on a foundation other than the

---

1. scarcely felicitous subtleties：不恰当的细微之处
2. Alonso Lopez Pinciano：阿隆索·洛佩兹·平恰诺（1547—1627），西班牙人文主义者、评论家，著有《诗学中的古代哲学》。
3. *no es la mentira, que seria coincider con la sophística, ni la historia que seria tomar la material al histórico; y no siendo historia porque toca fabúlas ni mentira porque toca historia, tiene por objeto el verisimil, que todo lo abraza. De aqui resulta que es un arte superior á la metaphysica, porqué comprende mucho mas, y se extiende a lo que es y á lo que no es.*：〈西〉诗不是夹杂了诡辩的谎言。也不是基于历史学家的素材写成的历史。它不是历史，因为它关乎寓言，它也不是谎言，因为它关乎历史。诗追求逼真，包罗万象。因此，诗是高于形而上学的艺术形式，它包含了更多的东西，扩展到它所是和所不是的东西。

probable, Francesco Patrizzi[1], the anti-Aristotelian, composed his *Poetica* between 1555 and 1586 in refutation of all Aristotle's main doctrines. Patrizzi notes that the word "imitation" is given many meanings by the Greek philosopher, who uses it now to denote a single word, now to describe a tragedy; at times it stands for a figure of speech, at others for a fiction: whence he draws the logical conclusion (from which, however, he shrinks alarmed) "that all philosophic and other kinds of writing and speaking are poetry, since they are made of words which themselves are imitations." He observes further that, according to Aristotle, it is impossible to distinguish between poetry and history (since both are imitations), or to prove that verse is not essential to poetry, or that history, science and art are unsuitable material for it; since Aristotle in several passages says that poetry may comprise "fable, actual occurrences, belief of others, duty, the best, necessity, the possible, the probable, the credible, the incredible, the suitable" as well as "all things worldly." After these objections, some sound, others sophistical, Patrizzi comes to the conclusion that "there is no truth in the dogma that poetry is wholly imitation; and even if it be imitation at all, it belongs not to poets alone, nor is it mere imitation of any kind, but something else not mentioned by Aristotle nor pointed out by any one else, nor yet borne into the mind of man. The discovery may possibly be made in course of time, or some one may hit upon the truth and bring it to light"; but up to the present "such discovery has not been made." [32]

Yet these confessions of ignorance, these endeavours, though vain, to escape from the Aristotelian circle of ideas, and the great literary controversies of the sixteenth century concerning the concept of poetic truth and the probable had their use in that they stimulated interest by

1. Francesco Patrizzi：弗朗切斯科·帕特里奇（1529—1597），意大利威尼斯科学家，哲学家，著有《论物理空间与数学》《通用哲学新论》《论诗学》等。

directing attention to a mystery still unsolved. Thought had once more begun to move upon the æsthetic problem, and this time it was not destined to be broken off or to lose itself.

[1] *Confess*, iv. x. ch. 13; *De Trinitate*, vi. ch. 10; *Epist.* 3, 18; *De civitate Dei*, xxii. ch. 19 (in *Opera*, ed. dei Maurini, Paris, 1679–1690, vols. i. ii. vii. viii.).

[2] *Summa theol.* I. 1. xxxix. 8; I. 11. xxvii. I (ed. Migne, i. cols. 794–795; ii. col. 219).

[3] Comparetti, *Virg. nel medio evo*, vol. i. *passim.*

[4] *De vulg. eloq.* (ed. Rajna), bk. ii. ch. 4.

[5] *Vita nuova*, ch. 25.

[6] *Convivio*, i. 1.

[7] *Prohemio al Condestable de Portugal*, 1445–1449 (in *Obras*, ed. Amador de los Rios, 1852), § 3.

[8] *De vulg. eloq.* bk. i. ch. 3.

[9] Lately reprinted under the editorship of padre M. Fernandez Garcia, Ad claras Aquas (Quarracchi), 1902.

[10] Windelband, *Gesch. d. Phil.* ii. pp. 251–270; De Wulf, *Philos, médiév.,* Louvain, 1900, pp. 317–320.

[11] *Dialogi di amore, composti per Leone, medico ...* , Rome, 1535.

[12] *Libro di natura e d' amore,* Venice, 1525 (Ven. 1563).

[13] *De divina proportione*, Venice, 1509.

[14] G. P. Lomazzo, *Trattato dell' arte della pittura, scultura ed architettura*, Milan, 1585, i. I, pp. 22–23.

[15] Aug. Niphi, *De pulcro el amore*, Rome, 1529.

[16] *Il Minturno o vero de la belleza* (in *Dialoghi*, ed. Guasti, vol. iii.).

[17] *Ration. philos.* part iv.; *Poeticor.* (Paris, 1638), art. vii.

[18] Fr. Robortelli, *In librum Arts, de arte poet, explicationes,* Florence, 1548; Lud. Castelvetro, *Poetica d'Aristotele vulgarizzata ed esposta*, 1570 (Basle, 1576), part i. particella iv. pp. 29–30.

[19] Bern. Segni, *Rettor. e poet. trad.* Florence, 1549; Vinc. Madii, *In Arist .... explanationes*, 1550; Petri Victorii, *Commentarii*, etc., Florence, 1560.

[20] *Poetica*, 1561 (ed. 3, 1586), i. I; vii. 3.

[21] *Annotationi net libro della Poetica*, Venice, 1575, preface.

[22] *Gerus. lib.* i. 3.

[23] *Poetic*, ch. I, art. 1.

[24] *Poetica trad.* preface.

[25] *Pensieri diversi*, bk. x. ch. 18.

[26] La Ménardière, *Poétique*, Paris, 1640; Le Bossu, *Traité du poème épique*, Paris, 1675.

[27] Borinski, *Poet. d. Renaiss.* p. 26.

[28] Hyeron. Frascatorii *Opera*, Venetian edition, Giunti, 1574, pp. 112–120.

[29] *Poet., ed. cit.* i. 1; ii. 1; iii. 7; v. I (pp. 64, 66, 71–72, 208, 580).

[30] *Annotationi*, preface.

[31] *Philosophia antiqua poetica*, Madrid, 1596 (reprinted Valladolid 1894).

[32] Francesco Patrici, *Della poetica, la Deca disputata*, "in which by history, by reason, by authority of the greatest worthies of antiquity, is shown the falsity of the most received opinions concerning Poetry down to our own day." Ferrara, 1586.

III FERMENTS OF THOUGHT IN THE SEVENTEENTH CENTURY

New words and new observations in the seventeenth century.

Interest in æsthetic investigation increased rapidly in the early years of the following century, owing either to the popularity acquired by certain new words or to the novel meanings given to words already familiar, which emphasized new aspects of artistic production and criticism, complicating the problem and rendering it thereby more puzzling and attractive. For example: wit, taste, imagination or fancy, feeling, and several others, which must be examined rather closely.

Wit

Wit (*ingegno*) differed somewhat from intellect. Free use of the word arose, if we mistake not, from its convenience in Rhetoric as conceived by antiquity; that is to say, a suave and facile mode of knowledge[1], as opposed to the severity of Dialectic; an "Antistrophe to Dialectic[2]," which substituted for reasons of actual fact those of probability or fancy; enthymemes[3] for syllogisms, examples for inductions; so much so that Zeno the Stoic[4]

1. suave and facile mode of knowledge: 简单易懂的认知
2. Antistrophe to Dialectic：辩证法的对应物
3. enthymemes：三段论省略式
4. Zeno the Stoic：斯多葛学派的芝诺（约公元前 490—前 430），古希腊哲学家，斯多葛学派的创始人，该学派认为世界理性决定事物的发展变化。芝诺被亚里士多德称为"辩证法之父"。

figured Dialectic with her fist clenched and Rhetoric with her hand open. The empty style of the decadent Italian authors in the seventeenth century found its complete justification in this theory of rhetoric; their prose and verse, Marinesque and Achillinesque, professed to exhibit not the true but the striking, subtly conceited, curious or nice. The word wit, *ingegno*, was now repeated much more frequently than in the preceding century; wit was hailed as presiding genius of Rhetoric; its "vivacities" were lauded to the skies[1]; "*belli ingegni*[2]" was a phrase seized upon by the French, who rendered it as "*esprit*" or "*beaux esprits*[3]." [1] One of the most noteworthy commentators on these matters (although opposed to the literary excesses of the times), Matteo Pellegrini of Bologna (1650), defines wit as "that part of the soul which in a certain way practises, aims, and seeks to find and create the beautiful and the efficacious"; [2] he considers the work of "wit" to be the "conceits" and "subtleties" noted by him in a previous pamphlet (1639). [3] Emmanuele Tesauro also descants at considerable length in his *Cannochiale Aristotelico*[4] (1654) upon wit and subtleties, not alone "verbal" and "lapidary" conceits, but also "symbolic" and "figurative" (statues, stories, devices, satires, hieroglyphs, mosaics, emblems, insignia, sceptres), and even "animated agents" (pantomimes[5], play-scenes, masques and dances): all things which may be grouped under "polite quibbling[6]" or rhetoric as distinct from "dialectic."

Amongst such treatises, product of their age, one written by the

1. its "vivacities" were lauded to the skies：其"活力"备受赞赏
2. *belli ingegni*：〈意〉美的巧智
3. *beaux esprits*：〈法〉美的巧智
4. Emmanuele Tesauro also descants at considerable length in his *Cannochiale Aristotelico*：伊曼纽尔·特索罗也在他的著作《亚里士多德的望远镜》中详细地阐述。特索罗（1592—1675），意大利修辞学家、戏剧家、诗人、历史学家。
5. pantomimes：哑剧
6. polite quibbling：文雅的遁词

Spaniard Baltasar Gracian[1] (1642) became celebrated throughout Europe. [4] Wit became in his hands the strictly inventive or artistic faculty, "genius"; *génie*, "genius" were now used as synonyms of wit, *ingegno* and *esprit*. In the following century Mario Pagano [5] [2] wrote: "Wit may be taken as equivalent to the *génie* of the French, a word now commonly used in Italy." To return to the seventeenth century, Bouhours[3], a Jesuit writer of dialogues on the *Manière de bien penser dans les ouvrages d'esprit* (1687), says that "'heart' and 'wit' are greatly in fashion just now, nothing else is spoken of in polite conversation, and all discourse is at last brought round to *l'esprit et le cœur*[4]." [6]

Taste.

The word *taste* or *good taste* was equally widespread and fashionable, signifying the faculty of judgement brought to bear on the beautiful, distinct to some extent from intellectual power, and sometimes divided into active and passive, so that it was usual to speak of one kind of taste as "productive" or "fertile" (thus coinciding with "wit"), and of another as "sterile."

Various meanings of the word taste.

From the rough notes which we possess as to the history of the concept

---

1. Baltasar Gracian：巴尔塔萨尔·格拉西安（1601—1658），西班牙耶稣会士、巴洛克散文作家与哲学家，著有《才智与创造性艺术》《英雄》《完美绅士》等。
2. Mario Pagano：马里奥·帕格诺（1748—1799），意大利法学家、作家，那不勒斯法学派创始人，温和派改革家，启蒙主义思想家，被尊为意大利统一运动先驱。
3. Bouhours：鲍赫斯（1628—1702），法国耶稣会牧师、散文家、语法学家以及新古典主义批评家，著有《批评的艺术》以及文中提到的《巧智工作中的良好思维方式》（*Manière de bien penser dans les ouvrages d'esprit*）。
4. *l'esprit et le cœur*：〈法〉巧智与心灵

of taste, several meanings of the word, not all of equal importance as indications of the development of ideas, detach themselves in a somewhat confused manner. "Taste," meaning "pleasure" or "delight," was an old-established word in Italy and Spain, as is shown in such phrases as "to have a taste for, to be to one's taste"; when Lope di Vega[1] and other Spaniards speak continually of the drama of their country as seeking to please the popular taste ("*deleita el gusto*"; "*para darle gusto*" [2]) they mean only the "pleasure" of the populace. In Italy there was a very ancient use of the word in the metaphorical sense of "judgement," either literary, scientific, or artistic; numberless examples of this use occur in writers of the sixteenth century (Ariosto, Varchi, Michæl Angelo, Tasso). To take but one of these: the lines in *Orlando Furioso*[3] where it is said of the Emperor Augustus[4], "*L'aver avuto in poesia buon gusto La proscrizione iniqua gli perdona*[5]" "For having had good taste in poetry he shall be forgiven his iniquitous proscriptions[6]"; or the remark of Ludovico Dolce[7] that some person "had such exquisite taste, he sang no verses save those of Catullus[8] and Calvus." [7] The word "taste," in the sense of a special faculty or attitude of mind, appears to have been used for the first time in Spain in the middle of the seventeenth century by Gracian, [8] the moralist and political writer already quoted. It is

1. Lope di Vega：洛佩·德·维加（1562—1635），文艺复兴时期西班牙最重要的诗人、剧作家，著有《羊泉村》《罗马奴隶》等。
2. *deleita el gusto; para darle gusto*：〈西〉欣赏或取悦
3. *Orlando Furioso*：《疯狂的奥兰多》，阿里奥斯托叙事诗。
4. Emperor Augustus：奥古斯都大帝（公元前 63—14），罗马帝国第一位元首，统治罗马长达 40 年，为罗马带来了两个世纪的和平与繁荣。
5. *L'aver avuto in poesia buon gusto La proscrizione iniqua gli perdona*：〈意〉鉴于他对诗的良好鉴赏力，就宽宥他吧。
6. iniquitous proscriptions：罪恶的流放
7. Ludovico Dolce：卢多维克·多尔塞（约 1508—1568），意大利人文主义者、多栖作家、绘画理论家、翻译家与编辑，著有《阿瑞替诺绘画谈话录》等。
8. Catullus：卡图卢斯 (约公元前 87—约前 54)，古罗马诗人，著有《诗集》等。

evidently to him that the Italian author Trevisano[1] alludes in a preface to a book by Muratori[2] (1708) when he speaks of "Spaniards, above all others cunning in metaphor," who express themselves in "that eloquent and laconic phrase, good taste"; touching further on taste and genius he quotes, "that ingenious Spaniard," Gracian, [9] who gave the word the sense of "practical wit," enabling one to perceive the "true signification" of things; his "man of good taste" becomes in our language "a man of tact" in the affairs of life. [10]

The transference of the word to the domain of æsthetic seems to have taken place in France during the last quarter of the century. "*Il y a dans l'art un point de perfection, comme de bonté ou de maturité dans la nature: celui qui le sent et qui l'aime a le goût parfait; celui qui ne le sent pas, et qui aime au deçà ou au delà, a le gout défectueux. Il y a donc un bon et un mauvais goût, et l'on dispute des goûts avec fondement,*[3]" writes La Bruyère[4] [11] (1688). As attributes or variants of taste it was usual to mention *delicacy* and *variety* or *variability*. Bearing its fresh critical — literary content, but not freed from the encumbrance of its earlier practical and moral significance, the word spread from France into other European countries. Thomasius[5] introduced it into Germany in 1687; [12] and in England it becomes "good taste." In Italy it appears as early as 1696 as title of a large book written by Camillo

---

1. Trevisano：特雷维萨诺（1669—1753），意大利巴洛克晚期画家。
2. Muratori：穆拉托里（1672—1750），意大利历史学家、学者，著有《宗教行业的克制能力》《外显道德哲学》等。
3. *Il y a dans l'art un point de perfection comme de bonté ou de maturité dans la nature: celui qui le sent et qui l'aime a le goût parfait; celui qui ne le sent pas, et qui aime au deçà ou au delà, a le gout défectueux. Il y a donc un bon et un mauvais goût, et l'on dispute des goûts avec fondement*：〈法〉艺术中存在着完美点，如同自然界中存在着善良和成熟：若能感受并爱上这种完美，则说明品味完美；反之，若感觉不到，或爱得太深或爱得不够，都说明品味有缺陷。品味自有高下，而我们对此的讨论有根有据。
4. La Bruyère：拉布吕埃（1645—1696），法国哲学家、道德家。
5. Thomasius：多马修（1655—1728），德国法学家、哲学家，著有《教会、国家与政治学论集》。

Ettori[1], the Jesuit, *Il buon gusto ne'componimenti rettorici*[2]. [13] The preface notes: "The expression 'good taste,' proper to those who rightly distinguish good from bad flavour in foods, is now in general use and claimed by every one as a title in connexion with literature and the humanities"; it reappears in 1708 at the beginning of Muratori's [14] book already quoted: Trevisano treats of it philosophically: Salvini[3] discusses it in his note upon the *Perfetta Poesia* of Muratori[4] above mentioned, where the subject of good taste occupies several pages, [15] and finally it gives its name to the Academy of Good Taste founded at Palermo[5] in 1718 [16] . Scholars of the day who took up the discussion of the theme, recollecting some passages scattered throughout the ancient classics, placed the new concept in relation with the "*tacitus quidam sensus sine ulla ratione et arte*[6]" of Cicero; and with the "*indicium*[7]" which "*nec magis arte traditur quam gustus aut odor*[8]" of Quintilian[9]. [17] More particularly Montfaucon de Villars (1671) [18] wrote a book on "Delicacy"; Ettori strove to find some definition more satisfactory than those current at the time (*e.g.* "it is the finest invention of wit, the flower of wit and extract of beauty's self," and similar conceits); [19] Orsi made it the subject of his *Considerazioni* written in reply to Bouhours' book.

---

1. Camillo Ettori：卡米洛・埃托里，生卒年月不详，意大利美学家，耶稣会会员。
2. *Il buon gusto ne'componimenti rettorici*：〈意〉《修辞作品中良好鉴赏力》，卡米洛・埃托里著。
3. Salvini：萨尔维尼（1653—1729），意大利自然主义者与古典主义者。
4. *Perfetta Poesia* of Muratori：穆拉托里的《论完美的诗》。
5. Palermo：巴勒莫，意大利西西里首府。
6. *tacitus quidam sensus sine ulla ratione et arte*：〈拉〉不与理性和艺术相伴的隐秘敏感
7. *indicium*：〈拉〉解释
8. *nec magis arte traditur quam gustus aut odor*：〈拉〉艺术鉴赏力和味觉、嗅觉一样，都是敏感的。
9. Quintilian：昆体良（约35—约100），古罗马著名教育家、演说家，著有《雄辩术原理》（即《论演说家的教育》）。

Fancy or Imagination.

In Italy in the seventeenth century we find imagination or fancy placed on a pinnacle. What do you mean by talking of probability and historical truth (asks Cardinal Sforza Pallavicino[1] in 1644), of false or true in connexion with poetry; which deals not with fiction, fact or historical probability but with primary apprehensions which assert neither truth nor falsehood? Following this line of argument, imagination takes the place of that probable, neither true nor false, advocated by some commentators of Aristotle; a theory strongly criticized by Pallavicino, here agreeing with Piccolomini[2], whom however he does not name, and in opposition to Castelvetro whom he explicitly mentions. He who goes to the play (continues Pallavicino) knows quite well that the scenes acted on the stage are not real; although he has no belief in them yet they please him greatly. For "if poetry desired to be mistaken for truth, the end she had in view would be a he, by the laws of nature and of God doomed inevitably to perish: for a lie is nothing but an untruth uttered in the hope that it may be mistaken for truth. How then should an art so tainted be allowed to flourish in the best-regulated republics? How should it be commended and used by the very writers of Holy Scripture?" *Ut pictura poësis*[3]: poetry is like painting, which is a "diligent imitation[4]" aiming at a close copy of the features, colours, acts, nay, even the hidden motives, of the objects it represents: and it "does not pretend that fiction is truth." The sole aim of poetic tales is "to adorn our understanding with imagery, that is to say, with sumptuous, novel, marvellous and

1. Pallavicino：帕拉维西诺（1672—1742），意大利诗人、歌剧编剧。
2. Piccolomini：皮克罗米尼（1405—1464），意大利学者、历史学家、诗人。1458 年当选教皇（1458—1464 年在位），即庇护二世。
3. *Ut pictura poësis*：〈拉〉诗如画
4. diligent imitation：精心描摹

splendid appearances. And this is known to diffuse so useful an influence on mankind that humanity insists on rewarding poets with praise more glorious than is bestowed on any other men; their books are protected from the ravages of time with greater solicitude than is shown to scientific treatises or productions of any other art; in the end the names of poets are crowned with adoring veneration. See how the world thirsts for beautiful first apprehensions, although these are neither laden with science nor are they vehicles of truth." [20]

Sixty years later these ideas, although expressed by a Cardinal, seemed all too daring to Muratori, who could not bring himself to allow poets so much latitude, or to enfranchize[1] them from their obligations to the probable. Nevertheless Muratori allows a large space to imagination, "an inferior apprehensive faculty" which, without caring whether things be false or true, confines itself to apprehending them, and "represents" the truth merely, leaving the task of "cognition" to the "superior apprehensive faculty" or intellect. [21] Even the stony heart of Gravina[2] yields to the charm of imagination: he admits it occupies a considerable place in the realm of poetry and suffers his own arid prose to describe it as "a sorceress, but beneficent," "a delirium[3] which cures madness." [22]

Earlier than either of these, Ettori commended it to the good rhetorician, "who in order that he may awaken images" must "familiarize himself with whatever is subject to bodily feeling" and "encounter the genius of imagination, which is a sensuous faculty," to these ends using "species rather than genera (since the latter, being more universal than the former, are less sensible), individuals rather than species, effects than causes, the number of the greater rather than

1. enfranchize：解放
2. Gravina：格拉维纳（1664—1718），意大利作家、法学家，著有《民法起源》《论罗马帝国》。
3. delirium：极度兴奋

the number of the less." [23]

As far back as 1578 the Spaniard Huarte[1] had maintained that eloquence is the product of imagination rather than of intellect or reason. [24] In England Bacon[2] (1605) ascribed science to intellect, history to memory and poetry to imagination or fancy: [25] Hobbes inquired into the procedure of poetry: [26] Addison[3](1712) devoted several numbers of his *Spectator*[4] to analysis of the "pleasures of imagination." [27] Somewhat later, the importance of imagination was felt in Germany, where it found advocates in Bodmer[5], Breitinger[6] and other writers of the Swiss school, who owed much to the influence of the Italians (Muratori, Gravina, Calepio) and the English: acting in their turn as teachers of Klopstock and the new German critical school. [28]

Feeling.

It was at this same period that opposition became clearly marked between those accustomed "*à juger par le sentiment*[7]" and those used to "*raisonner par principes*[8]." [29] The Frenchman, Du Bos[9], author of *Réflexions critiques sur la poésie et la peinture* (1719), upholds the theory of feeling; according to him art is simply a self-abandonment

1. Huarte：即胡安·瓦尔特（1529—1588），西班牙生理学家、心理学家。
2. Bacon：即弗朗西斯·培根（1561—1626），英国哲学家、政治家、科学家、法学家、散文家，"经验主义之父"，著有哲学著作《新工具》、散文集《培根随笔》等。
3. Addison：即约瑟夫·艾迪森（1672—1719），英国散文家、诗人、剧作家、政治家。
4. *Spectator*：《观察家》杂志，由艾迪森与友人创办。
5. Bodmer：博德莫尔（1698—1783），瑞士作家、批评家、学者、诗人，著有《论诗歌中的不可思议性》。
6. Breitinger：布莱廷格（1701—1776），瑞士语言学家和作家，与博德莫尔合著《论绘画》。
7. *à juger par le sentiment*：〈法〉用感觉去判断
8. *raisonner par principles*：〈法〉按原则去推理
9. Du Bos：杜博斯（1670—1742），法国作家。著有《诗歌与绘画评论集》（*Réflexions critiques sur la poésie et la peinture*）。

"*aux impressions que les objets étrangers font sur nous*[1]," setting aside all reflective labour. He laughs at those philosophers who deny the force of imagination, and Malebranche's[2] eloquent discourse founded on this denial draws from Du Bos the remark, "*c'est à notre imagination qu'il parle contre l'abus de l'imagination.*[3]" He refuses to see any intellectual nucleus in the productions of the arts, saying that art consists not in instruction but in style: nor is he too respectful towards the probable: he says he finds himself unable to set limits between it and the marvellous, and leaves to "born poets" the task of thus miraculously uniting opposites. For Du Bos there is no criterion of art save feeling, which he calls a "*sixième sens*[4]," against which dispute is vain since in such matters popular opinion invariably wins the day over the dogmatic pronouncements of artists and men of letters: all the ingenious conceits of the greatest metaphysicians, though unimpeachable in themselves, will not in the slightest degree diminish the lustre of poetry or despoil it of one single attraction. Attempts to discredit Ariosto and Tasso in the eyes of Italians were as vain as those made against the *Cid* in France. Other people's arguments can never persuade us of the contrary of what we feel. [30] These notions were adopted by many French writers: for example Cartaut de la Villate [31] observes, "*Le grand talent d'un écrivain qui veut plaire, est de tourner ses réflexions en sentiments*[5];" and Trublet[6], "*C'est un principe sûr, que la poésie*

1. *aux impressions que les objets étrangers font sur nous*：〈法〉我们对外界事物产生的印象
2. Malebranche：马勒伯朗士（1638—1715），法国牧师、理性主义哲学家，以其"上帝视野"、偶然主义和本体主义著称。
3. *c'est à notre imagination qu'il parle contre l'abus de l'imagination*：〈法〉说他反对滥用想象力是我们的想象。
4. *sixième sens*：〈法〉第六感
5. *Le grand talent d'un écrivain qui veut plaire, est de tourner ses réflexions en sentiments*：〈法〉欲取悦读者，优秀作家须有天分将反思转化为情感。
6. Trublet：特鲁布雷（1697—1770），法国牧师、道德主义者，与伏尔泰思想相左。

doit être une expression de sentiment.[1]" [32] Nor were the English slow in emphasizing the concept of "emotion" in their theories of literature.

Tendency to unite these terms.

In the writings of this period *imagination* was often identified with *wit*, *wit* with *taste*, *taste* with *feeling*, and *feeling* with first apprehensions or *imagination*; [33] we have already noted that taste is sometimes critical and sometimes productive: this fusion, identification and subordination of terms apparently distinct shows how they gravitate round one single concept.

Difficulties and contradictions in their definition.

A German critic, one of the very few who have sought to penetrate the darkness surrounding the origins of modern Æsthetic, considers the concept of taste (which we owe, he thinks, to Gracian) "the most important æsthetic doctrine which remained for modern times to discover." [34] But without going so far as to say that taste is the chief doctrine of the science, and the foundation of all the rest, instead of only a particular doctrine, and without recapitulating[2] what we have already said of Gracian's relation to the theory of taste, it is well to repeat that taste, wit, imagination, feeling, and so on, instead of new concepts scientifically grasped, were simply new words corresponding to vague impressions: at most they were problems, not concepts: apprehensions of ground still to be conquered, not yet annexed and brought into subjection. It must not be forgotten that the very men who made use of these terms could scarcely grope after the ideas

---

1. *C'est un principle sûr, que la poésie doit être une expression de sentiment.*：〈法〉确定的原则为，诗歌应表达情感。
2. recapitulating：概括

they suggested without falling back into the old traditions, the only ones on which they had an intellectual grasp. To them the new words were shades, not bodies: when they tried to embrace them their arms returned empty to their own breasts.

Wit and intellect.

Certainly wit differs to a certain extent from intellect. Yet Pellegrini and Tesauro, with other writers of treatises, never fail to point out that intellectual truth lies at the root of wit. Trevisano defines it as "an internal virtue of the soul which invents methods for expressing and executing its own concepts: it is recognizable now in the arrangement of things we invent, now in the clear expression of them: sometimes in cunning reconciliations of matters seemingly opposed, sometimes in tracing analogies but faintly discernible." To sum up, one must not "allow the actions of wit to go unaccompanied by those of intellect," or even by those of practical morality. [35] More ingenuously Muratori says, "Wit is that virtue and active force with which the intellect is able to assemble, unite and discover the similarities, relations and reasons of things." [36] In this manner wit, after having been distinguished from intellect, eventually becomes a part or a manifestation of it. By a somewhat different path the same conclusion is reached by Alexander Pope[1] when he counsels that wit be reined in like a mettlesome horse, and observes:

For wit and judgement often are at strife,
Though meant each other's aid like man and wife. [37]

Taste and intellectual judgement.

---

1. Alexander Pope：亚历山大·蒲柏（1688—1744），英国诗人、翻译家，荷马史诗《伊利亚特》和《奥德赛》的英译者，著有《人论》《道德论》等。

Similar vicissitudes[1] befell the word "taste," outcome of a metaphor (as was noted by Kant) whose effect was to stand in opposition to intellectualistic principles, as if to say that the judgement governing the choice of food destined solely for the delectation of the palate is of the same nature as that which decides opinions in matters of art. [38] Nevertheless, the very definition of this anti-intellectualistic concept contained a reference to intellect and reason; the implicit comparison with the palate was ultimately taken as signifying an anticipation of reflexion: as Voltaire wrote in the following century: "*De même que la sensation du palais anticipe la réflexion.*[2]" [39] Intellect and reason glimmer through all the definitions of taste belonging to this period. Mme. Dacier[3] wrote in 1684, "*Une harmonie, un accord de l'esprit et de la raison.*[4]" [40] "*Une raison éclairée qui, d'intelligence avec le cœur, fait toujours un juste choix parmi des choses opposées ou semblables*[5]," wrote the author of *Entretiens galants.* [41] According to another writer quoted by Bouhours, "taste" is "a natural feeling implanted in the soul, independent of any science that can possibly be acquired"; it is practically "an instinct of right reason." [42] The same Bouhours, whilst deprecating this interpretation of one metaphor by another, says, "Taste is more nearly allied to judgement than wit." [43] The Italian Ettori thinks that it may generally be described as "judgement regulated by art," [44] and Baruffaldi[6] (1710) identifies it with "discernment" reduced from theory to practice. [45]

---

1. Similar vicissitudes：同样的变迁
2. *De même que la sensation du palais anticipe la réflexion*：〈法〉就像味觉是反思的提前一样。
3. Dacier：达希尔（1654—1720），法国古典学者、翻译家、编辑家，译著有《贺拉斯》《亚里士多德诗学》《俄狄浦斯王》等。
4. *Une harmonie, un accord de l'esprit et de la raiso*：〈法〉这是一种和谐，一种心灵与理智的协调。
5. *Une raison éclairée qui, d'intelligence avec le cœur, fait toujours un juste choix parmi des choses opposées ou semblables*：〈法〉清晰的理性，用心灵的智慧，总能在完全相反或相似的事物中作出正确选择。
6. Baruffaldi：巴鲁法尔迪（1675—1755），意大利牧师、诗人、作家。

De Crousaz[1] (1715) observes: "*Le bon goût nous fait d'abord estimer par sentiment ce que la raison aurait approuvé, après qu'elle se serait donné le temps de l'examiner assez pour en juger par des justes idées.*[2]" [46] And somewhat prior to him Trevisano considered it "a sentiment always willing to conform to whatsoever reason accepts," and in conjunction with divine grace, a powerful help to man in revealing the true and good, no longer able to circulate freely among mankind owing to original sin. For König[3] (1727) in Germany taste was "a power of the intellect, product of a healthy mind and acute judgement which makes one able to feel the true, good and beautiful"; and for Bodmer in 1736 (after lengthy correspondence on the subject with his Italian friend Calepio[4]) "a practised reflexion, prompt and penetrating into the smallest details, by which intellect is able to distinguish the true from the false, the perfect from the imperfect." Calepio and Bodmer were opponents of pure feeling, and made a distinction between "taste" and "good taste." [47] Traversing the same intellectualistic path, Muratori speaks of "good taste" in "erudition" and others of "good taste in philosophy."

The "je ne sais quoi.[5]*"*

Perhaps those authors were wise who preferred to remain vague and to identify taste with an indefinable Something, a *je ne sais quoi*; a *nescio quid*[6]: a new expression which expressed nothing new, but at

---

1. De Crousaz：即让・皮埃尔・克鲁萨（1663—1750），瑞士神学家、哲学家，著有《逻辑新论》《论美》等。
2. *Le bon goût nous fait d'abord estimer par sentiment ce que la raison aurait approuvé, après qu'elle se serait donné le temps de l'examiner assez pour en juger par des justes idées.*：〈法〉好的品位让我们首先通过感觉理性会认可什么而进行判断——在理性给予自身充分的时间进行检验，并以正确思维加以判断之后。
3. König：乌尔里希・科尼西（1586—1642），德国画家，以铜版画闻名。
4. Calepio：卡雷比奥（1693—1762），意大利人文主义者，博德莫尔好友。
5. je ne sais quoi:〈法〉我说不出究竟
6. nescio quid：〈拉〉说不出究竟

least called attention to the problem. Bouhours (1671) discusses it at length: "*Les Italiens, qui font mystère de tout, emploient en toutes rencontres leur* ***non so che****: on ne voit rien de plus commune dans leurs poètes*[1]," and quotes Tasso and others in confirmation. [48] A note upon it is found in Salvini: "This 'good taste' has but recently come to the front; it seems a vague term applicable to nothing particular, and is equivalent to the *non so che*, to a happy or successful turn of wit." [49] Father Feijóo[2], who wrote on the *Razón del gusto* and on *El no se qué*[3] (1733), says very wisely: "*En muchas producciones no solo de la naturaleza, sino del arte, y aun mas del arte que de la naturaleza, encuentran los hombres, fuera di aquellas perfecciones sujetes á su comprehension racional, otro genero de primor misterioso que, lisonjeando el gusto, atormenta el entendemento. Los sentidos le palpan, pero no le puede dissipar la razon, y así, al querer explicarle, no se encuentran voces ni conceptos que cuadren á su idea, y salimos del paso con decir que hay un non se qué, que agrada, que enamora que hechiza, sin que pueda encontrarse revelacion mas clara da este natural misterio.*[4]" [50] And President Montesquieu[5]: "*Il y*

---

1. *Les Italiens, qui font mystère de tout, emploient en toutes rencontres leur* ***non so che****: on ne voit rien de plus commune dans leurs poètes*：〈法〉意大利人把任何事情都弄得神秘兮兮，总爱说“我说不出究竟：再稀松平常不过的诗作罢了”。
2. Father Feijóo：菲约神父（1676—1764），西班牙修士、学者，领导了西班牙启蒙运动。
3. *Razón del gusto* and on *El no se qué*：〈西〉《鉴赏和理性》和《我说不出究竟》，菲约神父著作。
4. *En muchas producciones no solo de la naturaleza, sino del arte, y aun mas del arte que de la naturaleza, encuentran los hombres, fuera di aquellas perfecciones sujetes á su comprehension racional, otro genero de primor misterioso que, lisonjeando el gusto, atormenta el entendemento. Los sentidos le palpa pero no le puede dissipar la razo y así, al querer explicarle, no se encuentran voces ni conceptos que cuadren á su idea, y salimos del paso con decir que hay un non se qué, que agrada, que enamora que hechiza, sin que pueda encontrarse revelacion mas clara da este natural misterio.*：〈西〉不仅在自然作品中，而且在艺术作品中，甚至更多是在艺术作品中，人们发现，除了与理性理解相关的完美外，还有一种神秘的完美，它们在提高鉴赏力的同时也折磨着理性。从情感上讲，人们觉得很清晰，从理性上讲却说不清道不明。所以在表达思想时会词不达意。于是人们总说“讲不出究竟是什么让我高兴，让我爱，让我着迷。换句话说，就是找不到对这自然的神秘进一步的描述”。
5. Montesquieu：孟德斯鸠（1689—1755），法国启蒙思想家、作家、政治哲学家，著有《法的精神》《理念体系》等。

a quelquefois dans les personnes ou dans les choses un charme invisible, une grâce naturelle, qu'on n'a pu définir, et qu'on a été forcé d'appeler le je ne sais quoi. Il me semble que c'est un effet principalement fondé sur la surprise.[1]" [51] Some writers rebelled against the subterfuge[2] of the *je ne sais quoi*, saying, rightly enough, that it was a confession of ignorance: but they knew not how to escape that ignorance without falling into confusion between taste and intellectual judgement.

Imagination and sensationalism. The corrective of Imagination.

If the attempt to define "wit" and "taste" usually resulted in intellectualism, it was easy to transform imagination and feeling into sensationalistic doctrines. We have seen how earnestly Pallavicino insisted on the non-intellectuality of the fantasies and inventions of the imagination. "Nothing presents itself to the admirer of the beautiful (he writes) to enable him to verify his cognition and satisfy himself that the object recognized is or is not that for which he takes it; if either by vision or by strong apprehension he is led to think it actually present by an act of judgement, his taste for beauty as beauty does not arise from such act of judgement, but from the vision or lively apprehension which might remain in ourselves even when the deception of belief was corrected"; just as happens when we are drowsy and know ourselves to be but half awake, yet are unwilling to tear ourselves from sweet dreams. For Pallavicino imagination cannot err; he assimilates it wholly to the sensations, which are incapable of truth or falsity. And if imaginative knowledge pleases, it is not because it holds a special

---

1. *Il y a quelquefois dans les personnes ou dans les choses un charme invisible, une grâce naturelle, qu'on n'a pu définir, et qu'on a été forcé d'appeler le je ne sais quoi. Il me semble que c'est un effet principalement fondé sur la surprise.*：〈法〉人或物，有着一种看不见的魅力，天然的优雅，难以名状。只能称其为“说不出究竟的”。我觉得，这好像主要是由惊奇引发的效果。
2. subterfuge：托词

truth (imaginative truth), but because it creates objects which "though false are pleasing": the painter makes not likenesses but images which, all resemblance apart, are pleasing to the sight: the poet awakens apprehensions "sumptuous, novel, marvellous, splendid." [52] His opinion coincides, if we mistake not, with Marino's sensationalism: "The poet should aim only at the marvellous ... he who cannot amaze his hearers is not worth a straw": [53] he applauds the oft-repeated dictum[1] of "Gabriel Chiabrera[2], that Pindar of Savona[3], that poetry should cause the eyebrows to arch themselves." [54] But in the *Treatise upon Style*[4] written later (1646) he repents of his youthful achievement and appears willing to return to the pedagogic theory: "And for as much as I theorized concerning poetry in the basest manner, treating it solely as a minister of that delight which the mind enjoys in the less noble operation of imagination or apprehension arising from imagination; and, therefore, in consequence I somewhat relaxed the strings which bind it to the probable: I now wish to demonstrate that poetry has other functions more exalted and fruitful, while remaining in strict servitude to the probable: which office is to guide our minds in the noble exercise of judgement; thus it becomes the nurse of philosophy which it nourishes with sweet milk." [55] The Jesuit Ettori, while inculcating the use of imagination and recommending orators to go to school with the "actors," points out that imagination should fulfil the simple office[5] of "interpreter" between intellect and truth, never assuming dominion, otherwise the orator would be treating his audience or readers "not as men, to whom intellect is proper, but as

1. dictum：名言
2. Gabriel Chiabrera：加布里埃洛·基亚布雷拉（1552—1638），意大利诗人。
3. Pindar of Savona：萨沃纳的品达（前 518—前 438），古希腊诗人，著有合唱琴歌、竞技胜利者颂等，有“抒情诗人之魁”之称。
4. *Treatise upon Style*：《论风格》
5. simple office：简单功能

beasts whom imagination satisfies." [56]

The conception of imagination as purely sensuous shows strongly in Muratori, who is so convinced that the faculty, if left to itself, would deteriorate into a riot of dreams and intoxication, that he links it to intellect as to "an authoritative friend" who shall influence the choice and combination of images. [57] The problem of the nature of imagination had strong attraction for Muratori, and, while traducing and vilifying[1], he returns to it again in his *Della forza della fantasia umana*[2]; [58] describing it as a material faculty essentially different from the mental or spiritual, and denying it the validity of knowledge. Although he had observed that the aim of poetry is distinct from that of science, in that the latter seeks to "know," and the former to "represent" truth, [59] he persisted in counting Poetry as an "art of delectation" subordinate to Moral Philosophy, of whom she was one of the three servants or ministers. [60] Very similarly Gravina held that along with novelty and delight in the marvellous, poetry should endow the mind of the vulgar with "truth and universal cognitions." [61]

Outside Italy the same movement was going on. Bacon, although he assigned poetry to imagination, yet considered it as something intermediary between history and science, approximating epic to history and the most lofty style, the parabolic[3], to science: ("*poēsis parabolica inter reliquas eminet*[4]" .) Elsewhere he calls poetry *somnium*[5] or declares absolutely that "*scientias fere non parit*[6]," and that "*pro lusu potius ingenii quam pro scientia est habenda*[7]": music, painting and

---

1. traducing and vilifying：贬低和中伤
2. *Della forza della fantasia umana*：〈意〉《论人类幻想的力量》，穆拉托里著。
3. parabolic：寓言诗
4. *poēsis parabolica inter reliquas eminet*：〈拉〉寓言诗最接近科学。
5. *somnium*：〈拉〉梦幻
6. *scientias fere non parit*：〈拉〉诗不产生科学
7. *pro lusu potius ingenii quam pro scientia est habenda*：〈拉〉诗是理性的游戏，不是科学。

sculpture are voluptuous arts. [62] Addison identified the pleasures of the imagination with those produced by visible objects or the ideas to which they give rise: such pleasures are not so strong as those of the senses nor so refined as those of the intellect: he groups together the pleasures experienced respectively in comparing imitations with the objects imitated, and in sharpening by this means the faculty of observation. [63]

Feeling and Sensationalism.

The sensationalism of Du Bos and other upholders of feeling appears very clearly. For Du Bos art is a pastime whose pleasantness consists in the fact that it occupies the mind without fatigue, and has affinities with[1] the pleasure provoked by gladiatorial contests[2], bullfights and tourneys[3]. [64]

For these reasons, whilst noting the importance, in the prehistory of Æsthetic, of these new words and the new views they express; and while recognizing their value as a ferment in the discussion of the æsthetic problem, taken up by thinkers of the Renaissance at the point at which it had been left by the ancients; we yet cannot discern in their apparition the true origin of our science. By these words and the discussions they aroused, the æsthetic fact clamoured even louder and more insistently for its own philosophical justification; but this it was not yet to attain either by this means or by any other.

[1] *E.g.* Molière, *Préc. ridic.* sc. i, 10.

[2] *I fonti dell' ingegno ridotti ad arte,* Bologna, 1650.

---

1. has affinities with：与……类似
2. gladiatorial contests：角斗竞赛
3. tourneys：骑马比武

[3] *Delle acutezze che altrimenti spiriti, vivezze e concetti volgarmenti si appellano*, Genova-Bologna, 1639.

[4] *Agudeza y arte de ingenio*, Madrid, 1642; enlarged, Huesca, 1649.

[5] *Saggio del gusto e delle belle arti,* 1783, ch. I, *note.*

[6] Ital. trans. in Orsi, *Considerazioni*, etc. (Modena, 1735), vol. i. dial. 1.

[7] *Orl. Furioso*, xxxv. 26; L. Dolce, *Dial. del pittura* (Venice, 1557); *ad init.*

[8] Borinski, *Poet. d. Renaiss.* p. 308 *seqq.*; *B. Gracian*, pp. 39–54.

[9] *Riflessioni sopra il buon gusto* (Venice, 1766), introd. pp. 72–84.

[10] Gracian, *Obras* (Antwerp, 1669); *El héroe, El discreto*, with introd. by A. Farinelli, Madrid, 1900. Cf. Borinski, *Poet. d. Renais, l.c.*

[11] *Les Caractères, ou les mœurs du siècle,* ch. I; *Des ouvrages de l'esprit.*

[12] In the programme: *Von der Nachahmung der Franzosen,* Leipzig, 1687.

[13] *Opera ... nella quale con alcune certe considerazioni si mostra in che consista il vero buon gusto ne'suddetti componimenti,* etc., etc., Bologna, 1696.

[14] *Delle riflessioni sopra il buon gusto nelle scienze e nell'arti,* 1708 (Venice, 1766).

[15] Muratori, *Della perfetta poesia italiana*, Modena, 1706, bk. ii. ch. 5.

[16] Mazzuchelli, *Scrittori d'Italia*, vol. ii. part iv. p. 2389.

[17] Cicero, *De oratore*, iii. ch. 50; Quintilian, *Inst. Orator*, vi. ch. 5.

[18] *De la délicatesse*, Paris, 1671.

[19] *Il buon gusto*, ch. 39, p. 367.

[20] *Del bene* (Naples, 1681), bk. i. part i. chs. 49–53. Cf. the same writer's *Arte della perfezion cristiana*, Rome, 1665, bk. i. ch. 3.

[21] *Perfetta poesia*, bk. i. chs. 14, 21.

[22] *Ragion poetica*, in *Prose italiane*, ed. De Stefano, Naples, 1839, i. ch. 7. 2 *Il buon gusto*, p. 10.

[23] *Esame degl'ingegni degl'huomini per apprender le scienze* (Ital. trans.

by C. Camilli, Venice, 1586), chs. 9–12.

[24] *De dignitate et augmentis scientiarum*, bk. ii. ch. 13.

[25] *De homine* (in *Opera phil.*, ed. Molesworth, vol. iii.), ch. 2.

[26] *Spectator*, Nos. 411–421 (*Works*, London, 1721, pp. 486–519).

[27] *Die Discourse der Mahlern*, 1721–1723; *Von dem Einflüss und Gebrauche der Einbildungskraft*, etc., 1727; and other writings of Bodmer and Breitinger.

[28] Pascal, *Pensées sur l'éloquence et le style*, § 15.

[29] *Réflexions critiques sur la poésie et la peinture*, 1719 (ed. 7, Paris, 1770) *passim.*; see especially sections 1, 23, 26, 28, 33, 34.

[30] *Réflexions critiques sur la poésie et la peinture*, 1719 (ed. 7, Paris, 1770) *passim.*; see especially sections 1, 23, 26, 28, 33, 34.

[31] Cartaut de la Villate, *Essais historiques et philosophiques sur le goût*, Aix, 1737; Trublet, *Essais sur divers sujets de literature et de morale*, Amsterdam, 1755.

[32] *Réflexions critiques sur la poésie et la peinture*, 1719 (ed. 7, Paris, 1770) *passim.*; see especially sections 1, 23, 26, 28, 33, 34.

[33] Cf. Du Bos, *op. cit.* § 33.

[34] Borinski, *B. Gracian*, p. 39.

[35] Trevisano, *op. cit.* pp. 82, 84.

[36] *Perfetta poesia*, bk. ii. ch. I (*ed. cit.* i. p. 299).

[37] A. Pope, *An Essay on Criticism*, 1709 (in *Poetical Works*, London, 1827), lines 81, 82.

[38] *Kritik der Urtheilskraft* (ed. Kirchmann), § 33.

[39] *Essai sur le goût* (in appendix to A. Gérard, *Essai sur le goût*, Paris, 1766).

[40] *Ibid.*

[41] Quoted in Sulzer, *Allg. Th. d. s. K.* ii. p. 377.

[42] *Manière de bien penser* (Ital. trans. *cit.*), dial. 4. 2 *Ibid.*

[43] *Op. cit.* chs. 2–4.

[44] *Osservazioni critiche* (in vol. ii. of Orsi's *Considerazioni*), ch. 8, p. 23.

[45] *Traité du beau* (Amsterdam ed., 1724), i. p. 170.

[46] J. Ulr. König, *Untersuchung von dem guten Geschmack in der Dicht- und Redekunst,* Leipzig, 1727, and (Calepio-Bodmer) *Briefwechsel von der Natur des poetischen Geschmackes*, Zürich, 1736; cf. for both Sulzer, ii. p. 380.

[47] *Les Entretiens d'Ariste et d'Eugène*, 1671 (Paris ed., 1734), conversation v.; "*Le je ne sçai quoi*"; cf. Gracian, *Oraculo manual,* No. 127, and *El héroe*, ch. 13.

[48] *Les Entretiens d'Ariste et d'Eugène*, 1671 (Paris ed., 1734), conversation v.; "*Le je ne sçai quoi*"; cf. Gracian, *Oraculo manual,* No. 127, and *El héroe*, ch. 13.

[49] In the notes to Muratori's *Perfetta poesia.*

[50] Feijóo, *Theatro critico*, vol. vi. Nos. 11–12.

[51] *Essai sur le goût dans les choses de la nature et de l'art.* Posthumous fragment (in appendix to A. Gérard, *op. cit.*).

[52] *Del bene, cap. cit.*

[53] Marino, in one of the sonnets in the *Murtoleide* (1608).

[54] *Del bene*, bk. i. part i. ch. 8.

[55] *Trattato dello stile* (Rome, 1666), ch. 30.

[56] *Il buon gusto*, pp. 12–13.

[57] *Perf. poesia*, i. ch. 18, pp. 232–233.

[58] Venice, 1745.

[59] *Perf. poesia*, i. ch. 6.

[60] *Op. cit.* i. ch. 4, p. 42.

[61] *Ragion poetica*, i. ch. 7.

[62] *De dignitate*, ii. ch. 13; iii. ch. I; iv. ch. 2; v. ch. 1.

[63] *Spectator, loc. cit.* esp. pp. 487, 503.

[64] *Op. cit.* § 2.

IV ÆSTHETIC IDEAS IN THE CARTESIAN AND LEIBNITIAN SCHOOLS, AND THE "ÆSTHETIC" OF BAUMGARTEN

Cartesianism[1] *and imagination.*

The obscure world of wit, taste, imagination, feeling and the *je ne sais quoi* was not selected for examination or even, so to speak, included in the picture of Cartesian philosophy. The French philosopher abhorred[2] imagination, the outcome, according to him, of the agitation of the animal spirits: and though not utterly condemning poetry, he allowed it to exist only in so far as it was guided by intellect, that being the sole faculty able to save men from the caprices of the *folle du logis*[3]. He tolerated it, but that was all; and went so far as not to deny it anything "*qu'un philosophe lui puisse permettre sans offenser sa conscience.*[4]" [1] It has been observed that the æsthetic parallel with Cartesian intellectualism is to be found in Boileau[5], [2] slave to rigid *raison* ("*Mais nous que la raison à ses règles engage* [6] ...") and enthusiastic partisan of allegory.

---

1. *Cartesianism*：笛卡尔主义。笛卡尔主义认为，精神和物质是两种绝对不同的实体，二者彼此完全独立，不能由一个决定或派生另一个。
2. abhorred：厌恶
3. *folle du logis*：〈法〉逻辑混乱
4. *qu'un philosophe lui puisse permettre sans offenser sa conscience*：〈法〉在不昧良心的前提下，哲学家可以接受的事情。
5. Boileau：布瓦洛（1636—1711），法国诗人、文学评论家，著有《讽刺诗》《诗的艺术》等。
6. *Mais nous que la raison à ses règles engage*：〈法〉是理性把我们束缚在它的规则之中了。

We have already had occasion to draw attention to the diatribe[1] of Malebranche against imagination. The mathematical spirit fostered in France by Descartes[2] forbade all possibility of a serious consideration of poetry and art. The Italian Antonio Conti[3], living in that country and witness of the literary disputes raging around him, thus describes the French critics (La Motte[4], Fontenelle[5] and their followers): "*Ils ont introduit dans les belles lettres l'esprit et la méthode de M. Descartes; et ils jugent de la poésie et de l'éloquence indépendamment des qualités sensibles. De là vient aussi qu'ils confondent le progrès de la philosophie avec celui des arts. Les modernes, dit l'Abbé Terrasson, sont plus grands géomètres que les anciens: donc ils sont plus grands orateurs et plus grands poètes.*[6]" [3] The fight against this mathematical spirit in the matters of art and feeling was still going on in France in the day of the encyclopædists; the din[7] of the battle was heard in Italy, as is shown by the writings of Bettinelli[8] and others. At the time when Du Bos published his daring book there was a counsellor in the parliament of Bordeaux, Jean-Jacques Bel by name,

1. diatribe：抨击
2. Descartes：笛卡尔（1596—1650），法国哲学家、数学家、科学家，被誉为"近代西方哲学之父"，开创了解析几何，著有《方法论》《几何学》《第一哲学沉思录》等。
3. Antonio Conti：安东尼奥·孔蒂（1677—1749），意大利物理学家、历史哲学家、数学家，著有素体诗《恺撒》等。
4. La Motte：拉莫特（1672—1731），法国作家、诗人、文艺评论家，著有《论悲剧》《论诗歌》《论荷马》等。
5. Fontenelle：丰特奈尔（1657—1757），法国哲学家，著有《关于宇宙多样性的对话》。
6. *Ils ont introduit dans les belles lettres l'esprit et la méthode de M. Descartes; et ils jugent de la poésie et de l'éloquence indépendamment des qualités sensibles. De là vient aussi qu'ils confondent le progrès de la philosophie avec celui des arts. Les modernes, dit l'Abbé Terrasso, sont plus grands géomètres que les anciens: donc ils sont plus grands orateurs et plus grands poètes.*：〈法〉他们把笛卡尔的精神和方法引入文学领域，他们对诗与口才的评价与感性品质无关。哲学的进步和艺术的进步由此被混为一谈。对此，特拉松修士写道：现代人是比古人更伟大的几何学家，所以现代人是更伟大更慷慨激昂的演说家和诗人。
7. din：喧嚣
8. Bettinelli：贝蒂内利（1718—1808），意大利耶稣会作家。

who composed a dissertation (1726) against the doctrine that feeling should be the judge of art. [4]

Crousaz and André[1].

Cartesianism was incapable of an Æsthetic of imagination. The *Traité du beau*[2] by the eclectic Cartesian J. P. de Crousaz (1715), maintained the dependence of beauty not upon pleasure or feeling, matters about which there can be no difference of opinion, but upon that which can be *approved* and therefore reduced to ideas. He enumerates five such ideas: variety, unity, regularity, order and proportion, observing, "*La variété tempérée par l'unité, la régularité, l'ordre et la proportion, ne sont pas assurément des chimères; elles ne sont pas du ressort de la fantaisie, ce n'est pas le caprice qui en décide*[3]": for him, that is to say, they were real qualities of the beautiful founded in nature and truth. He discovered similar characteristics of the beautiful in the individual beauties of the sciences (geometry, algebra, astronomy, physics, history), of virtue, eloquence and religion, finding in each the qualities laid down above. [5] Another Cartesian, the Jesuit André (1742), [6] distinguished between an *essential* beauty, independent of every institution, human and even divine; a *natural* beauty, independent of the opinions of mankind; and, lastly, a beauty to a certain extent *arbitrary* and of human invention: the first composed of regularity, order, proportion and symmetry (here André relied upon Plato and also as an afterthought brought in St. Augustine's definition): the

---

1. André：安德烈，即 Yves-Marie André（1675—1764），法国哲学家，数学家。著有《论美》。
2. *Traité du beau*：〈法〉《论美》，克鲁萨著作。
3. *La variété tempérée par l'unité, la régularité, l'ordre et la proportio ne sont pas assurément des chimères; elles ne sont pas du ressort de la fantaisie, ce n'est pas le caprice qui en decide*: 〈法〉 由统一，规律，秩序和比例调和而成的多样性绝非幻想，也不是随意形成的。

second having its principal measure in the light which generates colours (as a good Cartesian, he took full advantage of Newton's discoveries): the third belonging to fashion and convention, but never at liberty to violate essential beauty. Each of these three forms of beauty was subdivided into *sensible* beauty pertaining to bodies, and *intelligible* beauty of soul.

The English: Locke, Shaftesbury, Hutcheson and the Scottish School.

Like Descartes in France, Locke[1] in England (1690) is an intellectualist, and recognizes no form of spiritual elaboration save reflexion on the senses. None the less he takes over from contemporary literature the distinction between wit and judgement; according to him the former combines ideas with pleasing variety, discovering their similarities and relations and thus grouping them into beautiful pictures which divert and strike the imagination: the latter (judgement or intellect) seeks dissimilarities, guided by the criterion of truth. "The mind, without looking any further, rests satisfied with the agreeableness of the picture, and the gaiety of the fancy; and it is a kind of an affront to go about to examine it by the severe rules of truth and good reason; whereby it appears that it consists in something that is not perfectly conformable to them." [7] England produced philosophers who developed an abstract and transcendent Æsthetic, but one more tinged with sensationalism than that of the French Cartesians. Shaftesbury[2](1709) raises taste to a sense or instinct for the beautiful; a sense of order and proportion identical with moral sense and, with its preconceptions or presentations,

---

1. Locke：洛克（1632—1704），英国哲学家、生理学家，"自由主义之父"，著有《政府论》《人类理解论》等。
2. Shaftesbury：夏夫兹博里（伯爵）（1621—1683），英国伦理学家、美学家，新柏拉图派代表人物，著有《论特征》等。

anticipating the recognition of reason. Bodies, spirits, God are the three degrees of beauty. [8] Lineal descendant[1] of Shaftesbury was Francis Hutcheson[2](1723), who succeeded in popularizing the idea of an inward sense of beauty as something intermediate between sense and reason, and adapted to distinguish unity in variety, concord in the manifold, the true, the beautiful and the good in their substantial identity. Hutcheson maintains that from this sense springs the pleasure we take in art, in imitation and in the likeness between copy and original: the last a relative, as distinct from an absolute, beauty. [9] This view on the whole predominated in England during the eighteenth century and was adopted by Adam Smith[3] as well as by Reid[4], head of the Scottish school.

Leibniz[5]. Petites perceptions and confused knowledge.

Much more thoroughly and with much greater philosophical vigour Leibniz opened the door to that crowd of psychic facts from which Cartesianism recoiled in horror. In his conception of the real, governed by the law of continuity (*natura non facit saltus*[6]), presenting an uninterrupted scale of existence from the lowest beings to God, imagination, taste, wit and the like found ample room for shelter. The facts now called æsthetic were identified by Leibniz with Descartes'

---

1. Lineal descendant：直系后代
2. Francis Hutcheson：弗朗西斯·哈奇森（1694—1746），英国哲学家，生于爱尔兰，苏格兰启蒙运动创始人之一，著有《道德哲学体系》。
3. Adam Smith：亚当·斯密（1723—1790），英国经济学家、作家、哲学家，著有《道德情操论》《国富论》。
4. Reid：托马斯·里德（1710—1796），英国哲学家，苏格兰常识派创立者，著有《以常识原则探究人类理智》《人类主动性权力论集》。
5. Leibniz：莱布尼茨（1646—1716），德国哲学家、数学家，被誉为“17世纪的亚里士多德”。著有《形而上学论》《人类理智新论》《神义论》等。
6. *natura non facit saltus*：〈拉〉自然不会跳跃。

confused cognition, which might be *clear* without being *distinct*: scholastic terms borrowed, it would appear, from Duns Scotus, whose works were reprinted and widely read in the seventeenth century. [10]

In his *De cognitione, veritate et ideis*[1] (1684), after dividing *cognitio* into *obscura vel clara*, the *clara* into *confusa vel distincta*, and the *distincta* into *adaequata vel inadaequata*[2], Leibniz remarks that while painters and other artists are able to judge works of art very fairly they can give no reason for their decisions, and if questioned as to the reason of their condemnation of any work of art, they reply it lacks a *je ne sais quoi*: ("*at iudicii sui rationem reddere saepe non posse, et quaerenti dicere, se in re, quae displicet, desiderare nescio quid*[3]"). [11] They do possess, in fact, clear cognition, but confused and not distinct; what we should call today imaginative, not *ratiocinative*[4], consciousness: and indeed the latter does not exist in the case of art. There are things impossible to define: "*on ne les fait connaître que par des exemples, et, au reste, il faut dire que c'est un je ne sais quoi, jusqu'à ce qu'on en déchiffre la contexture.*[5]" [12] But these *perceptions confuses ou sentiments*[6] have "*plus grande efficacité que l'on ne pense: ce sont elles qui forment ce je ne sais quoi, ces goûts, ces images des qualités des sens.*[7]" [13] Whence it appears plainly that in his discussion of these

---

1. *De cognitione, veritate et ideis*：《论认知、真理与观念》，莱布尼茨著。
2. dividing *cognitio* into *obscura vel clara*, the *clara* into *confusa vel distincta*, and the *distincta* into *adaequata vel inadaequata*：将认知分为模糊的和清晰的，又将清晰认知分为混乱的和明确的，将明确认知再分为相应的和对应的。（斜体为拉丁文）
3. *at iudicii sui rationem reddere saepe non posse, et quaerenti dicere, se in re, quae displicet, desiderare nescio quid*：〈拉〉往往找不出审美判断的依据，如果有人问起，他们会说不喜欢的艺术品缺了点什么，但说不出究竟。
4. *ratiocinative*：推理的
5. *on ne les fait connaître que par des exemples, et, au reste, il faut dire que c'est un je ne sais quoi, jusqu'à ce qu'on en déchiffre la contexture.*：〈法〉只有通过实例才能解释清楚。而且必须说明，在解开谜底之前，它始终是"说不出究竟的"。
6. *perceptions confuses ou sentiments*：〈拉〉混乱的知觉或感情
7. *plus grande efficacité que l'on ne pense: ce sont elles qui forment ce je ne sais quoi, ces goûts, ces images des qualités des sens.*：〈法〉有着超乎想象的巨大能量：构成"说不出究竟的"，构成品鉴力和品质感的，正是这些混乱的知觉或感情。

perceptions Leibniz reposes upon the æsthetic theories we discussed in the preceding chapter; indeed at one point [14] he mentions Bouhours' book.

Intellectualism of Leibniz.

It might seem that by according *claritas*[1] and denying *distinction* to æsthetic facts Leibniz recognized that their peculiar character is neither sensuous nor intellectual. He might seem to have distinguished them by their "*claritas*" from pleasure or sense-motions, and from intellect by their lack of "*distinctio*." But the "*lex continui*[2]" and the Leibnitian intellectualism forbid this interpretation. In this case obscurity and clarity are quantitative degrees of one single consciousness, distinct or intellectual, towards which both converge and with which in the extreme case they unite.

To admit that artists judge with confused perceptions, clear but not distinct, does not involve denying that these perceptions may be capable of being connected and verified by intellectual consciousness. The self-same object that is confusedly though clearly recognized by imagination is recognized clearly and distinctly by the intellect; which amounts to saying that a work of art may be perfected by being determined by thought. In the very terminology adopted by Leibniz, who represents sense and imagination as obscure and confused, there is a tinge of contempt, as well as the suggestion of a single form of all cognition. This will help us to understand Leibniz' definition of music as "*exercitium arithmeticae occultum nescientis se numerare animi*[3]." Elsewhere he says: "*Le but principal de l'histoire, aussi bien que de la poésie,*

1. *claritas*:〈拉〉清晰性
2. *lex continui*：〈拉〉连续性法则
3. *exercitium arithmeticae occultum nescientis se numerare animi*：〈拉〉就像不识数的人朦朦胧胧做着算术练习

doit être d'enseigner la prudence et la vertu par des exemples, et puis de montrer le vice d'une manière qui en donne l'aversion et qui porte ou serve à l'éviter.[1]" [15]

The "*claritas*" attributed to æsthetic fact is not specifically different from, but rather a partial anticipation of, the "*distinctio*" of intellect. Undoubtedly this distinction of degree marks a great advance: but careful analysis shows that Leibniz does not differ fundamentally from those who, by inventing the new words and empirical distinctions examined above, called attention to the peculiarities of æsthetic facts.

Speculations on language.

We find the same invincible intellectualism in the speculations on language greatly in vogue at the time. When critics of the Renaissance and sixteenth century tried to rise above merely empirical and practical grammar and strove to reduce grammatical science to a systematic form, they fell into logicism and described grammatical forms by such terms as pleonastic, improper, metaphorical or elliptic[2]. Thus Julius Cæsar Scaliger (1540); thus, too, the most learned of all, Francisco Sanchez[3] (Sanctius or Sanzio), called Brocense, who, in his *Minerva* (1587), asserts that names are attached to things by reason, exclusive of interjections which are not parts of speech but merely sounds expressive of joy or sorrow; he denies the existence of heterogeneous and heteroclitic words[4], and works out a system

1. *Le but principal de l'histoire, aussi bien que de la poésie, doit être d'enseigner la prudence et la vertu par des exemples, et puis de montrer le vice d'une manière qui en donne l'aversion et qui porte ou serve à l'éviter.*：〈法〉历史的主要目标，同样也是诗歌的主要目标，是以实例教育人们谨言慎行、修身养德，并向人们展示背道而驰的恶果，进而厌恶并避免恶行。
2. pleonastic, improper, metaphorical or elliptic：赘述的，不规则的，比喻的，省略的
3. Francisco Sanchez：桑切斯（1550—1623），西班牙怀疑主义哲学家、生理学家，著有《密涅瓦》(Minerva)。
4. heterogeneous and heteroclitic words：变性名词和不规则名词

of syntax by means of four figures of construction, proclaiming the principle "*doctrinam supplendi esse valde necessarium*[1]," that is to say, that grammatical diversities must be explained as ellipsis, abbreviation or omission with reference to the typical logical form. [16] Gaspare Scioppio[2] follows him exactly, abusing the old grammar with his accustomed violence and crying up the "Sanctian" method, at that time still almost unknown, in his *Grammatica philosophica* (1628). [17] Amongst critics of the seventeenth century, Jacopo Perizonio must not be forgotten; he wrote a commentary on Sanchez' book (1687). Amongst recognized philosophers who studied the philosophy of grammar and noted the merits and defects of various tongues, we find Bacon. [18] In 1660 Claude Lancelot[3] and Arnauld[4] brought out the *Grammaire générale et raisonnée de Port-Royal*[5], a work applying the intellectualism of Descartes rigorously to grammatical forms, and dominated by the doctrine of the artificial nature of language. Locke and Leibniz both speculated about language, [19] but neither succeeded in creating a fresh point of view, although the latter did much to provoke inquiry into the historical origin of languages. All his life Leibniz cherished the notion of a universal language and of an "*ars characteristic universalis*[6]" as a combination likely to result in great scientific discoveries: prior to him, Wilkins[7] had fostered the same

---

1. *doctrinam supplendi esse valde necessarium*：〈拉〉词的替换很有必要。
2. Gaspare Scioppio：格斯派尔·肖坡（1576—1649），德国辩论家、语言学家，著有《哲学语法》(*Grammatica philosophica*)、《鞭打耶稣会》等。
3. Claude Lancelot：克劳德·兰斯洛特（1615—1695），法国修士、语法学家。
4. Arnauld：阿尔诺德（1612—1694），法国天主教神学家、哲学家、数学家。
5. *Grammaire générale et raisonnée de Port-Royal*：《波尔-罗亚尔普遍唯理语法》。
6. *ars characteristic universalis*：〈拉〉普遍特征的艺术
7. Wilkins：威尔金斯（1614—1672），英国牧师、自然哲学家、作家，著有《在月亮上发现世界》《数学的神奇》《论自然宗教的原理与责任》等。

hope, nor indeed, in spite of its utter absurdity, is it even yet wholly extinct.

J. C. Wolff.

In order to correct the æsthetic ideas of Leibniz it was necessary to alter the very foundations of his system, the Cartesianism upon which it rested. This could not be undertaken by disciples of his own personal school, in whom we notice rather an increase of intellectualism. Giving scholastic form to the brilliant observations of the master, Johann Christian Wolff's[1] system began with the theory of knowledge conceived as an "organon" or instrument, followed by systems of natural law, ethics and politics, together constituting the "organon" of practical activity: the remainder was theology and metaphysics, or pneumatology[2] and physics (doctrine of the soul and doctrine of phenomenal nature). Although Wolff distinguishes a productive imagination, ruled by the principle of sufficient reason, from the merely associative and chaotic, [20] yet a science of imagination considered as a new theoretical value could find no niche[3] in his schematism[4]. Knowledge of a lower order, as such, belonged to Pneumatology and was incapable of possessing its own "organon": at most it could be brought under the organon already existing, which corrected and transcended it by means of logical knowledge in the same way in which Ethics treats the "*facilitas appetitiva inferior*[5]." As in France the poetics of Boileau corresponded with the philosophy of

---

1. Johann Christian Wolff：约翰·克里斯丁·伍尔夫（1679—1754），德国哲学家，著有《经验主义心理学》《自然神学》《道德哲学》等。
2. pneumatology：灵物学
3. find no niche：无立锥之地
4. schematism：思想体系
5. *facilitas appetitiva inferior*：〈拉〉低等意欲官能

Descartes, so in Germany the rationalistic poetics of Gottsched[1] [21] reflect the Cartesian-Leibnitian theories of Wolff (1729).

Demand for an organon of inferior knowledge.

It was no doubt dimly seen that even in the inferior faculties some distinction was operative between perfect and imperfect, value and non-value. A passage in a book (1725) by the Leibnitian Bülffinger[2] has often been quoted where he says: "*Vellem existerent qui circa facultatem sentiendi, imaginandi, attendendi, abstrahendi et memoriam praestarent quod bonus ille Aristoteles, adeo hodie omnibus sordens, praestitit circa intellectum: hoc est ut in artis formant redigerent quicquid ad illas in suo usu dirigendas et iuvandas pertinet et conducid, quem ad modum Aristoteles in Organo logicam sive facultatem demonstrandi redegit in ordinem.*[3]" [22] But on reading the extract in its context one recognizes at once that the desired organon would have been merely a series of recipes for strengthening the memory, educating the attention, and so forth: a technique, in a word, not an æsthetic. Similar ideas had been spread in Italy by Trevisano (1708), who, by declaring that the senses might be educated through the mind, asserted the possibility of an *art of feeling* which should "endow manners with prudence and judgement

1. Gottsched：戈特谢德（1700—1766），德国哲学家、作家、评论家，著有《德语批评诗歌探究》。
2. Bülffinger：布尔芬格（1693—1750），德国学者，莱布尼茨主义者。
3. *Vellem existerent qui circa facultatem sentiendi, imaginandi, attendendi, abstrahendi et memoriam praestarent quod bonus ille Aristoteles, adeo hodie omnibus sordens, praestitit circa intellectum: hoc est ut in artis formant redigerent quicquid ad illas in suo usu dirigendas et iuvandas pertinet et conducid, quem ad modum Aristoteles in Organo logicam sive facultatem demonstrandi redegit in ordinem.*：〈拉〉说到感觉，想象，思索，抽象官能，人们会想到伟大的亚里士多德对于知性的定义，尽管如今他几乎无人问津了。人们仍会以艺术的方式再次重现这些官能，正如亚里士多德在他的《工具论》中所定义的逻辑性或秩序性官能一样。

with good taste." [23] We notice, moreover, that in his day Bülffinger was counted a depreciator of poetry, so much so that a tract against him was written in order to show that "poetry does not diminish the faculty of clear conception." [24] Bodmer and Breitinger were ready "to deduce all the parts of eloquence with mathematical precision" (1727), and the latter sketched a Logic of the Imagination (1740) to which he would have assigned the study of similitudes and metaphors; even had he carried out his project, it is difficult to see how it could have differed materially, from a philosophic point of view, from the treatises on the subject written by the Italian rhetoricians of the seventeenth century.

Alexander Baumgarten: his "Æsthetic."

These discussions and experiments filled the boyhood and helped to form the intellect of young Alexander Gottlieb Baumgarten of Berlin, a follower of the philosophy of Wolff and, at the same time, student and teacher of Latin rhetoric and poetry; these studies led him to reconsider the problem and search for some method by which the precepts of rhetoricians could be reduced to a rigorous philosophical system. On taking his doctor's degree in September 1735, when twenty-one years old, he published a thesis *Meditationes philosophicae de nonnullis ad poēma pertinentibus*[1]: [25] in which the word "Æsthetic" appears for the first time as name of a special science. [26] Baumgarten always remained much attached to his youthful discovery, and in 1742 when called to teach at the university of Frankfort-on-the-Oder, and again in 1749, he gave by request a course of lectures on Æsthetic (*quaedam consilia dirigendarum facultatum inferiorum novam per acroasis*

1. *Meditationes philosophicae de nonnullis ad poēma pertinentibus*：〈拉〉《完美诗歌的哲学沉思录》。

exposuit[1]). [27] In 1750 he printed a voluminous treatise wherein the word "Æsthetic" attained the honours of a title-page; [28] in 1758 he published a more slender second part: illness and finally death in 1762 prevented him from completing the work.

Æsthetic as science of sensory consciousness.

What was Æsthetic to Baumgarten? Its objects are sensible facts (ασθητά), carefully distinguished by the ancients from mental objects (νοητά); [29] hence it becomes *scientia cognitionis sensitivae, theoria liberalium artium, gnoseologia inferior, ars pulcre cogitandi, ars analogi rationis*[2]. [30] Rhetoric and Poetry constitute two special and interdependent disciplines which are entrusted by Æsthetic with the distinction between the various styles in literature and other mall differences, [31] for the laws she herself investigates are diffused throughout all the arts like guiding-stars for these various subsidiary arts (*quasi cynosura quaedam specialium*[3]) [32] and must be extracted not from isolated cases only, or from incomplete induction empirically, but from the totality of facts (*falsa regula peior est quant nulla.*[4]) [33] Nor must Æsthetic be confounded with Psychology, which furnishes its presuppositions only; an independent science, it gives the norm of sensitive cognition (*sensitive quid cognoscendi*[5]) and deals with "*perfectio cognitionis sensitivae, qua talis,*[6]" which is beauty (*pulcritudo*), just as the opposite, imperfection, is ugliness (*deformitas*) [34] . From the beauty of

1. *quaedam consilia dirigendarum facultatum inferiorum novam per acroasis exposuit*：〈拉〉有些人建议通过讲课培育低级认知功能。
2. *scientia cognitionis sensitivae, theoria liberalium artium, gnoseologia inferior, ars pulcre cogitandi, ars analogi rationis*：〈拉〉感性认知的科学、自由艺术的理论、低级认识论、以美的方式思考的艺术、理性类比的艺术
3. *quasi cynosura quaedam specialium*:〈拉〉像北极星
4. *falsa regula peior est quant nulla*：〈拉〉错误法则比没有法则更有害。
5. *sensitive quid cognoscendi*：〈拉〉感性认知
6. *perfectio cognitionis sensitivae, qua talis*：〈拉〉即感性认知的完善

sensitive cognition (*pulcritudo cognitionis*) we must exclude the beauty of objects and matter (*pulcritudo obiectorum et materiae*) with which it is often confused owing to habits of language, since it is easy to show that ugly things may be thought of in a beautiful manner and beautiful things in an ugly manner (*quacum ob receptam rei significationem saepe sed male confunditur; possunt turpia pulcre cogitare ut talia, et pulcriora turpiter*). [35] Poetical representations are confused or imaginative: distinctness, that is intellect, is not poetical. The greater the determination, the greater the poetry; individuals "*omnimode determinata*[1]" are highly poetical; poetical also are images or phantasms as well as all that appertains to the senses. [36] That which judges sensible or imaginary presentations is taste, or "*indicium sensuum*[2]." These, in brief, are the truths displayed by Baumgarten in his *Meditationes* and, with many distinctions and examples, in his *Æsthetic.* [37]

Criticism of judgements based on Baumgarten.

Nearly all German critics [38] are of opinion that from his own conception of Æsthetic as the science of sensitive cognition Baumgarten should have evolved a species of inductive Logic. But he can be cleared of this accusation: a better philosopher, perhaps, than his critics, he held that an inductive Logic must always be intellectual, since it leads to abstractions and the formation of concepts. The relation existing between "*cognitio confusa*[3]" and the poetical and artistic facts which belong to the realm of taste had been shown before his day, by Leibniz: neither he nor Wolff nor any other of their school ever dreamed of transforming a treatment of the "*cognitio confusa*" or "*petites*

---

1. *omnimode determinata*：〈拉〉极端具体的事物
2. *indicium sensuum*：〈拉〉感性的判断力
3. *cognitio confusa*：〈拉〉混乱的认知

perceptions[1]" into an inductive Logic. On the other hand, as a kind of compensation, these critics attribute to Baumgarten a merit he cannot claim, at least to the extent implied by their praises. According to them, he effected a revolution by converting [39] Leibniz' differences of degree or quantitative distinctions into a specific difference, and turning confused knowledge into something no longer negative but positive [40] by attributing a "*perfectio*" to sensitive cognition *qua talis*[2]; and by thus destroying the unity of the Leibnitian monad and breaking up the law of continuity, founded the science of Æsthetic. Had he really accomplished such a giant stride, his claim to the title of "father of Æsthetic" would have been placed beyond question. But, in order to win this appellation, Baumgarten ought to have been successful in unravelling all those contradictions in which he was involved no less than Leibniz and all intellectualists. It is not enough to posit a "*perfectio*"; even Leibniz did that when he attributed *claritas* to confused cognition, which, when devoid of clearness, remains obscure, that is to say, imperfect. It was imperative that this perfection "*qua talis*" should be upheld against the "*lex continui*[3]," and kept uncontaminated by any intellectualistic admixture. Otherwise he was bound to fall back into the pathless labyrinth[4] of the "probable" which is and is not false, of the wit which is and is not intellect, of the taste which is and is not intellectual judgement, of the imagination and feeling which are and are not sensibility and material pleasure. And in that case, notwithstanding the new name: notwithstanding (as we freely admit) the greater insistence than that of Leibniz upon the sensible nature of poetry, Æsthetic, as a science, would not have been born.

---

1. *petites perception*：〈法〉微小的感觉
2. sensitive cognition *qua talis*：〈拉〉感性认知本身
3. *lex continui*：〈拉〉连续性
4. pathless labyrinth：无法走出的迷宫

Intellectualism of Baumgarten.

Now Baumgarten overcame none of the obstacles above mentioned. Unprejudiced and continued study of his works forces one to this conclusion. Already in his *Meditationes* he does not seem able to distinguish clearly between imagination and intellect, confused and distinct cognition. The law of continuity leads him to set up a scale of more and less: amongst cognitions, the obscure are less poetical than the confused; the distinct are not poetical, but even those of the higher kinds (that is the distinct and intellectual) are to a certain extent poetical in proportion as they are lower in their nature; compound concepts are more poetical than simple; those of larger co mprehension are "*extensive clariores*[1]." [41] In the *Æsthetic* Baumgarten expounds his thought more fully and thereby exposes its defects. If the introduction of the book leads one to believe that he sees æsthetic truth to consist in consciousness of the individual, the belief is shattered by the explanations which follow. As a good objectivist he asserts that truth in the metaphysical sense has its counterpart in the soul, namely, subjective truth, logical truth in a wide sense, or æsthetico-logical. [42] And the complete truth lies not in the genus or species, but in the individual. The genus is true, the species more true, the individual most true. [43] Formal logical truth is acquired "*cum iactura*[2]," by jettisoning[3] much great material perfection: "*quid enim est abstractio, si iactura non est?*[4]" [44] So much being granted, logical truth differs from æsthetic in this: metaphysical or objective truth is presented now to the intellect, when it is logical truth in a narrow sense; now to the

1. *extensive clariores*：广义的明确性
2. *cum iactura*：〈拉〉通过削弱
3. jettisoning：抛弃
4. *quid enim est abstractio, si iactura non est?*：〈拉〉如果不是削弱，那抽象是什么呢？

analogy of reason and the lower cognitive faculties, when it is æsthetic; [45] a lesser truth in exchange for the greater which man is not always able to attain, thanks to the "*malum metaphysicum.*[1]" [46] Thus moral truths are comprehended in one fashion by a comic poet, in another by a moral philosopher; an eclipse is described in one way by an astronomer and in another by a shepherd speaking to his friends or his sweetheart. [47] Universals even are accessible, in part at least, to the inferior faculty. [48] Take the case of two philosophers, a dogmatic and a sceptic, arguing, with an æsthete listening to them. If the arguments of either party are so balanced that the hearer cannot determine which is true and which false, this appearance is to him æsthetic truth: if one adversary succeed in overbearing the other so that one argument is shown clearly to be wrong, the error just revealed is likewise æsthetic [49] falsity. Truths strictly æsthetic are (and this is the decisive point) those which appear neither entirely true nor entirely false: probable truths. "*Talia autem de quibus non complete quidem certi sumus, neque tamen falsitatem aliquam in iisdem appercipimus, sunt verisimilia. Est ergo veritas æsthetica, a potiori dicta verisimilitudo, ille veritatis gradus, qui, etiamsi non evectus sit ad completam certitudinem, tamen nihil contineat falsitatis observabilis.*[2]" [50] And especially the immediate sequel: "*Cujus habent spectator es auditor esve intra animum quum vident audiuntve, quasdam anticipationes, quod plerumque fit, quod fiery solet, quod in opinione positum est, quod habet ad haec in se quondam similitudinem, sive id falsum (logice et latissime), sive verum sit (logice et strictissime), quod non sit facile a nostris*

1. *malum metaphysicum*：〈拉〉有害的形而上学主义者
2. *Talia autem de quibus non complete quidem certi sumus, neque tamen falsitatem aliquam in iisdem appercipimus, sunt verisimilia. Est ergo veritas æsthetica, a potiori dicta verisimilitudo, ille veritatis gradus, qui, etiamsi non evectus sit ad completam certitudinem, tamen nihil contineat falsitatis observabilis.*：〈拉〉凡是我们从中看不出虚假性的，同时又没有确定把握的事实就是真的。所以，审美角度的真可称为可然性，它是这样一种程度的真：一方面未达到完全确定的程度，另一方面又不含有明显虚假性。

sensibus abhorrens: hoc illud ***est εἰκός*** *et verisimile quod, Aristotele et Cicerone assentiente, sectetur æstheticus.*[1]" [51] The probable embraces that which is true and certain to the intellect and the senses, that which is certain to the senses but not to the intellect, that which is probable logically and æsthetically, or logically improbable but æsthetically probable, or, finally, æsthetically improbable but on the whole probable or that whose improbability is not evident. [52] So we reach the admission of the impossible and absurd, the *αδύνατον* and ἄτοπον[2] of Aristotle.

If after reading these paragraphs, highly important as revealing the true thought of Baumgarten, we turn once more to the Introduction to his work, we notice at once his commonplace and erroneous conception of the poetic faculty. To a friend who suggested that there was no need for him to concern himself with confused or inferior consciousness both because "*confusio mater erroris*[3]" and because "*facilitate inferior es, caro, debellandae potius sunt quam excitandae et confirmandae*,[4]" Baumgarten replied that confusion is a condition wherein to find truth: that nature makes no sudden leap from obscurity to clarity: that noonday light is reached from night time through the dawn (*ex node per auroram meridies*[5]): that in the case of the

---

1. *Cujus habent spectator es auditor esve intra animum quum vident audiuntve, quasdam anticipationes, quod plerumque fit, quod fiery solet, quod in opinione positum est, quod habet ad haec in se quondam similitudinem, sive id falsum (logice et latissime), sive verum sit (logice et strictissime), quod non sit facile a nostris sensibus abhorrens: hoc illud* ***est εἰκός*** *et verisimile quod, Aristotele et Cicerone assentiente, sectetur æstheticus.*：〈拉〉这些事物对于听者和看者来说，在听或看的过程中，都存在心灵上的预感，也就是说它们通常总是会发生的。因此它们基于每个人的意见，建立在可然性之上。它们包含全部虚假（从逻辑上和广义上来说），也包含全部真实（从逻辑上和狭义上来说）。这是美学家所应遵循的，由亚里士多德和西塞罗提出的两个概念：表面显现和可然性。
2. *αδϋνατον* and *ἄτοπον*：〈希〉不可能和荒谬
3. *confusio mater erroris*：〈拉〉混乱是错误之母。
4. *facilitate inferior es, caro, debellandae potius sunt quam excitandae et confirmandae*：〈拉〉低级的官能和肉体应该被战胜，而不是被强化被训练。
5. *ex node per auroram meridies*：〈拉〉经过黎明

inferior faculties a guide, not a tyrant, is needed (*imperium in facilitates inferiores poscitur, non tyrannis*[1]) [53] . This is still the attitude of Leibniz, Trevisano and Bülffinger. Baumgarten is terrified lest he should be accused of treating subjects unworthy a philosopher. "*Quousque tandem*[2]" (says he to himself), "dost thou, professor of theoretic and moral philosophy, dare to praise lies and mixtures of true and false as though they were noble works?" [54] And if there is one thing above all others from which he is anxious to guard himself it is sensualism, unbridled and non-moralized. The sensitive perfection of Cartesianism and Wolffianism was liable to be confused with simple pleasure, with the feeling of the perfection of our organism: [55] but Baumgarten falls into no such confusion. When in 1745 one Quistorp[3] combated his æsthetic theory by saying that if poetry consisted in sensuous perfection it was a thing hurtful to men, Baumgarten answered disdainfully that he did not expect he should ever find time to reply to a critic of such calibre as to mistake his "*oratio perfecta sensitiva*[4]" for an "*oratio perfecte* (that is *omnino*) *sensitiva*[5]." [56]

New names and old meanings.

Save in its title and its first definitions Baumgarten's *Æsthetic* is covered with the mould of antiquity and commonplace. We have seen that he refers back to Aristotle and Cicero for the first principles of his science; in another instance he attaches his Æsthetic to the Rhetoric of antiquity, quoting the truth enunciated by Zeno the Stoic, "*esse duo cogitandi genera, alterum perpetuum et latius, quod Rhetorices sit, alterum*

1. *imperium in facilitates inferiores poscitur, non tyrannis*：〈拉〉低级感官需要管理，而不是暴政。
2. *Quousque tandem*：〈拉〉究竟为什么
3. Quistorp：奎斯托普（1718—1788），德国福音会神学家。
4. *oratio perfecta sensitiva*：〈拉〉完美的感觉言辞
5. *oratio perfecte* (that is *omnino*) *sensitiva*：〈拉〉完全的感觉言辞

concisum et contractius, quod Dialectices[1]," and identifying the former with the æsthetic horizon, the latter with the logical. [57] In his *Meditationes* he rests upon Scaliger and Vossius[2]; [58] of modern writers beside the philosophers (Leibniz, Wolff, Bülffinger) he quotes Gottsched, Arnold, [59] Werenfels, Breitinger [60] ; by means of these latter he is able to make acquaintance with discussions upon taste and imagination, even without direct acquaintance with Addison and Du Bos, as well as the Italians, whose writings had immense vogue in Germany in his day, and with whom his resemblances leap to the eye. Baumgarten always feels himself to be in perfect accord with his predecessors; never at variance with them. He never felt himself to be a revolutionary; and though some have been revolutionaries without knowing it, Baumgarten was not one of them. Baumgarten's works are but another presentation of the problem of Æsthetic still clamouring for solution in a voice so much the stronger as it uttered a commonplace: he proclaims a new science and presents it in conventional scholastic form; the babe about to be born receives the name of Æsthetic by premature baptism at his hands: and the name remains. But the new name is devoid of new matter; the philosophical armour covers no muscular body. Our good Baumgarten, full of ardour and conviction, and often curiously brisk and vivacious in his scholastic Latinism, is a most sympathetic and attractive figure in the history of Æsthetic: of the science in formation, that is to say, not of the science brought to completion: of Æsthetic *condenda* not *condita*[3].

---

1. *esse duo cogitandi genera, alterum perpetuum et latius, quod Rhetorices sit, alterum concisum et contractius, quod Dialectices*：〈拉〉有两种认识方法，一种是永恒的、广义的，被称为修辞学；一种是简明的，用于揭示矛盾的，被称为辩证法。
2. Vossius：福修斯（1577—1649），荷兰古典主义学者、人文主义者和神学家，著有《古代诗人的时代》《希腊史第四卷》等。
3. Æsthetic *condenda* not *condita*：美学在建立中，但尚未建立起来。（斜体为拉丁文）

[1] Letters to Balzac and the Princess Elizabeth.

[2] *Art poétique* (1669–1674).

[3] Letters to Marquis Maffei, about 1720, in *Prose e poesie*, Venice, 1756, ii. p. cxx.

[4] Sulzer, *op. cit.* i. p. 50.

[5] *Traité du beau* (2nd ed., Amsterdam, 1724; Paris ed., 1810).

[6] *Essai sur le beau*, Paris, 1741.

[7] *An Essay concerning Human Understanding* (French trans. In *Œuvres*, Paris, 1854), bk. ii. ch. 11, § 2.

[8] *Characteristics of Men, Manners, Opinions, Times*, 1709–1711.

[9] *Enquiry into the Original of our Ideas of Beauty and Virtue*, London, 1723.

[10] See above, p. 211.

[11] *Opera philosophica* (ed. Erdmann), p. 78.

[12] *Ibid*, preface.

[13] *Nouveaux Essais*, ii. ch. 22.

[14] *Op. cit.* ii. ch. 11.

[15] *Essais de Théodicée*, part. ii. § 148.

[16] Francisci Sanctii, *Minerva seu de causis linguæ latinæ commentarius*, 1587 (ed. with add. by Gaspare Scioppio, Padua, 1663); cf. bk. i. chs. 2, 9, and bk. iv.

[17] Gasperis Sciopii, *Grammatica philosophica*, Milan, 1628 (Venice, 1728).

[18] *De dignitate*, etc., bk. vi. ch. i.

[19] Locke, *Essay*, etc., bk. lii.; Leibniz, *Nouveaux Essais*, bk. iii.

[20] *Psychol. empirica* (Frankfurt and Leipzig, 1738), §§ 138–172.

[21] Joh. Chr. Gottsched, *Versuch einer critischen Dichtkunst*, Leipzig, 1729.

[22] *Dilucidationes philosophicæ de Deo, anima humana et mundo*, 1725 (Tübingen, 1768), § 268.

[23] Preface to *Rifless. sul gusto, ed. cit.* p. 75.

[24] Borinski, *Poetik d. Renaiss.* p. 380 note.

[25] Halæ Magdeburgicæ, 1735 (reprinted, ed. B. Croce, Naples, 1900).

[26] *Med.* § 116.

[27] *Æsthetica*, i. pref.

[28] *Æsthetica. Scripsit* Alex. Gottlieb Baumgarten, *Prof. Philosoph., Traiecti eis Viadrum, Impens. Ioannis Christiani Kleyb*, 1750; 2nd part, 1758.

[29] *Med.* § 116.

[30] *Æsth.* § i.

[31] *Med.* 117.

[32] *Æsth.* § 71.

[33] *Ibid.* § 53.

[34] *Med.* § 115.

[35] *Æsth.* § 14.

[36] *Ibid.* § 18.

[37] *Med.* § 92.

[38] Ritter, *Gesch. d. Philos.* (Fr. trans., *Hist, de la phil. mod.* iii. p. 365); Zimmermann, *Gesch. d. Æsth.* p. 168; J. Schmidt, *L. u. B.* p. 48.

[39] Danzel, *Gottsched*, p. 218; Meyer, *L. u. B.* pp. 35–38.

[40] Schmidt, *op. cit.* p. 44.

[41] *Med.* § § 19, 20, 23.

[42] *Æsth.* § 424.

[43] *Op. cit.* § 441.

[44] *Op. cit.* § 560.

[45] *Æsth.* § 424.

[46] *Op. cit.* § 557.

[47] *Op. cit.* § § 425, 429.

[48] *Op. cit.* § 443.

[49] *Op. cit.* § 448.

[50] *Op. cit.* § 483.

[51] *Op. cit.* § 484.

[52] *Æsth.* § § 485, 486.

[53] *Op. cit.* § § 7, 12.

[54] *Op. cit.* § 478.

[55] Cf. Wolff, Psych, empir. § 511, and the passage there quoted from Descartes; also § § 542, 550.

[56] Th. Joh. Quistorp, in *Neuen Bücher-Saal*, 1745, fasc. 5; *Erweis dass die Poesie schon für sie selbst ihre Liebhaber leichtlich unglücklich machen könne*; and A. G. Baumgarten, *Metaphysica*, 2nd ed., 1748, preface; cf. Danzel, *Gottsched*, pp. 215, 221.

[57] *Æsth.* § 122.

[58] *Med.* § 9.

[59] *Op. cit.* § § 111, 113.

[60] *Æsth.* § 11.

V GIAMBATTISTA VICO

Vico as inventor of æsthetic science.

The real revolutionary who by putting aside the concept of probability and conceiving imagination in a novel manner actually discovered the true nature of poetry and art and, so to speak, invented the science of Æsthetic, was the Italian Giambattista Vico.

Ten years prior to the publication in Germany of Baumgarten's first treatise, there had appeared in Naples (1725) the first *Scienza nuova*[1], which developed ideas on the nature of poetry outlined in a former work (1721), *De constantia iurisprudentis*[2], outcome of "twenty-five years' continuous and harsh meditation." [1] In 1730 Vico republished it with fresh developments which gave rise to two special books (*Della sapienza poetica*[3] and *Della discoperta del vero Omero*[4]) in the second *Scienza Nuova.* Nor did he ever tire of repeating his views and forcing them upon the attention of his hostile contemporaries at every opportunity, seizing such occasion even in prefaces and letters, poems on the occasion of weddings or funerals, and in such press notices as fell to his duty as public censor of literature.

And what were these ideas? Neither more nor less, we may say, than the solution of the problem stated by Plato, attacked but not solved

---

1. *Scienza nuova*：〈拉〉《新科学》
2. *De constantia iurisprudentis*：〈拉〉《论法的永恒》
3. *Della sapienza poetica*：〈拉〉《论诗歌的智慧》
4. *Della discoperta del vero Omero*：〈拉〉《论真正荷马的发现》

by Aristotle, and again vainly attacked during the Renaissance and afterwards: is poetry rational or irrational, spiritual or brutal? and, if spiritual, what is its special nature and what distinguishes it from history and science?

As we know, Plato confined it within the baser part of the soul, the animal spirits. Vico re-elevates it and makes of it a period in the history of humanity: and since history for him means an ideal history whose periods consist not of contingent facts but of forms of the spirit, he makes it a moment in the ideal history of the spirit, a form of consciousness. Poetry precedes intellect, but follows sense; through confusing it with the latter, Plato failed to grasp the position it should really occupy and banished it from his Republic[1]. "Men at first feel without being aware; next they become aware with a perturbed and agitated soul; finally they reflect with an undisturbed mind. This Aphorism is the Principle of poetical sentences which are formed by the sense of passions and affections; differing thereby from philosophical sentences which are formed by reflexion through ratiocination; whence the latter approach more nearly to truth the more they rise towards the universal, while the former have more of certainty the more they approach the individual." [2] An imaginative phase of consciousness, but one possessed of positive value.

Poetry and Philosophy: imagination and intellect.

The imaginative phase is altogether independent and autonomous with respect to the intellectual, which is not only incapable of endowing it with any fresh perfection but can only destroy it. "The studies of Metaphysics and Poetry are in natural opposition one to the other; for the former purges the mind of childish prejudice and the

---

1. banished it from his Republic：把诗驱逐出他的理想国

latter immerses and drowns it in the same: the former offers resistance to the judgement of the senses, while the latter makes this its chief rule: the former debilitates[1], the latter strengthens, imagination: the former prides itself in not turning spirit into body, the latter does its utmost to give a body to spirit: hence the thoughts of the former must necessarily be abstract, while the concepts of the latter show best when most clothed with matter: to sum up, the former strives that the learned may know the truth of things stripped of all passion: the latter that the vulgar may act truly by means of intense excitement of the senses, without which stimulant they assuredly would not act at all. Hence from all time, in all languages known to man, never has there been a strong man equally great as metaphysician and poet: such a poet as Homer, father and prince of poetry." [3] Poets are the senses, philosophers the intellect, of mankind. [4] Imagination is "stronger in proportion as reason is weaker." [5]

No doubt "reflexion" may be put in verse; but it does not become poetry thereby. "Abstract sentences belong to philosophers, since they contain universals; and reflexions concerning such passions are made by poets who are false and frigid[2]." [6] Those poets "who sing of the beauty and virtue of ladies by reflexion ... are philosophers arguing in verses or in love-rhymes." [7] One set of ideas belongs to philosophers, another to poets: these latter are identical with those of painters, from which "they differ only in colours and words." [8] Great poets are born not in epochs of reflexion but in those of imagination, generally called barbarous: Homer, in the barbarism of antiquity: Dante in that of the Middle Ages, the "second barbarism of Italy." [9] Those who have chosen to read philosophic reason into the verse of the great father of Greek poetry have transferred the character of a later age into an earlier, since the era of poets precedes that of philosophers

1. debilitates：削弱
2. poets who are false and frigid：虚伪薄情的诗人

and countries in infancy were sublime poets. Poetic locutions arose before prose, "by the necessity of nature" not "by caprice of pleasure"; fables or imaginative universals were conceived before reasoned, *i.e.* philosophical universals. [10]

With these observations Vico justified and at the same time corrected the opinion of Plato in the *Republic*, denying to Homer wisdom, every kind of wisdom; the legislative of Lycurgus[1] and Solon[2], the philosophic of Thales[3], Anacharsis[4] and Pythagoras, the strategic of military commanders. [11] To Homer (he says) belongs wisdom, undoubtedly, but poetic wisdom only: the Homeric images and comparisons derived from wild beasts and the elements of savage nature are incomparable; but "such success does not spring from talent imbued with domesticity and civilized with any philosophy." [12]

When anybody takes to writing poetry in an era of reflexion, it is because he is returning to childhood and "putting his mind in fetters"; no longer reflecting with his intellect, he follows imagination and loses himself in the particular. If a true poet dallies with philosophical ideas, it is not "that he may assimilate them and dismiss imagination," but merely "that he may have them in front of him, to examine as though on a stage or public platform." [13] The New Comedy which made its appearance after Socrates is undeniably impregnated with philosophic ideas, with intellectual universals, with "intelligible kinds of human conduct"; but its authors were poets in so far only as they knew how to transform logic into imagination and their ideas into portraits. [14]

---

1. Lycurgus：莱克格斯（约公元前 9 世纪），传说中的斯巴达立法家。
2. Solon：梭伦（约公元前 638—前 558），古希腊雅典政治改革家、立法者、诗人。
3. Thales：泰勒斯（公元前 624—前 564），古希腊哲学家、数学家和天文学家，古希腊七大先贤之一。
4. Anacharsis：阿纳卡西斯（约公元前 6 世纪），赛西亚哲学家，犬儒主义先驱。

Poetry and History.

The dividing line between art and science, imagination and intellect, is here very strongly drawn: the two distinct activities are repeatedly contrasted with a sharpness that leaves no room for confusion. The line of demarcation between poetry and history is hardly less firm. While not quoting Aristotle's passage, Vico implicitly shows why poetry seemed to Aristotle more philosophical than history, and at the same time he dispels the erroneous opinion that history concerns the particular and poetry the universal. Poetry joins hands with science not because it consists in the contemplation of concepts but because, like science, it is ideal. The most beautiful poetic story must be "wholly ideal": "by means of idea, the poet breathes reality into things otherwise unreal; masters of poetry claim that their art must be wholly compact of imagination, like a painter of the ideal, not imitative like a portrait-painter: whence, from their likeness to God the Creator, poets and painters alike are called divine." [15] And against those who blame poets for telling stories which, they say, are untrue, Vico protests: "The best stories are those approximating most nearly to ideal truth, the eternal truth of God: it is immeasurably more certain than the truth of historians who often bring into play caprice, necessity or fortune; but such a Captain as, for instance, Tasso's Godfrey[1] is the type of a captain of all times, of all nations, and so are all personages of poetry, whatever difference there may be in sex, age, temperament, custom, nation, republic, grade, condition or fortune; they are nothing save the eternal properties of the human soul, rationally discussed by politicians, economists and moral philosophers, and painted as portraits by the poet." [16] Referring to an observation made by Castelvetro, and approving it in part, to the effect that if poetry

1. Tasso's Godfrey：塔索笔下的戈弗雷。塔索在《解放了的耶路撒冷》记述了法兰克骑士戈弗雷解放圣地的过程。

is a presentiment of the possible it should be preceded by history, imitation of the real, yet finding himself confronted by the difficulty that, nevertheless, poets invariably precede historians, Vico solves the problem by identifying history with poetry: primitive history was poetry, its plot was narration of fact, and Homer was the first historian; or rather "he was a heroic character amongst Greek men, in so far as they poetically narrated their own history." [17] Poetry and history, therefore, are originally identical; or rather, undifferentiated. "But inasmuch as it is not possible to give false ideas, since falsity arises from an embroiled combination of ideas, so is it impossible to give a tradition, however fabulous, that has not had, at the beginning, a basis of truth." [18] Hence we gain an entirely new insight into mythology: it is no longer an arbitrary calculated invention, but a spontaneous vision of truth as it presented itself to the spirit of primitive man. Poetry gives an imaginative vision; science or philosophy intelligible truth; history the consciousness of certitude[1].

Poetry and language.

Language and poetry are, in Vico's estimation, substantially the same. In refuting the "vulgar error of grammarians" who maintain the priority of the birth of prose over that of verse, he finds "within the origin of Poetry, so far as it has been herein discovered," the "origin of languages and the origin of letters." [19] This discovery was made by Vico after "toil as disagreeable and overwhelming as we should undergo had we to strip off our own nature and enter into that of the primæval men of Hobbes, Grotius, or Puffendorf; creatures possessing no language at all, by whom were created the languages of the ancient world." [20] But his painful labour was richly repaid

1. certitude：确信

by his refutation of the erroneous theory that languages sprang from convention or, as he said, "signified at will," whereas it is evident that "from their natural origin words must have had natural meanings; this is plainly seen in common Latin ... wherein almost all words have arisen by natural necessity, either from natural properties or from their sensible effects; and in general, metaphor forms the bulk of language in the case of every people." [21] This argument strikes a blow at another common error of the grammarians, "that the language of prose writers is correct, that of poets incorrect." [22] The poetic tropes grouped under the heading of metonymy[1] seem to Vico to be "born of the nature of primitive peoples, not of capricious selection by men skilled in poetic art"; [23] stories told "by means of similitudes, imagery and comparisons," result "from lack of the genera and species required to define things with propriety," and "are therefore, by reason of natural necessities, common to entire peoples." [24] The earliest languages must have consisted of "dumb gestures and objects which had natural connexions with the ideas to be expressed." [25] He observes very acutely that to these figurate languages belong not only hieroglyphics but the emblems, knightly bearings, devices and blazons which he calls "mediæval hieroglyphics." [26] In the barbarous Middle Ages "Italy was forced to fall back on the mute language ... of the earliest gentile nations in which men, before discovering articulate speech, were obliged like mutes to use actions or objects having natural connexions with the ideas, which at that time must have been exceedingly sensuous, of the things which they wished to signify; such expressions, clad in almost vocal words, must have had all the lively expressiveness of poetic diction." [27] Hence arise three kinds or phases of language: dumb show, the language of the gods; heraldic language[2], or that of the heroes; and spoken language. Vico also looked forward

1. metonymy：借喻
2. heraldic language：歌颂丰功伟绩的语言

to a universal system of etymology[1], a "dictionary of mental words common to all nations."

Inductive and formalistic Logic

A man with ideas of this sort about imagination, language and poetry could not say he was satisfied with formalistic and verbal Logic, whether Aristotelian or scholastic. The human mind (says Vico) "makes use of intellect when from things which it feels by sense it gathers something that does not fall under sense: this is the true meaning of the Latin *intelligere*[2]." [28] In a rapid outline of the history of Logic, Vico wrote: "Aristotle came and taught the syllogism, a method more suited to expound universals in their particulars than to unite particulars by the discovery of universals: then came Zeno with his sorites[3], which corresponds with modern philosophic methods and refines, without sharpening, the wits; and no advantage whatever was reaped from either by mankind at large. With great reason, therefore, does Verulam[4], equally eminent as politician and philosopher, propound, commend and illustrate induction in his Organum[5]: he is followed by the English with excellent results to experimental philosophy." [29] From this source is derived his criticism of mathematics, which have always, but especially in his day, been considered as the type of perfect science.

Vico opposed to all formal theories of poetry.

In all this, Vico is not only a thorough revolutionary, but is quite

1. universal system of etymology：词源学普遍体系
2. *intelligere*：〈拉〉心智
3. sorites：多段论法
4. Verulam: 维鲁兰男爵，即英国哲学家培根。
5. Organum:《新工具》，培根著。

conscious of being so: he knows himself to be in opposition to all previous theories on the subject. He says that his new principles of poetry "are wholly opposed to, and not merely different from, all which have been imagined from the time of Plato and his disciple Aristotle to Patrizzi, Scaliger and Castelvetro among the moderns; poetry is now discovered to have been the first language used by all nations alike, even the Hebrew." [30] In another passage he says that by his theories "is overthrown all that has ever been said of the origin of poetry, beginning from Plato and Aristotle, right down to our own Patrizzi, Scaliger and Castelvetro; and it is found that poetry arising through defect of human ratiocination is as sublime as any which owes its being to the later rise of philosophy and the arts of composition and criticism; indeed, that these later sources never gave rise to any poetry that could equal, far less surpass it." [31] In the Autobiography he boasts of having discovered "other principles of poetry than those found by Greeks and Latins and all others from those times down to the present day; on these are founded other views on mythology." [32]

These ancient principles of poetry "laid down first by Plato and confirmed by Aristotle" had been the anticipation or prejudice which had misled all writers on poetic reason (among whom he cites Jacopo Mazzoni[1]). Statements "even of most serious philosophers such as Patrizzi and others" upon the origin of song and verse are so inept that he "blushes even to mention them." [33] It is curious to see him annotating the *Ars Poetica* of Horace, with a view to finding some plausible sense in it by applying the principles of the *Scienza nuova*. [34]

It is probable that he was familiar with the writings of Muratori among contemporaries, for he quotes him by name, and of Gravina, who was a personal acquaintance; but if he read the *Perfetta Poesia*

---

1. Jacopo Mazzoni：雅科博·马佐尼（1548—1598），意大利哲学家，伽利略好友，著有《论人类生活的三种方式：行动、思想与宗教》。

and the *Forza della fantasia*[1] he could not have been satisfied by the treatment meted out to the faculty of imagination, so highly valued and respected by himself; and if Gravina influenced him at all it must have been by provoking him to contradiction. In this latter (if not directly in such French writers as Le Bossu) he may have met with the fallacy of regarding Homer as a repository of wisdom, a fallacy which he combated with vigour and pertinacity[2]. In his estimation, among the gravest faults of the Cartesians was their inability to appreciate the world of imagination and poetry. Of his own times he complained they were "benumbed by analytical methods[3] and by a philosophy which sought to deaden every faculty of soul which reached it through the body, especially that of imagination, now held to be mother of all human error 'times' of a wisdom which freezes the generous soul of the best poetry," and prevents all understanding of it. [35]

Vico's judgements of the grammarians and linguists who preceded him

It is just the same with the theory of language. "The manner of birth and the nature of languages has been the cause of much painful toil and meditation: nor, from the *Cratylus* of Plato, in which in our other works we have falsely delighted and believed" (he alludes to the doctrine followed by him in his own first book, *De antiquissima Italorum sapientia*[4]), "down to Wolfgang Latius[5], Julius Cæsar Scaliger, Francisco Sanchez and others, can we find anything to satisfy our understanding; so much that in discussing matters of this kind Signor

1. *Perfetta Poesia* and the *Forza della fantasia*：《论完美诗歌》和《想象的力量》，穆拉托里著（斜体为意大利文）。
2. combated with vigour and pertinacity：猛烈而无休止地抗争
3. benumbed by analytical methods：被分析法折磨得麻木不仁
4. *De antiquissima Italorum sapientia*：〈意〉《论古代意大利人民的智慧》
5. Wolfgang Latius：沃尔夫冈·拉齐乌斯（1514—1565），奥地利人文主义者、绘图师、历史学家、生理学家，著有《奥地利其他类型地方地图》。

Giovanni Clerico[1] says there is nothing in philology involved in such a maze of doubt and difficulty." [36] The chief grammarian-philosophers do not escape criticism. Grammar, says he, lays down rules for speaking correctly: Logic for speaking truly; "and since in the order of nature we must speak truly before learning to speak correctly, Giulio Cesare della Scala[2], followed by the best grammarians, employs all his magnificent energy to reason to the causes of the Latin language from the principles of logic. But his great design ended in failure for this reason, that he attached himself to the logical principles of a single philosopher, namely Aristotle, whose principles are too universal to explain the almost infinite particulars which naturally beset him who would reason concerning a language. Whence it happened that Francisco Sanchez, who followed him with admirable zeal, attempting in his *Minerva* to explain the innumerable particles which are found in Latin by his famous principle of ellipsis, and trying thereby, though without success, to vindicate[3] the logical principles of Aristotle, fell into the most cumbrous clumsinesses[4] among an almost innumerable host of Latin phrases whereby he meant to make good the slight and subtle omissions employed by Latin in expressing its meaning." [37] The origin of parts of speech and syntax is wholly different from that assigned to them by folk who fancied that "the people who invented language must first have gone to school to Aristotle." [38] The same criticism undoubtedly must have extended to the logico-grammarians of Port-Royal, for Vico remarked that the Logic of Arnauld was built "on the same plan as that of Aristotle." [39]

1. Signor Giovanni Clerico：乔万尼·克莱里科先生（生卒年不详），意大利语文学家，曾发表关于《新科学》的评论。
2. Giulio Cesare della Scala：朱利奥·塞萨尔·戴拉·斯卡拉（1484—1558）意大利作家、哲学家、物理学家。
3. vindicate：维护
4. cumbrous clumsinesses：笨重累赘

Influence of seventeenth-century writers on Vico.

It may well be granted that Vico was more in sympathy with the seventeenth-century rhetoricians, in whom we have detected a premonition of æsthetic science. For Vico, as for them, wit (referring to imagination and memory) was "the father of all invention": judgement concerning poetry was for him a "judgement of the senses," a phrase equivalent to "taste" or "good taste," expressions never used by him in this connexion. There is no doubt he was familiar with the writers of treatises on wit and conceits[1], for, in a dry rhetorical manual written for the use of his school (in which one looks in vain for a shadow of his own personal ideas), he quotes Paolo Beni[2], Pellegrini, Pallavicino[3] and the Marquis Orsi[4]. [40] He highly esteems Pallavicini's treatise on *Style* and has knowledge of the book *Del bene* by the same author; [41] perhaps too his mind was not unaffected by the flash of genius which had enabled the Jesuit for one instant to perceive that poetry consists of "first apprehensions." He does not name Tesauro, but here is no doubt he knew him; indeed the *Scienza nuova* includes a section, besides that on poetry, upon "blazons," "knightly bearings," "military banners," "medals," and so forth, precisely similar in method to that of Tesauro when he treats of "figurate conceits" in his *Cannochiale aristotelico*. [42] For Tesauro such conceits are merely metaphorical ingenuities, like any other; for Vico they are wholly the work of imagination, for imagination expresses itself not in words only, but in the "mute language" of lines and colours. He knew something also of Leibniz; the great German and Newton[5] were by him described as "the greatest wits of the time" [43];

1. wit and conceits：妙思奇喻
2. Paolo Beni：保罗·贝尼（1552—1625），意大利语文学家、文学家、耶稣会士，著有《评亚里士多德诗学》《历史四书》等。
3. Pallavicino：帕拉维奇诺（1607—1667），意大利主教、哲学家。
4. Marquis Orsi：奥尔西侯爵乔万尼·朱塞佩（1652—1733）。
5. Newton：牛顿（1643—1727），英国物理学家、数学家，著有《自然哲学的数学原理》《光学》等。

but he seems to have remained in complete ignorance of the æsthetic attempts of the Leibnitian school in Germany. His "Logic of poetry" was a discovery independent of, and earlier than, Bülffinger's Organon of the inferior faculties, the *Gnoseologia inferior*[1] of Baumgarten, and the *Logik der Einbildungskraft*[2] of Breitinger. In truth, Vico belongs on one side to the vast Renaissance reaction against formalism and scholastic verbalism, which, beginning with the reaffirmation of experience and sensation (Telesio, Campanella, Galileo, Bacon), was bound to go on by reasserting the function of imagination in individual and social life: on the other side he is a precursor[3] of Romanticism.

Æsthetic in the "Scienza nuova."

The importance of Vico's new poetic theory in his thought as a whole as well as in the organism of his *Scienza nuova* has never been fully appreciated, and the Neapolitan philosopher is still commonly regarded as the inventor of the Philosophy of History. If by such a science is meant the attempt to deduce concrete history by ratiocination and to treat epochs and events as if they were concepts, the only result of Vico's efforts to solve the problem could have been failure; and the same is true of his many successors. The fact is that his philosophy of history, his ideal history, his *Scienza nuova d' intorno alia comune natura delle nazioni*[4], does not concern the concrete empirical history which unfolds itself in time: it is not history, it is a science of the ideal, a Philosophy of the Spirit. That Vico made many discoveries in history proper which have been to a great extent confirmed by

1. *Gnoseologia inferior*：〈拉〉《低级认识论》，鲍姆加登著。
2. *Logik der Einbildungskraft*：〈德〉《想象力的逻辑》，布莱廷格著。
3. precursor：先驱
4. *Scienza nuova d' intorno alia comune natura delle nazioni*：〈意〉《关于各民族共同本质的新科学》

modern criticism (*e.g.* on the development of the Greek epic and the nature and genesis of feudal society in antiquity and in the Middle Ages) certainly deserves all emphasis; but this side of his work must be kept distinctly apart from the other, strictly philosophical, side. And if the philosophical part is a doctrine expounding the ideal moments of the spirit, or in his own words "the modifications of our human mind," of these moments or modifications Vico undertakes especially to define and fully describe not the logical, ethical and economic moments (though on these too he throws much fight), but precisely the imaginative or poetic. The larger portion of the second *Scienza nuova* hinges on the discovery of the creative imagination, including the "new principles of Poetry," the observations on the nature of language, mythology, writing, symbolic figures and so forth. All his "system of civilization, of the Republic, of laws, of poetry, of history, in a word, of humanity at large" is founded upon this discovery, which constitutes the novel point of view at which Vico places himself. The author himself observes that his second book, dedicated to Poetic Wisdom, "wherein is made a discovery totally opposed to Verulam's," forms "nearly the whole body of the work"; but the first and third books also deal almost exclusively with works of the imagination. It might be maintained, therefore, that Vico's "New Science" was really just Æsthetic; or at least the Philosophy of the Spirit with special emphasis upon the Philosophy of the Æsthetic Spirit.

Vico's mistakes.

Among so many luminous points, or rather in such a general blaze of light, there are yet dark nooks in his mind; corners that remain in shadow. By not maintaining a rigid distinction between concrete history and the philosophy of the spirit, Vico allowed himself to suggest historical periods which do not correspond with the real periods, but are rather allegories, the mythological expression of his philosophy of the spirit. From the same source arises the multiplicity of those periods (usually three in number) which Vico finds in the history of civilization in general, in poetry and language and practically every subject. "The first peoples, who were the children of the human race, founded first the world of the arts: next, after a long interval, the philosophers, who were therefore the aged among nations, founded the world of the sciences: with which humanity attained completion." [44] Historically, understood in an approximate sense, this scheme of evolution has some truth; but only an approximate truth. In consequence of the same confusion of history and philosophy he denied primitive peoples any kind of intellectual logic, and conceived not only their physics, cosmology, astronomy and geography as poetic in character, but their morals, their economy and their politics as well. But not only has there never been a period in concrete human history entirely poetic and ignorant of all abstraction or power of reasoning, but such a state cannot even be conceived. Morals, politics, physics, all presuppose intellectual work, however imperfect they may be. The ideal priority of poetry cannot be materialized into a historical period of civilization.

Linked with this error is another into which Vico often falls when he asserts that "the chief aim of poetry" is to "teach the ignorant vulgar to act virtuously" and to "invent fables adapted with the popular understanding capable of producing strong emotion." [45] Having regard to the clear explanations he himself gave of the inessentiality of abstractions and intellectual artifice in poetry; when we remember that for him poetry makes her own rules for herself without consulting anybody, and that he clearly established the peculiar theoretical nature of the imagination, such a proposition cannot be taken as a return to the pedagogic and heteronomous theory of poetry which in substance he had left far behind: therefore, without doubt, it follows from his historical hypothesis of a wholly poetical epoch of civilization, in which education, science and morality were administered by poets.

Another consequence is that "imaginative universals" are apparently sometimes understood by him as imperfect universals (empirical or representative concepts as they were subsequently called); although, on the other hand, individualization is so marked in them and their unphilosophical nature so accentuated that their interpretation as purely imaginative forms may be taken s normal. In conclusion, we remark that fundamental terms are not always used by Vico in the same sense: it is not always clear how far "sensation," "memory," "imagination," "wit" are synonymous or different. Sometimes "sensation" seems outside the spirit, at others one of its chief moments; poets are sometimes the organ of "imagination," sometimes the "sensation" of humanity; and imagination is described as "dilated memory[1]." These are the aberrations of a thought so virgin and original that it was not easy to regulate.

Progress still to be achieved.

To sever the Philosophy of the Spirit from History, the modifications of the human mind from the historic vicissitudes[2] of peoples, and Æsthetic from Homeric civilization, and by continuing Vico's analyses to determine more clearly the truths he uttered, the distinctions he drew and the identities he divined; in short, to purge Æsthetic of the remains of ancient Rhetoric and Poetics as well as from some over-hasty schematisms imposed upon her by the author of her being: such is the field of labour, such the progress still to be achieved after the discovery of the autonomy of the æsthetic world due to the genius of Giambattista Vico.

[1] *Scienza nuova prima*, bk. iii. ch. 5 (*Opere di G. B. Vico*, edited by G.

1. dilated memory：膨胀的记忆
2. historic vicissitudes：历史变迁

Ferrari, 2nd ed., Milan, 1852–1854).

[2] *Scienza nuova seconda, Elementi*, liii.

[3] *Scienza nuova pr.* bk. iii. ch. 26.

[4] *Scienza nuova sec.* bk. ii. introd.

[5] *Op. cit. Elem.* xxxvi.

[6] *Op. cit.* bk. ii.; *Sentenze eroiche.*

[7] Letter to De Angelis of December 25, 1725.

[8] Letter to De Angelis, *cit.*

[9] *Scienza nuova sec.* bk. iii.; Letter to De Angelis, *cit.*; *Giudizio su Dante.*

[10] *Scienza nuova sec.* bk. ii.; *Logica poetica.*

[11] *Republica*, x.

[12] *Scienza nuova sec.* bk. iii. *ad init.*

[13] Letter to De Angelis, *cit.*

[14] *Scienza nuova sec.* bk. iii. *passim.*

[15] *Scienza nuova pr.* bk. iii. ch. 4.

[16] Letter to Solla, January 12, 1729; cf. *Scienza nuova sec. Elem.* xliii.

[17] *Scienza nuova sec.* bk. iii.

[18] *Scienza nuova pr.* bk. iii. ch. 6.

[19] *Scienza nuova sec.* bk. ii., *Corollari d'intorno all'origine della locuzion poetica,* etc.

[20] *Scienza nuova pr.* bk. iii. ch. 22.

[21] *Scienza nuova sec.* bk. ii., *Corollari d'intorno all'origini delle lingue*, etc.

[22] *Op. cit.* bk. ii., *Corollari d'intorno a'tropi,* etc., § 4.

[23] *Scienza nuova pr.* bk. iii. ch. 22.

[24] *Scienza nuova sec.* bk. iii., *Pruove filosofiche.*

[25] *Scienza nuova pr.* bk. iii.-ch. 22.

[26] *Op. cit.* bk. iii. chs. 27–33.

[27] Letter to De Angelis, *cit.*

[28] *Scienza nuova sec.* bk. ii. introd.

[29] *Scienza nuova sec.* bk. ii., *Ultimi corollari,* § vi.

[30] *Scienza nuova pr.* bk. iii. ch. 2.

[31] *Scienza nuova sec.* bk. ii., *Della metafisica poetica,* etc.

[32] *Scienza nuova pr.* bk. iii. ch. 37.

[33] *Scienza nuova pr.* bk. iii. ch. 37.

[34] *Note all'Arte poetica di Orazio*, in *Opere, ed. cit.* vi. pp. 52–79.

[35] Letter to De Angelis, *cit.*

[36] *Scienza nuova pr.* bk. iii. ch. 22; cf. the review of Clerico (Le Clerc) in *Opere*, iv. p. 382.

[37] *Giudizio intorno alia gram. d'Antonio d'Aronne*, in *Opere*, vi. pp. 149–150.

[38] *Scienza nuova sec.* bk. ii., *Corollari d'intorno all'origini delle lingue*, etc.

[39] *Vita, cit.* p. 343.

[40] *Instituzioni oratorie e scritti inediti,* Naples, 1865, pp. 90 *seqq.*: *De senteniiis, vulgo del ben parlare in concetti.*

[41] Letter to the Duke of Laurenzana, March 1, 1732; and cf. letter to Muzio Gæta.

[42] Cf. p. 190.

[43] *Scienza nuova sec.* bk. i., *Del metodo.*

[44] *Scienza nuova sec., Ultimi corollari,* § 5.

[45] *Scienza nuova sec.* bk. iii. ch. 3; *Scienza nuova sec.* bk. ii., *Della metafisica poetica*; and bk. iii. *ad init.*

VI MINOR ÆSTHETIC DOCTRINES OF THE EIGHTEENTH CENTURY

The influence of Vico.

This step in advance had no immediate effect. The pages in the *Scienza nuova* devoted to æsthetic doctrine were actually the least read of any in that marvellous book. Not that Vico exercised no influence at all; we shall see that several Italian authors both of his own time and of the generation immediately following show traces of his æsthetic ideas; but these traces are all external and material and therefore sterile. Outside Italy the *Scienza nuova* (already announced by a compatriot in 1726 in the *Acta* of Leipzig with the graceful comment that *magis indulget ingenio quam veritati*[1] and the pleasing information that *ab ipsis Italis taedio magis quam applausu excipitur*[2]) [1] was mentioned toward the end of the century, as is well known, by Herder[3], Goethe, and some few others. [2] In connection with poetry, especially with the Homeric question, Vico's book was quoted by Friedrich August Wolf[4], to whom

1. *magis indulget ingenio quam veritati*：〈拉〉（《新科学》）谈论更多的是精巧，而非真。
2. *ab ipsis Italis taedio magis quam applausu excipitur*：〈拉〉与其说（《新科学》）在意大利广受欢迎，不如说是被迫接受。
3. Herder：赫尔德（1744—1803），德国哲学家、神学家、诗人、文学评论家，著有《论语言的起源》《人性的哲学史大纲》（未完成）等。
4. Friedrich August Wolf：弗里德里希·奥古斯特·沃尔夫（1759—1824），德国古典主义者，被称为现代语文学先驱。

it had been recommended by Cesarotti[1] [3] after the publication of the *Prolegomena ad Homerum*[2] (1795), but without any suspicion of the importance of its general doctrine of poetry, of which the Homeric hypothesis was a mere application. Wolf (1807) imagined himself in the presence of a talented forerunner in an isolated problem, instead of a man of intellectual stature towering above any philologist, however great.

Italian writers: A. Conti.

Neither by reliance on the works of Vico, who founded no real school, nor, it must be added, by any independent effort along new lines, did thought succeed in maintaining or improving upon the position already attained. A notable attempt to establish a philosophical theory of poetry and the arts was made by the Venetian A. Conti, who left numerous sketches for essays on imagination, the faculties of the soul, poetic imitation and similar subjects, designed for inclusion in a large treatise on the Beautiful and Art. Conti had started by professing ideas very like those of Du Bos, affirming that the poet must "put everything in images"; that taste is as indefinable as feeling, and that there are persons without taste just as there are blind and deaf persons; he also wrote polemical tracts[3] against the Cartesians. Later he abandoned his sensationalistic or sentimentalist theories, [4] and, inquiring into the nature of poetry, declared himself ill-satisfied with Castelvetro, Patrizzi, and even Gravina. "Had Castelvetro," he observes, "who writes so subtly of Aristotle's *Poetics*, given two or three chapters to a philosophical explanation of the

1. Cesarotti：切萨罗蒂（1730—1808），意大利诗人、作家、翻译家、文学评论家，著有《论美》《论悲剧的快感》《论语言的哲学》。
2. *Prolegomena ad Homerum*：〈拉〉《荷马绪论》，沃尔夫著。
3. polemical tracts: 论战性著作

idea of imitation, he would have solved many questions raised but not clearly answered by himself concerning poetic theories. In his *Poetica* and in his controversy against Torquato Tasso, Patrizzi never succeeded in clearly defining the philosophical idea of imitation; he collected much useful information about the history of poetry, but wilfully lost the Platonic doctrine by allowing it to mingle with the historical detail instead of gathering it up without sophistry into a single point, when it would have appeared in a very different guise. The *Ragion poetica*[1] of Gravina shadows forth a sort of philosophical idea of imitation; but so wholly engrossed is he in deducing therefrom rules for lyrical, dramatical and epic poetry, and illustrating each with examples from the most celebrated poets, Greek, Latin and Italian, that he is too busy to question the sufficiency of the fertile idea he has propounded." [5] A close follower of contemporary European thought, Conti was familiar with Hutcheson[2], whose theories he vigorously repudiated, observing, "Why this multiplication of faculties?" The soul is one, and for scholastic convenience only has been divided into three faculties: sense, imagination, intellect; the first "concerns herself with objects present before her; imagination with those afar into which memory gradually merges: but the object of sense and imagination is always particular; it is only the mind, the intellect, the spirit, that by comparing particulars apprehends the universal." "Before introducing a new sense for the pleasure of beauty" Hutcheson should have "assigned limits to these three faculties of cognition and demonstrated that the pleasure occasioned by beauty does not arise from the three pleasures of these three faculties, or from intellectual pleasure alone, to which they all reduce, if the functions of the soul be carefully analysed." Thus it would appear that the mistake of

---

1. *Ragion poetica*：《诗性之理》，格拉维纳著。
2. Hutcheson：哈奇森（1694—1746），爱尔兰哲学家，著有《道德哲学体系》。

the Scotchman [6] arose from his habit of separating pleasure from the cognitive faculties, placing the former apart in a special empty "sense of beauty." [7] On the other hand, when rewriting the history of the opinions of various critics upon the Aristotelian doctrine of the universal in poetry, Conti gave much weight to the dialogue *Naugerius seu De poëtica*[1] of Fracastoro[2]; [8] for an instant he seems on the point of grasping the essence of the poetic universal and identifying it with the characteristic, which makes us call even horrible things wholly beautiful. "In all his journeys Balzac never saw a beautiful old woman: in the poetic or picturesque sense an old woman is highly beautiful, if depicted as having suffered all the dilapidations of age": immediately after, however, he identifies the characteristic with Wolff's concept of perfection: "It does not differ from being, nor does being differ from the truth which the schoolmen call transcendental and which is the object of all arts and all sciences; we call it the object of poetry when by means of imaginary presentations it ravishes the intellect and moves the wall, transporting both these faculties into the ideal and archetypal world of which, following S. Augustine, Father Malebranche discourses at length in his *Recherche de la vérité*[3]." [9] In the same way Fracastoro's universal gives place to the universal of science: "Owing to the infinity of their determinations all we can know of particulars is their common properties, which is merely another manner of saying that we have no science save of universal. Thus it is precisely the same if we say the object of poetry is science or the universal; which is the doctrine of Navagero[4], following Aristotle." [10] The "imaginative universals of Signor Vico" (with whom he had interchanged some

---

1. *Naugerius seu De poëtica*：〈拉〉《瑙格吕斯或论诗艺》，弗拉卡斯特罗著。
2. Fracastoro：弗拉卡斯特罗（约 1476—1553），意大利生理学家、诗人、数学家、地理学家和天文学家。
3. *Recherche de la vérité*：〈法〉《探索真理》，马勒布朗士著。
4. Navagero：纳瓦杰罗（1483—1529），意大利诗人、政治家、历史学家和植物学家。

letters) opened no new views for him: he notes that Signor Vico "talks a great deal about them" and "holds that the most uncivilized men, having framed them not from any wish to please or serve others, but from the necessity of expressing their feelings as nature taught them, spoke in poetical language the elements of a theology, a physics, and an ethics wholly poetical." Conti excuses himself from immediate examination of "this critical question" and only opines[1] that "it can be shown in many ways that these imaginative universals are the material or object of poetry, in so far as they contain within them sciences or things considered in themselves" [11] — a conclusion diametrically opposed to that which "Signor Vico" meant to express. Conti is next obliged to ask himself how it is possible that poetry's object should be not the true but the probable, when the universal of poetry is the same as that of science. He answers by coming down to the commonplace level of a Baumgarten: "When sciences receive a particular colouring, we pass from the true to the probable." Imitation means giving the impression of truth; that is done by selecting a few of its features only; and this is the procedure in which the probable just consists. If you wish to describe the rainbow poetically, a great part of the Newtonian optics must be thrown overboard; thus "many circumstances of mathematical demonstration" will be neglected in poetical descriptions, and the rest, which is utilized, will form the probable or that particular "which awakens the universal idea, slumbering in the minds of the learned." The great art of poetry consists "in selection of the image containing the greatest number of points of universal doctrine which, by being inserted in the example, may so colour the precept that I may find it without seeking it, or recognize it through its connexion with events described." [12] Hence poetry cannot be content with imitation; allegory too is needed: "in

1. opines：强调

ancient poetry one thing is read and another is meant." Here follows the inevitable instance of the Homeric poems, in which Conti certainly finds elements which cannot be reduced to instruction and allegory and therefore to some extent deserve the Platonic condemnation. [13] He recognizes a species of imagination differing from passive sensibility, "which Father Malebranche calls active imagination, and Plato the art of imagery; it comprises all that is meant by wit, sagacity, judgement and good taste, which teach a poet to use or not to use at a given time or place the rules and licences of art, and to control the extravagance of his imagery." [14] On the question of literary taste he follows the opinion of Trevisano and decides that it consists in "setting in mutual harmony, that is to say restraining within limits, the soul's cognitive faculties, memory, imagination and intellect, allowing none to overwhelm another." [15]

Quadrio[1] *and Zanotti*[2].

By assiduous travail[3] of thought and perpetual search for the best, Conti kept himself at the highest level of æsthetic speculation in contemporary Europe (Vico always excepted); at the same level as Baumgarten in Germany. We pass rapidly over other Italian writers such as Quadrio (1739), author of the first great encyclopædia of universal literature, in which he defines poetry as "the science of things human and divine, presented in pictures to the populace, and written in words connected by measure"; [16] and Francesco Maria Zanotti (1768), who describes poetry as "the art of versification

1. *Quadrio*：夸德里奥（1695—1756），意大利历史学家、文学家，著有《论诗的历史与理性》。
2. *Zanotti*：扎诺提（1692—1777），意大利哲学家、作家，著有《理念的吸引力》《论诗艺》等。
3. assiduous travail：辛勤劳作

in order to give pleasure": [17] the first is worthy of a mediæval anthologist, the second of a no less mediæval composer of handbooks on rhythm and methods of composition. The only serious student of æsthetic was Melchior Cesarotti.

M. Cesarotti

Cesarotti called attention to popular and primitive poetry: he translated Ossian[1] and illustrated the text with dissertations; he unearthed antique Spanish poems and even the folk-songs of Mexico and Lapland[2]; he studied Hebrew poetry; he dedicated the greater part of his life to the Homeric poems, examining all the theories of critics past and present, encountering Vico in this connexion and discussing his views. Besides this, he debated the origin of poetry, the pleasure given by tragedy, taste, the beautiful, eloquence, style, in short every problem belonging to æsthetics which had been raised up to his time. [18] One seems to catch an echo of Vico as one listens to his words on La Motte: "He had logic, but knew not that the logic of poetry differs somewhat from ordinary logic: he was a man of great talent, but he recognized talent only, and was incapable of feeling the immeasurable distance between judicious prose and poetry: the real Homer with his attractive faults will always be more beloved than his reformed Homer with his cold, affected virtue." [19] Cesarotti purposed (1762) bringing out a great theoretico-historical book in whose first part "we shall suppose the non-existence of poetry and poetic art and try to trace by what path a man of illuminated reason can have reached the idea of the possibility of such an art and how he can have attained perfection by these means: every one will be able to see poetry growing up under

---

1. Ossian：《莪相》，凯尔特神话中，莪相是古爱尔兰著名的英雄人物，传说是一位优秀的诗人。
2. Lapland：拉普兰（北欧接近北极圈的地区）

his eyes, so to speak, and attest the truth of theory by the testimony of his own personal feelings." [20] Although celebrated throughout Italy in his day as one who "with the most pure torch of philosophy has thrown beams of light into the darkest recesses of poetry and eloquence," [21] it does not appear that the distinguished scholar, the pleasing and desultory[1] philosopher, offered any profound or original solutions. In 1797 he defined poetry as "the art of representing and perfecting nature by means of picturesque, animated, imaginative and harmonious discourse." [22]

Bettinelli[2] *and Pagano.*

The fashion of the day in philosophy made men impatient of the ideas found in writers of treatises of former times. Arteaga praises Cesarotti for "that fine tact, that impartial criticism, that logical spirit derived not from the trickling streamlets of Sperone[3], Castelvetro, Casa[4] and Bembo[5], but from the profound and inexhaustible springs of Montesquieu, Hume[6], Voltaire, d'Alembert[7], Sulzer[8], and writers of like temper." [23] Writing to Saverio Bettinelli,

---

1. desultory: 随意的
2. *Bettinelli*：贝蒂内里（1718—1808），意大利耶稣会作家，著有《米尔之后的艺术与风俗》《学习的复兴》等。
3. Sperone：斯佩罗尼（1500—1588），意大利文艺复兴时期人文主义者、剧作家，著有《修辞学谈话录》。
4. Casa：卡萨（1503—1556），意大利佛罗伦萨诗人、作家、外交家、法官。
5. Bembo：本博（1470—1547），意大利学者、诗人、文学理论家，著有《论平民语言》《维尼塔历史》等。
6. Hume：休谟（1711—1776），英国苏格兰哲学家、历史学家、经济学家，散文家，著有《人性论》《英格兰史》《宗教的自然历史》等。
7. d'Alembert：达朗贝尔（1717—1783），法国数学家、物理学家、哲学家、音乐理论家，著有《动力学》《数学手册》等。
8. Sulzer：舒尔策（1508—1585），瑞士神学家、宗教改革家。

who was preparing a work on *Enthusiasm*, Paradisi[1] hoped it would prove "a metaphysical history of enthusiasm which shall outweigh all those Poetics which are only fit to be burned," and would "make waste paper of Castelvetro, the 'Mintumo,' and that stupid creature, Quadrio." [24] In spite of these aspirations Bettinelli's book (1769) contains little beyond vivacious and eloquent empirical observations concerning the psychology of poets, "poetic enthusiasm," to which he assigns six degrees, namely, elevation, vision, rapidity, novelty and surprise, passion and transfusion. Equally empirical was Mario Pagano in his two fragments, *Gusto e le belle arti* and *Origine e natura della poesia* (1783–1785), in which he grotesquely combines some ideas from Vico with the current sensationalism. Theoretico-imaginative form and sensuous pleasure are presented by him as two historical periods of art. "In their cradle the fine arts are directed towards making a true imitation of nature rather than towards loveliness. Their first steps are towards expression rather than charm ... . In the most ancient poetry, even in the ballads of barbarous ages, there lives a most compelling pathos: passions are expressed naturally, even the sound of the words is alive with the expression of the things described." But "the period of perfection is reached at the moment when exact imitation of nature is coupled with complete beauty, accord and harmony," when "the taste is refined and society reaches its most complete form of culture." Fine arts "precede by a short time the dawn of philosophy, that is to say, the time of the most intense perfection of society"; indeed, certain modes of art, such as tragedy, must necessarily come later than philosophy whose aid must be invoked to further "the purgation of manners[2]." [25]

---

1. Paradisi：帕拉迪西（1736—1783），意大利诗人、经济学家、教育家，著有文中提到的《品鉴力与美术》（*Gusto e le belle arti*），《诗歌的起源与本质》（*Origine e natura della poesia*）等。
2. purgation of manners：移风易俗

German disciples of Baumgarten: G. F. Meier.

The compatriots and successors of Baumgarten, like those of Vico, did little by way of understanding or improving upon his work. An enthusiastic admirer and disciple of Baumgarten who had attended his lectures at Frankfort-on-the-Oder, Georg Friedrich Meier[1], came forward in 1746 to defend the *Meditationes* against the attacks of Quistorp to whom the master had deigned no reply; [26] already in 1748, prior to the publication of the *Æsthetic*, he had published the first volume of his *Principles of all the Beautiful Sciences*, [27] followed in 1749 and 1750 by the second and third volumes. This book, which is a complete exposition of Baumgarten's theory, is divided, according to the master's method, into three parts: invention of beautiful thoughts (heuristic), æsthetic method (methodic), and the beautiful signification of thoughts (semiotic); the first of these (occupying two and a half volumes) is subdivided into three sections: beauty of sense-apprehension (æsthetic richness, grandeur, verisimilitude, vivacity, certainty, sensitive life and wit), sensitive faculties (attention, abstraction, senses, imagination, subtlety, acumen, memory, poetic power, taste, foresight, conjecture, signification and the minor appetitive faculties), and the diverse kinds of beautiful thought (æsthetic concepts, judgements, and syllogisms). Elsewhere than in this book, which was reprinted many times (in 1757 an epitome was issued [28]), Meier discusses Æsthetic in several of his numerous works, especially in a little tract, *Considerations on the First Principles of all Fine Arts and Sciences.* [29] Who was more tenderly inclined than he towards the science so recently born and baptized? He was ardent

1. Georg Friedrich Meier：乔治・弗里德里希・梅尔（1718—1777），德国哲学家、美学家，鲍姆加登的追随者，著有《形而上学》《试论通用设计艺术》《论所有美术与科学的基本原理》（*Considerations on the First Principles of all Fine Arts and Sciences*）等。

in her defence against those who denied both her possibility and her utility, and against those who admitted these yet complained, not unreasonably, that she was substantially the same as that which in former days had been treated as Poetics and Rhetoric. He parried[1] this accusation, of which he recognized the partial truth, by asserting that it was impossible for one writer to have perfect knowledge of all the arts: another of his excuses was to the effect that Æsthetic was a science too young to show the perfection reached by other sciences after the cultivation of centuries; in one place he says he has no intention of arguing "with those enemies of Æsthetic who will not or cannot see the true nature and aim of this science, but have built for themselves in its place a deformed and miserable image against which, when they fight, they fight against themselves." With philosophic resignation he concludes that the same fate is in store for Æsthetic as for every science: "At first when almost unknown they encounter enemies and detractors who ridicule them through ignorance and prejudice; but later they meet persons of intellect who, by working at them conjointly, carry them on to their proper perfection." [30]

Confusions of Meier.

Students of the new science flocked to Halle University[2] to hear Meier lecture on Æsthetic whose "chief author" or "inventor" (*Haupturheber, Erfinder*), as Meier never tired of repeating, was "Herr Professor Baumgarten"; at the same time warning them that his own *Anfangsgründe*[3] were no mere transcription of Baumgarten's lectures. [31] Still, while recognizing the great gifts of Meier as publicity-agent, the facility, clarity and wealth of his eloquence, and his shrewdness in

1. parried: 避开
2. Halle University: 德国哈勒大学。1649 年，由骑士学院改制而成立。
3. *Anfangsgründe*：〈德〉《原理》，即《论所有美术与科学的基本原理》。

polemic, one cannot altogether deny the justice of the remark upon "Professor Baumgarten of Frankfort and his ape[1] (*Affe*) Professor Meier of Halle." [32] Every defect of Baumgarten's Æsthetic reappears accentuated in Meier; the limits of the inferior cognitive faculties, alleged as the domain of poetry and the arts, are laid down by him most strangely. It is curious to note how, for example, he interprets the difference between the confused (æsthetical) and the distinct (logical), and the proposition that beauty disappears when made the object of distinct thought. "The cheeks of a beautiful girl whereon bloom the roses of youth are lovely so long as they are looked at with the naked eye. But let them be examined with a magnifying glass. Where is their beauty? One can hardly believe that such a disgusting surface, scaly, all mounts and hollows, the pores full of dirt, with hairs sprouting here and there, can be the seat of that amorous attraction which subdues the heart." [33] That is described as "æsthetically false" whose truth the inferior faculty is unable to grasp: for example, the theory that bodies are composed of monads[2]. Once they have become intelligible to these faculties, general concepts possess great æsthetic richness, since they include infinite consequences and particular cases. [34] Æsthetic also comprehends those things which cannot be thought distinctly or, if so thought, might be capable of upsetting philosophic gravity: a kiss may be an excellent subject for a poet; but whatever would be thought of a philosopher who sought to demonstrate its necessity by the mathematical method? [35] Moreover, Meier includes the whole theory of observation and experiment in Æsthetic, to which this theory belongs, he says, by right of its connexion with the senses, [36] and also the whole theory of the appetitive faculties, because "æsthetic requires not only a fine wit but a noble heart as well." [37] He comes near truth sometimes, when, for

---

1. ape：模仿者
2. monads：单细胞

example, he observes that the logical form presupposes the æsthetic and that our first concepts are sensitive, later becoming distinct by the help of logic; [38] and when he condemns allegory as "among the most decadent forms of beautiful thinking." [39] But, on the other hand, he thinks that logical distinctions and definitions, although not necessarily sought after by genius, are very useful in poetry; they are even indispensable as regulators of beautiful thinking and make up, as it were, the skeleton of the body poetic: great care, however, must be taken not to judge æsthetical general concepts, *notiones æstheticæ universales*[1], with the rigorous exactitude[2] demanded by philosophical. And since such concepts, taken singly, may be likened to unstrung jewels, they must be connected by the string of æsthetic judgement and syllogism, the theory of which is identical with that presented by Logic, setting aside that part which is of little or no use to genius, but belongs exclusively to the philosopher. [40] In his *Considerations* of 1757 Meier, having combated the principle of imitation (which appeared to him at once too broad, since science and morals are also imitations of nature, and too narrow, since art does not imitate natural objects solely nor should it imitate them all, for the immoral must be excluded), reaffirmed the thesis that the æsthetic principle consists in the "greatest possible beauty of sense-perception." [41] He upheld this by condemning as erroneous the belief that this sense-perception is wholly sensuous and confused, without any gleam of distinctness or rationality. The perception of sweet, bitter, red, etc., is wholly sensuous; but there is another perception which is both sensuous and intellectual, confused and distinct, in which both faculties, the higher and the lower, collaborate. When intellectuality prevails in this consciousness, then we have science: when sensibility, then we have poetry. "From our explanation it will be gathered that the inferior

1. *notiones æstheticæ universales*：〈拉〉审美的一般性概念
2. rigorous exactitude: 严格意义

cognitive faculties must collect all the material of a poem, and all its parts. Intellect and judgement, on the other hand, watch and ensure that these materials are placed side by side in such a way that in their connexion distinction and order may be observed." [42] Here a plunge into sensationalism, there a fugitive glimpse of truth: most often, and in conclusion, an adherence to the old mechanical, ornamental, pedagogic theory of poetry: this is the impression left on us by the æsthetic writings of Meier.

M. Mendelssohn[1] *and other followers of Baumgarten. Vogue of Æsthetic.*

Another disciple of Baumgarten, Mendelssohn, conceiving beauty as "indistinct image of a perfection," deduced that God can have no perception of beauty, as this is merely a phenomenon of human imperfection. According to him a primary form of pleasure is that of the senses, arising from "the bettered state of our bodily constitution"; a secondary form is the æsthetic fact of sensible beauty, that is to say, unity in variety; a third form is perfection, or harmony in variety. [43] He too repudiates[2] Hutcheson's *deus ex machina*[3], the sense of beauty. Sensible beauty, perfection such as can be apprehended by the senses, is independent of the fact that the object represented is beautiful or ugly, good or bad by nature; it suffices that it leaves us not indifferent: whence Mendelssohn agrees with Baumgarten's definition, "a poem is a discourse sensibly perfect." [44] Elias Schlegel[4] (1742) conceived

1. *M. Mendelssohn*：门德尔松（1729—1786），德国犹太裔哲学家，18 至 19 世纪犹太人启蒙运动发起人。
2. repudiates：拒绝接受
3. *deus ex machina*：〈拉〉机械降神，戏剧术语。当剧情发展到难以继续的情节，就用滑轮从舞台上方降下神来解决困境。
4. Elias Schlegel：伊利亚斯 · 施莱格尔（1719—1749），德国评论家、诗人，著有诗剧《妇女的胜利》《哑巴美人》及文中提到的《论品鉴力》（*Considerations on Taste*）等。

art as imitation, not so servile as to seem a copy, but having similarity rather than identity with nature: he considered the duty of poetry was first to please and only afterwards to instruct. [45] Treatises on Æsthetic, university lectures or slender volumes for use of the public, *Theories of the Fine Arts and Letters, Manuals, Sketches, Texts, Principles, Introductions, Lectures, Essays,* and *Considerations on Taste* poured down thick and fast on Germany during the second half of the eighteenth century. There are at least thirty full or complete treatises and many dozens of minor tracts or fragments. After the Protestant universities, the Catholic took up the new science, which was taught by Riedel[1] at Vienna, Herwigh at Würzburg, Ladrone at Mainz, Jacobi at Freiburg, and by others at Ingolstadt after the expulsion of the Jesuits. [46] A pretty little volume on the First Principles of the Fine Arts [47] was written (1790) for Catholic schools by the notorious Franciscan friar Eulogius Schneider[2], who, after being unfrocked, terrorised Strasburg in the days of the Convention, and met his end under the guillotine. The frenzied output of these German *Æsthetics* resembles that of *Poetics* in Italy in the sixteenth century, after the rise to popularity of Aristotle's treatise. Between 1771 and 1774 the Swiss Sulzer brought out his great æsthetic encyclopædia, *The General Theory of the Fine Arts*[3], in alphabetical order, with historical notes upon each article, which were greatly enlarged in the second edition of 1792, edited by a retired Prussian captain, von Blankenburg. [48] In 1799, one J. Koller published a first *Sketch of the History of Æsthetic*[4], [49] in which he observes not unjustly, "Patriotic youth will be pleased to recognize that Germany has produced more literature on this subject than any

1. Riedel：即弗雷德里希·雷德尔（1742—1785）德国作家，美学家。
2. Eulogius Schneider：欧洛基乌斯·施耐德（1756—1794），德国方济会修士、哲学家，波恩大学教授，著有《论女性教育》《美术的基本原理通论及具体写作中的美的风格》等。
3. *The General Theory of the Fine Arts*：《美术的通用理论》，舒尔策著。
4. *Sketch of the History of Æsthetic*：《美学史草稿》，科勒尔著。

other country." [50]

Eberhard[1] *and Eschenburg*[2].

Confining ourselves to bare mention of the works of Riedel (1767), Faber (1767), Schütz (1776–1778), Schubart[3] (1777–1781), Westenrieder[4] (1777), Szerdahel (1779), König (1784), Gang (1785), Meiners (1787), Schott (1789), Moritz (1788), [51] we will select from the crowd the *Theory of Fine Arts and Letters*[5] (1783) of Johann August Eberhard, successor to Meier in the Chair at Halle, [52] and the *Sketch of a Theory and Literature of Letters*[6] (1783) by Johann Joachim Eschenburg, one of the most popular books of the day for students. [53] Both these authors are followers of Baumgarten, with inclinations towards sensationalism; amongst other things Eberhard considered the beautiful as "that which pleases the most distinct senses," that is to say, of sight and hearing.

J. G. Sulzer.

A word must be accorded to Sulzer, in whom we find the most curious alternation of new and old, the romantic influence of the new Swiss school and the utilitarianism and intellectualism of his day. He asserts that beauty exists wherever unity, variety and order are found:

---

1. *Eberhard*：艾伯哈德（1739—1809），德国神学家、哲学家，著有《美术与美学的理论》。
2. *Eschenburg*：即约翰·约阿希姆·爱森堡（1743—1820），德国评论家与文学史学家，著有《古典文学手册》《美学的理论与文学草稿》等。
3. Schubart：舒巴特（1739—1791），德国诗人、作曲家、演奏家、记者。
4. Westenrieder：韦斯顿里德尔（1748—1829），德国著名作家、历史学家，著有《美学导论》等。
5. *Theory of Fine Arts and Letters*：《美术与文学理论》，艾伯哈德著
6. *Sketch of a Theory and Literature of Letters*：《文学理论草稿》，约翰·约阿希姆·爱森堡著。

the work of an artist is strictly in the form, in lively expression (*lebhafte Darstellung*): the material is irrelevant to art, but the duty of every reasonable and sensible man is to make judicious selection. The beauty which is used to clothe the good as well as the bad is not the ineffable, celestial Beauty, offspring of the alliance between the beautiful, the good and the perfect, which awakens more than mere pleasure, a veritable joy which ravishes and beatifies our soul. Such is the human face when, by filling the eye of the beholder with the pleasure of form arising from the variety, proportion and order of the features, it proceeds to arouse the imagination and intellect by its suggestion of interior perfection; of the same nature is the statue of a great man carved by Phidias[1], or a patriotic oration by Cicero. If truth lie outside art and belong to philosophy, the most noble use to which art may be put is to make us feel the important truths which lend her strength and energy, not to mention that truth itself enters into art in the shape of truthful imitation or representation. Sulzer also repeats (and he is not the last) that orators, historians and poets are intermediaries between speculative philosophy and the people. [54]

K. H. Heydenreich.

Karl Heinrich Heydenreich[2] returns to a sounder tradition when he defines art (1790) as “a representation of a determinate state of sensibility,” and observes that man, as a cognitive being, is impelled to enlarge the sphere of his cognitions and impart his discoveries to his fellows, while as a sensitive being he is impelled to represent

1. Phidias：菲狄亚斯（公元前 480—前 430），古希腊著名雕塑家、建筑设计师。
2. Karl Heinrich Heydenreich：卡尔·海因里希·海顿雷希（1764—1801），德国哲学家、诗人，著有《论门德尔松如何表现斯宾诺莎》《美学体系》等。

and communicate his sensations; whence arise science and art. But Heydenreich does not clearly grasp the cognitive character of art; for in his opinion sensations become objects of artistic representation either because they are pleasing or, when not pleasing, because they are useful to further the moral aims of man as a social being; the objects of sensibility which enter into art must be possessed of intrinsic excellence and value and bear reference not to a single individual but to the individual as a rational being: hence the objectivity and necessity of taste. Like Baumgarten and Meier, he divides Æsthetic into three parts: a doctrine of *inventio*, another of *methodica*, a third of the *ars significandi*.[1] [55]

J. G. Herder.

Another disciple of Baumgarten is J. G. Herder, who had an unbounded admiration for the old Berlin master, whom he calls "the Aristotle of his day," and defends him warmly against those who think fit to describe him as a "stupid and obtuse syllogizer[2]" (1769). On the other hand he had slight esteem for subsequent Æsthetic, for example Meier's work, which he stigmatized[3] accurately enough as "in part a re-mastication of Logic, in part a patchwork of metaphorical terms, comparisons and examples." "O Æsthetic!" he cries with emphasis, "O Æsthetic! the most fertile, the most beautiful and by far the most novel of all abstract sciences, in what cavern of the Muses is sleeping the youth of my philosophic nation destined to bring thee to perfection?" [56] He denied Baumgarten's claim to have established an *Ars pulchre cogitandi* instead of limiting himself to a simple *Scientia de*

1. *inventio ... methodica ... ars significandi*: 〈拉〉思维的发明，审美的方法，艺术的意义
2. stupid and obtuse syllogizer：愚钝的三段论者
3. stigmatized：指责

pulchro et pulchris philosophice cogitans[1], and ridiculed the scruple which held Æsthetic to be unworthy of the dignity of Philosophy. [57] To compensate for this, however, he accepted the fundamental definition cf poetry as *oratio sensitiva perfecta*[2]: gem of definitions (says he), the best that has ever been invented, that penetrates to the heart of the matter, touches the true poetic principles and opens the most extended view over the entire philosophy of the beautiful, "coupling poetry with her sisters, the fine arts." [58] Like Cesarotti the Italian, but with much less vivacity and brilliance, Herder the German had studied primitive poetry, Ossian and the songs of ancient peoples, Shakespeare (1773), popular love-songs (1778), the spirit of Hebrew poetry (1782), and oriental poetry; these studies powerfully impressed upon his mind the sensitive nature of poetry. His friend Hamann[3] (1762) had written these memorable words, which read like an extract from one of Vico's aphorisms: "Poetry is the mother-tongue of mankind: in the same way that the garden is older than the ploughed field, painting than writing, song than declamation, barter than trade. The repose of our most ancient progenitors was a slumber deeper than ours; their motion a tumultuous dance. They spent seven days in the silence of thought or of stupor; and opened their mouths to pronounce winged words. Their speech was sensation and passion, and they understood nothing but images. Of images is composed all the treasure of human knowledge and felicity." [59] Although Herder, who knew and admired Vico, [60] does not mention him by name when treating of language and poetry, one might suppose him to be influenced by the great Neapolitan at least in the final consolidation of his theories; but, on

1. *Scientia de pulchro et pulchris philosophice cogitans*：〈拉〉用美和美的哲学思维的科学
2. *oratio sensitiva perfecta*：〈拉〉完美的感性言辞
3. Hamann：哈曼（1730—1788），德国哲学家，著有《论宗教的力量与犹太教》。

the contrary, the authors whom he chiefly quotes in this connexion are Du Bos, Goguet[1] and Condillac[2], and observes "the first beginnings of human speech in tone, gesture, expression of sensations and thoughts by means of images and signs, can only have been a kind of crude poetry, and so it is among every savage nation in the world." Not a speech with punctuation and a sense of syllable, like ours, learning as we do to read and write, but an unsyllabled melody which gave birth to the primitive epic. "Natural man depicts what he sees and as he sees it, alive, powerful, monstrous; in order or disorder, as he sees and hears, so he reproduces. Not alone did barbarous tongues thus arrange their images, but Greek and Latin do the same. As the senses offered material, so the poets utilized it; especially in Homer we see how closely nature is followed in images which glow and fade perpetually and inimitably[3]. He describes things and events line by line, scene by scene; and, in the same way, he paints men in their very bodies, actually as they speak and move." Later we distinguish epic from what we call history; because the former "not only describes what has happened but describes the event in its entirety, showing how it occurred in the only possible way, having regard to surrounding circumstance of body and spirit": this is the reason of the more philosophical character of poetry. As for pleasure, no doubt we do find poetry pleasant; but the idea that the poet's motive is merely to excite pleasure cannot be condemned too strongly. "Homer's gods were as essential and indispensable to the poet's world as the forces of motion are to the world of matter. Without the deliberations and activities of Olympus, none of the necessary events which happen on this earth

---

1. Goguet：即安东尼·高古埃（1716—1758），法国法学家、历史学家，著有《法律、艺术与科学的起源》《古代人的发展历程》等。
2. Condillac：孔狄亚克(1714—1780)，法国哲学家、知识论学家，著有《论人类认识的起源》《论体系》等。
3. perpetually and inimitably：永无止境地和无以伦比地

could take place. Homer's magic island in the western sea belongs to the map of his hero's wanderings by the same necessity which placed it on the map of the world: it was necessary to the plan of his poem. It is the same with the severe Dante and his circles of Hell and Heaven." Art is formative: she disciplines, orders and governs the imagination and every faculty of man: not only did she generate history, "but, earlier yet, she created gods and heroes and purified the uncouth imaginations and fables of peoples with their Titans, monsters and Gorgons, reducing to limit and law the riotous imagination of ignorant men which knows no bounds or rule." [61]

Notwithstanding these intuitions, so like those of Vico early in the same century, Herder as a philosopher is inferior to his Italian predecessor, and in point of fact does not rise superior to Baumgarten. By application of Leibniz' law of continuity, he too arrived at the opinion that the pleasing, the true, the beautiful and the good are degrees of one single activity. For instance, sensible pleasure " is a participation in the true and the good, so far as the senses may comprehend them; the feeling of pleasure and pain is no other than the feeling of the true and the good, that is to say, the consciousness that the aim of our organism, the conservation of our well-being and the avoidance of our hurt, has been attained." [62] Fine arts and letters are all instructive (*bildend*): hence the terms *humaniora*[1], the Greek *καλόν*[2], the Latin *pulchrum*[3], the *gentle* arts of days of chivalry, *les belles lettres et les beaux arts*[4] of the French. A group of them (gymnastic, dance, etc.) educates the body; a second group (painting, plastic, music) educates the nobler senses of man, the eye, the ear, the hand and tongue; a third (poetry) touches the

---

1. *humaniora*：〈拉〉人文科学
2. *Καλόν*：〈希〉美
3. *pulchrum*：〈拉〉美
4. *les belles lettres et les beaux arts*：〈法〉文学与艺术

intellect, the imagination and the reason: a fourth group governs human tendencies and inclinations. [63] Herder disapproved of the facile theorists of art who began straight away with a definition of beauty, a complex and involved concept. He held that the theory of fine arts should be subdivided into three theories, each to be built up from the foundations, the theory of sight, of hearing and of touch, that is to say of painting, music and sculpture, *i.e.* into æsthetical Optics, æsthetical Acoustics and æsthetical Physiology. "Fairly well elaborated in the psychological and subjective aspects, Æsthetic is sadly undeveloped in all that belongs to the object and to the sensation of beauty, without which there can never be a fertile theory of the Beautiful capable of influencing all the arts." [64] Taste is not "a fundamental faculty of the soul but a habitual application of our judgement (intellectual judgement) to objects of beauty"; an acquired facility of the intellect (of which Herder outlines the genesis). [65] The poet is poet not only in his imagination but in his intellect. In 1782 he writes: "The barbarous name Æsthetic of recent invention indicates nothing beyond a section of Logic: that which we call taste is neither more nor less than a quick and rapid judgement which does not exclude truth and profundity, but rather presupposes and promotes them. All didactic poetry is nothing more than philosophy rendered sensible: the fable as exposition of a general doctrine is truth in act, in activity ... . When expounded and applied to human affairs, Philosophy is not only a fine art in herself (*schöne Wissenschaft,*) but the mother of Beauty: it is only through her that Rhetoric and Poetry can ever be educational, useful, or in the truest sense pleasant." [66]

Philosophy of language.

Herder and Hamann deserve our gratitude for having brought a current of fresh air into the study of the philosophy of language.

The lead given by the Port-Royal authors had been followed since the beginning of the century by many writers of logical or general grammars. According to the French Encyclopædia, "*La grammaire générale est la science raisonnée des principes immuables et généraux de la parole prononcée ou écrite dans toutes les langues*[1]," [67] and d'Alembert spoke of grammarians of invention and grammarians of memory, assigning to the former the duty of studying the metaphysics of grammar. [68] General grammars had been written by Du Marsais[2], De Beauzée[3], and Condillac in France; Harris[4] in England; and many others. [69] But what was the relation between general grammar and particular grammars? If logic be one, how comes it that languages are many? Is the variety of tongues but a deviation on their part from one single model? And, if there be no such deviation or error, what is the explanation of the fact? What is language, and how was it born? If language be external to thought, how can thought exist if not in language? "*Si les hommes*," says Jean-Jacques Rousseau[5], "*ont eu besoin de la parole pour apprendre à penser, ils ont eu bien plus besoin encore de savoir penser pour trouver l'art de la parole*" [6]; appalled at the difficulty, he declares his conviction "*de l'impossibilité presque démontrée que les langues*

1. *La grammaire générale est la science raisonnée des principes immuables et généraux de la parole prononcée ou écrite dans toutes les langues*：〈法〉通用语法是所有语言的口语或书面语不变的、通行的规则的理性学科。
2. Du Marsais：杜马塞（1676—1756），法国哲学家、语法学家，著有《学习拉丁文的合理方法》《语法原理》，参与编写《百科全书》。
3. De Beauzée：迪博泽（1717—1789），法国语言学家，著有《通用语法》，参与编写《百科全书》。
4. Harris：哈里斯（1709—1780），英国政治家、语法学家，著有《赫尔墨斯，语言与通用语法的哲学探究》。
5. Jean-Jacques Rousseau：让·雅克·卢梭（1712—1788），法国哲学家、作家，其政治哲学影响了欧洲启蒙运动，著有《爱弥儿》《忏悔录》《社会契约论》等。
6. *Si les hommes ... ont eu besoin de la parole pour apprendre à penser, ils ont eu bien plus besoin encore de savoir penser pour trouver l'art de la parole*：〈法〉如果需要通过语言学习如何思考，那就先要学会思考怎么才能找到语言的艺术。

aient pu naître et s'établir par des moyens purement humains.[1]" [70] Such questions became fashionable; books on the origin and formation of language were written by de Brosses[2] (1765) and Court de Gébelin[3] (1776) in France, by Monboddo[4] (1774) in England, Süssmilch[5] (1766) and Tiedemann[6] in Germany, and Cesarotti(1785) in Italy, and by others who had some slight acquaintance with Vico, but profited little by it. [71] None of the above-named writers was able to free himself of the notion that speech was either natural and mechanical, or else a symbol attached to thought: whereas in fact it was impossible to solve the difficulties under which they were labouring except by dropping the notion of a sign or symbol and attaining the conception of the active and expressive imagination, verbal imagination, language as the expression not of intellect but of intuition. An approach towards this explanation was made by Herder in a brilliant and imaginative thesis in 1770 upon this subject of the origin of language, chosen for discussion by the Berlin Academy. In it he says that language is the reflexion or consciousness (*Besonnenheit*[7]) of man. "Man shows reflexion when he puts forth freely such force of mind as enables him to make selection from amongst the crowd of sensations by which he

---

1. *de l'impossibilité presque démontrée que les langues aient pu naître et s'établir par des moyens purement humains*：〈法〉纯粹靠人类自身能力产生和建立语言，几无可能，这一点已几乎被证明。
2. de Brosses：布罗塞（1709—1777），法国作家、语言学家，著有《论语言的机械形式与词源的物理原理》。
3. Court de Gébelin：科特·德·吉贝林（1725—1784），法国新教牧师、语言学家，著有《论语言的起源与通用语法》。
4. Monboddo：蒙博多（爵士）詹姆斯·伯内特（1714—1799），英国苏格兰法官、语言学家、哲学家、自然神论者，著有《语言的起源与发展历程》《古代形而上学》等。
5. Süssmilch：苏斯米尔西（1707—1767），德国新教牧师、统计学家、人口学家。
6. Tiedemann：狄德曼（1748—1803），德国哲学家、哲学史学家，著有《斯多葛哲学体系》《试解语言的起源》等。
7. *Besonnenheit*:〈德〉悟性

is assailed: from the ocean of the senses, so to speak, to select a single wave and consciously to watch it. He shows reflexion when, amidst the thronging chaos of images which pass before him as in a dream, he can in a waking moment collect himself and fasten his attention upon a single image, examine it calmly and clearly, and separate it from its neighbours. Once again, man shows reflexion when he is able not merely to grasp vividly and clearly all the properties of an image, but also to recognize one or more of its distinctive properties." The language of man "does not depend on the organization of the mouth, for even he who is dumb from birth has, if he reflects, a language; it is not a cry of the senses, since it resides in a reflective creature, not in a breathing machine; it is not an affair of imitation, since imitation of nature is a means, and we are here trying to explain the end: much less is it an arbitrary convention; a savage in the depths of the forest would have had to create a language for himself even though he never used it. Language is an understanding of the soul with herself, necessary just in so far as man is man." [72] Here language begins to show itself no longer as purely mechanical or as something derived from arbitrary choice and invention, but as a creative activity and a primary affirmation of the activity of the human mind. Herder's essay may not state such a view unequivocally, but it points forward to such a conclusion in a striking way for which its author has not received the credit he deserves. Hamann, in reviewing his friend's theories, agreed with him in denying the origin of language by invention or arbitrary choice; while dwelling also on the liberty of man, he regarded language as something which man could only have learned by means of a mystical *communicatio idiomatum*[1] from God. [73] That, too, was one way of recognizing that the mystery of language is not to be solved except by placing it in the forefront of the problem of the spirit.

---

1. *communicatio idiomatum*：〈拉〉属性互通

[1] Vico, *Opere, ed. cit.* iv. p. 305.

[2] Herder, *Briefe zur Beförderung der Humanität*, 1793–1797, Letter 59; Goethe, *Italien. Reise*, Mar. 5, 1787.

[3] Letters from Wolf to Cesarotti, June 5, 1802; in Cesarotti, *Opere*, vol. xxxviii. pp. 108–112; cf. *ibid.* pp. 43–44, and vol. xxxvii. pp. 281, 284, 324; cf. on the question of the relations between Wolf and Vico, Croce, *Bibliografia vichiana,* pp. 51, 56–58, and *Supplem.* pp. 12–14.

[4] Letter in French to Mme. Ferrant (1719), and to the Marquis Maffei in *Prose e poesie*, vol. ii. (1756), pp. lxxxv.-civ., cviii.-cix.

[5] *Prose e poesie,* vol. i., 1739, pref.

[6] Francis Hutcheson (1694–1746) was an Irishman. Croce's mistake is probably due to the fact that he studied and taught at Glasgow, or that his family was ultimately of Scottish origin. — TR.

[7] *Prose e poesie*, vol. ii. pp. clxxi.-clxxvii.

[8] See above, pp. 216–217.

[9] *Prose e poesie*, vol. ii. pp. 242–246.

[10] *Op. cit.* ii. p. 249.

[11] *Op. cit.* ii. pp. 252–253.

[12] *Prose e poesie*, vol. ii. pp. 233–234.

[13] *Op. cit.* i. pref.

[14] *Op. cit.* ii. p. 127.

[15] *Op. cit.* i. p. xliii.

[16] Fr. Sav. Quadrio, *Della storia e della ragione d'ogni poesia*, Bologna, 1739, vol. i. part i. dist. i. ch. 1.

[17] Fr. M. Zanotti, *Dell' arte poetica, ragionamenti cinque*, Bologna, 1768.

[18] On Ossian, *Opere*, vols, ii.-v.; on Homer, vols, vi.-x.; *Saggio copra il diletto della tragedia,* vol. xxix. pp. 117–167; *Saggio sul bello*, vol. xxx. pp. 13–70; on *Filosofia del gusto*, vol. i.; on *Eloquenza*, lecture, vol. xxxi.

[19] *Opere*, vol. xl. p. 49.

[20] *Ibid.* p. 55.

[21] Letter from Corniani to Cesarotti, November 21, 1790, in *Opere*, vol. xxxvii. p. 146.

[22] *Saggio sopra le istituzioni scolastiche, private e pubbliche*, in *Opere*, vol. xxix. pp. 1–116.

[23] Letter of March 30, 1764, in *Opere*, vol. xxxv. p. 202.

[24] Saverio Bettinelli, *Dell'entusiasmo nelle belle arti, 1769,* in *Opere*, iii. pp. xi.-xiii.

[25] Fr. M. Pagano, *De'saggi politici*, Naples, 1783–1785, vol. i. Appendix to § 1, "Sull'origine e natura della poesia"; vol. ii. § 6, "Del gusto e delle belle arti."

[26] See above, p. 217.

[27] *Anfangsgründe aller schönen Wissenschaften*, Halle, 1748–1750.

[28] *Auszug aus den Anfangsgründe*, etc., *ibid.* 1758.

[29] *Betrachtungen über den ersten Grundsätzen aller schönen Künste u. Wissenschaften, ibid.* 1757.

[30] Preface to 2nd ed. (1768) of vol. ii. of *Anfangsgründe*, and *Betrachtungen, cit.*, esp. §§ 1, 2, 34.

[31] Preface to vol. i., and cf. § 5.

[32] In a letter to Gottsched, 1747, in Danzel, *Gottsched*, p. 215.

[33] *Anfangsgründe*, § 23.

[34] *Op. cit.* § 49.

[35] *Op. cit.* § 55.

[36] *Op. cit.* §§ 355–370.

[37] *Op. cit.* §§ 529–540.

[38] *Op. cit.* § 5.

[39] *Op. cit.* § 413.

[40] Anfangsgründe, §§ 541–670.

[41] Betrachtungen, § 20.

[42] Op. cit. § 21.

[43] *Briefe über die Empfindungen*, 1755 (in *Opere filosofiche*, Ital. trans.,

Parma, 1800, vol. ii.). Letters 2, 5, 11.

[44] *Betrachtungen üb. d. Quellen d. sch. Wiss. u. K.*, 1757, later entitled *Über die Hauptgrundsätze*, etc., 1761, in *Opere, ed. cit.* ii. pp. 10, 12–15, 21–30.

[45] J. E. Schlegel, *Von der Nachahmung*, 1742; cf. Braitmaier, *Gesch. d. poet. Th.* i. p. 249 *sqq.*

[46] Koller, *Entwurf*, p. 103.

[47] *Die ersten Grundsätze der schönen Kunst überhaupt, und der schönen Schreibart insbesondere*, Bonn, 1790; cf. Sulzer, i. p. 55, and Koller, pp. 55–56.

[48] See Bibliographical Appendix.

[49] *Entwurf zur Geschichte u. Literatur d. Ästhetik*, etc., Regensburg. 1799; see Bibl. App.

[50] Koller, *op. cit.* p. 7.

[51] Notices and extracts in Sulzer and Koller, *opp. citt.*

[52] Joh. Aug. Eberhard, *Theorie der schönen Künste u. Wissenschaften*, Halle, 1783; reprinted 1789, 1790.

[53] Joh. Joach. Eschenburg, *Entwurf einer Theorie u. Literatur d. s. W.*, Berlin, 1783; reprinted 1789.

[54] Allgem. Th. d. sch. Künste, on words Schön, Schönheit, Wahrheit, Werke des Geschmacks, etc.

[55] Karl Heinrich Heydenreich, *System der Ästhetik*, vol. i., Leipzig, 1790, esp. pp. 149–154. 367–385. 385–392.

[56] *Kritische Wälder oder Betrachtungen über die Wissenschaft und Kunst des Schönen*, Fourth Forest, 1769, in *Sämmtliche Werke*, ed. B. Suphan, Berlin, 1878, vol. iv. pp. 19, 21, 27.

[57] *Kritische Wälder, loc. cit.* pp. 22–27.

[58] Fragment, *Von Baumgarten Denkart*; and cf. *op. cit.* pp. 132–133.

[59] *Æsthetica in mice*, in *Kreuzzüge des Philologen*, Königsberg, quoted in Herder, *Werke*, xii. 145.

[60] See above, p. 267.

[61] *Kaligone*, 1800, in *Werke, ed. cit.*, xii. pp. 145–150.

[62] *Kaligone*, pp. 34–55.

[63] *Ibid.* pp. 308–317.

[64] *Kritische Wälder, loc. cit.* iv. pp. 47–127.

[65] *Op. cit.* pp. 27–36.

[66] *Sophron*, 1782, § 4.

[67] *Encyclopédie, ad verb.*

[68] *Éloge de Du Marsais*, 1756 (introd. to *Œuvres de Du Marsais*, Paris, 1797, vol. i.).

[69] Du Marsais, *Méthode raisonnée*, 1722; *Traité des tropes,* 1730; *Traité de grammaire générale* (in *Encyclopédie*); De Beauzée, *Grammaire générale pour servir de fondement à l'étude de toutes les langues*, 1767; Condillac, *Grammaire française*, 1755; J. Harris, *Hermes, or a Philosophical Enquiry concerning Language and Universal Grammar*, 1751.

[70] *Discours sur l'origine de l'inégalité parmi les hommes*, 1754.

[71] De Brosses, *Traité de la formation mécanique des langues*, 1765; Court de Gébelin, *Histoire naturelle de la parole*, 1776; Monboddo, *Origin and Progress of Language*, 1774; Süssmilch, *Beweis dass der Ursprung der menschlichen Sprache göttlich sei*, 1766; Tiedemann, *Ursprung der Sprache*; Cesarotti, *Saggio sulla filosofia delle lingue*, 1785 (in *Opere*, vol. i.); D. Colao Agata, *Piano, ovvero icerche filosofiche sulle lingue*, 1774; Soave, *Ricerche intorno all'istituzione naturale d'una società e d'una lingua,* 1774.

[72] *Abhandlung über den Ursprung der Sprache*, in a small book *Zwei Preisschriften*, etc. (2nd ed., Berlin, 1789), esp. pp. 60–65.

[73] Steinthal, *Ursprung der Sprache*, 4th ed., pp. 39–58.

VII OTHER ÆSTHETIC DOCTRINES OF THE SAME PERIOD

Other writers of the eighteenth century: Batteux.

A great medley of heterogeneous ideas is noticeable among other writers on Æsthetic during the same period. In 1746 appeared a little volume by Abbé Batteux[1] bearing the attractive title of *The Fine Arts reduced to a Single Principle* in which the author attempted a unification of all the different rules laid down by the writers of treatises. All such rules (says Batteux) are branches emerging from one trunk; he who possesses the simple principle will be able to deduce the rules one by one without entangling himself in their mass, which can but involve him in endless coils. The author had passed in review the *Ars Poetica* of Horace and that of Boileau, and the works of Rollin[2], Dacier, le Bossu and d'Aubignac[3]; but had found real help only in Aristotle's principle of imitation, which he thought could be easily and strikingly applied to poetry, painting, music and the art of gesture. But suddenly the Aristotelian principle of imitation yields place to a wholly new rendering, namely the "imitation of natural *beauty.*" The business of art is to "select the most beautiful parts of nature in order to frame

---

1. Abbé Batteux：查理·巴托神父（1713—1780），法国哲学家、美学家，著有《文学原理》和文中提到的《简化为单一原则的美术》（*The Fine Arts reduced to a Single Principle*）等。
2. Rollin：罗兰（1661—1741），法国历史学家、教育家。
3. d'Aubignac：奥比尼亚克神父（1604—1676），法国作家、牧师，著有《戏剧的实践》等。

them into an exquisite whole which shall be more beautiful than nature's self, without ceasing to be natural." Now, what may this greater perfection, this beautiful nature, be? On one occasion Batteux identifies it with truth: but "with the truth which may be; with beauty-truth, which is represented as though it really existed with all the perfections it could possibly receive," recalling one example from the ancients in the Helen of Zeuxis[1], and one from the moderns in the *Misanthrope*[2] of Molière . In another place he explains that beautiful nature, "*tum ipsius (obiecti) naturæ, tum nostræ convenit*[3]," i.e. that it has the closest connexion with our own perfection, our advantage and our interest, and is, at the same time, perfect in itself. The aim of imitation is "to please, to move, to soften, in one word, to delight"; so beautiful nature must be interesting and furnished with unity, variety, symmetry and proportion. Embarrassed by the question of artistic imitation of things naturally ugly or objectionable, Batteux falls back on saying, as Castelvetro had said before him, that displeasing objects please when imitated, since imitation, being always imperfect, in comparison with the reality, cannot excite the horror and disgust aroused by the latter. From pleasure he deduces the other aim of utility: if the aim of poetry be to give pleasure, and "pleasure by moving the passions, then in order to give a perfect and enduring pleasure it ought to rouse such passions only as it is well to excite, not those inimical[4] to goodness." [1]

The English: W. Hogarth[5].

It is difficult to string together a more insubstantial mass of

---

1. Helen of Zeuxis：《海伦》，宙西斯画作。
2. Misanthrope：《愤世者》，莫里哀剧作。
3. *tum ipsius (obiecti) naturæ, tum nostræ convenit*：〈拉〉时而适合事物自身的自然，时而适合我们的自然。
4. inimical：敌意的
5. Hogarth：贺加斯（1697—1764），英国画家和社会评论家。

contradictions. But Batteux is rivalled and outdone by the English philosophers or rather scribblers on Æsthetic or rather on things in general which sometimes accidentally include æsthetic facts. Happening to find in Lomazzo[1] some words attributed to Michæl Angelo on the beauty of shapes, Hogarth the artist took into his head the idea that the figurative arts can be regulated by a special principle which can be expressed in a particular fine. [2] Filled with this discovery, in 1745 he designed a frontispiece[2] for a volume of his engravings; it depicted a painter's palette scored across with an undulating line[3] and the words *The Line of Beauty*. Public curiosity was immediately aroused by this hieroglyphic, to be satisfied a little later by the publication of his book *The Analysis of Beauty* (1753). [3] In this he combated the mistake of judging pictures either by the subject or the excellence of the imitation instead of by their form, which is the true essential of art and is composed "of symmetry, variety, uniformity, simplicity, intricacy and quantity; all things which co-operate in the production of beauty, correcting and restraining each other as required." [4] But immediately afterwards Hogarth proclaims that there must also be correspondence and agreement with the thing copied; for "regularity, uniformity and symmetry give pleasure in so far only as they serve to give the illusion of faithful correspondence." [5] Further on, the reader learns that "amongst the immense variety of undulating lines which may be conceived, there is but one which truly merits the name of the Line of Beauty, and this is a precisely serpentine line which may be called the Line of Grace." [6] Again, we are told that intricacy of lines is beautiful because "the active mind likes to be engaged," and the eye delights in being "guided in a sort of hunt." [7] A straight line has no beauty, and the pig, the bear, the

1. Lomazzo：洛马佐，意大利科莫省下属城镇。
2. frontispiece：卷首插画
3. undulating line：曲线

spider and the toad are ugly because devoid of undulating lines. [8] The ancients showed much judgement in the management and grouping of lines, "varying from the precise line of grace only on those occasions when the character or action demanded." [9]

E. Burke[1].

With similar indecision Edmund Burke wavers between the principle of imitation and other heterogeneous or imaginary principles in his book, *An Enquiry into the Origin of our Ideas of the Sublime and the Beautiful* (1756). He observes, "Natural properties contained in an object give pleasure or displeasure to the imagination: beyond this, however, imagination may delight in the likeness of a copy to its original"; he asserts that from "these two reasons" arises the whole pleasure of imagination. [10]

Without dwelling further on the second, he proceeds to a lengthy discussion of the natural qualities which should be found in an object of sensible beauty: "Firstly, comparative smallness; secondly, smooth surface; thirdly, variety in disposition of the parts; fourthly, that it have no angularity, all lines fusing one in another; fifthly, a structure of great delicacy betraying no signs of violence; sixthly, vivid colouring without glare or harshness; seventhly, if it have any glaring colour, let it be different from the background." These are the properties of beauty working in harmony with nature and least liable to suffer from caprice and differences of taste. [11]

H. Home.

These books of Hogarth and Burke are generally described as

1. Burke：艾德蒙·柏克（1730—1797），英国作家、政治家和哲学家，著有《自然社会辨》《壮美和秀美观念的哲学探源》等。

classical; if so, they belong to the type of classic that fails to convince. To a somewhat higher type belongs the *Elements of Criticism* (1761) of Henry Home, Lord Kaimes[1], who seeks "the true principles of the fine arts" with the object of converting criticism into "a rational science," and to this end chooses "the upward path of facts and experiments." Home confines himself to feelings derived from objects of sight and hearing, which, in so far as unaccompanied by desires, are more truly described as simple feelings (emotions, not passions). These occupy a middle position between mere sense-impressions and intellectual or moral ideas, and are therefore akin to both; and it is from these that the pleasures of beauty are derived. Beauty is divided into beauty of relation and intrinsic beauty. [12] Of the latter, Home's only account is that regularity, simplicity, uniformity, proportion, order and other pleasing qualities have been "so disposed by the Author of nature in order to increase our happiness here on earth which, as is clearly shown in numberless instances, is not foreign to his care." This notion is confirmed when he reflects that "our taste for such details is not accidental, but uniform and universal, being a very part of our nature"; adding that "regularity, uniformity, order and simplicity help to facilitate perception and make it possible for us to form clearer conception of objects than it would be possible to gain by the most earnest attention were such qualities not present." Proportions are often combined with a view to utility, "as we see that the best proportioned amongst animals are also the strongest; but there are also many examples in which this conjunction does not hold good[2]"; wherefore the wisest plan "is to rest content with the final cause just

1. Henry Home, Lord Kaimes：亨利·霍姆即凯姆斯勋爵（1696—1782），英国苏格兰法官、哲学家、作家、农业改良家，著有《历史上的法律线索》《人的历史手稿》及文中提到的《批评的要素》（*Elements of Criticism*）等。
2. hold good：适用

mentioned: that of the increase of our happiness intended by the Author of nature." [13] In his *Essay on Taste* (1758) and on *Genius* (1774) Alexander Gérard[1] employs by turns, according to the various forms of art, the principles of association, of direct pleasure, of expression, and even of moral sense: the same kind of explanation reappears in another *Essay on Taste* by Alison[2] (1792).

Eclecticism and sensationalism: E. Platner[3].

It is impossible to classify works of such calibre, almost wholly lacking as they are in scientific method; on each page their writers pass from physiological sensationalism to moralism; from the imitation of nature to mysticism and transcendent finalism[4] without the slightest sense of incongruity[5]. It would be absurd to take them seriously; in comparison it is almost refreshing to come across a frank hedonist in the German, Ernst Platner, who interpreted Hogarth's inquiry into lines after a fashion of his own and was unable to see anything in æsthetic facts except a reverberation of sexual pleasure. Where can we find a beauty, he asks, that is not derived from the female figure, the centre of all beauty? Undulating lines are beautiful because found in a woman's body; beautiful are all movements distinctively feminine; beautiful the tones of music melting one into another; beautiful the poem where one thought embraces another with tenderness and facility. [14] Condillac's sensationalism had already shown itself wholly

---

1. Alexander Gérard：亚历山大・杰拉德（1728—1795），苏格兰政治家、学者、哲学家，著有文中提到的《论品鉴力》（*Essay on Taste*）《论天才》（*Genius*）。
2. *Essay on Taste* by Alison：阿里森的《论品鉴力》
3. *E. Platner*：恩斯特・普拉特钠（1744—1818），德国人类学家、生理学家、理性主义哲学家，著有《迟到的启蒙》《新人类学》等。
4. transcendent finalism：超验目的论
5. incongruity：不相称

incapable of understanding æsthetic productivity; the associationism especially promoted by the work of Hume fared no better.

Fr. Hemsterhuis[1].

The Dutchman Hemsterhuis considered beauty as a phenomenon born of the meeting between sensibility, which gives multiplicity, and the internal sense, which tends to unity; hence the beautiful is "that which exhibits the greatest number of ideas in the shortest time." Man, to whom it is not permitted to attain ultimate unity, finds in beauty an approximate unity which gives him a pleasure somewhat analogous with the joy of love. This theory of Hemsterhuis, in which elements of mysticism and sensationalism mingle with glimpses of truth, developed later into the sentimentalism of Jacobi, for whom the totality of Truth and Goodness and even the Supersensible itself are sensibly present to the soul in the form of beauty. [15]

Neo-Platonism and mysticism: Winckelmann[2].

Platonism or, more accurately, neo-Platonism was revived by the creator of the history of figurative art, Winckelmann (1764). Contemplation of the masterpieces of antique plastic art, and the impression of superhuman loftiness and divine indifference which they create all the more irresistibly because we cannot reawaken the life they once possessed or understand their real significance, led Winckelmann, and others with him, to the conception of a Beauty which, descending from the seventh heaven of the divine Idea,

---

1. *Fr. Hemsterhuis*：弗朗索瓦·亨姆斯特胡伊斯（1721—1790），荷兰美学家、道德哲学家，著有《论雕塑》《论无神论》等。
2. *Winckelmann*：温克尔曼（1717—1768），德国艺术史学家、考古学家，著有《古迹中的艺术史》。

embodied itself in works of this description. Baumgarten's follower Mendelssohn had denied the enjoyment of beauty to God: the neo-Platonist Winckelmann gave it back to him and lodged it in his bosom.

Beauty and lack of significance.

"Wise men who have meditated upon the causes of universal Beauty, seeking her amongst created things and trying to gain the contemplation of Supreme Beauty, have placed it in the perfect harmony of creatures with their ends and of their parts with one another. But as this is equivalent to perfection, which man is incapable of attaining, our concept of universal beauty remains indeterminate, and arises by means of particular cognitions which, when accurately collected and fitted together, give us the highest idea we can attain of human beauty, which we elevate in proportion as we raise it above matter. But, again, since the Creator deals out[1] perfection to all his creatures in the proportion that befits them, and since every concept rests on some cause which must be sought outside the concept itself, the cause of Beauty which is to be found in every created thing cannot be sought in anything outside these created things. For this reason, and because our cognitions are comparative concepts, whereas Beauty cannot be compared with anything higher, it is difficult to attain a distinct and universal cognition of Beauty." [16] The only way out of this difficulty and others like it is the recognition that "supreme beauty resides in God": "the concept of human beauty becomes the more perfect in proportion as it can be thought more in conformity and agreement with supreme Being, which is distinguished from matter by its own unity and indivisibility. This conception of Beauty is as a spirit which, freed by fire from the prison of matter, strives to

1. deals out：分发；分配

conjure up a creature in the likeness of the first reasonable creature formed by the divine intelligence. The forms of such an image are simple and continuous and within this unity they are varied and for that very reason harmonious." [17] To these characteristics is added "lack of significance" (*Unbezeichnung*[1]), since supreme beauty cannot be described with points or lines different from those which alone can constitute that beauty; its form "is not peculiar to this or that determinate person, neither does it express any state of feeling or sensation of passion, things which disturb unity and overcloud beauty." Winckelmann concludes: "We look upon Beauty as a purest water drawn from the centre of the spring; the less taste it has the higher it is esteemed because free from all impurities." [18]

To perceive pure beauty, a special faculty is required, which certainly is not sense, but may perhaps be intellect or even, as Winckelmann says, "a fine internal sense" free from all intentions or passions of instinct, inclination or pleasure. Having asserted beauty to be something supersensible, it is not surprising that Winckelmann should wish, if not wholly to exclude colour[2], at least to reduce it to a minimum, and treat it not as a constitutive element in beauty but as secondary and ancillary[3]. [19] True beauty is given in form: by which he means line and surface, forgetting that these are only apprehended by the senses, and could not be seen without being in some way coloured.

Winckelmann's contradictions and compromises.

When error refuses to retire, hermit-like, to the narrow cell of a brief aphorism[4], it finds itself condemned to self-contradiction in order

---

1. *Unbezeichnung*：〈德〉缺乏意义
2. wholly to exclude colour：完全排除色彩因素
3. secondary and ancillary：次要的和附加的
4. brief aphorism：简短的名言，警句

to live at all in the world of concrete facts and problems. Although composed with a view to stating a theory, the work of Winckelmann always led him among concrete historical facts clamouring to be brought into relation with his formally stated idea of supreme beauty. In his admission of line-drawing and his further admission, on a lower plane, of colour, we have two compromises already; to which a third is added in his principle of Expression. "Since human nature has no state intermediate between pain and pleasure" and as living creature without such feelings is inconceivable, "the human figure must be represented in a condition of action and passion, which artists call expression." Hence Winckelmann, after dealing with Beauty, goes on to treat of Expression. [20] He then found himself obliged to effect a fourth compromise between the single constant supreme beauty and individual beauties; for while he preferred the male to the female body as a completer embodiment of perfect beauty, he could not shut his eyes to the obvious fact that we know and admire beautiful women's bodies and even beautiful animals' bodies.

A. R. Mengs[1].

Friend and, in a sense, collaborator of Winckelmann was Raphæl Mengs the artist, no less eager than his archæological fellow-countryman to understand the nature of that beauty which the one studied as a critic while the other produced it as a painter. Remarking, writes Mengs, that of the two chief duties of a painter, the imitation of appearances and the selection of the most beautiful objects, much has been written on the former, while the latter "has scarcely been touched by the modems, who would have been ignorant of the art of drawing were it not for the statues of ancient Greece"; [21] pondering this,

---

1. *A. R. Mengs*：拉斐尔·孟斯（1728—1779），德国画家，新古典主义绘画的先驱。

"I read, asked and looked at everything likely to throw light on the subject, but never was I satisfied; either they spoke of beautiful things or of qualities which are the attributes of beauty, or they pretended to explain, as the saying is, the obscure by the more obscure, or even confused the beautiful with the pleasing: so that finally I determined to search for the nature of beauty on my own account." [22] One of his works on this subject was published during his lifetime by the advice and assistance of Winckelmann (1761); many others appeared posthumously[1] (1780), all were reprinted several times and translated into several languages. In his *Dreams of Beauty*[2] he says, "I have been sailing a long time on a vast sea seeking the understanding of beauty, and still I am far from any shore and in great doubt how to shape my course: gazing around, my sight is confounded by the immensity of the subject." [23] In truth it seems as though Mengs never arrived at a formula satisfactory to himself, although he conformed more or less to Winckelmann's doctrine that "beauty consists in material perfection according to our ideas; and since God alone is perfect, beauty is divine"; it is the "visible idea of perfection" and stands in the same relation to it as does a visible to a mathematical point. Our ideas proceed from the purposes which the Creator has willed to fulfil in various things; hence the multiplicity of beauties. In general, Mengs finds the types of things in natural species: *e.g.* "a stone, of which we have the idea that it should be uniform in colour"; which "is called ugly if it happen to be spotted"; or a child "would be ugly if he were like a man of mature age, just as a man is ugly when shaped like a woman, and a woman when she is like a man." He adds surprisingly, "As among stones there is but one perfect species, the diamond; among metals, gold; and among animated creatures, man only; so there is difference and distinction in every order, and very rarely is

1. appeared posthumously：作者身后出版的
2. *Dreams of Beauty*：《美之梦想》，孟斯著。

there perfection." [24] In his *Dreams of Beauty* he considers beauty as "a middle disposition, including perfection on the one hand and the pleasing on the other"; in reality it is a third thing, differing from perfection and the pleasing, and deserving a special name for itself. [25] The art of painting arises from four sources: beauty, significant or expressive character, the pleasing united to harmony, and colouring. Mengs finds the first amongst the ancients, the second in Raphæl, the third in Correggio[1] and the fourth in Titian. [26] From this empirical studio-gossip[2] he rouses himself to exclaim, "The force of beauty so transports me that I will tell thee, reader, what I feel. All nature is beautiful, and so is virtue; beautiful are forms and proportions; beautiful are appearances and beautiful the causes thereof; more beautiful is reason, most beautiful of all is the great first cause." [27]

G. E. Lessing[3].

An attenuated, that is to say, a less metaphysical, echo of Winckelmann's theory is found in Lessing (1766), who infused a new spirit into the literature and social life of the Germany of his time. According to Lessing the aim of art is "delight"; and since delight is a "superfluous thing" it seems reasonable that the legislator should not allow to art that liberty which is indispensable to science in her search for truth, the soul's necessity. For the Greeks painting was what by its nature it ought to be, "the imitation of beautiful bodies." "Its (Hellenic) cultivator represented nothing but the beautiful: common beauty of a low grade served him as an accidental subject, an exercise,

1. Correggio：柯雷乔（1499—1534），意大利文艺复兴时期画家之一，作品包括帕尔玛大教堂圆屋顶画的巨幅壁画《圣母升天》。
2. empirical studio-gossip：经验性的画室闲扯
3. Lessing：莱辛（1729—1781），德国作家、哲学家、戏剧家和艺术评论家，启蒙主义时代杰出代表之一，著有《拉奥孔》及文中提到的《汉堡剧评》（*Dramaturgie*）等。

a diversion. The attractiveness of his work must depend simply and solely on the perfection of his subject: he was far too true an artist to wish his audience to content itself with the barren pleasure arising from mere resemblance or from the inspection of skilful workmanship: nothing in his art was dearer to him, nothing seemed more noble, than the end at which it aimed." [28] Pictorial representation must exclude everything unpleasing or ugly; "painting as imitation may express ugliness: painting as a fine art will refuse to do so: all visible objects belong to art taken under the former title: the latter may claim only such objects as awaken pleasing sensations." If, on the contrary, ugliness may be represented by the poet, the reason is this: poetic description "conveys a less displeasing sense of bodily malformation which, in the end, almost loses its character as such; unable to use it for itself, the poet uses it as a means to provoke certain mixed feelings (the ridiculous, the terrible), in which we are content to remain, in the absence of any purely pleasant feelings." [29] In his *Dramaturgie* (1767) Lessing takes his stand upon the Aristotelian *Poetics*: it is well known that not only did he approve of rules in general but he believed those laid down by Aristotle to be as incontrovertible as the theorems of Euclid. His polemic[1] against French writers and critics is waged in the name of probability, not to be confounded with historical accuracy. He understood the universal as a sort of average of what appears in individuals, and catharsis as a conversion of passions into virtuous dispositions, asserting it as beyond doubt that the aim of all poetry is to inspire a love for virtue. [30] He follows the example of Winckelmann in introducing the concept of ideal beauty into the doctrine of figurative art: "expression of corporeal beauty is the aim of painting: therefore supreme beauty of body is the supreme aim of art. But this supreme beauty of body is found in man only, and for him

1. polemic：激烈的争辩

it exists only through the ideal. This ideal may be found among the brute creation in inferior degree; but is entirely absent from vegetable or inanimate nature." Landscape and flower painters are not really artists because "they imitate beauties possessed of no ideal: whereby they work by eye and hand alone, genius having little or no part in their compositions." Nevertheless, Lessing prefers a landscape painter to "the painter of historic pieces who, instead of making beauty his aim, merely depicts a crowd in order to show his cunning in simple expression, not in expression subordinate to beauty." [31] The ideal of bodily beauty then consists "chiefly in the ideal of form, but also in that of texture of the flesh, and in that of permanent expression. Mere colouring and transitory expression have no ideal since nature herself has placed no indelible seal upon them." [32] At the bottom of his heart Lessing dislikes colour; and when he finds the pen-sketches of painters showing "a life, a freedom, a brilliancy never to be found in their painted pictures," he asks himself "whether the most marvellous colouring can compensate so heavy a loss," and whether it is not to be wished "that painting in oils had never been invented"? [33]

Theorists of ideal Beauty.

Ideal beauty, that curious alliance between God and the subtle outline traced with pen or graver, that cold academical mysticism, came into fashion. In Italy (the home of Winckelmann and Mengs, who published many of their works in Italian) it was much discussed by artists, antiquaries and connoisseurs[1]. The architect Francesco Milizia[2] professed himself a follower of "the principles of Sulzer

1. connoisseurs：〈法〉艺术爱好者；鉴赏家
2. Francesco Milizia：弗兰塞斯克 · 米丽奇亚（1725—1798），意大利艺术史学家、建筑家。

and Mengs"; [34] the Spaniard d'Azara[1], living in Italy, edited and annotated Mengs, adding his own definition of beauty: "The union of the perfect and the pleasing made visible"; [35] another Spaniard, Arteaga[2], one of the many Jesuit refugees in Italy, wrote a treatise on *Ideal Beauty* (1789); [36] the Englishman Daniel Webb[3] on coming to Rome and making the acquaintance of Mengs seized upon the ideas he heard him express on beauty, collected them and actually published them in a book anticipating Mengs' own. [37]

G. Spalletti[4] and the characteristic.

The first voice of dissent from this doctrine of ideal beauty was raised in 1764 by a small circle of Italians who asserted the characteristic to be the principle of art. As such appears to be the necessary interpretation of the little *Essay on Beauty* written by Guiseppe Spalletti in the form of a letter to Mengs, with whom Spalletti had discussed the subject "in the solitudes of Grottaferrata[5]," and who had urged him to put all his thoughts in writing. [38] Its polemical character, though not openly asserted, is discernible in every page. "Truth in general, conscientiously rendered by the artist, is the object of Beauty in general. When the soul finds those characteristics which wholly converge upon the matter which the work of art claims to represent, it judges that work beautiful. The same is true of the works

---

1. d'Azara：达萨拉（1730—1804），西班牙外交家、古董收藏家。
2. Arteaga：阿尔泰加（1747—1799），西班牙戏剧评论家、音乐评论家，著有《意大利音乐剧的演化》《阿拉伯人对诗歌起源的影响》《论理想美》（*Ideal Beauty*）等。
3. Daniel Webb：丹尼尔·韦伯（1718—1798），爱尔兰美学家，著有《绘画美之探究》《评诗歌之美》《音乐与诗歌相通性研究》等。
4. *G. Spalletti*：朱塞佩·斯帕莱蒂（生卒年不详），意大利美学家，著有《论美》（*Essay on Beauty*）。
5. in the solitudes of Grottaferrata：格罗塔费拉塔的寂寥。格罗塔费拉塔系一意大利市镇，拥有圣玛丽亚迪格罗塔费拉塔修道院。

of nature: if the soul perceives a man of fine proportions having the face of a lovely woman, which causes it to doubt whether the object before it be man or woman, it esteems that man ugly rather than the reverse, through deficiency of the characteristic of truth[1]; if this can be said of natural Beauty, how much more can it be said of the Beauty of art." The pleasure given by Beauty is intellectual, that is to say, it is the pleasure of apprehending truth: when confronted by ugly things represented characteristically, man "delights in having increased his cognitions": Beauty, "with its property of supplying to the soul likeness, order, proportion, harmony and variety, provides it with an immense field for the construction of innumerable syllogisms, and by reasoning in this manner it will take pleasure in itself, in the object which arouses such pleasure, and in the feeling of its own perfection." Finally, the beautiful may be defined as "the inherent modification of the object under observation which presents it in the inevitably characteristic manner in which it is bound to appear." [39] In contrast to the fallacious profundity[2] of Winckelmann and Mengs we welcome the sound good sense of this obscure Spalletti, upholder of the Aristotelian position against the revived neo-Platonism of the æstheticians.

Beauty and the characteristic: Hirt[3], Meyer[4], Goethe.

Many years went by before a similar rebellion arose in Germany; at length in 1797 the art-historian Aloys Hirt, basing his case on ancient works of art which depicted all things, even things utterly vulgar and

1. deficiency of the characteristic of truth：缺乏真实特征
2. fallacious profundity：虚妄的深刻
3. *Hirt*：阿洛伊斯·希尔特（1759—1837），德国艺术史学家、考古学家，著有《古代绘画艺术史》《古代建筑艺术史》《神话、考古与艺术画册》等。
4. *Meyer*：海因里希·梅尔（1760—1832），瑞士画家、艺术史学家、艺术评论家。

ugly, ventured to deny the view that ideal beauty is the principle of art, and that expression has only a secondary place, above which it must not rise for fear of disturbing ideal beauty. For the ideal he substituted the characteristic, as a principle to be applied equally to gods, heroes or animals. Character is "that individuality by which form, movement, signs, physiognomy and expression, local colour, fight, shade and chiaroscuro[1] are distinguished and represented in the manner demanded by the object." [40] Another historian of art, Heinrich Meyer, who started from the position of Winckelmann and went on by adopting a series of compromises, finally asserting an ideal of trees and landscape side by side with the ideal of man and various other animals, tried to find an intermediate position between this doctrine and Hirt's, in the course of controversy with the latter. And Wolfgang von Goethe, forgetful of his youthful days when he chanted the praises of Gothic architecture, returning home from an Italian tour impregnated with Greece and Rome in 1798, also sought a middle term between Beauty and Expression; dwelling on the thought of certain characteristic contents which should supply the artist with forms of beauty to be by him remodelled and developed into complete beauty. The characteristic was thus the mere point of departure, and beauty was simply the result of the artist's elaboration: "we must start from the characteristic" (says he) "in order to attain the beautiful." [41]

[1] *Les Beaux Arts réduits à un même principe,* Paris, 1746; see esp. part i. ch. 3; part ii. chs. 4, 5; part iii. ch. 3.

[2] See above, p. 142.

[3] *Analysis of Beauty*, London, 1753 (Ital. trans., Leghorn, 1761).

[4] *Op. cit.* p. 47.

---

1. chiaroscuro：明暗对比

[5] *Op. cit.* p. 57.

[6] *Op. cit.* p. 93.

[7] *Op. cit.* pp. 61, 65.

[8] *Analisi della bellezza*, p. 91.

[9] *Op. cit.* p. 176.

[10] *Enquiry into the Origin of our Ideas of the Sublime and the Beautiful*, 1756 (Ital. trans., Milan, 1804); cf. the preliminary discourse on "Taste."

[11] *Enquiry into the Origin of our Ideas of the Sublime and the Beautiful*, part iii. § 18.

[12] *Elements of Criticism*, 1761, vol. i. introd. and chs. 1–3.

[13] *Elements of Criticism*, i. ch. 3, pp. 201–202.

[14] *Neue Anthropologie*, Leipzig, 1790, § 814, and the lectures on Æsthetic published posthumously in 1836; cf. Zimmermann, *op. cit.* p. 204.

[15] Zimmermann, *op. cit.* pp. 302–309; v. Stein, *Entstehung d. n. Ästh.* p. 113.

[16] *Geschichte der Kunst des Altertums,* 1764 (in *Werke*, Stuttgart, 1847, vol. i.), bk. iv. ch. 2, § 51, p. 131.

[17] *Op. cit.* § 22, pp. 131–132.

[18] *Op. cit.* § 23, p. 132.

[19] *Geschichte,* § 19, pp. 130–131.

[20] *Op. cit.* bk. iv. ch. ii. § 24.

[21] *Geschichte*, bk. v. chs. ii. and vi.

[22] Letter of January 2, 1778, *Opere*, Rome, 1787 (reprinted Milan, 1836), ii. pp. 315–316.

[23] *Opere*, i. p. 206.

[24] *Riflessioni sulla bellezza e sul gusto della pittura*, in *Opere*, i. pp. 95, 100, 102–103.

[25] *Opere*, i. p. 197.

[26] *Ibid.* p. 161.

[27] *Ibid.* p. 206.

[28] *Laokoon*, § 2.

[29] *Op. cit.* § § 23, 24.

[30] *Hamburg. Dramaturgie* (ed. Göring, vols. xi. and xii.), passim, esp. Nos. 11, 18, 24, 78, 89.

[31] *Laokoon*, appendix, § 31.

[32] *Op. cit.* § § 22, 23.

[33] *Op. cit. ad fin.* p. 268.

[34] *Dell' arte di vedere nelle belle arti del disegno secondo i principi di Sulzer e di Mengs*, Venice, 1871.

[35] D'Azara, in Mengs, *Opere*, i. p. 168.

[36] *Investigaciones filosóficas sobre la belleza ideal, considerada como objeto de todas las artes de imitación*, Madrid, 1789.

[37] *Ricerche su le bellezze della pittura* (Ital. trans., Parma, 1804); cf. D'Azara, *Vita del Mengs*, in *Opere*, i. p. 27.

[38] *Saggio sopra la bellezza*, dated "Grottaferrata, July 14, 1764," and published at Rome, 1765, anonymously.

[39] *Saggio*, esp. § § 3, 12, 15, 17, 19, 34.

[40] *Über das Kunstschöne*, in the review *Die Horen*, 1797; cf. Hegel, *Vorles. ii. Ästh.* i. p. 24; and Zimmermann, *Gesch. d. Ästh.* pp. 356–357.

[41] Goethe, *Der Sammler und die Seinigen* (in *Werke*, ed. Goedecke, vol. xxx.)

VIII IMMANUEL KANT

I. Kant.

Of all these writers, Winckelmann and Mengs, Home and Hogarth, Lessing and Goethe, none was a philosopher in the true sense of the word: not even those who like Meier laid claim to the title, nor those who had some gifts for philosophy like Herder or Hamann. After Vico, the next European mind of real speculative genius is Immanuel Kant, who now comes before us in his turn.

Kant and Vico.

That Kant took up the problem of philosophy where Vico laid it down (not, of course, in a directly historical, but in an ideal, sense) has already been noted by others. [1] How far he made an advance upon his predecessor and how far he failed to reach the same level it is not here our business to inquire; we must confine ourselves strictly to the consideration of Æsthetic questions.

Summarizing the results of such a consideration, we may say at once that though Kant holds an immensely important place in the development of German thought; though the book containing his examination of æsthetic facts is among his most influential works; and though in histories of Æsthetic written from the German point of view, which ignore practically the whole development of European thought from the sixteenth to the eighteenth century, Kant can pose as the man who discovered the problem of Æsthetic or solved it or brought it within sight of solution; yet in an unprejudiced and complete history whose aim is to take broad views and to consider not the popularity of a book or the historical importance of a nation but the intrinsic value of ideas, the judgement passed on Kant must be very different. Like Vico in the serious tenacity with which he reflected upon æsthetic facts, more fortunate than he in having a much larger stock of material gathered from preceding discussion and argument, Kant was at once unlike and less successful than Vico in that he was unable to attain a doctrine substantially true, and unable also to give his thoughts the necessary system and unity.

Identity of the concept of art in Kant and Baumgarten.

In fact, what was Kant's idea of art? Strange as our reply may seem to those who recollect the explicit and insistent war waged by him against the school of Wolff, and the concept of beauty as a perfection confusedly perceived, we must assert that Kant's idea of art was fundamentally the same as that of Baumgarten and the Wolffian school. [2] In that school his mind had been trained; he always had a great respect for Baumgarten whom in the *Critique of Pure Reason*[1] he calls "that excellent analyst"; he chose the text of Baumgarten for two of his University lectures on Metaphysics, and that of Meier for his lecture on Logic (*Vernunftlehre*). Kant, like them, therefore considered Logic and Æsthetic (or theory of art) as conjoined sciences. They were thus described by him in his *Scheme of Lectures* in 1765, when he proposed, while expounding the critique of reason, to "throw a glance at that of taste, that is to say, at Æsthetic, since the rules of one apply to the other and each throws light upon the other."

---

1. *Critique of Pure Reason*: 《纯粹理性批判》，康德著。

Kant's "Lectures."

In his University lectures he distinguished æsthetic truth from logical truth in the style of Meier; even citing the example of the beautiful rosy face of a girl which, when seen distinctly, *i.e.* through a microscope, ceases to be beautiful. [3] It is æsthetically true (said he) that a man once dead cannot come to life again, although this is in opposition to logic and moral truth: it is æsthetically true that the sun plunges into the sea, but it is false logically and objectively. To what degree it is necessary to combine logical truth with æsthetic the learned have never yet been able to decide; not even the greatest æstheticians. In order to become accessible, logical concepts must assume æsthetic forms; a garb to be abandoned only in the rational sciences which seek profundity. Æsthetic certainty is subjective: it is content with authority, *i.e.* the citation of the opinions of great men. On account of our weakness, for we are strongly attached to the sensible, æsthetic perfection often helps us to render our thoughts distinct. In this, examples and images co-operate; æsthetic perfection is the vehicle for logical perfection; taste is the analogue of intellect. There are logical truths which are not æsthetic truths: and on the other hand we must exclude from abstract philosophy exclamations and other sentimental commotions proper to the other truth. Poetry is a harmonious play of thoughts and sensations. Poetry and eloquence differ in this: in the former, thoughts adapt themselves to sensations; in the latter the contrary is the case. In these lectures Kant sometimes taught that poetry is anterior to eloquence because sensations come before thoughts; and he observed (perhaps under Herder's influence) that the poetry of Eastern peoples, lacking concepts, is wanting in unity and taste although rich in imaginative detail. Poetry formed out of the pure play of sensibility is doubtless a possibility, *e.g.* love-poems: but true poetry disdains such productions, concerned as they are with sensations which every one knows ought to be expelled from our breasts. True poetry must strive to present virtue and intellectual truth in sensible form, as has been done by Pope in his *Essay on Man*[1], in which he attempts to vivify poetry by means of reason. On other occasions Kant definitely says that logical perfection is the basis of every other, æsthetic perfection being merely an adornment of the logical; something of the latter may be omitted in order to appeal to the audience, but it must never be disguised or falsified. [4]

This is Baumgartenism pure and simple; unless we are prepared to look on these Lectures as representing a pre-critical period of thought, or an exoteric doctrine superseded eventually by Kant's own original esoteric ideas[2] in his *Critique of the Judgment*[3] (1790). Not to open such a controversy, let us put these Lectures on one side (although they often throw no little light on the signification of Kantian phrases and formulæ), and refuse to raise the question what pages of the *Critique of the Judgment* are derived from Baumgarten and Meier; he who reads the works of these disciples of Wolff and passes immediately to the *Critique of Judgment* often has the impression that the atmosphere surrounding him is unchanged. But if the *Critique of Judgment* itself be examined without prejudice it will be seen that Kant always adhered to Baumgarten's conception of art as the sensible and imaginative vesture of an intellectual concept.

Art in the "Critique of Judgment."

According to Kant, art is not pure beauty wholly detached from the concept, it is adherent beauty, which presupposes and attaches itself to a concept. [5] This is the work of genius, the faculty of representing æsthetic ideas. An æsthetic idea is "a representation of the imagination

---

1. *Essay on Man*：《论人》，蒲柏著。
2. original esoteric ideas：原有的深奥学说
3. *Critique of the Judgment*：《判断力批判》，康德著。

which accompanies a given concept: a representation conjoined with such truthful representation of particulars as to be unable to find for it any expression that may mark a determinate concept, thereby endowing the given concept with something of the ineffable[1]; a feeling which stimulates the cognitive faculties and reinforcing the tongue, which is simply the letter, with the spirit." Genius, then, has two constitutive elements, imagination and intellect; it consists in "that happy disposition, which no science can teach or diligence attain, to find ideas for a given concept and, also, to select the expression by which the subjective commotion it excites as accompaniment to a concept may be communicated to others." No concept is adequate to the æsthetic idea, as no representation of the imagination can ever possibly be adequate to the concept. Examples of æsthetic attributes are found in the eagle of Jupiter with the thunderbolt in its claws, and the peacock of the proud Queen of Heaven: "they do not, like logical attributes, represent that which is contained in our concepts of the sublimity or majesty of creation, but something else which gives occasion to the imagination to run riot over[2] a multitude of kindred representations which make us think more than we can express in a given concept by means of words, and give us an æsthetic idea, which serves to this rational idea instead of a logical representation, precisely with the aim of quickening our feelings by throwing open to them a view over a vast field of kindred representations." There are a *modus logicos*[3] and a *modus æstheticus*[4] of expressing our thoughts: the first consists in following determinate principles: the other in the mere feeling of the unity of the representation. [6] To imagination, to intellect and to spirit (*Geist*) we must add taste, the link between

1. something of the ineffable：不可言喻之物
2. run riot over：自由驰骋
3. *modus logicos*：〈拉〉逻辑方式
4. *modus æstheticus*：〈拉〉审美方式

imagination and intellect. [7] Art may therefore represent natural ugliness: artistic beauty "is not a beautiful *thing* but a beautiful representation of a thing": although the representation of ugliness has limits varying with the individual arts (a reminiscence of Lessing and Winckelmann), and an absolute limit at the disgusting and nauseating[1], which kill representation itself. [8] In natural things, too, there is adherent beauty which cannot be judged by the æsthetic judgement alone but demands a concept. Nature thus appears as a work of art, though superhuman art: "the teleological[2] judgement is the basis and condition of the æsthetic." When we say "this is a beautiful woman," we merely mean that "nature beautifully represents in the form of this woman her purpose in the construction of the female body": it is necessary therefore, besides noting simple form, to aim at a concept, "so that the object may be apprehended through an æsthetic judgement logically conditioned." [9] By this means is formed the ideal of beauty in the human face, the expression of moral life. [10] Kant admits that there may also be artistic productions without a concept, comparable with the free beauties of nature, flowers and some birds (parrot, humming-bird, bird of paradise, etc.): ornamental drawings, cornice-mouldings[3], musical fantasies without words, represent nothing, no object reducible to a determined concept, and must be reckoned among free beauties. [11] But does not this necessitate their exclusion from true and proper art, from the operation of genius in which fancy and intellect must both, according to Kant, have a place?

Imagination in Kant's system.

This is Baumgartenism transposed into a higher key, more

1. nauseating：令人作呕的
2. teleological：目的论的
3. cornice-mouldings：飞檐装饰线条

concentrated, more elaborated, more suggestive, until from moment to moment it seems about to burst into a wholly different conception of art. But it is still Baumgartenism, from whose intellectualistic bonds it never escapes. Nor was escape possible. A profound concept of imagination was entirely lacking to Kant's system and his philosophy of the spirit. Glancing over the table of faculties of the spirit which precedes his *Critique of Judgment*, we see that Kant co-ordinates with it the cognitive faculty, the feeling of pleasure and pain, and the appetitive faculty; to the first corresponds intellect, to the second, judgement (teleological and æsthetic), to the third, reason; [12] he finds no place for imagination amongst powers of the spirit but places it among the facts of sensation. He knows a reproductive imagination and an associative, but he knows nothing of a genuinely productive imagination, imagination in the proper sense. [13] We have seen that, in his doctrine, genius is the co-operation of several faculties.

The forms of intuition and the Transcendental Æsthetic.

Yet sometimes Kant had an inkling that intellectual activity is preceded by something which is not mere sensational material, but is an independent non-intellectual theoretical form. He obtained a glimpse of this latter form not when he was reflecting on art in the strict sense but when he was examining the process of knowledge: he does not treat of it in his *Critique of Judgment*, but in the first section of his *Critique of Pure Reason*, in the first part of the *Transcendental Doctrine of Elements*. He says here that sensations only enter the spirit when the latter itself gives them form; a form not identical with that which intellect gives to sensations, but much simpler, namely pure intuition, the totality of the *a priori* principles of sensibility. There must therefore be "a science which forms the first part of the transcendental doctrine of elements, distinct from that which contains the principles of pure thought and is named transcendental Logic." Now, what name does

Kant confer upon this science whose existence he has deduced? None other than Transcendental Æsthetic (*die transcendentale Ästhetik*). In a note he even insists that this is the right name for the new science of which he treats, and censures[1] the Germans for their habit of applying it to the Critique of Taste, which, as he thought at that time, could never become a science. Thus, he concludes, we approach more closely to the usage of the ancients, among whom the distinction between *αἰσθητὰ καὶ νοητά*[2] [14] was well known.

Nevertheless, after having so rightly postulated the necessity for a science of the forms of sensation or pure intuition, purely intuitive knowledge, Kant went on, simply because he had no exact idea of the nature of the æsthetic faculty and of art, to fall into an intellectualistic error by reducing the form of sensibility or pure intuition into the two categories or functions of space and time, and by asserting that the spirit emerges from the chaos of sensation by organizing its sensations in space and time. [15] But space and time as such are very far from being primitive categories; they are relatively late and complex formations. [16] As examples of the matter of sensation Kant quoted hardness, impenetrability, colour and so forth. But the mind only recognizes colour and hardness in so far as it has already given form to its sensations; considered as brute matter, sensations fall outside the cognitive spirit, they are a limit; colour, hardness, impenetrability and so on, when recognized, are already intuitions, spiritual elaborations, the æsthetic activity in its rudimentary manifestation[3]. The characterizing or qualifying imagination which is æsthetic activity ought to have occupied in the *Critique of Pure Reason* the pages devoted to the discussion of space and time, and would thus have constituted a real Transcendental Æsthetic, a real prologue to the transcendental

1. censures：指责
2. *αἰσθητὰ καὶ νοητά*：〈希〉感性和理性
3. rudimentary manifestation：原始体现

Logic. In this manner Kant would have achieved the truth aimed at by Leibniz and Baumgarten and would have joined hands with Vico.

Theory of Beauty distinguished by Kant from that of Art.

His repeatedly announced opposition to the school of Wolff concerns not the concept of art but that of Beauty; two concepts for Kant entirely distinct. First of all, he did not admit that sensation could be called "confused knowledge," a confused form, that is, of intellectual cognition; rightly judging this to be a false account of sensibility, since a concept, however confused, is always a concept or a rough sketch of a concept, never an intuition. [17] But he further denied that pure beauty contained a concept, and therefore denied that it was a perfection sensibly apprehended. These reflexions have no doubt some connexion with those concerning the nature of art in the *Critique of Judgment;* but the connexion is far from close, still less are they actually fused into a single whole. That Kant was minutely familiar with eighteenth-century writers who had discussed beauty and taste is shown by his Lectures, wherein they are all quoted and used. [18] Of these the greater part, especially the English, were sensationalists, others intellectualists; some few, as we have noted, were inclined towards mysticism. Kant began by tending towards sensationalism in æsthetic problems, then became the adversary of sensationalists and intellectualists alike. This development can be traced in his *Observations on the Beautiful and Sublime,* as well as in his Lectures; its final expression is reached in the *Critique of Judgment.*

Of the four moments, as he calls them, *i.e.* the four determinations, he accords to Beauty, the two negative are directed, one against the sensationalists, the other against the intellectualists. "That is beautiful which pleases *without interest*[1]": "That is beautiful which pleases *without*

---

1. *without interest*：没有利害关系的

concepts." [19] Here he asserts the existence of a spiritual region, distinct on one side from the pleasurable, the useful and the good, and on the other from truth. But this region, as we know very well, is not that of art, which Kant attaches to the concept: it is the region of a special activity of feeling which he calls judgement or, more exactly, æsthetic judgement.

Mystical features in Kant's theory of Beauty.

The other two moments give some kind of a definition of this region: "That is beautiful which has the form of finality without the representation of an end": "That is beautiful which is the object of universal pleasure." [20] What is this mysterious sphere? What this disinterested pleasure we experience in pure colours and tones, in flowers, and even in adherent beauty[1] when we make abstraction from the concept to which it adheres?

Our answer is: there is no such sphere; it does not exist; the examples given are instances either of pleasure in general or of facts of artistic expression. Kant, who so emphatically criticizes the sensationalists and the intellectualists, does not show the same severity towards the neo-Platonic line of thought whose revival we remarked in the eighteenth century. Winckelmann in particular exercised strong influence over his mind. In one course of his Lectures we find him making a curious distinction between form and matter: in music melody is matter and harmony form: in a flower the scent is material and the shape (*Gestalt*) is form (*Form*). [21] This reappears slightly modified in the *Critique of Judgment.* "In painting, statuary and all the figurative arts in architecture and gardening, so far as they are fine arts, the drawing is the essential; in which the foundation of taste lies not in

1. adherent beauty：依存美

what gratifies (*vergnügt*) in sensation, but in that which pleases (*gefällt*) by its form. The colours which illuminate the drawing belong to sensuous stimulus[1] (*Reiz*) and may bring the object more vividly before the senses, but do not render it worthy of contemplation as a thing of beauty; they are, moreover, often limited by the exigencies of the beautiful form, and even where their sensuous stimulus is legitimate, they are ennobled only by the beautiful form." [22] Continuing in pursuit of this phantasm of beauty which is not the beauty of art nor yet the pleasing, and is equally detached from expressiveness and pleasure, Kant loses himself in insoluble contradictions. Little inclined to submit himself to the charm of imagination, abhorring "poetic philosophers" like Herder, [23] he makes statements and refuses to commit himself to them, affirms and immediately criticizes his affirmations, and wraps up Beauty in a mystery which, at bottom, was nothing more than his own individual incertitude and inability to see clearly the existence of an activity of feeling which, in the spirit of his sane philosophy, represented a logical contradiction. "Necessary and universal pleasure" and "finality without the idea of an end" are the organized expression in words of this contradiction.

By way of clearing up the contradiction he arrives at the following thought: "The judgement of taste is founded on a concept (the concept of a general foundation of the subjective teleology of nature through judgement); but it is a concept by which it is impossible to know or demonstrate anything of the object, because the object in itself is indeterminable and unsuited to cognition; on the other hand, it has validity for every one (for every one, I say, in so far as it is an individual judgement, immediately accompanying intuition), since its determining reason reposes, perhaps, in the concept of that which may be regarded as the supersensible substrate of mankind." Beauty,

1. sensuous stimulus：感官刺激

then, is a symbol of morality. "The subjective principle alone, that is the indeterminate idea of the supersensible in us, can be considered the only key able to unlock this faculty springing from a source we cannot fathom: excepting by its aid, no comprehension of it can possibly be reached." [24] These cautious words, and all others here used by Kant to conceal his thoughts, do not hide his tendency to mysticism. A mysticism without conviction or enthusiasm, almost in spite of himself, but very evident nevertheless. His inadequate grasp of the æsthetic activity led him to see double, even triple, and caused the unnecessary multiplication of his explanatory principles. Although he was always ignorant of the genuine nature of the æsthetic activity, he was indebted to it for suggesting to him the pure categories of space and time as the Transcendental Æsthetic; it caused him to develop the theory of imaginative embellishment of intellectual concepts by the work of genius; finally it forced him to acknowledge a mysterious faculty of feeling, midway between theoretical and practical activity, cognitive and yet not cognitive, moral and indifferent to morality, pleasing yet wholly detached from the pleasure of the senses. Great use of this power was made by Kant's immediate successors in Germany who were delighted to find their daring speculations supported by that severe critic of experience, the philosopher of Königsberg[1].

[1] B. Spaventa, *Prolus. ed introd. alle lezioni di filosofia*, Naples, 1862 pp. 83–102; *Scritti filosofici*, ed. Gentile, pp. 139–145, 303–307.
[2] *Kritik d. rein. Vernunft* (ed. Kirchmann), i. 1, § 1, note.
[3] See above, p. 276.
[4] Extract from Kant's lectures of 1764 and later, in O. Schlapp, *Kant's Lehre vom Genie, passim*, esp. pp. 17, 58, 59, 79, 93, 96, 131–

1. Königsberg：哥尼斯堡，康德的故乡。

134, 136–137, 222, 225, 231–232, etc.

[5] *Kritik d. Urtheilskraft* (ed. Kirchmann), § 16.

[6] *Kritik d. Urth.* § 49.

[7] *Op. cit.* § 50.

[8] *Op. cit.* § 48.

[9] *Krit. d. Urth.* § 48.

[10] *Op. cit.* § 17.

[11] *Op. cit.* § 16.

[12] For the historical genesis of this tripartition, cf. remarks in Schlapp, *op. cit.* pp. 150–153.

[13] See also *Anthropol.* (ed. Kirchmann), § § 26–31; cf. Schlapp, *op. cit.* p. 296.

[14] *Kritik d. rein. Vernunft*, i. I, § 1 and note.

[15] *Op. cit.* § § 1–8.

[16] See above, pp. 36–37.

[17] *Krit. d. r. Vern.* § 8, and introd. to § ii.; cf. *Krit. d. Urth.* § 15.

[18] See catalogue in Schlapp, *op. cit.* pp. 403–404, and *passim.*

[19] *Krit. d. Urth.* § § 1–9.

[20] *Op. cit.* § § 10–22.

[21] Schlapp, *op. cit.* p. 78.

[22] *Krit. d. Urth.* § 14.

[23] For Kant's judgement of Herder, see Schlapp, *op. cit.* pp. 320–327, note.

[24] *Kritik d. Orth.* § § 57–59.

IX THE ÆSTHETIC OF IDEALISM: SCHILLER, SCHELLING, SOLGER, HEGEL

The "Critique of Judgment" and metaphysical idealism.

It is well known that Schelling[1] held the *Critique of Judgment* to be the most important of the three Kantian *Critiques*, and that Hegel together with the great majority of the followers of metaphysical idealism had a special affection for the book. According to them the third *Critique* was the attempt to bridge the gulf, to resolve the antitheses between liberty and necessity, teleology and mechanism, spirit and nature: it was the correction Kant was preparing for himself, the concrete vision which dispelled the last traces of his abstract subjectivism.

F. Schiller.

The same admiration and an opinion even more favourable were extended by them to Friedrich Schiller[2], the first to elaborate that part of Kant's philosophy and to study the third sphere which united sensibility to reason. "It was the artistic sense dwelling in his also profoundly philosophical mind," says Hegel, "which, against the

---

1. Schelling：谢林（1775—1854），德国哲学家，著有《超验唯心主义体系》《人类自由的本质之哲学探究》《宗教与哲学》等。
2. Friedrich Schiller：弗里德里希・席勒（1759—1805），德国著名诗人、剧作家、哲学家、历史学家，启蒙主义文学代表人物之一，作品有戏剧《阴谋与爱情》、美学论著《审美教育书简》等。

abstract infinity[1] of Kant's thought, against his living for duty, against his conception of nature and reality, and of sense and feeling as utterly hostile to intellect, asserted the necessity and enunciated the principle of totality and reconciliation, even before it had been recognized by professed philosophers: to Schiller must be allowed the great merit of having been the first to oppose the subjectivity of Kant, and of having dared try to go beyond it." [1]

Relations between Schiller and Kant.

Discussion has raged around the true relation between Schiller and Kant, and it has lately been maintained that his Æsthetic was not, as would seem to be the case, derived from Kant, but from the pandynamism[2] which, starting from Leibniz, had propagated itself in Germany through Creuzer[3], Ploucket[4] and Reimarus[5] down to Herder, who had conceived a wholly animated nature. [2] There can be no doubt that Schiller shared Herder's conception, as may be seen from the theosophical tone[6] of the fragment of correspondence between Julius and Raphæl[7] and in other writings. It cannot be denied, however, that whatever personal feelings Kant may have had towards Herder, or Herder towards his former teacher (against whose

1. abstract infinity：抽象的无限性
2. pandynamism：泛物力论
3. Creuzer：克罗伊策（1771—1858），德国语言学家、考古学家，著有《希腊之外其他古代民族的象征与神话》《希腊人的历史性艺术》等。
4. Ploucket：即戈特弗里德·普洛克特（1716—1790），德国哲学家、逻辑学家，著有《物质与表象的原理》《思辨哲学的基础》《形而上学中的无限论证法》等。
5. Reimarus：赖马鲁斯（1694—1768），德国哲学家、启蒙主义作家，著有《自然宗教的基本真理论集》《正确使用理性的指南》等。
6. theosophical tone：神智论。神智学认为通过直接认识、哲学思辨就能洞悉神和世界的本性，把上帝看作是一切存在和善的超越的源泉。
7. fragment of correspondence between Julius and Raphæl：《尤里乌斯与拉斐尔通信集》残卷

Critique of Judgment he published his *Kaligone*[1], as he had replied to the *Critique of Pure Reason* with his *Metacritica*[2]), when Kant in somewhat dubious manner made the first step towards a reconciliation, the breach was at all events partially healed. The dispute is therefore of small importance: we shall find it more useful to observe that Schiller introduced an important correction of Kant's views when he obliterated[3] every trace of the double theory of art and the beautiful, giving no weight to the distinction drawn between pure and adherent beauty, and finally abandoning the mechanical conception of art as consisting in beauty joined to the intellectual concept. It was certainly his own experience of active artistic work that led him to this simplification.

The æsthetic sphere as the sphere of Play.

Schiller defined the æsthetic sphere as the sphere of play (*Spiel*); the unfortunate term, suggested to him partly by some phrases of Kant, partly, perhaps, by an article on card-games by one Weisshuhn which he published in his review *The Hours* (*Die Horen*)[4], [3] has given rise to the belief that he anticipated certain modern doctrines of artistic activity as the overflow of exuberant spirits[5], analogous with the play of children and animals. Schiller did not fail to warn his readers against such a mistaken interpretation (to which, however, he lent himself) when he begged them not to think of "games in real life, which are usually concerned with wholly material things," nor yet of the idle dreaming of the imagination left to itself. [4] The activity of the

1. *Kaligone*：《卡利戈尼》，赫尔德著。赫尔德在书中批驳了康德所说的无内容的纯形式、无概念的情感、无目的的合目的性。
2. *Metacritica*：《大批判》，赫尔德著。
3. obliterated：消除
4. *The Hours* (*Die Horen*)：《季节女神》杂志，席勒主编。
5. exuberant spirits：精力过剩

play of which he treated held the mean between the material activity of the senses, of nature, of animal instinct or passion as it is called, and the formal activity of intellect and morality. The man who plays, *i.e.* contemplates nature æsthetically and produces art, sees all natural objects as animated; in such a phantasmagoria[1] mere natural necessity gives place to the free determination of the faculties; spirit appears as spontaneously reconciled with nature, form with matter. Beauty is life, the living form (*lebende Gestalt*[2]); not life in the physiological sense, since beauty does not extend throughout all physiological life, nor is it restricted to that alone: marble when worked by an artist may have a living form; and a man, although possessed of life and form, need not be a living form. [5] Wherefore art must conquer nature with form: "in an artistic work of true beauty the content ought to be nil, the form everything: by form man is influenced in his entirety; by content in his separate faculties only. The true secret of great artists is that they cancel matter through form (*den Stoff durch die Form vertilgt*[3]); the more imposing, overwhelming or seductive the matter is in itself, the greater its obstinacy in striving to emphasize its own particular effect, the more the spectator inclines to lose himself immediately in the matter, so much the more triumphant is the art which brings it into subjection and enforces its own sovereign power. The mind of hearer or spectator should remain perfectly free and calm; from the magic circle of art it should issue as pure and perfect as when it left the hands of the Creator. The most frivolous object should be treated in such a manner as to enable us to pass at once to the most serious matters; and the most serious in such a way that we may pass from them to the lightest game." There is a fine art of passion; a passionate fine art would be a

---

1. phantasmagoria：幻想
2. *lebende Gestalt*：〈德〉活的形象，席勒用来指感性生活和理性形式（形象）的统一。
3. *den Stoff durch die Form vertilgt*：〈德〉能全销材质于形式之中

contradiction in terms. [6] "So long as man in his early physical state passively absorbs the world of senses and simply feels it, he is one with it; and precisely because he merely is a world there is for him as yet no world at all. Only when in his æsthetic state he places the world outside himself and contemplates it, does he detach his personality from the rest; then a world appears to him, since he is no longer one with the world." [7]

Æsthetic education.

Schiller ascribed high educational value to art thus conceived as at once sensible and rational, material and formal. Not that it teaches moral precepts or excites to good actions; if it acted thus, or when it acted thus, it would at once cease, as we have seen, to be art. Determination in whatsoever direction, to the good or the bad, to pleasure or to duty, destroys the character of the æsthetic sphere, which is rather indeterminism. By means of art man frees himself from the yoke of the senses; but before putting himself spontaneously under that of reason and duty, he takes as it were a little breathing-space by staying in a region of indifference and serene contemplation. "While having no claim to promote exclusively any special human faculty, the æsthetic condition is favourable to each and all without favouritism; and the reason why it favours none in particular is that it is the foundation of the possibility of all alike. Every other exercise gives some inclination to the soul, and therefore presupposes a special limit; æsthetic activity alone is unlimited." This indifference, which if not yet pure form is not pure matter, confers its educational value on art; it opens a way to morality, not by preaching and persuading, that is to say, determining, but by making determination possible. Such is the fundamental concept of his celebrated *Letters on the Æsthetic Education of Man*[1] (1795), in

1. *Letters on the Æsthetic Education of Man*：《审美教育书简》，席勒著。

which Schiller took his cue from the conditions of his times and from the necessity of finding a middle way between supine acquiescence[1] in tyranny and savage rebellion as exemplified by the revolution then raging in France.

Vagueness and lack of precision in Schiller's Æsthetic.

The defects of Schiller's æsthetic doctrine are its lack of precision and its generality. Who has given a better description of certain aspects of art, the catharsis produced by artistic activity, the serenity and calm resulting from the domination over natural impressions? Equally just is his remark that art, although wholly independent of morality, is in some way connected with it. But what precisely this connexion may be, or what the exact nature of æsthetic activity, Schiller does not succeed in explaining. Conceiving the moral and intellectual as the only formal activities (*Formtrieb*[2]) and denying as a convinced anti-sensationalist in opposition to Burke and philosophers of his type that art can belong to the passionate and sensuous nature (*Stofftrieb*[3]), he cut himself off from the means of recognizing the general category to which artistic activity belongs. His own concept of the formal is too narrow: too narrow, also, his concept of the cognitive activity, in which he is able to see the logical or intellectual form, but not that of the imagination. What for him was this art he describes as an activity neither formal nor material, neither cognitive nor moral? Was it for him, as for Kant, an activity of feeling, a play of several faculties at once? It would seem so, since Schiller distinguishes four points of view or relations of man with things: the physical, in which these affect our senses: the logical, in which they excite knowledge: the moral, in which they appear to us

1. supine acquiescence：浑浑噩噩
2. *Formtrieb*：〈德〉形式冲动
3. *Stofftrieb*：〈德〉感性冲动

as an object of rational volition: and the æsthetic "in which they refer to our powers in entirety without becoming the determinate object of any one faculty." For example, a man is pleased æsthetically when his feeling depends in no way on the pleasure of the senses and when he is not conscious of thinking about any law or end. [8] We look in vain for any more conclusive reply.

It must not be overlooked that Schiller delivered a course of lectures on Æsthetic in Jena University[1] in 1792, and that his writings on the subject intended for reviews were couched in a popular style: no less popular, in his own opinion, was the style of the book quoted above, which grew out of a series of letters actually sent to his patron the Duke of Holstein-Augustenburg. But the great work to be entitled *Kallias*[2], which he intended writing upon Æsthetic, was never completed; the only fragments which have reached us are contained in the correspondence with Körner[3] (1793–1794). From the discussions between the two friends we gather that Körner was not satisfied with Schiller's formula and desired something objective, something more precise, a positive characteristic of the beautiful: and one day Schiller told him that he had definitely discovered such a characteristic. But what it was that he had discovered we do not know; no mention of it occurs in any further document, and we are left in doubt as to whether we have lost an integral part of his thought or merely the momentary illusion of a discovery.

Schiller's caution and the rashness of the Romanticists.

The uncertainty and vagueness of Schiller's theory seem almost

---

1. Jena University：耶拿大学。始建于 1558 年，是德国最为传统，古老的大学之一。为纪念席勒，1934 年改名为“弗里德里希 · 席勒大学”。
2. *Kallias*：《论美》。席勒在该著作中着重探讨了美的本质问题。
3. Körner：考尔纳（1756—1831），德国法学家、作家，著有《美学观点》《德国的希望》等。

a merit in contrast with that which followed. He had constituted himself guardian of the teaching of Kant and refused to abandon the realm of criticism; faithful disciple of his master, he conceived the third sphere not as real but as an ideal, a concept not constitutive but regulative, an imperative. "From transcendental motives, reason here demands that communion be established between formal and material activity; that is to say, there must be an activity of play, since the concept of humanity can be complete only by the union of reality with form, the accidental with the necessary, passivity with liberty. This demand must be made because reason, in conformity with her essence, aims at perfection and at sweeping away all obstacles; and every exclusive operation of one or other activity leaves humanity incomplete and confined within limits." [9] Schiller's thought, as it appears in his correspondence with Körner, has been well represented as follows: "The union of sensibility with liberty in the Beautiful, which does not actually take place but is supposed to do so, suggests to man an intuition of the union of these elements within himself: a union which does not take place actually but ought to do so." [10] The times which followed had no such nice scruples. Kant had given new vigour to the production of works on æsthetic, and, as in the days following Baumgarten, every new year saw a number of new treatises. It was the fashion. "Nothing swarms like æstheticians" (wrote Jean Paul Richter[1] in 1804 when preparing his own book on the subject for publication): "it is rare for a youth who has paid his fees for a course of lectures on Æsthetic not to produce a book on some point of the science in the hope that the public may refund him his expenses by buying his book: some there are indeed who pay their professor's fees out of their author's royalties." [11] It was hoped, not unreasonably, that the exploration of the obscure region of æsthetic might throw

1. Jean Paul Richter：让·保罗·里希特（1763—1825），德国重要诗人。

some light on metaphysics, and the procedure of artists seemed to offer a good example to philosophers seeking to create a world for themselves: so philosophy modelled itself upon art and, as though to render the transition easier, the concept of art was brought as close as possible to that of philosophy. Romanticism, gaining vogue daily, was a renewal or continuation of that "age of genius" in which the youth of Goethe and Schiller had been passed; and as the period of *Sturm und Drang*[1] had zealously worshipped the genius who breaks all rules and oversteps all limitations, so did Romanticism hail the domination of a faculty called Fancy, or more frequently Imagination, to which were attributed the most diverse characteristics and the most miraculous effects.

Ideas on Art: J. P. Richter.

The Romantic theorists, artists themselves for the most part, abounded in truthful and subtle observations concerning artistic procedure. Jean Paul Richter makes many excellent remarks about productive imagination, which he distinguishes clearly from the reproductive and asserts to be shared by all men as soon as they are able to say "This is beautiful"; for "how could a genius be acclaimed or even tolerated for a single month, not to mention thousands of centuries, by the common herd, if he had not a strong connecting-link of relationship with the herd?" He also describes how imagination is variously divided among individuals: as simple talent, as passive or feminine genius, and in the highest degree as the active or masculine genius, formed by reflexion and instinct, in which "all faculties flourish simultaneously and fancy is no isolated flower, but the goddess Flora herself who, in order to produce new combinations, crosses with

1. *Sturm und Drang*：〈德〉狂飙突进运动，18 世纪德国文学界的一次文艺形式从古典主义向浪漫主义过渡的文学解放运动。

each other those blossoms whose conjunction is fertile, and is, so to speak, a faculty full of faculties." [12] This latter sentence betrays a tendency on Richter's part to exaggerate the functions of imagination and to construct upon it a kind of mythology.

Romantic Æsthetic and idealistic Æsthetic.

Contemporary systems of philosophy are partly impregnated with, and partly the source of, such mythologies: the Romantic conception of art may be said to have found its most complete expression in German idealism, where this attained its most coherent and systematic form.

J. G. Fichte.

It did not attain this form with Fichte[1], the first great pupil of Kant; for though Fichte regarded imagination as the activity which creates the universe, effects the synthesis of the ego and the non-ego, posits the object and therefore precedes consciousness, he does not connect it with art. [13] In his æsthetic notions Fichte is influenced by Schiller, with the addition of a moralism imposed upon him by the general character of his system; hence the ethical sphere, midway between the cognitive and the æsthetic, becomes from his point of view a mere appurtenance[2] of morality, as being the representation of, and hence reverence for, the moral ideal. [14] His subjective idealism eventually produced an æsthetic doctrine through the work of Friedrich Schlegel[3]

1. Fichte：费希特（1762—1814），德国作家、哲学家，古典主义哲学的主要代表人物之一。
2. appurtenance：附庸
3. Friedrich Schlegel：弗里德里希·施莱格尔（1772—1829），德国诗人、作家、哲学家、文艺理论家，著有《希腊诗歌的研究》《特征与批评》《欧洲文学史》等。

and Ludwig Tieck[1]; the doctrine of Irony as the basis of art.

Irony: Schlegel, Tieck, Novalis.

The ego which created the universe can also destroy it; the universe is an empty appearance at which the only true reality, the ego, can smile, holding itself aloof, like an artist or a creative god, from creatures of its own which it does not take seriously. [15] Friedrich Schlegel described art as a perpetual parody of itself and a "transcendental farce.[2]" Tieck defined irony as "a power which allows the poet to dominate the matter which he handles." Another Romantic Fichtian, Novalis[3], dreamed of a magical idealism, an art of creation by the instantaneous act of the ego and of realizing our dreams.

F. Schelling.

But it is only to the *System of Transcendental Idealism*[4] (1800) of Schelling, to his *Bruno*(1802), to his celebrated course of lectures on the *Philosophy of Art* given at Jena in 1802–1803 (repeated at Würzburg, and distributed subsequently in manuscript notes all over Germany), to the no less celebrated lecture on the *Relation between the Figurative Arts and Nature* (1807), as well as to other works of this eloquent and enthusiastic philosopher that we owe the first great philosophical affirmation of Romanticism, and of a renewed and conscious neo-Platonism in Æsthetic.

Beauty and character.

---

1. Ludwig Tieck：路德维西·蒂克（1773—1853），德国诗人、小说家、翻译家、评论家，著有《施特恩巴尔德的游历》《关于艺术的遐想》等。
2. transcendental farce：先验的闹剧
3. Novalis：诺瓦利斯（1772—1801），德国诗人、作家、哲学家，著有《夜的赞美诗》等。
4. *System of Transcendental Idealism*：《先验唯心论体系》，谢林著。

Like all the other idealistic philosophers, Schelling held firmly to the fusion of the theories of art and the beautiful already effected by Schiller. From this point of view it is interesting to note his explanation of the condemnation of art by Plato: this condemnation, says Schelling, was directed against the art of his time, the natural and realistic art of antiquity in general, with its character of finitude[1]: Plato could not have uttered such a condemnation (as we moderns are unable to utter it) if he had known Christian art, whose characteristic is infinity. [16] The pure abstract beauty of Winckelmann is not enough; no less inadequate, false and negative is that concept of the characteristic which would try to make art something dead, hard and ugly by imposing upon it the limitations of the individual. Art is beauty and characteristic in one; characteristic beauty, character from which beauty is evolved, according to Goethe's saying; it is therefore not the individual but the living concept of the individual. When the artist's eye recognizes the creative idea of the individual and draws it forth, he transforms the individual into a world in itself, into a species (*Gattung*), an eternal idea (*Urbild*), and fears no more the limitation or hardness which is the condition of life: characteristic beauty is that plenitude of form which kills form; it does not inflame passion, it regulates it, like the banks of a river which are filled but not overflowed by the waters. [17] In all of this we feel the influence of Schiller, with something added which Schiller could never have expressed.

Art and Philosophy.

Indeed, whilst gratefully acknowledging the excellent contributions to the theory of art made by the writers who succeeded Kant, Schelling laments that in none of them can he find exact scientific method

---

1. finitude：有限

(*Wissenschaftlichkeit*[1]), [18] The true point of departure in his theory is in the philosophy of nature, *i.e.* in that criticism of the teleological judgement which Kant places directly after that of the æsthetic judgement in his third *Critique*. Teleology is the union of theoretical and practical philosophy; but the system would be incomplete but for the possibility of demonstrating in the subject itself, in the ego, the identity of the two worlds, theoretical and practical; an activity which has, and at the same time has not, consciousness; unconscious as nature, conscious as spirit. This activity is precisely the æsthetic activity: "the general organ of philosophy, keystone of the whole edifice." [19] There are but two ways open to one who is desirous of escaping from common realities: poetry, which transports into the ideal world; and philosophy which annihilates the real world. [20] Strictly speaking, "there is but one sole absolute work of art; it may exist in various exemplars, but in itself it is one, although it may not yet possess existence in its original form." True art is not the impression of one moment, but the representation of infinite life; [21] it is transcendental intuition become objective, and is therefore not only the organ but the document of philosophy. A time will come when philosophy will return to poetry, from which she has detached herself; and from the new philosophy a new mythology will arise. [22] The Absolute is thus the object of art as well as of philosophy (as Schelling insists elsewhere in greater detail): the first represents it in idea (*Urbild*), the second in its reflexion (*Gegenbild*): "philosophy portrays ideas, not realities: so is it with art: those same ideas of which real things, as philosophy demonstrates, are imperfect copies, themselves appear in the objective arts as ideas, *i.e.* in all their perfection, and represent the intellectual world in the world of reflexion." [23] Music is the "very ideal rhythm of Nature and the Universe, which by means of

1. *Wissenschaftlichkeit*：〈德〉科学性

this art makes itself felt in the derivative world"; perfect creations of statuary are "the very ideas of organic nature represented objectively"; the Homeric epic, "the very identity constituting the foundation of history in the Absolute." [24] But while philosophy gives an immediate representation of the Divine, of absolute Identity, art can but give the immediate representation of Indifference; and "since the degree of perfection or reality in a thing becomes higher in proportion as it approaches nearer to the absolute Idea and the fulness of infinite affirmation and in proportion as it comprehends within itself other powers, it is clear that art, above everything else, is in closest relation with philosophy, from which it is distinguished merely by the character of its specification: in everything else it may be considered as the highest power in the ideal world." [25] To the three powers of the real and ideal world correspond in a rising scale the three ideas of Truth, Goodness and Beauty. Beauty is neither the mere universal (truth), nor mere reality (action), but the perfect interpenetration of both: "beauty exists when the particular (the real) is so adequate to its concept that the latter, as infinite, enters the finite and presents itself to our contemplation in concrete form. With the appearance of the concept, the real becomes truly similar and equal to the idea, wherein the universal and the particular find their absolute identity. Without ceasing to be rational, the rational becomes at the same time apparent and sensible." [26] But as above the three powers is poised God, their point of union, so Philosophy stands supreme over the three ideas; concerning itself not with truth or morality or even beauty alone, but with that which belongs to all the three in common, deduced from one common source. If philosophy assumes the character of science and truth, while yet remaining superior to truth, this is made possible by the fact that science and truth are its formal determination; "philosophy is science in the sense that truth, goodness and beauty, *i.e.* science, virtue and art, interpenetrate each other; therefore it is also not science but is that which is common to science, virtue and art."

This interpenetration distinguishes philosophy from all other sciences; for instance, if mathematics can dispense with morality and beauty, philosophy cannot do so. [27]

Ideas and the gods. Art and mythology.

In Beauty are contained truth and goodness, necessity and liberty. When beauty appears to be in conflict with truth, the truth in question is a finite truth with which beauty ought not to agree, because, as we have seen, the art of naturalism and of the merely characteristic is a false art. [28] The individual forms of art, being in themselves representatives of the infinite and the universe, are called Ideas. [29] Considered from the point of view of reality, Ideas are gods; their essence, their "in-itself," is in fact equivalent to God; every idea is an idea so far as it is God in a particular form; every idea, therefore, is equal to God, but to a particular god. Characteristic of all the gods is pure limitation and indivisible absoluteness: Minerva is the idea of wisdom united with strength, but she is lacking in womanly tenderness; Juno[1] is power without wisdom and without the sweet attraction of love, for which she is forced to borrow the cestus of Venus[2]; Venus again has not the weighty wisdom of Minerva. What would become of these ideas if deprived of their limitations? They would cease to be objects of Imagination. [30] Imagination is a faculty which has no connexion with pure intellect or with reason (*Vernunft*) and is distinct from fancy (*Einbildungskraft*) which collects and arranges the products of art, whereas imagination intuits them, forms them out of itself, represents them. Imagination is to fancy as intellectual intuition is to reason: it is therefore the intellectual intuition of art. [31]

1. Juno：朱诺，罗马神话中的天后，主神丘比特（Jupiter）之妻，主司婚姻生育等。
2. cestus of Venus：维纳斯的皮腰带，具有爱情的魔力。

"Reason" no longer suffices in a philosophy such as this: intellectual intuition, which for Kant was a limiting concept, is now asserted as really existing: intellect sinks to a subordinate place: even the genuine imagination which operates in art is overshadowed by this new-fangled Imagination[1], twin with intellectual Intuition, who sometimes changes places with this sister of hers. Mythology is proclaimed a necessary condition of all art: mythology which is not allegory, for in the latter the particular signifies only the universal, while the former is already itself the universal; which explains how easy it is to allegorize[2], and how fascinating are such poems as those of Homer which lend themselves to such interpretations. Christian, as well as Hellenic, art has its mythology: Christ; the persons of the Trinity; the Virgin mother of God. [32] The line between mythology and art is as shadowy as that between art and philosophy.

K. W. Solger[3].

The year 1815 saw the publication of Solger's principal work, *Erwin*, a long philosophical dialogue on the beautiful; subsequently in 1819 he gave a course of lectures on Æsthetic which were published posthumously. He was one of those who found but a glimpse of truth in Kant and held the post-Kantians in very slight estimation, particularly Fichte; in Schelling, who begins from the original unity of the subjective and the objective, he detects for the first time a speculative principle not adequately developed, since Schelling had never triumphed dialectically over the difficulties of intellectual intuition. [33]

---

1. new-fangled Imagination：新鲜出炉的幻想
2. allegorize：讽喻
3. Solger：佐尔格（1780—1819），德国哲学家，著有《索福克勒斯的悲剧》《埃尔温——关于美学和艺术的四问答》《哲学谈话录》。

Fancy and Imagination.

Solger was one of those who conceived of Imagination as totally distinct from Fancy: fancy (says he) belongs to common cognition and is none other than "the human consciousness, in so far as it continues, in temporal succession, infinitely reasserting an original intuition"; it presupposes the distinctions between common cognition, abstraction and judgement, concept and representation, amongst which "it acts as mediator by giving to the general concept the form of individual representation; and to the latter the form of a general concept; in this manner it has its being among the antitheses of the ordinary understanding." Imagination is totally different; proceeding "from the original unity of the antitheses in the Idea, it acts so that the elements in opposition, separated as they are from the idea, find themselves united in the reality; by its means we are capable of apprehending objects higher than those of common cognition and of recognizing in them the idea itself as real: also, in art, it is the faculty of transforming the idea into reality." It presents itself in three modes or degrees: as Imagination of the Imagination, which conceives the whole as idea, and activity as nothing more than the development of the idea in reality; as Sensibility of the Imagination, in so far as it expresses the life of the idea in the real and reduces the one to the other; lastly (and here we have the highest grade of artistic activity, corresponding with Dialectic in philosophy) as Intellect of the Imagination or artistic Dialectic, conceiving idea and reality in such a way that one passes over into the other, that is to say, into reality. Other divisions and subdivisions are made on which it is not necessary to dwell. Imagination is said to produce the Irony essential to true art: this is the Irony of Tieck and Novalis, of whom Solger is in a sense a follower. [34]

Art, practice and religion.

Solger joins Schelling in placing beauty in the region of the Idea, inaccessible to common consciousness. It is distinct from the idea of Truth, because instead of dissolving the appearances of common consciousness after the manner of truth, art accomplishes the miracle of making appearance dissolve itself while still remaining appearance; artistic thought, therefore, is practical, not theoretical. Furthermore, it is distinct from the idea of Goodness, with which at first sight it would seem to be closely related, because in the case of Goodness the union of ideal with real, of the simple with the multiple, of the infinite with the finite, is not real and complete, but remains ideal, a mere ought-to-be. It is related more closely to Religion, which thinks the Idea as the abyss of life where our individual conscience must lose itself in order to become "essential" (*wesentlich*), while in beauty and art the Idea manifests itself by gathering into itself the world of distinctions between universal and particular and placing itself in their place. Artistic activity is more than theoretical, it is of a practical nature, but realized and perfected; art, therefore, belongs not to theoretical philosophy (as Kant thought, according to Solger), but to practical. Necessarily attached on one side to infinity, it cannot have common nature as its object; for example, art is absent from a portrait, and the ancients showed their discrimination in selecting gods and heroes for objects in sculpture since every deity — even in limited and particular form — always signifies a determinate modification of the Idea. [35]

G. W. F. Hegel.

The same concept of art appears in the philosophy of Hegel, whatever may be the minor differences which he felt to separate himself from his predecessors. Little concerned as we are with the shades and varieties of mystical Æsthetic exhibited by each of these thinkers, we are chiefly concerned to lay bare[1] the substantial

1. lay bare：使显露

underlying identity, the mysticism of arbitrarism which gives them their historic place in Æsthetic.

Art in the sphere of absolute spirit.

Opening the *Phenomenology* and the *Philosophy of Spirit*[1], one need not expect to find any discussion of art in the analysis of the forms of the theoretical Spirit, among definitions of sensibility and intuition, language and symbolism, and various grades of imagination and thought. Hegel places Art in the sphere of absolute Spirit, together with Religion and Philosophy, [36] and in this he regards Kant, Schiller, Schelling and Solger as his precursors, for like them he strongly denies that art has the function of representing the abstract concept, but not that it represents the concrete concept or Idea. Hegel's whole philosophy consists in the affirmation of a concrete concept, unknown to ordinary or scientific thought. "Indeed," says he, "no concept has in our day been more mishandled than the concept in itself and for itself; for by concept is generally meant the abstract determinateness or one-sidedness of representation and intellectualistic thought, with which it is naturally impossible to think either the entirety of truth or concrete beauty." [37] To the realm of the concrete concept belongs art, as one of the three forms wherein the freedom of the spirit is achieved; it is the first form, namely that of immediate, sensible, objective knowledge (the second is religion, a representative consciousness *plus* worship, an element extraneous to mere art: the third is philosophy, free thought of the absolute spirit). [38]

Beauty as sensible appearance of the Idea.

Beauty and truth are at the same time one yet distinct. "Truth is

1. *Phenomenology* and the *Philosophy of Spirit*：《精神现象学》和《精神哲学》，黑格尔著。

Idea as Idea, according to its being-in-itself and its universal principle, and so far as it is thought as such. There is no sensible or material existence in Truth; thought contemplates therein nothing but universal idea. But the Idea must also realize itself externally and attain an actual and determinate existence. Truth also as such as existence; but when in its determinate external existence it is immediately for consciousness, and the concept remains immediately one with the external appearance, the Idea is not only true but beautiful. In this way Beauty may be defined as the sensible appearance of the Idea." [39] The Idea is the content of art: its sensible and imaginative configuration; its form: two elements which must interpenetrate and form a whole, hence the necessity that a content destined to become a work of art should show itself capable of such transformation; otherwise we have but an imperfect union of poetic form with prosaic and incongruous content[1]. [40] An ideal content must gleam through the sensible form; the form is spiritualized by this ideal light; [41] artistic imagination does not work in the same way as the passive or receptive fancy, it does not stop at the appearances of sensible reality but searches for the internal truth and rationality of the real. "The rationality of the object selected by him should not be alone in awakening the consciousness of the artist: he should have well meditated upon the essential and the true in all their extension and profundity, for without reflexion a man cannot become conscious of that which is within himself, and all great works of art show that their material has been thought again and again from every side. No successful work of art can issue from light and careless imagination." [42] It is a delusion to fancy that poet and painter need nothing beyond intuitions: "a true poet must reflect and meditate before and during the execution of his poem." [43] But it is always understood that the

1. an imperfect union of poetic form with prosaic and incongruous content：不完美的诗歌形式，内容平淡不协调

thought of the poet does not take the form of abstraction.

Æsthetic in metaphysical idealism and Baumgartenism.

Some critics [44] affirm that the æsthetic movement from Schelling to Hegel is a revived Baumgartenism on the ground that this movement regarded art as a mediator of philosophical concepts; they mention the fact that a follower of Schelling, one Ast[1], was moved by the trend of his system to substitute didactic poetry for drama as the highest form of art.[2] [45] Putting aside some isolated and accidental deviations, there is no truth in this affirmation: these philosophers are hostile to intellectualistic and moralistic views, frequently entering upon definite and explicit polemic against them. Schelling wrote: "Æsthetic production is in its origin an absolutely free production ... . This independence on any extraneous purpose[3] constitutes the sanctity and purity of art, enabling it to repel all connexion with mere pleasure, a connexion which is a mark of barbarism, or with utility, which cannot be demanded of art save at times when the loftiest form of the human spirit is found in utilitarian discoveries. The same reasons forbid an alliance with morality and hold even science at arm's length, although nearest by reason of her disinterestedness; having her aim, however, outside herself, she must restrict herself definitely to serve as means to something higher than herself: the arts." [46] Hegel says, "Art contains no universal as such." "If the aim of instruction is treated as an aim, so that the nature of the content represented appears for itself directly, as an abstract proposition, prosaic reflexion, or general

---

1. Ast：阿斯特（1778—1841），德国哲学家、语文学家，著有《柏拉图的生平与著作》《艺术学体系》《美学基础》等。
2. substitute didactic poetry for drama as the highest form of art：用说教诗取代戏剧作为艺术的最高形式
3. extraneous purpose：外在目的

theory, and is not merely contained indirectly and implicitly in the concrete artistic form, the result of such a separation is to reduce the sensible and imaginative form, the true constituent of a work of art, to an idle ornament, a covering (*Hülle*[1]) presented simply as a covering, an appearance maintained as mere appearance. The very nature of the work of art is thus completely altered, for a work of art must not present to intuition a content in its universality, but this universal individualized and converted into a sensible individual." [47] It is a bad sign, he adds, when an artist sets himself about his work from a motive of abstract ideas instead of that of the fulness of life (*Überfülle des Lebens*). [48] The aim of art lies in itself, in presentation of truth in a sensible form; any other aim is altogether extraneous. [49] It would not be hard to prove, certainly, that by separating art from pure representation and imagination and making it in some sense the vehicle of the concept, the universal, the infinite, these philosophers were facing in the direction of the road opened by Baumgarten. But to prove this would mean accepting as a presupposition the dilemma that if art be not pure imagination, it must be sensuous and subordinate to reason; and it is just this presupposition and dilemma that the metaphysical idealists denied. The road they tried to follow was to conceive a faculty which should be neither imagination nor intellect but should partake of both; an intellectual intuition or intuitive intellect, a mental imagination after the fashion of Plotinus.

Mortality and decay of art in Hegel's system.

In a greater degree than any of his predecessors Hegel emphasized the cognitive character of art. But this very merit brought him into a difficulty more easily avoided by the rest. Art being placed in the

---

1. *Hülle*：〈德〉外壳

sphere of absolute Spirit, in company with Religion and Philosophy, how will she be able to hold her own in such powerful and aggressive company, especially in that of Philosophy, which in the Hegelian system stands at the summit of all spiritual evolution? If Art and Religion fulfilled functions other than the knowledge of the Absolute, they would be inferior levels of the Spirit, but yet necessary and indispensable. But if they have in view the same end as Philosophy and are allowed to compete with it, what value can they retain? None whatever; or, at the very most, they may have that sort of value which attaches to transitory historical phases in the life of humanity. The principles of Hegel's system are at bottom rationalistic and hostile to religion, and hostile no less to art. A strange and painful consequence for a man like Hegel, endowed with a warmly æsthetic spirit and a fervid lover of the arts[1]; almost a repetition of the hard fate endured by Plato. But as the Greek philosopher, in obedience to the presumed command of religion, did not hesitate to condemn the mimetic art and the Homeric poetry he loved, so the German refused to evade the logical exigencies[2] of his system and proclaimed the mortality, nay, the very death, of art. "We have assigned," he says, "a very high place to art: but it must be recollected that neither in content nor in form can art be considered the most perfect means of bringing before the consciousness of the mind its true interests. Precisely by reason of its form, art is limited to a particular content. Only a definite circle or grade of truth can be made visible in a work of art; that is to say, such truth as may be transfused into[3] the sensible and adequately presented in that form, as were the Greek gods. But there is a deeper conception of truth, by which it is not so intimately allied to the sensible as to permit of its being received or expressed suitably in material

1. fervid lover of the arts：狂热的艺术爱好者
2. logical exigencies：逻辑需要
3. transfuse into：灌输

fashion. To this class belongs the Christian conception of truth; and, furthermore, the spirit of our modern world, more especially that of our religion and our mental evolution, seems to have passed the point at which art is the best road to the apprehension of the Absolute. The peculiar character of artistic production no longer satisfies our highest aspirations ... . Thought and reflexion have superseded fine art." Many reasons have been adduced in order to account for the moribund condition of modern art[1]; in especial, the prevalence of material and political interests; the true reason, says Hegel, consists of the inferiority in grade of art in comparison with pure thought. "Art in its highest form is and for us must remain a thing of the past"; and just because the thing has vanished, one can reason about it philosophically. [50] The Æsthetic of Hegel is thus a funeral oration: he passes in review the successive forms of art, shows the progressive steps of internal consumption and lays the whole in its grave, leaving Philosophy to write its epitaph[2].

Romanticism and metaphysical idealism had elevated art to such a fantastic height among the clouds that at last they were obliged to admit that it was so far away as to be absolutely useless.

[1] *Vorles. über die Ästhetik* (2nd ed., Berlin, 1842), vol. i. p. 78.
[2] Sommer, *Gesch. d. Psych. u. Ästh.* pp. 365–432.
[3] Danzel, *Ges. Aufs.* p. 242.
[4] *Briefe ü. d. Ästh. Erzieh.* (in Werke, ed. Goedecke), Letters 15, 27.
[5] *Op. cit.* Letter 15.
[6] *Briefe*, Letter 22.
[7] *Op. cit.* Letter 25.
[8] *Briefe*, Letter 20.

---

1. moribund condition of modern art：当代艺术奄奄一息的状态
2. epitaph：墓志铭

[9] *Briefe*, Letter 15.

[10] Danzel, *Ges. Aufs.* p. 241.

[11] *Vorschule der Ästh.*, 1804 (French trans., *Poétique ou introduction à l'Esth.*, Paris, 1862), preface.

[12] *Vorschule d. Ästh.* chs. 2, 3.

[13] *Grundl. der Wissenschaftslehre,* in *Werke* (Berlin, 1845), vol. i. pp. 214–217.

[14] Danzel, *Ges. Aufs.* pp. 25–30; Zimmermann, *G. d. A.* pp. 522–572.

[15] Hegel, *Vorles. üb. d. Ästh.* introd. vol. i. pp. 82–88.

[16] *Vorles. üb. d. Methode d. akadem. Stud.* (1803), lecture 14; in *Werke* (Stuttgart, 1856–1861), vol. v, pp. 346–347.

[17] *Üb. d. Verhältniss d. bild. Künste, z. d. Natur* in *Werke*, vol. vii. pp. 299–310.

[18] *Philos, d. Kunst,* posthumous, introd. in *Werke*, v. p. 362.

[19] *System d. transcend. Idealismus,* in *Werke*, § i. vol. iii. introd. § 3, p. 349.

[20] *Op. cit.* § 4, p. 351.

[21] *System d. transcend. Idealismus*, in *Werke*, part vi. § 3, p. 627.

[22] *Op. cit.* § 3, pp. 627–629.

[23] *Phil. d. Kunst*, pp. 368–369.

[24] *Op. cit.* p. 369.

[25] *Op. cit.* General Part, p. 381.

[26] *Phil. d. Kunst,* p. 382.

[27] *Op. cit.* p. 383.

[28] *Op. cit.* p. 385.

[29] *Op. cit.* pp. 389–390.

[30] *Phil. d. Kunst*, pp. 390–393.

[31] *Op. cit.* p. 395.

[32] *Op. cit.* pp. 405–451.

[33] *Vorles. üb. Ästhetik*, Heyse, Leipzig, 1829, pp. 35–43.

[34] Vorles. üb. Ästh. pp. 186–200.

[35] *Op. cit.* pp. 48–85.

[36] *Encykl. d. phil. Wiss.* § § 557–563.

[37] *Vorles. üb. Ästh.* (*ed. cit.*) i. p. 118.

[38] *Op. cit.* i. pp. 129–133.

[39] *Vorles. üb. Ästh.* i. p. 141.

[40] *Op. cit.* i. p. 89.

[41] *Op. cit.* i. pp. 50–51.

[42] *Op. cit.* i. pp. 354–355.

[43] *Encykl.* § 450.

[44] Danzel, *Ästh. d. hegel. Sch.* p. 62; Zimmermann, *G. d. A.* pp. 693–697; J. Schmidt, *L. u. B.* pp. 103–105; Spitzer, *Krit. St.* p. 48.

[45] Fr. Ast, *System der Kunstlehre*, Leipzig, 1805; cf. Spitzer, *op. cit.* p. 48.

[46] *System d. transcend. Idealismus* (1800), part vi. § 2; in *Werke*, § I, vol. iii. pp. 622–623.

[47] *Vorles. üb. d. Ästh.* i. pp. 66–67.

[48] *Vorles. üb. d. Ästh.* i. p. 353.

[49] *Op. cit.* i. p. 72.

[50] *Vorles. üb. d. Ästh.* i. pp. 13–16.

X SCHOPENHAUER AND HERBART

Æsthetic mysticism in the opponents of Idealism.

Nothing, perhaps, shows more clearly how well this imaginative conception of art suited the spirit of the times (not only a particular fashion in philosophy, but the psychological conditions expressed by the Romantic movement) than the fact that the adversaries of the systems of Schelling, Solger and Hegel either agreed with this conception in general or, while believing themselves to be departing widely from it, actually returned to it involuntarily.

A. Schopenhauer.

Everybody knows with what lack, shall we say, of *phlegma philosophicum*[1] Arthur Schopenhauer fought against Schelling, Hegel and all the "charlatans" and "professors" who had divided amongst themselves the heritage of Kant. But what was the artistic theory accepted and developed by Schopenhauer?

Ideas as the object of art.

His theory, like Hegel's own, turns upon the distinction between the concept which is abstraction and the concept which is concrete,

1. *phlegma philosophicum*：〈拉〉黏液质的（指性情沉静、平稳）

or Idea; although Schopenhauer's Ideas are by himself likened to Plato's, and in the particular form in which he presents them more nearly resemble those of Schelling than the Idea of Hegel. They have something in common with intellectual concepts, for like them they are unities representing a plurality of real things: but "the concept is abstract and discursive, entirely indeterminate in its sphere, rigorously precise within its own limits only; the intellect suffices to conceive and understand it, speech expresses it without need for other intermediary, and its own definition exhausts its whole nature; the idea, on the contrary (which may be defined clearly as the adequate representative of the concept) is absolutely intuitive, and although it represents an infinite number of individual things, it is not for that any the less determined in all its aspects. The individual, as individual, cannot know it; in order to conceive it he must strip himself of all will, of all individuality, and raise himself to the state of a pure knowing subject[1]. The idea, therefore, is attained by genius only, or by one who finds himself in a genial disposition attained by that elevation of his cognitive powers inspired usually by genius." "The idea is unity become plurality by means of space and time, forms of one intuitive apperception; the concept, on the contrary, is unity extracted from plurality by means of abstraction, which is the procedure of our intellect: the concept may be described as *unitas post rem*[2]; the idea, *unitas ante rem.*[3]" [1] Schopenhauer is in the habit of calling ideas the genera of things; but on one occasion he remarks that ideas are of species, not genera[4]; that genera are simply concepts, and that there are natural species, but only logical genera. [2] This psychological illusion as to the existence of ideas for types originates (as we find elsewhere in

1. pure knowing subject：纯粹认知主体
2. *unitas post rem*:〈拉〉后于事物的普遍性
3. *unitas ante rem*：〈拉〉先于事物的普遍性
4. ideas are of species, not genera：理念是属类而非种类

Schopenhauer) in the habit of converting the empirical classifications of the natural sciences into living realities. "Do you wish to see ideas?" he asks; "look at the clouds which scud across the sky; look at a brooklet leaping over rocks; look at the crystallization of hoar-frost[1] on a window-pane with its designs of trees and flowers. The shapes of the clouds, the ripples of the gushing brook, the configurations of the crystals exist for us individual observers, in themselves they are indifferent. The clouds in themselves are elastic vapour; the brook is an incompressible fluid, mobile, transparent, amorphous[2], the ice obeys the laws of crystallization: and in these determinations their ideas consist." [3] All these are the immediate objectification of will in its various degrees; and it is these, not their pale copies in real things, that art delineates[3]; whence Plato was right in one sense and wrong in another, and is justified and condemned by Schopenhauer exactly in the same way as by Plotinus of old, as well as by Schopenhauer's worst enemy, the modern Schelling. [4] In consequence, each art has a special category of ideas for its own dominion. Architecture, and in some cases hydraulics, facilitate the clear intuition of those ideas which constitute the lower degrees of objectification — weight, cohesion, resistance, hardness, the general properties of stone and some combinations of light; gardening and (most curious association) landscape painting represent the ideas of vegetable nature; sculpture and animal painting those of zoology; historical painting and the higher forms of sculpture that of the human body; poetry the very idea of man himself. [5] As for music, that (let him who can justify the logical discontinuity) is outside the hierarchy of the other arts. We have seen how Schelling considered it to be representative of the very rhythm of the universe; [6] differing but slightly from this,

1. hoar-frost：白霜
2. amorphous：无固定形状的
3. delineates：刻划

Schopenhauer affirms that music does not express ideas but, parallel with ideas, Will itself. The analogies between music and the world, between the fundamental bass and crude matter, between the scale and the series of species, between melody and conscious will, led him to the conclusion that music was not, as Leibniz thought, an arithmetic but a metaphysic: *exercitium metaphysices occultum nescientis se philosophari animi*[1]. [7]

Æsthetic catharsis.

To Schopenhauer, no less than his idealistic predecessors, art beatifies; it is the flower of life; he who contemplates art is no longer an individual but a pure knowing subject, at liberty, free from desire, from pain, from time. [8]

Signs of a better theory in Schopenhauer.

Schopenhauer's system no doubt contains here and there premonitions[2] of a better and more profound treatment of art. Schopenhauer, who was capable on occasion of clear and keen analysis, constantly insists that the forms of space and time must not be applied to the idea or to artistic contemplation, which admits of the general form of representation only. [9] From this he might have inferred that art, so far from being a superior and extraordinary level of consciousness, is actually its most immediate level, namely that which in its primitive simplicity precedes even common perception with its reference of objects to a position in the spatial and temporal series. To free oneself from common perception and to live in

1. *exercitium metaphysices occultum nescientis se philosophari animi*：〈拉〉在没有哲学思维的人那里，形而上学式的练习。
2. premonitions：预告

imagination does not mean rising to a Platonic contemplation of the ideas, but descending once more into the region of immediate intuition, becoming children again, as Vico had seen. On the other hand Schopenhauer had begun to examine the categories of Kant with an unprejudiced eye; he was not satisfied with the two forms of intuition, and wished to add to them a third, causality. [10] In conclusion, we note that, like his predecessors, he makes a comparison between art and history, with this difference and advantage over the idealist authors of the philosophy of history, that for him history was irreducible to concepts; it was contemplation of the individual, and therefore not science. Had he persevered in his comparison between art and history, he would have arrived at a better solution than that at which he stopped; that is to say, that the matter of history is the particular in its particularity and contingency, while that of art is that which is, and is always identical. [11] But instead of pursuing these happy ideas Schopenhauer preferred to play variations on the themes fashionable in his day.

J. F. Herbart.

Most astounding of all is the fact that a dry intellectualist, the avowed enemy of idealism, of dialectic and of speculative constructions, head of the school calling itself realistic or the school of exact philosophy, Johann Friedrich Herbart, when he turns his attention to Æsthetic, turns mystic too, though in a slightly different way. How weightily he speaks when expounding his philosophical method! Æsthetic must not bear the blame of the faults into which metaphysic has fallen; we must make it an independent study, and detach it from all hypothesis about the universe. Nor must it be confounded with psychology or asked to describe the emotions awakened by the content of works of art, such as the pathetic or the comic, sadness or joy; its duty is to determine the essential character of art and beauty. In the analysis of particular cases of beauty and in registering what they reveal lies the way of salvation. These proposals and promises have misled numbers of people as to the nature of Herbart's Æsthetic. But *ce sont là jeux de princes*[1]; by paying attention we shall see what Herbart meant by analysis of particular case; and how he held himself aloof from metaphysics.

Pure Beauty and relations of form.

Beauty, for him, consisted in relations: relations of tone, colour, line, thought and will; experience must decide which of these relations are beautiful, and æsthetic science consists solely in enumerating the fundamental concepts (*Musterbegriffe*) in which are summarized the particular cases of beauty. But these relations, Herbart thought, were not like physiological facts; they could not be empirically observed, *e.g.* in a psycho-physical laboratory. To correct this error it is only necessary to observe that these relations include not only tones, lines and colours, but also thoughts and will, and that they extend to moral facts no less than to objects of external intuition. He declares explicitly "No true beauty is sensible, although it frequently happens that sense-impressions precede and follow the intuition of beauty." [12] There is a profound distinction between the beautiful and the pleasant; for the pleasant needs no representation, while the beautiful consists in representation of relations, followed immediately in consciousness by a judgment, an appendix (*Zusatz*) which expresses unqualified approbation ("*es gefällt!*"). And while the pleasant and the unpleasant "in the progress of culture gradually become transient and unimportant, Beauty stands out more and more as something permanent and possessed of undeniable value." [13] The judgment of

---

1. *ce sont là jeux de princes*：〈法〉这里存在原则性批判

taste is universal, eternal, immutable: "the complete representation (*vollendete Vorstellung*) of the same relations is always followed by the same judgment; just as the same cause always produces the same effect. This happens at all times and in all circumstances, conditions and complications, which gives to the particularity of certain cases the appearance of a universal rule. Granted that the elements of a relation are universal concepts, it is plain that although in judging we think only of the content of these concepts, the judgment must have a sphere as large as that common to the two concepts." [14] Herbart considers æsthetic judgements as a general class comprising ethical judgements as a subdivision: "amongst other beauties is to be distinguished morality, as a thing not only of value in itself but as actually determining the unconditioned value of persons"; within morality in the narrowest sense is distinguished in turn justice. [15] The five ethical ideas guiding moral life (internal liberty, perfection, benevolence, equity and justice) are five æsthetic ideas or rather æsthetic concepts applied to relations of will.

Art as sum of content and form.

Herbart looks on art as a complex fact, the combination of an extra-æsthetic element, content, which may have logical or psychological or any other kind of value, and a purely æsthetic element, form, which is an application of the fundamental æsthetic concepts. Man looks for that which is diverting, instructive, moving, majestic, ridiculous; and "all these are mingled with the beautiful in order to procure favour and interest for the work. The beautiful thus assumes various complexions, and becomes graceful, magnificent, tragic, or comic; it can become all these because the æsthetic judgement, in itself calmly serene, tolerates the company of the most diverse excitations of the soul which are no part of itself." [16] But all these things have nothing to do with beauty. In order to discover the objectively beautiful or ugly, one must make abstraction from every predicate[1] concerning the content. "In order to recognize the objectively beautiful or ugly in poetry, one must show the difference between this and that thought, and the discussion will concern itself with thoughts; to recognize it in sculpture, one must show the difference between this and that outline, and the discussion will turn upon outlines; to recognize it in music, one should show the difference between this and that tone, and the discussion will turn upon tones. Now, such predicates as 'magnificent, charming, graceful' and so forth contain nothing whatever about tones, outlines or thoughts, and therefore tell us nothing about the objectively beautiful in poetry, sculpture, or music; indeed they rather lead us to believe in the existence of an objective beauty to which thought, outline, or tone are equally accidental, which may be approached by receiving impressions from poetry, sculpture, music and so forth, obliterating the object and giving oneself up to the pure emotion of mind." [17] Very different is the æsthetic judgement, the "cold judgement of the connoisseur" who considers exclusively form, *i.e.* objectively pleasant formal relations. This abstraction from the content in order to contemplate pure form is the catharsis produced by art. Content is transitory, relative, subject to moral law and liable to moral judgement: form is permanent, absolute, free. [18] Concrete art may be the sum of two or more values; but the æsthetic fact is form alone.

Herbart and Kantian thought.

The reader who goes behind appearances and discounts diversities of terminology will not fail to observe the close similarity of the æsthetic doctrine of Herbart to that of Kant. In Herbart we again

---

1. predicate：断言

find the distinction between free and adherent beauty, and between form and the sensuous stimulus (*Reiz*) attached to form: we find an affirmation of the existence of pure beauty, the object of necessary and universal, but not discursive, judgements; lastly, we find a certain connexion between beauty and morality, between Æsthetic and Ethics. In these matters Herbart is perhaps the most faithful follower and propagator of the thought of Kant, whose doctrine contains the germ of his own. In one passage he describes himself as "a Kantian, but of the year 1828"; and he is quite right, even in pointing out the exact difference in date. Amidst the errors and uncertainties of his æsthetic thought, Kant is rich in suggestion and scatters fertile seed; he belongs to a period when philosophy was still young and impressionable. Herbart, coming later, is dry and one-sided; he takes whatever is false in Kant's doctrine and hardens it into a system. If they had done little else, the Romanticists and idealists had at least united the theory of beauty to that of art, and destroyed the rhetorical and mechanical view; and they had brought into relief (frequently exaggerating, doubtless) various important characteristics of artistic activity. Herbart re-states the mechanical view, restores the duality, and presents a capricious, narrow, barren mysticism, devoid of all breath of artistic feeling.

[1] Welt als Wille u. Vorstellung, 1819 (in *Sämmtl. Werke*, ed. Grisebach, vol. i.). bk. iii. § 49.

[2] *Ergänzungen* (ed. Grisebach, vol. ii.), ch. 29.

[3] *Welt a. W. u. V.* iii. § 35.

[4] See above, p. 291.

[5] *Welt a. W. u. V.* iii. §§ 42–51.

[6] See above, p. 325.

[7] *Welt a. W. u. V.* § 53.

[8] *Op. cit.* § 34.

[9] *Welt a. W. u. V.* § 32.

[10] *Kritik d. kantischen Philosophie,* in append, to *op. cit.* pp. 558–576.

[11] *Ergänzungen*, ch. 38.

[12] *Einleitung in die Philosophie,* 1813, in *Werke*, ed. Hartenstein, vol. 1. p. 49.

[13] *Einleitung in die Philosophie*, pp. 125–128.

[14] *Allgemeine praktische Philosophie*, in *Werke*, viii. p. 25.

[15] *Einleitung*, p. 128.

[16] *Einleitung*, p. 162.

[17] *Op. cit.* pp. 129–130.

[18] *Op. cit.* p. 163.

XI FRIEDRICH SCHLEIERMACHER

Æsthetic of content and Æsthetic of form: meaning of the contrast.

We have now reached a point when we are able to give ourselves an exact account of the signification and importance of the celebrated war waged for over a century in Germany between the Æsthetic of content (*Gehaltsästhetik*) and the Æsthetic of form (*Formästhetik*); a war which gave birth to vast works on the history of Æsthetic undertaken from one or other point of view, and sprang from Herbart's opposition to the idealism of Schelling, Hegel, and their contemporaries and followers. "Form" and "Content" are among the most equivocal words in the whole philosophical vocabulary, particularly in Æsthetic; sometimes, indeed, what one calls form, others call content. The Herbartians were specially given to quoting in their own defence Schiller's dictum, that the secret of art consists in "cancelling content by form." But what is there in common between Schiller's concept of "form," which placed the æsthetic activity side by side with the moral and intellectual, and Herbart's "form," which does not penetrate or enliven, but clothes and adorns a content? Hegel, on the other hand, often gives the name "form" to what Schiller would call "matter" (*Stoff*), that is, the sensible matter which it is the business of spiritual energy to dominate. Hegel's "content" is the idea, the metaphysical truth, the constituent element of beauty: Herbart's "content" is the emotional and intellectual element which falls outside beauty. The Æsthetic of "form" in Italy is an æsthetic of expressive activity; the form is neither a clothing nor a metaphysical idea nor sensible matter, but a representative or imaginative faculty with the power of framing impressions; yet there have been attempts to confute this Italian æsthetic formalism with the same arguments that are used against German æsthetic formalism, a totally different thing in every respect. And so forth. Having given a plain account of the thoughts of the post-Kantian æstheticians, we shall be able to appreciate their opponents without seeking light from their obscure terminology or allowing ourselves to be misled by the banners they wave. The antithesis between the Æsthetic of content and that of form, the Æsthetic of idealism and that of realism, the Æsthetic of Schelling, Solger, Hegel and Schopenhauer and that of Herbart, will appear in its true light, as the family quarrel between two conceptions of art united by a common mysticism, although one is destined almost to meet with truth during its long journey, while the other wanders ever further away.

The first half of the nineteenth century was for Germany a period of many fine-sounding philosophical formulæ: subjectivism, objectivism, subjective — objectivism; abstract, concrete, abstract-concrete; idealism, realism, idealism — realism; between pantheism[1] and theism[2] Krause inserted his pan-en-theism[3]. In the midst of this uproar, in which the second-rate men shouted down the first-rate and made good their claim to their only true property, namely words, it is not surprising that a few modest clear thinkers, philosophers who preferred to think about realities, should have the worst of it and remain unheard and unnoticed, lost among the roaring crowd or labelled with a false ticket.

---

1. pantheism：泛神论，认为神存在于自然界一切事物之中，没有另外的超自然主宰或精神力量。
2. theism：一神论，相信一个神的理论或宗教。
3. pan-en-theism：万有在神论，克劳泽提出的理论，主要观点为世界是神的一部分。

Friedrich Schleiermacher[1].

This, at least, seems to have been the lot of Friedrich Schleiermacher, whose æsthetic doctrine is amongst the least known although it is perhaps the most noteworthy of the day.

Wrong judgements concerning him.

Schleiermacher delivered his first lectures on Æsthetic at Berlin University in 1819, and from that date he began to study the subject seriously with a view to writing a book on it. He repeated his lectures on two occasions, in 1825 and 1832–1833; but his death, which occurred in the following year, prevented him from carrying out his plan, and all we know of his thoughts on Æsthetic comes from his lectures, as collected by his pupils and published in 1842. [1] A Herbartian historian of Æsthetic, Zimmermann[2], attacks the posthumous work of Schleiermacher with real ferocity[3]; after twenty pages of invective and sarcasm[4] he concludes by asking, how could his pupils so dishonour their great master by publishing such a mass of waste paper, "all play upon words, sophistical conceits and dialectical subtleties"? [2] Nor was the idealistic historian Hartmann much more benevolent when he describes the work as "a confused mess in which, among much that is merely trivial, many half-truths and exaggerations, one can detect a few acute observations"; and says that, in order to make bearable "such unctuous afternoon sermons

---

1. *Friedrich Schleiermacher*：弗里德里希·施莱尔马赫（1768—1834），德国哲学家、神学家与《圣经》学者，通常被称为“现代自由神学之父”，著有《宗教演讲录》《基督教信仰》。
2. Zimmermann：齐默尔曼（1807—1878），德国神学家和历史学家，著有《农民大战争通史》。
3. real ferocity：极为粗暴
4. invective and sarcasm：讽刺谩骂

delivered by a preacher in his dotage[1]," it must be shortened by three-quarters; and that, "as regards fundamental principles," it is simply useless, offering no innovations upon concrete idealism as presented by Hegel and others; and that, in any case, it seems impossible "to attach it to any line of thought except the Hegelian, to which Schleiermacher's contribution is only of second-rate importance." He further observes that Schleiermacher was primarily a theologian, and in philosophy more or less an amateur. [3] Now it cannot be denied that Schleiermacher's doctrine has reached us in a hazy form, by no means free from uncertainties and contradictions; and, which is more important, it is here and there affected for the worse by the influence of contemporary metaphysics. But, side by side with these defects, what excellent method, really scientific and philosophical; what a number of cornerstones well and truly laid; what wealth of new truths, and of difficulties and problems not suspected or discussed before his day!

Schleiermacher contrasted with his predecessors.

Schleiermacher considered Æsthetic as an essentially modern line of thought, and drew a sharp distinction between the *Poetics* of Aristotle, which never shakes itself free from the empirical standpoint of the maker of rules, and what Baumgarten tried to do in the eighteenth century. He praised Kant for having been the first truly to include Æsthetic among the philosophical sciences, and recognized that in Hegel artistic activity had attained the highest elevation by being brought into connexion and almost into equality with religion and philosophy. But he was not satisfied either with the followers of Baumgarten when they degenerated into the absurd attempt to construct a science or theory of sensuous pleasure, or with the

---

1. such unctuous afternoon sermons delivered by a preacher in his dotage：昏庸老朽的牧师做的巧言令色的午后布道词

Kantian point of view which made its principal aim the consideration of taste; or with the philosophy of Fichte, in which art became a means of education; or with the more widely received opinion which placed at the centre of Æsthetic the vague and equivocal concept of Beauty. Schiller pleased him by having called attention to the moment of artistic spontaneity or productiveness, and he praised Schelling for having laid stress on the importance of the figurative arts, which lend themselves less easily than poetry to facile and illusory moralistic interpretations[1]. [4] Having with the utmost clearness excluded from Æsthetic the study of practical rules as empirical, and therefore irreducible to a science, he assigned to Æsthetic the task of determining the proper position of artistic activity in the scheme of ethics. [5]

Place assigned to Æsthetic in his Ethics.

To avoid falling into error over this terminology, we must call to mind that the philosophy of Schleiermacher followed the ancient traditions in its tripartite division into Dialectic, Ethics and Physics. Dialectic corresponds with ontology; Physics embraces all the sciences of natural facts; Ethics includes the study of all free activities of mankind (language, thought, art, religion and morality). Ethics represented to him not only the science of morality but what others name Psychology or, better still, the Science or Philosophy of the Spirit. This explanation once given, Schleiermacher's point of departure seems to be the only one just and permissible, and we shall not be surprised when he talks of will, of voluntary acts and so on, where others would have simply spoken of activity or spiritual energy; he even endows such expressions with a broader meaning than that

---

1. facile and illusory moralistic interpretations：廉价且虚无缥缈的道德说教

conferred upon them by practical philosophy.

Æsthetic activity as immanent and individual.

A double distinction may be made amongst human activities. In the first place, there are activities which we presume to be constituted in the same manner in all men (such as the logical activity) and are called activities of identity; and others whose diversity is presumed, which are called activities of difference or individual activities. Secondly, there are activities which exhaust themselves in the internal life, and others which actualize themselves in the external world: immanent activities[1] and practical activities. To which of the two classes in each of the two orders does artistic activity belong? There can be no doubt of its different modes of development, if not actually in each individual person, at least in different peoples and nations; therefore it belongs properly to activities of difference or individual activities. [6] As for the other distinction, it is true that art does realize itself in the external world, but this fact is something superadded[2] ("*ein später Hinzukommendes*") "which stands to the internal fact as the communication of thought by means of speech or writing stands to thought itself": art's true work is the internal image[3] ("*das innere Bild ist das eigentliche Kunstwerk*"). Exceptions to this might be adduced[4], such as mimicry; but they would be apparent only. Between a really angry man and the actor who plays the part of an angry man on the stage there is this difference: in the second case anger appears as controlled and therefore beautiful; that is, the internal image is in the actor's soul interposed between the fact of passion and its physical

1. immanent activities：内心活动
2. superadded：附加的
3. art's true work is the internal image：真正的艺术作品是内在的意象
4. adduced：引证

manifestation. [7] Artistic activity "belongs to those human activities in which we presuppose the individual in its differentiation; it belongs equally to those activities developing essentially within themselves and not completing themselves in any external world. Art, therefore, is an immanent activity in which we presuppose differentiation." Internal, not practical: individual, not universal or logical.

Artistic truth and intellectual truth.

But if art be one form of thought, there must be one form of thought in which identity is presupposed, and another in which difference is presupposed. We do not look for truth in poetry; or, rather, we do look for truth, but for one that is totally different from that objective truth to which there must correspond some being, either universal or individual (scientific and historical truth). "When a character in a poem is said to be devoid of truth, a slur[1] is cast on the given poem; but if the character is said to be a pure invention, corresponding with no reality, that is quite a different matter." The truth of a poetic character consists in the coherence with which a single person's divers modes of thinking and acting are represented: even in portraits it is not an exact correspondence with an objective reality that makes the thing a work of art. From art and poetry "springs no iota of knowledge[2]" (*das Geringste vom Wissen*); "it expresses but the truth of the single consciousness." There are then "productions of thought and of sensible intuitions, opposed to the other productions because they do not presuppose identity, and they express the singular as such." [8]

Difference of artistic consciousness from feeling and religion.

---

1. slur：指责
2. springs no iota of knowledge：不产生丝毫的知识

The domain of art is immediate self-consciousness[1] (*unmittelbare Selbstbewusstsein*), which must be carefully distinguished from the thought or concept of the ego or of the determinate ego. This latter is the consciousness of identity in the diversity of moments; immediate self-consciousness is "diversity itself, of which one must be aware, since life in its entirety is but the development of consciousness." In this domain art has often been confused with two facts which accompany it: sensuous consciousness (the feeling of pleasure and pain), and religion. A double confusion, of which the sensationalists fall into the first half and Hegel into the second; Schleiermacher clears it up by proving that art is free productivity, whereas sensuous pleasure and religious feeling, however different in other ways, are both determined by an objective fact (*äussere Sein*). [9]

Dreams and art: inspiration and deliberation.

The better to understand this free productivity, we must further circumscribe the domain of immediate consciousness. In this we can find nothing more helpful than comparing it with the images produced by dreams. The artist has his own dreams: he dreams with open eyes, and from among the thick-thronging images[2] of this dream-state those having sufficient energy alone become works of art, the rest remaining a mere background from which the others stand out. All the essential elements of art are found in the dream-state, which is the production of free thoughts and sensuous intuitions consisting of mere images. Certainly something is lacking in dreams, and they differ from art not only in their absence of technique, which has already been excluded as irrelevant to art, but in another way, viz. that a dream is a chaotic fact, without stability, order, connexion or measure. But when some sort of

1. immediate self-consciousness：直接的自我意识
2. thick-thronging images：大量的意象

order is introduced into the chaos the difference at once disappears, and the likeness to art merges in identity. This internal activity which introduces order and measure, fixes and determines the image, is that which distinguishes art from a dream or transforms a dream into art. It often involves struggle, labour, the obligation to stem the involuntary flood of internal images; in a word, it means reflexion or deliberation. But the dream and the cessation of dreaming[1] are equally indispensable elements of art. There must be production of thoughts and images and, together with such production, there must be measure, determination and unity, "otherwise each image would be confused with its neighbour and have no definiteness." The instant of inspiration (*Begeisterung*) is as essential as that of deliberation (*Besonnenheit*). [10]

Art and the typical.

But in order to arrive at artistic truth it is also I necessary (here Schleiermacher's thought becomes less clear and accurate) that the singular be accompanied by consciousness of the species; consciousness of the self as individual man is impossible without consciousness of mankind; nor is a single object true unless referred to its universal. In a pictured landscape "every tree must possess natural truth, that is to say, it must be contemplated as a specimen of a given kind; similarly, the whole complex of natural and individual life must have effective truth of nature and constitute a single harmony. Just because in art we do not strive after the production of individual figures in themselves and for themselves, but their internal truth as well, we commonly assign to them a high place as being a free realization of that in which all cognition has its value, that is to say,

---

1. cessation of dreaming：梦境中止

in the principle that all forms of being are inherent in the human spirit. If this principle fails, truth is no longer possible; scepticism only remains." The productions of art are the ideal or typical figures which real nature would create were it not impeded by external influences. [11] "The artist creates a figure on the basis of a general scheme, rejecting whatever may hinder or impede the play of the living forces of reality; such a production, founded on a general scheme, is what we call the Ideal." [12]

In spite of all these determinations, Schleiermacher did not apparently intend to limit the artist's scope. He remarks, "When an artist represents something really given, whether portrait, landscape or single human figure, he renounces the freedom of productivity and adheres to the real." [13] There is a twofold tendency at work in the artist: towards perfection of type, and towards representation of natural reality. An artist must not fall into the abstractness of the type or into the unmeaningness of empirical reality. [14] If in flower-painting it is necessary to bring out the specific type, a much more complete individualisation is demanded when representing man, owing to the lofty position which he occupies. [15] Representation of the ideal in the real does not exclude "an infinite variety, such as is found in actual reality." "For instance, the human face wavers between the ideal and caricature[1], in its moral conformation no less than in its physical. Every human face contains elements of disfigurement (*Verbildung*,) but it has also something by which it is a determinate modification of human nature; this does not appear openly, but a practised eye can seize it and ideally complete the face in question." [16] Schleiermacher is keenly aware of the difficulties and perplexities of such problems as the question whether there exists one or many ideals of the human face. [17] He observes that the two views which strive for mastery in

1. wavers between the ideal and caricature：摇摆于完美画作和讽刺漫画之间

the field of poetry may be extended to art as a whole. Some assert that poetry and art should represent the perfect, the ideal, that which would have been produced by nature, had she not been prevented by mechanical forces; others reject the ideal as incapable of realisation and prefer that the artist should depict man as he really is, with those perturbing elements[1] which in reality belong to him no less than his ideal qualities. Each view is a half-truth: it is the duty of art to represent the ideal as well as the real, the subjective as well as the objective. [18] The comic element, that is the unideal and the faulty ideal, is included in the circle of art. [19]

Independence of art.

In respect to morality, art is free just as philosophical speculation is free: its essence excludes practical and moral effects. This leads to the proposition that "there is no difference between various works of art, except in so far as they can be compared in respect of artistic perfection" (*Vollkommenheit in der Kunst.*) "Given an artistic object perfect of its kind, it has an absolute value which cannot be increased or diminished by anything else. If motions of the will could truly be described as consequences of works of art, a different standard of values would apply to works of art: and since the objects which an artist may depict are not all equally adapted to influence volition, a scale of values would exist which did not depend on artistic perfection." Nor must we confound the judgement passed upon the varied and complex personality of the artist himself with the strictly æsthetic judgement passed upon his work. "In this respect the biggest, most complicated canvas is on a level with the smallest arabesque[2], the longest poem with the shortest: the value of a work of art depends

1. perturbing elements：不确定因素
2. arabesque：阿拉伯式蔓草花纹

on the perfect manner in which the external corresponds to the internal." [20]

Schleiermacher rejects the doctrine of Schiller because in his opinion it makes art a sort of game or pastime in contrast to the serious affairs of life: a view, he says, for business men to whom their business is the only serious thing. Artistic activity is universally human, a man devoid of it is inconceivable; although, of course, there are in this respect great differences betwixt man and man, running from the mere desire to enjoy art to real taste, and from this again to productive genius. [21]

Art and language.

The artist makes use of instruments which, by their nature, are framed not for the individual but for the universal; of this kind is language. But it is the business of poetry to extract the individual from language which is universal without giving to its productions the form of the antithesis between individual and universal which is proper to science. Of the two elements of language, the musical and the logical, the poet claims the first for his own ends and constrains the other to awaken individual images. In comparison with pure science as in comparison with the individual image, there is something irrational about language: but the tendencies of speculation and of poetry are always contrary, even in their use of language; the former tends to make language approximate to mathematical formulæ; the latter to imagery (*Bild*). [22]

Schleiermacher's defects.

Leaving out many details which will be touched on in their proper places, the foregoing is a fair summary of the heads of Schleiermacher's æsthetic thought. Adding up the accounts of the whole statement of views, on the side of error and oversight we find: first, ideas or types are not wholly excluded, in spite of all Schleiermacher's care and anxiety to safeguard artistic individualisation and to make the ideas and types superfluous. Secondly, there is still, undefeated and unexpelled, a certain residue of abstract formalism, visible at various points of his theories. [23] Thirdly, the definition of art as an activity of mere difference may be diluted but is not destroyed by making art a difference of complexes of individuals, a national difference. A closer reflexion on the history of art, a recognition of the possibility of appreciating the art of various nations and various times, a more patient investigation into the moment of artistic reproduction, even an examination of the relation between science and art, would have led Schleiermacher to treat this difference as empirical and surmountable, still holding firmly to the distinctive character (individual as opposed to universal) he assigned to art in comparison with science. Fourthly, he did not recognize the identity of æsthetic activity with linguistic, and failed to make it the basis of all other theoretic activity. It would seem, moreover, that Schleiermacher had no clear ideas concerning that artistic element which enters into the constitution of historic narrative and is indispensable as the concrete form of science; or concerning language, taken not as a complex of abstract means of expression but as expressive activity.

Schleiermacher's services to Æsthetic.

These defects and uncertainties may perhaps be attributable in part to the fact that his thoughts on æsthetic have reached us in an inchoate form[1], very far from a mature development. But if on the other hand we wish to cast up the sum of his very striking merits, it will suffice to

1. inchoate form：未经深思熟虑的形式

run over the list of accusations heaped upon him by the two historians before mentioned, Zimmermann and Hartmann. Schleiermacher has denuded Æsthetic of its imperative character[1]; he recognizes in it a form of thought differing from logical thought; he gives this science a non-metaphysical and merely anthropological character; he denies the concept of beauty, substituting that of artistic perfection, and actually affirms the æsthetic equivalence of small and great works of art, so long as each is perfect in its own sphere; he considers the æsthetic fact as pure human productivity: and so on and so forth. All these criticisms are meant for blame and are really praise; for what is blame to the mind of a Zimmermann or a Hartmann, is to ours praise. In the metaphysical orgy[2] of his day, in the perpetual building and pulling down of more or less arbitrary systems, Schleiermacher the theologian, with philosophic acumen, fixed his eye upon what was really characteristic of the æsthetic fact and succeeded in defining its properties and connexions; when he failed to see clearly and wandered from the track, he never abandoned analysis for fantastic caprice. By his discovery that the obscure region of immediate consciousness is also that of the æsthetic fact, he seems to bid his distracted contemporaries listen to the old adage[3]: *Hic Rhodus, hic salta.*[4]

[1] *Vorlesungen üb. Ästhetik* published by Lommatsch, Berlin, 1842 (*Werke*, sect. iii. vol. vii.).

[2] Zimmermann, *G. d. A.* pp. 608–634.

---

1. denuded Æsthetic of its imperative character：否定了美学至高无上的特点。
2. metaphysical orgy：形而上学的纵情狂欢
3. old adage：古老格言
4. *Hic Rhodus, hic salta.*：〈拉〉“此地便是罗德岛，你就跳一下试试。”源自《伊索寓言》，某运动员称自己在罗德岛能跳得很远，听者说道：“这儿就是罗德岛，你跳一下试试吧！”

[3] E. von Hartmann, *Deutsche Ästh. s. Kant,* pp. 156–169.

[4] *Vorles. üb. Ästhetik* pp. 1–30.

[5] *Op. cit.* pp. 35–51.

[6] *Vorles. üb. Ästh.* pp. 51–54.

[7] *Vorles. üb. Ästh.* pp. 55–61.

[8] *Op. cit.* pp. 61–66; cf. *Dialektik*, ed. Halpern, pp. 54–55, 67.

[9] *Vorles. üb. Ästh.* pp. 67–77.

[10] *Vorles. üb. Ästh.* pp. 79–91.

[11] *Op. cit.* pp. 123, 143–150.

[12] *Op. cit.* p. 505; cf. p. 607.

[13] *Op. cit.* p. 505.

[14] *Vorles. üb. Ästh.* pp. 506–508.

[15] *Op. cit.* pp. 156–157.

[16] *Op. cit.* pp. 550–551.

[17] *Op. cit.* p. 608.

[18] *Op. cit.* pp. 684–686.

[19] *Op. cit.* pp. 191–196; cf. pp. 364–365.

[20] *Vorles. üb. Ästh.* pp. 209–219; of. pp. 527–528.

[21] *Op. cit.* pp. 98–111.

[22] *Vorles. üb. Ästh.* pp. 635–648.

[23] Cf. *e.g.* p. 467 *seqq.*

XII THE PHILOSOPHY OF LANGUAGE: HUMBOLDT AND STEINTHAL

Progress of Linguistic.

About the time when Schleiermacher was meditating on the nature of the æsthetic fact, a movement of thought was gaining ground in Germany which, tending as it did to overthrow the old concept of language, might have proved a powerful aid to æsthetic science. But not only had the æsthetic specialists — if we may so call them — no notion of the existence of this movement, the new philosophers of language never brought their ideas into relation with the æsthetic problem, and their discoveries languished imprisoned within the narrow scope of Linguistic, condemned to sterility.

Linguistic speculation at the beginning of the nineteenth century.

Research into the relations between thought and speech, between the unity of logic and the multiplicity of languages, had been promoted, like many other things, by the *Critique of Pure Reason*: the earliest Kantians often tried to apply the Kantian categories of intuition (space and time) and of intellect to language. The first to make the attempt was Roth[1] [1] in 1795; the same who wrote an essay twenty years later on *Pure Linguistic*. Many other noteworthy books on this

1. Roth：洛特（1748—1814），德国语言学家，著有《对赫尔墨斯的批判或关于人类语言的理性概念之哲学探讨和普通语言学》。

subject appeared in quick succession: those of Vater[1], Bernhardi[2], Reinbeck[3] and Koch[4] were published one after another in the first ten years of the nineteenth century. In all these treatises the dominating subject is the difference between language and languages; between the universal language, corresponding with Logic, and concrete, historical languages disturbed by feeling and imagination or whatever other name was applied to the psychological element of differentiation. Vater distinguishes a general Linguistic (*all gemeine Sprachlehre*), constructed *a priori* by means of the analysis of the concepts contained in the judgement, from a comparative Linguistic (*vergleichende Sprachlehre*) which attempts by means of induction to reach probable laws through the study of a number of languages. Bernhardi considers language to be an "allegory of intellect[5]" and distinguishes it as functioning either as the organ of poetry or that of science. Reinbeck speaks of an Æsthetic Grammar and a Logical. Koch, more energetic than the others, asserts positively that the character of language is "*non ad Logices sed ad Psychologiae rationem revocanda.*[6]" [2] Some few philosophers speculated on language and mythology: for example Schelling considered them to be the products of a pre-human consciousness[7] (*vormenschliche Bewusstsein*,) presenting them, in a fantastic allegory, as diabolic suggestions[8] which precipitate the ego from the infinite to the

---

1. Vater：法特尔（1771—1826），德国神学家、圣经学者、语言学家，著有《希伯来语语法》《波兰语语法》《俄语语法》等。
2. Bernhardi：伯恩哈迪（1769—1820），德国语言学家、作家，著有《拉丁文语法大全》《希腊语语法大全》《语言学基本原理》等。
3. Reinbeck：莱茵贝克（1766—1849），德国人文主义者，语言学家、文学家、美学家、教育家，著有《语言学手册》《莱茵贝克戏剧全集》《德国当下的教育》等。
4. Koch：科赫（1749—1816），德国音乐家。
5. allegory of intellect：知性的寓言
6. *non ad Logices sed ad Psychologiae rationem revocanda.*：〈拉〉（语言）不应与逻辑学而应与心理学相关。
7. pre-human consciousness：先人类意识
8. diabolic suggestions：魔鬼的暗示

finite. [3]

Wilhelm von Humboldt[1]*: relics of intellectualism.*

Even the famous philologist, Wilhelm von Humboldt, was unable to detach himself entirely from the prejudice of the substantial identity and the purely historical, accidental diversity between logical thought and language. His celebrated dissertation, *On the Diversity of Structure of Human Languages*[2] (1836), [4] is based on the notion of a perfect language split up and distributed amongst particular tongues according to the linguistic or intellectual capacity of various nations. "For," says he, "since disposition towards speech is general in mankind, and all men must necessarily carry within themselves the key to the comprehension of all languages, it follows that the form of all languages must be substantially equal and all must attain the same general end. Diversity can exist solely in the means, and within the bounds permitted by the attainment of the end." Yet this same diversity becomes a real divergence not only in sounds, but in the use of sound made by the linguistic sense in respect to the form of language, or rather, in respect to its own idea of the form of the determinate language. "Languages being merely formal, the operation of the linguistic sense by itself should produce mere uniformity; the linguistic sense must exact from every tongue the same right and legitimate construction that is found in one of them. In practice, however, the facts are quite otherwise, partly owing to the reaction of sounds, and partly by reason of the individual aspect assumed by

---

1. *Wilhelm von Humboldt*：威廉·冯·洪堡（1767—1838），德国哲学家、语言学家，创立了柏林洪堡大学，著有《论国家的作用》《论语言的多样性及其对人类理智发展的影响》。
2. *On the Diversity of Structure of Human Languages*：《论人类语言结构的多样性》。

the same internal meaning in phenomenal reality." Linguistic force "cannot maintain its equality everywhere or show the same intensity, vivacity or regularity; it cannot be supported by an exactly equal tendency towards the symbolic treatment of thought or by exactly equal pleasure in richness and harmony of sound." These, then, are the causes which produce in human languages that diversity which manifests itself in every branch of the civilization of nations. But reflexion on languages "ought to reveal to us a form which of all possible forms best fits the purpose of language" and approaches most closely to its ideal; and "the merits and defects of existing languages must be estimated by their nearness or remoteness from this form." Humboldt finds the nearest approximation to such an ideal in the Sanskrit tongues[1], which can therefore be used as a standard of comparison. Setting Chinese apart in a class by itself, he proceeds to the division of the possible forms of language into inflective, agglutinative and incorporative[2]; types which are found combined in various proportions in every real language. [5] He also inaugurated the division of languages into inferior and superior, unformed and formed, according to the way in which verbs are treated. He was never able to rid himself of a second prejudice connected with the first, namely that language exists as something objective outside the talking man, unattached and independent, and waking up when needed for use.

Language as activity. Internal form.

But Humboldt opposes Humboldt: amongst the old dross[3] we detect the brilliant gleams of a wholly new concept of language. Certainly

---

1. Sanskrit tongues：梵语
2. inflective, agglutinative and incorporative：屈折语的、粘着语的，综合语的
3. old dross：老生常谈

his work is for this very reason not always free from contradictions and from a kind of hesitation and awkwardness which appear characteristically in his literary style and make it at times labored and obscure[1]. The new man in Humboldt criticizes the old man when he says, "Languages must be considered not as dead products but as an act of production ... Language in its reality is something continually changing and passing away. Even its preservation in writing is incomplete, a kind of mummification[2]: it is always necessary to render the living speech sensible. Language is not a work, *ergon*[3], but an activity, *energeia*[4] ... It is an eternally repeated effort of the spirit in order to make articulated tones capable of expressing thought." Language is the act of speaking. "True and proper language consists in the very act of producing it by means of connected utterance; that is the only thing that must be thought of as the starting-point or the truth in any inquiry which aims at penetrating into the living essence of language. Division into words and rules is a lifeless artifice of scientific analysis." [6] Language is not a thing arising out of the need of external communication; on the contrary, it springs from the wholly internal thirst for knowledge and the struggle to reach an intuition of things. "From its earliest commencement it is entirely human, and extends without intention to all objects of sensory perception or internal elaboration ... Words gush spontaneously from[5] the breast without constraint or intention: there is no nomad tribe in any desert without its songs. Taken as a zoological species, man is a singing animal which connects its thoughts with its utterances." [7] The new man leads Humboldt to discover a fact hidden from the authors of

1. labored and obscure：佶屈聱牙
2. mummification：木乃伊般的干巴巴
3. *ergon*：〈希〉尔格（功和能量的单位）
4. *energeia*：〈拉〉活力
5. gush spontaneously from：自发地喷薄而出

logico-universal grammars: namely the internal form of language (*innere Sprachform*), which is neither logical concept nor physical sound, but the subjective view of things formed by man, the product of imagination and feeling, the individualization of the concept. Conjunction of the internal form of language with physical sound is the work of an internal synthesis; "and here, more than anywhere else, language by its profound and mysterious operation recalls art. Sculptor and painter also unite the idea with matter, and their efforts are judged praiseworthy or not according as this union, this intimate interpenetration, is the work of true genius, or as the idea is something separate, painfully and laboriously imposed upon the matter by sheer force of brush or chisel[1]." [8]

Language and art in Humboldt.

But Humboldt was content to regard the procedure of artist and speaker as comparable by analogy, without proceeding to identify them. On the one hand, he was too one-sided in his view of language as a means for the development of thought (logical thought); on the other, his own æsthetic ideas, always vague and not always true, prevented his perception of the identity. Of his two principal writings on Æsthetic, that on *Beauty Masculine and Feminine*[2] (1795) seems to be wholly under the influence of Winckelmann, whose antithesis between beauty and expression is revived, and the opinion expressed that specific sexual characters diminish the beauty of the human body and that beauty asserts itself only by triumphing over differences of sex. His other work, which is inspired by Goethe's *Hermann und Dorothee*[3], defines art as "representation of nature by means of fancy;

1. by sheer force of brush or chisel：纯粹靠画笔和美工刀
2. *Beauty Masculine and Feminine*：《论阳刚美与阴柔美》，洪堡著。
3. *Hermann und Dorothee*：〈德〉《赫尔曼与多罗泰》，歌德诗作。

the representation being beautiful, just because it is the work of fancy," a metamorphosis[1] of nature carried to a higher sphere. The poet reflects the pictures of language, itself a complex of abstractions. [9] In his dissertation on Linguistic, Humboldt distinguishes poetry and prose, treating the two concepts philosophically, not by the empirical distinction between free and measured or periodic and metric language. "Poetry gives us reality in its sensible appearance, as it is felt internally and externally; but is indifferent to the character which makes it real, and even deliberately ignores that character. It presents the sensuous appearance to fancy and, by this means, leads towards the contemplation of an artistically ideal whole. Prose, on the contrary, looks in reality for the roots which attach it to existence, the cords which bind her to it: hence it fastens fact to fact and concept to concept according to the methods of the intellect, and strives towards the objective union of them all in an idea." [10] Poetry precedes prose: before producing prose, the spirit necessarily forms itself in poetry. [11] But, beside these views, some of which are profoundly true, Humboldt looks on poets as perfecters of language, and on poetry as belonging only to certain exceptional moments, [12] and makes us suspect that after all he never recognized clearly or maintained firmly that language is always poetry, and that prose (science) is a distinction not of æsthetic form but of content, that is, of logical form.

H. Steinthal[2]. *The linguistic function independent of the logical.*

Humboldt's contradictions about the concept of language lost him his principal follower, Steinthal. With the help of his master, Steinthal

---

1. metamorphosis：改造
2. *H. Steinthal*：斯坦因塔尔（1823—1899），德国语言学家和哲学家。著有《洪堡的语言学与黑格尔的哲学》《语法、逻辑学与心理学：各自的基本原理与相互关系》《普通伦理学》。

restated the position that language belongs not to Logic but to Psychology, [13] and in 1855 waged a gallant war against the Hegelian Becker[1], author of *The Organisms of Language*, one of the last logical grammarians, who pledged himself to deduce the entire body of the Sanskrit languages from twelve cardinal concepts. Steinthal declares it is not true that one cannot think without words: the deaf-mute thinks in signs; the mathematician in formulæ. In some languages, as in Chinese, the visual element is as necessary to thought as the phonetic, if not more so. [14] In this he may have overshot the mark, and failed to establish the autonomy of expression with regard to logical thought; for his examples only confirm the fact that if we can think without words, we cannot think without expressions. [15] But he successfully demonstrates that concept and word, logical judgement and proposition, are incommensurable. The proposition is not the judgement but the representation (*Darstellung*) of a judgement; and all propositions do not represent logical judgements. It is possible to express several judgements in a single proposition. The logical divisions of judgements (the relations of concepts) find no counterpart in the grammatical divisions of propositions. "A logical form of the proposition is just as much a contradiction as the angle of a circle or the circumference of a triangle." He who talks, in so far as he talks, possesses not thoughts but language. [16]

Identity of the problems of the origin and the nature of language.

Having thus freed language from all dependence on Logic, having repeatedly proclaimed the principle that language produces its forms independently of Logic and in the fullest autonomy, [17] and having

1. Becker：卡尔·贝克尔（1775—1849），德国语言学家、生理学家、教育家，著有《德语构词法》《德语语法大全》和文中提到的《语言的有机体》（*The Organisms of Language*）等。

purified Humboldt's theory from the taint of the logical grammar[1] of Port Royal, Steinthal seeks the origin of language, recognizing, with his master, that the question of its origin is identical with that of nature of language, its psychological genesis or rather the position it occupies in evolution of the spirit. "In the matter of language there is no difference between its original creation (*Urschöpfung*) and the creation which is daily repeated." [18] Language belongs to the vast class of reflex movements[2]; but to say that is to look at it from one side only and to omit its own essential peculiarity. Animals have reflex movements and sensations like man; but in animals the senses "are wide gates through which external nature rushes to the assault with such impetus[3] as to overwhelm the mind and deprive it of all independence and freedom of movement." In man, however, language can arise because man is resistance to nature, conqueror of his own body, freedom incarnate: "language is liberation: even to-day we feel our mind lightened and freed from a weight when we speak." In the situation immediately preceding the production of speech man must be conceived as "accompanying all his sensations and all the intuitions received by his mind with the most lively contortions of body[4], attitudes of mimicry, gestures, and above all tones, articulate tones." What element of speech did he lack? One only, but a most important one: the conscious conjunction of reflex bodily movements with the excitations of his mind. If sensuous consciousness is already consciousness, it lacks the consciousness of being conscious; if it is already intuition, it is not intuition of intuition; what it lacks is in a word the internal form of speech. When that arises, there arises too its inseparable accompaniment, words. Man does not select sound: it is

1. taint of the logical grammar：逻辑语法残余
2. vast class of reflex movements：反应运动的广义范畴
3. assault with such impetus：藉此力量征服
4. most lively contortions of body：最活跃的身体运动

given him, and he takes it of necessity, instinctively, without intention or choice. [19]

Steinthal's mistaken ideas on art: his failure to unite Linguistic and Æsthetic.

This is not the place for detailed examination of the whole of Steinthal's theory and the various phases, not always progressive, through which he travelled, especially after the beginning of his spiritual collaboration with Lazarus[1], with whom he studied ethnopsychology[2] (*Völkerpsychologie*), of which they both took Linguistic to be a part. [20] But, while giving him full credit for bringing Humboldt's ideas into coherent order, and for clearly differentiating, as had never before been done, between linguistic activity and the activity of logical thought, it must be noted that Steinthal never recognized the identity of the internal form of language (which he also called the intuition of intuition, or apperception) with the æsthetic imagination. The Herbartian psychology to which he clung afforded him no clue to such a discovery. Herbart and his followers divorced psychology from logic as a normative science and never succeeded in discerning the true connection between feeling and spiritual formation, soul and spirit; they never understood that logical thought is one of these spiritual formations: an activity, not a code of external laws. The domain allotted by them to Æsthetic we already know; for them Æsthetic too was only another code of beautiful formal relations. Under the influence of these doctrines Steinthal was led to regard Art as the embellishment of thoughts[3], Linguistic as the science of speech, and Rhetoric or Æsthetic as a thing differing from

1. Lazarus：莫里茨·拉扎鲁斯（1824—1903）德国犹太裔哲学家，种族心理学创始人。
2. ethnopsychology：种族心理学
3. embellishment of thoughts：思想的美化

Linguistic since it is science of fine or beautiful speaking. [21] In one of his innumerable tracts he says, "Poetics and Rhetoric both differ from Linguistic, since they are obliged to touch on many important topics before reaching language. These sciences therefore have but one section devoted to Linguistic, which is the concluding section of Syntax[1]. Moreover Syntax has a character entirely different from Rhetoric and from Poetics; the former is occupied solely with correctness (*Richtigkeit*) of language; the latter two sciences study beauty or grace of expression (*Schönheit oder Angemessenheit des Ausdrucks*): the principles of the first are merely grammatical, the others must consider matters outside language; for example, the disposition of the orator and so forth. To speak plainly, Syntax is to Stylistic[2] as is the grammatical measure of the quantity of vowels to the theory of metre." [22] That speaking invariably means good or beautiful speaking, since speech that is neither good nor beautiful is not really speech, [23] and that the radical renewal of the concept of language inaugurated by Humboldt and himself must produce far-reaching effects on the cognate sciences[3] of Poetics, Rhetoric and Æsthetic and, by transforming, unify them, never entered Steinthal's head. After all this labour and all this minute analysis, the identification of language and poetry, and of the science of language with the science of poetry, the identification of Linguistic with Æsthetic, still found its least faulty expression in the prophetic aphorisms of Giambattista Vico.

[1] *Antihermes oder philosophische Untersuchung üb. d. reine Begriff d. menschl. Sprache und die allgemeine Sprachlehre*, Frankfurt and Leipzig, 1795.

[2] For these writers, see accounts and quotations in Loewe,

1. Syntax：句法
2. Stylistic：文体学
3. cognate sciences：同源科学

Hist, crit. gramm. univ., passim, and Pott, introd. to Humboldt, pp. clxxi.–ccxii.; cf. also Benfey, *Gesch. d. Sprachwiss.,* introd.

[3] In *Philos, der Mythologie*: cf. Steinthal, *Urspr.* pp. 81–89.

[4] *Üb. d. Verschiedenheit d. menschl. Sprachbaues*, posthumous work (2nd ed. by A. F. Pott, Berlin, 1880).

[5] *Verschiedenheit*, etc. pp. 308–310.

[6] *Verschiedenheit*, etc., pp. 54–56.

[7] *Verschiedenheit*, etc., pp. 25, 73–74, 79.

[8] *Op. cit.* pp. 105–118.

[9] Zimmermann, *G. d. A.* pp. 533–544.

[10] *Verschiedenheit*, etc., pp. 326–328.

[11] *Op. cit.* pp. 239–240.

[12] *Op. cit.* pp. 205–206, 547, etc.

[13] *Grammatik, Logik und Psychologie, ihre Principien u. Ihr Verhältn. z. einand.*, Berlin, 1855.

[14] *Gramm., Log. u. Psych.* pp. 153–158.

[15] See above, pp. 60–62.

[16] *Gramm., Log. u. Psych*, pp. 183, 195.

[17] *Einleitung i. d. Psych, u. Sprachwissenschaft* (2nd ed., Berlin, 1881), p. 62.

[18] *Gramm., Log. u. Psych*, p. 231.

[19] *Op. cit.* pp. 285, 292, 295–306.

[20] Steinthal, *Ursprung d. Sprache* (4th ed. Berlin, 1888), pp. 120–124. M. Lazarus, *Das Leben der Seele,* 1855 (Berlin, 1876–1878), vol. ii. *Zeitschrift f. Völkerpsych. u. Sprachwiss.* from 1860 onwards, edited by Steinthal and Lazarus together.

[21] *Gramm., Log. u. Psych*, pp. 139–140, 146.

[22] *Einleit.* pp. 34–35.

[23] See above, pp. 100–101.

XIII MINOR GERMAN ÆSTHETICIANS

Minor æstheticians in the metaphysical school.

When we turn from the pages of methodical and serious thinkers such as Schleiermacher, Humboldt and Steinthal, we are filled with distaste by the books written in enormous quantities during the first half of the nineteenth century by disciples of Schelling and Hegel. We are fatigued and almost disgusted as we pass from this illuminating and scientific study to something which oscillates between vapid fancies and charlatanism[1]; between the vanity of empty formulæ and the attempt, not always free from dishonesty, to employ them in order to amaze and overwhelm the reader or student.

Krause, Trahndorff[2]*, Weisse*[3] *and others.*

Why should we encumber a general History of Æsthetic (which ought, certainly, to take account of aberrations from the truth[4], but only in so far as they indicate the general trend of contemporary

---

1. oscillates between vapid fancies and charlatanism：在胡思乱想和喋喋不休之间纠缠不清
2. *Trahndorff*：特兰多尔夫（1782—1863），德国作家、神学家、哲学家、美学家，著有《美学或世界观与艺术的演讲录》《世界史的疑问或：上帝只是理念，还是客观的真实？》等。
3. *Weisse*：韦塞（1801—1866），德国新教哲学家、美学家，莱比锡大学哲学教授，著有《美学体系》《哲学犬儒主义或基督教哲学》等。
4. aberrations from the truth：与真理之间的偏差

thought) with the theories of such men as Krause, Trahndorff, Weisse, Deutinger[1], Oersted[2], Zeising[3], Eckardt[4] and the crowd of manipulators of manuals and systems? The only one who obtained a hearing outside his native Germany was Krause, who was imported into Spain; we are justified, therefore, in leaving them to the memory or forgetfulness of their compatriots. For Krause, [1] the humanitarian, the freethinker, the theosophist[5], everything is organism, everything is beauty; beauty is organism, and organism is beauty: Essence, that is to say God, is one, free and entire; one, free and entire is Beauty. There is but one artist, God; but one art, the divine. The beauty of finite things is the Divinity, or rather the likeness of Divinity manifested in the finite. Beauty brings into play reason, intellect and imagination in a mode conforming to their laws, and awakens disinterested pleasure and inclination in the soul. Trahndorff, [2] describing the various degrees by which the individual seeks to grasp the essence or form of the universe (the degrees of feeling, intuition, reflexion and presentiment), and noting the insufficiency of simple theoretical knowledge till supplemented by the Will, the Will which is power (*Können*), in its three degrees of Aspiration, Faith and Love, places the Beautiful in the highest grade, in Love: it would seem, therefore, that Beauty is Love which comprehends itself. Christian Weisse [3] attempted, like Trahndorff, to reconcile the God of Christianity with the Hegelian philosophy: in his estimation the æsthetic Idea is superior to the logical, and leads to religion, to God; the idea of beauty, existing

1. Deutinger：多伊廷格尔（1815—1864），德国哲学家、宗教作家。
2. Oersted：奥斯特（1777—1851），丹麦物理学家、化学家、哲学家，发现奥斯特定律，构建了后康德哲学，著有《对自然科学普遍的爱》《论自然的精神》《再论自然的精神》。
3. Zeising：蔡辛格（1810—1876），德国心理学家、哲学家、美学家，著有《美学研究》《艺术与偏好》等。
4. Eckardt：艾卡尔特（1827—1871），奥地利文学家、美学家，著有《美学基础》《对比泛神论对美学进行一神论推理》等。
5. theosophist：神智学者

outside the sensible universe, is the reality of the concept of beauty, and, as the idea of divinity is absolute Love, so must that of Beauty be found truly in Love. The same reconciliation was attempted by the Catholic theologian Deutinger; [4] beauty, for him, is born of power (*Können*), an activity parallel with those of the knowledge of truth and the doing of good but (differing in this from knowledge, which is receptive) realizing itself in an outward movement from within, mastering the world of matter and imprinting upon it the seal of personality. An internal ideal intuition, the Idea: an external shapable matter: the power of interpenetrating internal with external, invisible with visible, ideal with real: such is Beauty. Oersted [5] (the celebrated Danish naturalist whose works were translated into German and gained him a considerable reputation in Germany) defines beauty as the objective Idea in the moment of subjective contemplation: the Idea expressed in things in so far as it reveals itself to intuition. Zeising [6] turned his attention partly to exploration of the mysteries of the golden section, and partly to speculations on Beauty, which he considered as one of the three forms of the Idea; first, the Idea which expresses itself in object and subject; secondly, the Idea as intuition; and thirdly, the Absolute which appears in the world and is conceived intuitively by the spirit. Eckardt, [7] intent on creating a theistic Æsthetic[1] which should avoid the one-sided transcendence of deism[2] on the one hand and the one-sided immanence of pantheism[3] on the other, maintained that its principles must be sought not in the feelings of the contemplator, not in works of art, not in the idea of the beautiful, not in the concept of art, but in the creative spirit of the artist, the original fount of beauty; and since a creative artist cannot be conceived except as derived from the highest creative genius which is God, Eckardt invokes aid from a

1. theistic Æsthetic：有神论的美学
2. transcendence of deism：自然神论的超验性
3. immanence of pantheism：泛神论的内在性

psychology of God (*eine Psychologie des Weltkünstlers*).

Fried. Theodor Vischer[1].

If quantity is as important as quality, we must devote some space to Friedrich Theodor Vischer, the bulkiest of all German æstheticians, indeed the German æsthetician *par excellence*: after publishing a book on *The Sublime and the Comic, a contribution to the Philosophy of the Beautiful*, [8] in 1837, he produced four huge tomes on *Æsthetic as Science of the Beautiful* between 1846 and 1857, [9] where, in hundreds of paragraphs and long observations and sub-observations, is massed a stupendous[2] amount of æsthetic material, of matter foreign to Æsthetic, and of subjects taken haphazard from the whole thinkable universe. Vischer's work is divided into three parts: a Metaphysic of the Beautiful, which investigates the concept of Beauty in itself, no matter where and how it is realized; a treatise on concrete Beauty, which inquires into the two one-sided modes of realization, Beauty of nature and Beauty of imagination, one lacking subjective, the other lacking objective, existence; lastly, a theory of the arts, which studies the synthesis in art of the two artistic moments, the physical and psychical, the objective and subjective. It is easy to sum up Vischer's concept of æsthetic activity; it is Hegel's concept, debased. For Vischer, Beauty belongs neither to the theoretical nor to the practical activity, but is placed in a serene sphere, superior to these antitheses; that is to say in the sphere of absolute Spirit, in company with Religion and Philosophy; [10] but,

1. *Fried. Theodor Vischer*：弗里德里希·提奥多·费肖尔（1807—1887），德国小说家、诗人、剧作家、美学家，著有文中提及的《美学》（*Æsthetic as Science of the Beautiful*），《论崇高与喜剧性兼论美的哲学》（*The Sublime and the Comic, a contribution to the Philosophy of the Beautiful*）。
2. stupendoust：大得惊人的

in contradistinction to[1] Hegel, Vischer assigns the first place in this sphere to Religion, the second to Art, and the third to Philosophy. Much ingenuity was devoted in those days to moving these words about like pieces on a chess-board; it has been observed that of the six possible combinations of the three terms Art, Religion and Philosophy, four were actually adopted: by Schelling, *P.R.A.*; by Hegel, *A.R.P.*; by Weisse, *P.A.R.*; and by Vischer, *R.A.P.* [11] But Vischer himself [12] states that Wirth[2], author of a *System of Ethics*, [13] opted for the fifth combination, *R.P.A.*, which leaves us but the sixth, *A.P.R.*, unclaimed, unless (as is not improbable) some unrecognized genius seized upon it and made it the text of his system. Beauty, therefore, as the second form of the absolute Spirit, is the realization of the Idea, not as abstract concept but as union of concept and reality; and the Idea determines itself as species (*Gattung*), and every idea of a species, even on the lowest degree, is beautiful as being an integral part in the totality of Ideas; although the higher the degree of the idea the greater is its beauty. [14] Highest of all degrees is that of human personality: "in this spiritual world the Idea attains its true significance; the name of idea is given to the great moral motive powers to which the concept of species may also be applied in the sense that they stand to their restricted spheres in the same relation in which the genus stands to its species and individuals." At the head of all is the Idea of morality: "the world of moral and autonomous ends is destined to furnish the most important, the most worthy content of the Beautiful"; with the warning, however, that Beauty, in actualizing this world through intuition, excludes art having a moral tendency. [15] So Vischer proceeds now to degrade Hegel's Idea to the simple class-concept, now to couple it with the idea of the Good; now,

1. in contradistinction to：有别于
2. Wirth：维尔特（1810—1879），德国新教神学家、哲学家，著有《思辨伦理学体系》《神的思辨理念》《哲学研究》等。

in accord with the teaching of his master, to make it different from, yet superior to, intellect and morality.

Other tendencies.

From the first, the Herbartian formalism was little studied and less followed: two writers, Griepenkerl[1] in 1827 and Bobrik[2] in 1834, made some attempt to develop and apply the cursory[3] notes with which Herbart contented himself. [16] Schleiermacher's lectures, even before their appearance in book form, had served as basis for a series of elegant dissertations by Erich Ritter (1840) [17] (better known as a historian of philosophy); his work is of little value, for instead of dwelling on the important points of the master's doctrine Ritter brings into prominence secondary matters relating to sociability and the æsthetic life. A penetrating critic of German Æsthetic from Baumgarten to the post-Kantian school was Wilhelm Theodor Danzel[4], who lived about this time and very properly rebelled against the claim to find "thought" in works of art: "Artistic thought:" he writes; "unhappy phrase, which helped to condemn an entire epoch to the Sisyphean labour[5] of trying to reduce art to intellectual and rational thinking! The thought of a work of art is nothing save that which is contemplated in a definite way; it is not represented, as is commonly asserted, in a work of art, it is the work of art itself. Artistic thought

1. Griepenkerl：格里彭克尔（1782—1849），德国语文学家、教育家、音乐学家、美学家，著有《美学教程》《逻辑学教程》等。
2. Bobrik：伯布里克（1802—1870），德国语文学家、哲学家、美学家，著有《美学的自由合约》《逻辑学的新时间体系》《共济会的历史、基本观念与规则》等。
3. cursory：粗略的
4. Wilhelm Theodor Danzel：威廉·提奥多·丹泽尔（1818—1850），德国文学史学家、哲学家，著有《论歌德的斯宾诺莎主义》《论黑格尔哲学中的美学》《莱辛的生平与著作》等。
5. Sisyphean labour：徒劳无益

can never be expressed by concepts and words." [18] By his early death Danzel ended the hopes he raised by his original views on the science and history of Æsthetic.

Theory of the Beautiful in nature, and that of the Modifications of Beauty.

The post-Hegelian metaphysical Æsthetic is chiefly noteworthy for the fuller development of two theories or, to speak more accurately, of two very curious combinations of arbitrary assertion and fanciful caprice: the so-called theory of Natural Beauty, and the theory of Modifications of the Beautiful. Neither of the two had any intimate or necessary connexion with this philosophical movement, to which they are rather linked by historical or psychological causes; by the relationship between facts of pleasure and pain and the inclination towards mysticism; by the confusion arising from the really æsthetic (imaginative) quality of some representations wrongly described as observation of natural beauties; or by the scholastic and literary tradition of discussing these cases of pleasure and pain and extra-æsthetic natural beauties in books devoted to the discussion of art. [19] These metaphysicians were sometimes rather grotesque and remind one of the story told of Paisiello[1], that in the fury of composition he set even the stage directions of his libretto[2] to music; bitten with the rage for construction and dialectic, they did not spare even the indexes of chaotic old books, but seized on them as suitable material for a dialectical exercise.

Development of the first theory. Herder.

---

1. Paisiello：帕伊谢洛（1740—1816），意大利古典主义作曲家，作品有《磨坊女》《妮娜》《吉普赛集市》等。
2. libretto：（歌剧）剧本

Beginning with the theory of Natural Beauty, observations on beautiful natural objects are found among the inquiries of the ancient philosophers on beauty, and especially among the mystical effusions of neo-Platonists and their followers in the Middle Ages and the Renaissance. [20] Less frequently such questions were introduced into treatises on Poetics: Tesauro (1654) is among the first who, in his *Cannochiale aristotelico*, discusses not only the conceits of men, but also of God, the angels, nature and animals; and somewhat later (1707) Muratori speaks of "the beauty of matter," of which examples are "the gods, a flower, the sun, a rivulet[1]." [21] Observations on that which is outside art and is merely natural, are made by Crousaz, by André, and especially by those authors of the eighteenth century who wrote on Beauty and Art in an empirical and gallant style. [22] It was the influence of these persons that led Kant, as we have seen, to sever the theory of beauty from that of art, specially connecting free beauty with objects of nature and those productions of man which reproduce natural beauties. [23] When the adversary of Kant's theory of Æsthetic, Herder (1800), in his sketch of an ethical system united spirit and nature, pleasure and value, feeling and intellect, he inevitably made much of natural beauty, and affirmed that everything in nature has its own beauty, the expression of its own greatest content, and that this accounts for the ascending scale of beautiful objects: beginning with outlines, colours and tones, light and sound, and proceeding by way of flowers, water and sea, to birds, terrestrial animals, and man himself. For instance "a bird is the sum of the properties and perfections of its element, a representation of its potency[2], a creature of light, song and air"; amongst terrestrial animals, the ugliest are those resembling man, as the melancholy moping monkey; the most beautiful, those of perfect build, well proportioned, noble, free in action; those which

1. rivulet：小河
2. potency：力量

express sweetness; those, in fine, which live in harmony and happiness, endowed with a perfection of their own, harmless to man. [24]

Schelling, Solger, Hegel.

Schelling, on the contrary, utterly, denies the concept of beauty in nature, and considers that such beauty is purely accidental and that art alone supplies the norm by which it can be discovered and judged. [25] Solger also excludes natural beauty; [26] so does Hegel, who distinguishes himself not by denying it but by proceeding with the utmost inconsequence to deal at length with the beautiful in nature. It is in fact not clear whether he means that really no beauty exists in nature and that man introduces it in his vision of things, or whether natural beauty really exists though inferior in degree to the beauty of art. "The beauty of art," he says, " stands higher than that of nature; it is beauty born and reborn by the work of the spirit, and spirit alone is truth and reality; hence beauty is truly beauty only when it participates in spirit and is produced therefrom. Taken in this sense, the beauty of nature appears as a mere reflexion of the beauty appertaining to spirit, as an imperfect and incomplete mode, which substantially is contained within the spirit itself." In confirmation, he adds that nobody has attempted a systematic exposition of natural beauties, whereas there actually is, from the point of view of the utility of natural objects, a *materia medica*[1]. [27] But the second chapter of the first part of his Æsthetic is devoted precisely to natural Beauty on the ground that, in order to grasp the idea of artistic beauty in its entirety, three stages must be traversed: beauty in general, natural beauty (whose defects show the necessity for art), and, lastly, the Idea; "the first existence of the Idea is nature, and its first beauty is natural beauty." This beauty, which is beauty for us

1. *materia medica*：〈拉〉药物学

and not for itself, has several phases, from that in which the concept is immersed in matter to the point of disappearing, such as physical facts and isolated mechanisms, to that higher phase in which physical facts are united in systems (*e.g.* the solar system); but the Idea first reaches a true and real existence in organic facts, in the living creature. And even the living creature is liable to the distinction between beautiful and ugly; for example, among animals, the sloth[1], trailing itself laboriously and incapable of animation or activity, displeases us by its apathetic somnolence[2]; nor can beauty be found in amphibians[3] or in many kinds of fish, or in crocodiles, or toads, as well as in many insects and especially in those equivocal creatures which express a transition from one class to another, such as the ornithorhyncus, a mixture of bird and beast. [28] These samples may suffice to show the general trend of Hegel's doctrine of natural beauty; elsewhere he discusses the external beauty of abstract form, regularity, symmetry, harmony, etc., which are; precisely the concepts which the formalism of Herbart placed in the heaven of the Ideas of the Beautiful.

Schleiermacher.

Schleiermacher, who praised Hegel for his attempt to exclude natural beauty from his Æsthetic, excluded it from his own not verbally but actually, by confining his attention to the artistic perfection of the internal image formed by the energy of the human spirit. [29] But the so-called Feeling for Nature which came in with Romanticism, and the *Cosmos*[4] and other descriptive works of Humboldt, [30] directed attention increasingly to the impressions

1. sloth：树懒
2. apathetic somnolence：提不起神的昏昏欲睡模样
3. amphibians：两栖动物
4. *Cosmos*：《宇宙》，洪堡著。

awakened by natural facts.

Alexander von Humboldt.

This led to the compilation of those systematic lists of natural beauties whose impossibility had been proclaimed by Hegel, though he himself had furnished an example of them; amongst others, Bratraner published an *Æsthetic of the Vegetable World*[1]. [31]

Vischer's "Æsthetic Physics[2]*."*

The best-known and most widely circulated treatment of the subject was contained in this very work of Vischer's; who following Hegel's example devoted a section of his *Æsthetic*, as we have seen, to the objective existence of Beauty, *i.e.* to the Beauty of nature, and entitled it by the perhaps new and certainly characteristic name of Æsthetic Physics (*ästhetische Physik*). This Æsthetic Physics comprised the beauty of inorganic nature (light, heat, air, water, earth); organic nature, with its four vegetable types and its animals vertebrate and invertebrate; and beauty of human beings, divided into generic and historic. The generic was subdivided into sections on the beauty of general forms (age, sex, conditions, love, marriage, family); of special forms (races, peoples, culture, political life); and of individual forms (temperament and character). Historical beauty included that of ancient history (Oriental, Greek, Roman), of Mediæval or Germanic, and of modern times; because, according to Vischer, it was the duty of Æsthetic to cast a glance over universal history before summing up the different degrees of the beautiful according to the varying phases of the struggle for freedom against nature. [32]

---

1. *Æsthetic of the Vegetable World*：《植物界的美学》，布拉特拉内著。
2. *Æsthetic Physics*：《美学物理学》

The theory of the modifications of Beauty: from antiquity to the eighteenth century.

As regards the Modifications of Beauty, it should be remembered that the ancient manuals of Poetics, and more frequently those of Rhetoric, contained more or less scientific definitions of psychological states and facts; Aristotle attempted in his *Poetics* to determine the nature of a tragic action or personality, and sketched a definition of the comic; in his Rhetoric he writes at considerable length of wit; [33] sections of the *De oratore* of Cicero and the *Institutions* of Quintilian [34] are devoted to wit and the comic; the lofty style was the subject of a lost treatise of Cæcilius[1], which anticipated that attributed to Longinus[2], whose title was translated in modern times as *De sublimitate* or *On the Sublime*. Following the example of the ancients, this kind of medley was perpetuated by writers of the sixteenth and seventeenth centuries; whole treatises on the comic are incorporated in, for instance, the *Argutezza* of Matteo Pellegrini (1639) and the *Cannochiale* of Tesauro. La Bruyère treated of the sublime [35] and Boileau by his translation gave a fresh vogue to Longinus: the following century saw Burke inquiring into the origin of our ideas of the beautiful and the sublime, and deriving the former from the instinct for sociability, the latter from that of self-preservation; he also tried to define ugliness, grace, elegance and extraordinary beauty; Home, in his celebrated *Elements of Criticism*[3], discussed grandeur, sublimity, the ridiculous, wit, dignity and grace: Mendelssohn discussed sublimity, dignity and grace in fine art, and described some of these facts as due to mixed feelings, in which he was followed by Lessing [36] and others: Sulzer welcomed

---

1. Cæcilius：凯西琉斯，约公元 1 世纪生活在罗马庞贝城，古罗马银行家。
2. Longinus：朗吉努斯，古罗马时代希腊文艺评论家，约生活在公元 1 世纪或 3 世纪，著有《论崇高》。
3. *Elements of Criticism*：《批评要素》，霍姆著。

all these various concepts into his æsthetic encyclopædia and collected round them an elaborate bibliography. A new and curious meaning of the word humour reached the continent from England at this time. Its original meaning was simply "temperament," and sometimes "spirit," or "wit" ("*belli umori*" in Italy; in the seventeenth century there was in Rome an Academy of *Umoristi*[1]). Voltaire introduced it into France and wrote in 1761, "*Les Anglais ont un terme pour signifier cette plaisanterie, ce vrai comique, cette gaieté, cette urbanité, ces saillies, qui échappent à un homme sans qu'il s'en doute; et ils rendent cette idée par le mot* humour[2] ... "; [37] in 1767 Lessing distinguishes humour from the German *Laune*[3] (caprice, whim), [38] a distinction maintained by Herder in 1769 in opposition to Riedel who had confused the terms. [39]

Kant and the post-Kantians.

Accustomed to find all these subjects treated in the same book, philosophers at first theorized about them all without attempting to link them up together by introducing an artificial logical connexion. Kant, who had already in imitation of Burke written in 1764 a dissertation on the beautiful and the sublime, ingenuously remarked in the course of his lectures on Logic in 1771 that the beautiful and the æsthetic are not identical, because "the sublime also belongs to Æsthetic"; [40] and in his *Critique of Judgment*, while treating of the comic in a mere digression[4] (a magnificent piece of psychological

---

1. Academy of *Umoristi*：幽默者学会
2. *Les Anglais ont un terme pour signifier cette plaisanterie, ce vrai comique, cette gaieté, cette urbanité, ces saillies, qui échappent à un homme sans qu'il s'en doute; et ils rendent cette idée par le mot* humour：〈法〉人们不知不觉中流露出来的俏皮话，那种真正喜剧式的，欢乐的，文雅的，让人捧腹的语言。对此，英国人专门有个词来表达：幽默。
3. *Laune*：〈德〉情绪
4. in a mere digression：作为题外话

analysis) [41] places side by side with and as if on an equality with the "Analytic of Beauty," an "Analytic of the Sublime." [42] We may note in passing that, before the publication of the third Critique, Heydenreich arrived at the same doctrine of the sublime which is contained in Kant's book. [43] Did Kant ever think of uniting the beautiful and the sublime and deducing them from a single concept? Apparently not. By his declaration that the principle of beauty must be sought outside ourselves, and that of the sublime within us, he tacitly assumes that the two objects are wholly disparate[1]. In 1805 Ast, a follower of Schelling, declared the necessity of overcoming what he called the Kantian dualism of the beautiful and the sublime: [44] others reproached Kant with having treated the comic by the psychological, not the metaphysical, method. Schiller wrote a series of dissertations on the tragic, the sentimental, the ingenuous, the sublime, the pathetic, the trivial, the low, the dignified and the graceful, and their varieties, the fascinating, the majestic, the grave, and the solemn. Another artist, Jean Paul Richter, discoursed at great length on wit and humour, described by him as the romantic comic, or the sublime reversed (*umgekehrte Erhabene*). [45]

Herbart, in virtue of his formalistic principle, asserts that all these concepts are irrelevant to Æsthetic; he attributes them to the work of art, not to pure beauty; [46] Schleiermacher comes to the same conclusion, but for much better reasons, as a result of his sane conception of art. Amongst other things he observes: "It is usual to describe the beautiful and the sublime as two kinds of artistic perfection; and so accustomed have we grown to the union of these two concepts that we must make an effort to convince ourselves how very far they are from being co-ordinate or from together exhausting the concept of artistic perfection"; he regrets that even the best

1. wholly disparate：截然不同

æstheticians should give rhetorical descriptions of them instead of demonstrating them. "The thing," says he, "is not right and just" (*hat keine Richtigkeit*), and he proceeds to exclude the whole subject from his Æsthetic, [47] as he had done previously in the case of natural beauty. Other philosophers, however, clung persistently to their search for a connexion between these various concepts, and called in dialectic to help them. The habit of applying dialectic to empirical concepts affected everybody at that time; even the great enemy of dialectic, Herbart, showed the cloven hoof, when in order to explain the union of different æsthetic ideas in the beautiful he appealed to the formula "they lose regularity in order to regain it." [48] Schelling asserted that the sublime is the infinite in the finite, and the beautiful the finite in the infinite, adding that the absolutely sublime includes the beautiful, and the beautiful the sublime; [49] and Ast, whom we have mentioned already, spoke of a masculine, positive element, which is the sublime, and a feminine, negative element which is the graceful and pleasing: between which there is a contrast and a struggle.

Culmination of the development.

These exercises in dialectical system-building developed and increased till about the middle of the nineteenth century they assumed two distinct forms whose history must here be shortly outlined.

Double form of the theory. The overcoming of the ugly. Solger, Weisse and others.

The first form may be called the Overcoming of the Ugly. This theory conceives the comic, the sublime, the tragic, the humorous, and so forth, as so many engagements in the war between the Ugly and the Beautiful, wherein the latter was invariably victorious, and arose by means of this war to more and more lofty and complex manifestations. The second form of the theory may be described as the Passage from Abstract to Concrete; it held that Beauty cannot emerge from the abstract, cannot become this or that concrete beauty, except by particularizing itself in the comic, tragic, sublime, humorous, or some other modification. The first form was already well developed in Solger, an adherent of the romantic theory of Irony: but historically it presupposes the æsthetic theory of the Ugly, first sketched by Friedrich Schlegel in 1797. We have already noted that Schlegel considered the characteristic or interesting, not the beautiful, to be the principle of modern art; hence the importance attached by him to the piquant, the striking (*frappant*), the daring, the cruel, the ugly. [50] Solger found here the basis for his dialectic; amongst other things he maintains that the finite, earthly element may be dissolved and absorbed in the divine, which constitutes the tragic: or else the divine element may be entirely corrupted by the earthly, producing the comic. [51] These methods of Solger were followed by Weisse (1830), and by Ruge[1] (1837); for the former, ugliness is "the immediate existence of beauty" which is overcome in the sublime and the comic; for the latter, the effort to achieve the Idea, or the Idea searching for itself, generates the sublime; when the Idea loses instead of discovering itself, ugliness is produced; when the Idea rediscovers itself and rises out of ugliness to new life, the comic. [52] A whole treatise entitled *The Æsthetic of the Ugly*[2] [53] was published by Rosenkranz[3] in 1853, presenting this concept as intermediate between the beautiful and the comic, and tracing it from its first origin to that "sort of perfection" it attains in the satanic. Passing from the common (*Gemeine*) which is the petty, the

---

1. Ruge：鲁格（1802—1880），德国哲学家、政论作家、剧作家、翻译家，著有《鲁格集》《当代史》，译著有巴克尔的《文明史》。
2. *The Æsthetic of the Ugly*：《丑的美学》，罗森克兰茨著。
3. Rosenkranz：罗森克兰茨（1805—1879），德国哲学家、教育家、美学家，著有《心理学：主观精神的科学》《丑的美学》《逻辑理念的科学》等。

weak, the low, and the sub-species of the low, viz. the usual, the casual, the arbitrary and the crude, Rosenkranz goes on to describe the repugnant, trisected into[1] the awkward, the dead and empty, and the horrible: thus he proceeds from tripartition to tripartition, dividing the horrible into the absurd, the nauseating and the wicked: the wicked into criminal, spectral and diabolical[2]: the diabolical into demoniac, magical and satanic. He opposes the childish notion that ugliness acts as a foil to beauty in art, and justifies its introduction by the necessity for art to represent the entire appearance of the Idea; on the other hand he admits that the ugly is not on the same level as the beautiful, for, if the beautiful can stand by itself alone, the other cannot do so and must always be reflected by and in the beautiful. [54]

Passage from abstract to concrete: Vischer.

The second form prevailed with Vischer. The following extract will serve as an illustration of his manner: "The Idea arouses itself from the tranquil unity in which it was fused with the appearance and pushes onward, affirming, in face of its own finitude, its infinity"; this rebellion and transcendence is the sublime. "But Beauty demands full satisfaction for this disruption of its harmony: the violated right of the image must be reasserted: this can be accomplished only by means of a fresh contradiction, that is to say by the negative position now taken up by the image towards the Idea by rejecting all interpenetration with it and by affirming its own separate existence as the whole"; this second moment is the comic, negation of a negation. [55] The same process is further enriched and complicated by Zeising, who compares the modifications of Beauty to the refraction of colours[3]:

---

1. trisected into：一分为三
2. criminal, spectral and diabolical：犯罪的，鬼魅的和恶魔的
3. refraction of colours：色彩的折射

the three primary modifications, the sublime, the attractive and the humorous, correspond with the primary colours violet, orange and green; the three secondary, pure beauty, comic and tragic, to the colours red, yellow and blue. Each of these six modifications (exactly like the degrees of the Ugly in Rosenkranz) branches out, like fireworks, into three rays: pure beauty into the decorous, noble and pleasing: the attractive into graceful, interesting and piquant: the comic into buffoonery, the diverting and burlesque[1]: the humorous into the quaint, capricious and melancholy: the tragic into the moving, pathetic and demoniac: the sublime into the glorious, majestic and imposing. [56]

The Legend of Sir Purebeauty.

All the works of this period on Æsthetic are filled in this way with the *gest, chanson* or romaunt[2] of the knight Sir Purebeauty (*Reinschon*) and his extraordinary adventures, recounted in two conflicting versions. According to one story, Sir Purebeauty is constrained to abandon his beloved leisure by the Mephistophelean devices of the temptress Ugliness, who leads him into countless dangers from which he invariably emerges victorious; his victories and successes (his Marengo, Austerlitz and Jena[3]) are called the Sublime, the Comic, the Humorous and so forth. The other story tells how the knight, bored by his life of loneliness, sallies[4] forth purposely to seek adversaries and occasions for fighting; he is always vanquished, but even in his

---

1. the comic into buffoonery, the diverting and burlesque：喜剧分为滑稽的，有趣的和戏谑的
2. *gest, chanson* or romaunt：〈法〉功绩，赞歌或者传奇
3. Marengo, Austerlitz and Jena：马伦哥战役、奥斯特里茨战役、耶拿战役，皆为拿破仑大获全胜的战役。
4. sallies：出击

overthrow *ferum victorem capit*[1], he transforms and irradiates[2] the enemy. Beyond this artificial mythology, this legend composed without the least imagination or literary skill, this miserably dull tale, it is vain to look for anything whatever in the much elaborated theory of German æstheticians known as the Modifications of Beauty.

[1] *Abriss der Ästhetik*, post. 1837; *Vorlesung üb. Ästh.* (1828–1829), post. 1882.

[2] *Ästhetik*, Berlin, 1827.

[3] *Ästhetik*, Leipzig, 1830; *System d. Ästh.*, lectures, post. Leipzig, 1872.

[4] *Kunstlehre*, Ratisbon, 1845–1846 (*Grundlinien einer positiven Philosophie*, vols. iv. v.).

[5] *Der Geist in der Natur*, 1850–1851; *Neue Beitrage z. d. Geist i.d. Natur*, post. 1855.

[6] *Ästhetische Forschungen*, Frankfurt a. M. 1855.

[7] *Die theistische Begründung d. Ästhetik im Gegensatz z. d. pantheistichen*, Jena, 1857; same author, *Vorschule d. Ästh.*, Karlsruhe, 1864–1865.

[8] *Üb. d. Erhabene u. Komische*, Stuttgart, 1837.

[9] *Ästhetik oder Wissenschaft d. Schönen*, Reutlingen, Leipzig and Stuttgart, 1846–1857, 3 parts in 4 vols.

[10] *Ästh.* introd. §§ 2–5.

[11] Hartmann, *Dtsch. Ästh. s. Kant*, p. 217, note.

[12] *Ästh.* introd. § 5.

[13] *System der spekulativen Ethik*, Heilbronn, 1841–1842.

[14] *Ästh.* §§ 15–17.

[15] *Op. cit.* §§ 19–24.

---

1. *ferum victorem capit*：〈拉〉征服了野蛮的胜利者
2. irradiates：启发

[16] Griepenkerl, *Lehrb. d. Ästh.*, Brunswick, 1827. Bobrik, *Freie Verträge üb. Ästh.*, Zürich, 1834.

[17] *Üb. d. Principien d. Ästh.*, Kiel, 1840.

[18] *Ges. Aufs.* pp. 216–221.

[19] See above, pp. 119–125.

[20] See above, pp. 211–212.

[21] *Cannochiale arist.* ch. 3: *Perfetta poesia*, bk. I. chs. 6, 8.

[22] See above, pp. 237–238, 290–293.

[23] See above, pp. 307–309.

[24] *Kaligone, op. cit.* pp. 55–90.

[25] *System d. transcend. Ideal*, part vi. § 2.

[26] *Vorles. üb. Ästh.* p. 4.

[27] *Vorles. üb. Ästh.* I. pp. 4–5.

[28] *Vorles. üb. Ästh.* I. pp. 148–180.

[29] *Op. cit.* introd.

[30] *Ansichten der Natur*, 1088; *Kosmos*, 1845–1858.

[31] *Ästhetik. Pflanzenwelt*, Leipzig, 1853.

[32] *Ästh.* § 341.

[33] *Poet.* 5. 13–14; *Rhet.* iii. 10, 18.

[34] *De orat.* ii. 54–71; *Inst. orat.* vi. 3.

[35] *Caractères*, I.

[36] *Hamb. Dramat.* Nos. 74–75.

[37] Letter to abbé d'Olivet, August 20, 1761.

[38] *Hamb. Dramat.* No. 93; in *Werke, ed. cit.* xii. pp. 170–171, note.

[39] *Kritische Wälder*, in *Werke, ed. cit.* iv. pp. 182–186.

[40] Schlapp, *op. cit.* p. 55.

[41] *Kr. d. Urth., Anmerkung,* § 54.

[42] *Op. cit.* bk. ii. §§ 23–29.

[43] *System d. Ästh.* introd. p. xxxvi *n.*

[44] *System der Kunstlehre* cf. Hartmann, *op. cit.* p. 387.

[45] *Vorschule d. Ästh.* chs. 6–9.

[46] See above, pp. 341–342.

[47] *Vorles. üb. Ästh.* p. 240 *seqq.*

[48] Cf. Zimmermann, *G. d. Ästh.* p. 788.

[49] *Philos, d. Kunst,* § § 65–66.

[50] Cf. Hartmann, *Deutsch. Ästh. s. Kant,* pp. 363–364.

[51] *Vorles üb. Ästh.* p. 85.

[52] *Neue Vorschule d. Ästh.* Halle, 1837.

[53] K. Rosenkranz, *Ästhetik des Hässlichen*, Kœnigsberg, 1853.

[54] *Ästh. d. Hässl.* pp. 36–40.

[55] *Ästh.* § § 83–84, 154–155.

[56] *Ästh. Forsch.* p. 413.

XIV ÆSTHETIC IN FRANCE, ENGLAND AND ITALY DURING THE FIRST HALF OF THE NINETEENTH CENTURY

Æsthetic movement in France: Cousin, Jouffroy.

In the last quarter of the eighteenth century and the first half of the nineteenth century German thought, notwithstanding the glaring errors which vitiated[1] it, and were soon to bring about a violent and indeed exaggerated reaction[2], must on the whole be awarded the foremost place in the general history of European thought as well as in the individual study of Æsthetic, the contemporary philosophy of other countries standing on an inferior level of the second and third degree. France still lay under the dominion of the sensationalism of Condillac[3] and, at the opening of the century, was quite incapable of grasping the spiritual activity of art. A faint gleam of Winckelmann's abstract spiritualism[4] just appears in the theories of Quatremère de Quincy[5], who, in criticism of Émeric-David[6] (in his turn a critic of ideal beauty and an adherent of the imitation of nature), [1] maintained that the

---

1. vitiated：削弱
2. exaggerated reaction：过激反应
3. sensationalism of Condillac：孔狄亚克感觉论
4. abstract spiritualism：抽象唯心论
5. Quatremère de Quincy：夸特梅尔·德·昆西（1755—1849），法国考古学家、建筑理论家、艺术评论家，著有《三论法国艺术设计》《建筑史词典》《论埃及建筑》等。
6. Émeric-David：艾莫瑞克-大卫（1755—1839），艺术评论家、法国考古学家，续编了《法国文学史》并著有《古代与现代雕塑艺术之研究》。

arts of design have pure beauty, devoid of individual character[1], as their objective; they depict man and not men. [2] Some sensationalists, such as Bonstetten[2], vainly endeavoured to trace the peculiar processes of imagination in life and in art. [3] Followers of the orthodox spiritualism[3] of the French universities date the beginning of a new era, and the foundation of Æsthetic in France, to 1818, the year when Victor Cousin[4] first delivered at the Sorbonne[5] his lectures on the True, the Beautiful and the Good, which later formed his book with the same name, frequently reprinted. [4] These lectures of Cousin are but poor stuff, although some scraps of Kant are to be found in them here and there; he denies the identity of the beautiful with the pleasant or useful, and substitutes the affirmation of a threefold beauty, physical, intellectual and moral, the last being the true ideal beauty, having its foundations in God; he says that art expresses ideal Beauty, the infinite, God, that genius is the power of creation, and that taste is a mixture of fancy, sentiment and reason. [5] Academic phrases all of them; pompous and void[6] and, for that very reason, well received. Of much greater value were the lectures on Æsthetic delivered by Théodore Jouffroy[7] in 1822, before a small audience, and published posthumously in 1843. [6] Jouffroy allowed a beauty of expression, to be found alike in art and nature: a beauty of imitation, consisting in

---

1. devoid of individual character：缺乏个人色彩
2. Bonstetten：邦斯泰唐（1745—1832），瑞士自由主义作家、思想家，深受卢梭影响，著有《论国民教育》《人之研究，或思想与思维能力的研究》等。
3. orthodox spiritualism：正教唯心论
4. Victor Cousin：维克多·库辛（1792—1867），法国哲学家，折中主义创始人，著有《洛克的哲学》《哲学札记》《为大学与哲学辩护》等。
5. Sorbonne：索邦大学，即巴黎第四大学，以人文教育著称。
6. pompous and void：华丽而空洞的
7. Théodore Jouffroy：提奥多·约夫里埃（1796—1842），法国哲学家，著有《道德哲学概论》《伦理学导论（含道德体系批判研究）》《美学教程》等。

the perfect accuracy with which a model is reproduced: a beauty of idealisation, which reproduces the model, accentuating a particular quality in order to give it greater significance: and, finally, a beauty of the invisible or of content, reducible to force (physical, sensible, intellectual, moral), which, as force, awakens sympathy. Ugliness is the negation of this sympathetic beauty; its species or modifications are the sublime and the graceful. One sees that Jouffroy did not succeed in isolating the strictly æsthetic fact in his analysis and gave, instead of a scientific system, little beyond explanations of the use of words. He could not see or understand that expression, imitation and idealization are identical with each other and with artistic activity. Moreover he had many curious ideas, chiefly concerning expression. He said that if we were to see a drunkard with all the most disgusting symptoms of intoxication[1] on a road where there was also an unhewn rock[2], we should be pleased by the drunken man, since he had expression, and not by the rock, since it had none. Beside Jouffroy, whose theories, crude and immature though they be, reveal an inquiring mind, it is hardly worth while to cite Lamennais[3] [7] , who like Cousin regarded art as the manifestation of the infinite through the finite, of the absolute through the relative. French Romanticism in de Bonald[4], de Barante[5] and Mme. de Staël[6] had defined literature as "the expression of society," had honoured, under German influence, the characteristic

1. symptoms of intoxication：醉酒症状
2. unhewn rock：未经打磨的石头
3. Lamennais：拉梅内（1782—1854），法国天主教牧师、哲学家、政论家、翻译家，著有《从教会传统来看主教机构》《论宗教题材的区别》等。
4. de Bonald：德·博纳尔德（1754—1840），法国天主教哲学家、政治家，著有《政权与教权的理论》《社会秩序的自然法则分析论》等。
5. de Barante：德·巴兰特（1782—1866），法国政治家、历史学家，著有《论公众与贵族》《法兰西共和国督政府历史》等。
6. Mme. de Staël：斯达尔夫人（1766—1817），瑞士裔法国女作家，著有《对自杀的反思》《幻想女士》《论卢梭的品格与著作》等。

and the grotesque, [8] and had proclaimed the independence of art by means of the formula "art for art's sake"; but these vague affirmations or aphorisms did not supersede, philosophically speaking, the old doctrine of the "imitation of nature."

English Æsthetic.

In England associationistic psychology[1] still flourished (and has continued to flourish uninterruptedly), unable to emancipate itself wholly from sensationalism or to understand imagination. Dugald Stewart[2] [9] had recourse to the wretched expedient of establishing two forms of association: one of accidental associations, the other of associations innate[3] in human nature and therefore common to all mankind. England did not escape German influence, as appears, for example, in Coleridge[4], to whom we owe a saner concept of poetry and the difference between it and science [10] (in collaboration with the poet Wordsworth[5]), and in Carlyle[6], who placed intellect lower than imagination, "organ of the Divine." The most noteworthy English æsthetic essay of this period is the *Defence of Poetry* by Shelley[7](1821), [11]

---

1. associationistic psychology：联想心理学
2. Dugald Stewart：杜加尔德·斯图亚特（1753—1828），苏格兰哲学家、数学家，任教于爱丁堡大学，著有《人类理智哲学的要素》《道德哲学大纲》《主动力与道德力的哲学》等。
3. innate: 与生俱来的
4. Coleridge：柯勒律治（1772—1834），英国诗人、文学家、哲学家、神学家，湖畔派诗人之一，著有《古舟子咏》《忽必烈汗》《哲学史讲义》《逻辑学》等。
5. Wordsworth：华兹华斯(1770—1850)，英国桂冠诗人。其诗歌理论动摇了英国古典主义诗学的统治，推动了英国诗歌的革新和浪漫主义运动的发展。
6. Carlyle：卡莱尔（1795—1881），英国哲学家、历史学家、讽刺作家，著有《论历史上的英雄、英雄崇拜和英雄业绩》《法国大革命史》等。
7. *Defence of Poetry* by Shelley：雪莱的《为诗一辩》。雪莱（1792—1822），英国浪漫主义诗人之一，代表作有《西风颂》《致云雀》《解放的普罗米修斯》等。

containing profound, if not very systematic, views on the distinctions between reason and imagination, prose and poetry; on primitive language and the faculty of poetic objectification which enshrines and preserves "the record of the best and happiest moments of the happiest and best minds."

Italian Æsthetic.

In Italy, where neither Parini[1] nor Foscolo[2] [12] had been able to shake off the fetters of the old doctrines (although the latter, in his later writings, was in several ways an innovator in literary criticism), many treatises and essays on Æsthetic were published during the earlier decades of the century, the greater part showing the influence of Condillac's sensationalism, which had a great vogue in Italy. Such authors as Delfico[3], Malaspina[4], Cicognara[5], Talia[6], Pasquali, Visconti[7] and Bonacci[8] belong more exclusively to the special, or rather, the anecdotal, history of Italian philosophy. Now and then,

---

1. Parini：即朱塞佩·帕里尼（1729—1799），意大利启蒙主义讽刺作家，新古典主义诗人，著有《一天：早晨、中午、傍晚、夜晚》。
2. Foscolo：福斯科洛（1778—1827），意大利作家、革命家、诗人，代表作有长诗《墓地哀思》、小说《奥提斯的最后几封信》、诗歌《致黑夜》等。
3. Delfico：德尔菲科（1744—1835），意大利经济学家，著有《婚姻哲学论》《对美的新研究》等。
4. Malaspina：即马拉斯皮纳·迪·桑纳扎罗（1754—1835），意大利收藏家、博物学家、美学家，著有《美的规律在绘画与建筑中的应用》《希腊建筑的历史框架》等。
5. Cicognara：奇科尼亚拉（1767—1834），意大利考古学家、收藏家、艺术史学家，著有《法拉里斯文人与艺术家的历史回忆》《法拉里斯伟大画家与雕塑家的生平》《论美术》等。
6. Talia：詹巴蒂斯塔·塔利亚，生卒年不详，19 世纪意大利美学家。
7. Visconti：维斯孔蒂（1751—1818），意大利古籍学家、艺术史学家，当时的古罗马雕塑领域顶级专家，著有《皮奥·克莱蒙提奥博物院》《哥贝尼石碑》等。
8. Bonacci：即格拉蒂利亚诺·博纳西，生卒年不详，19 世纪意大利美学家。

however, one comes across remarks that are not wholly contemptible, as in Melchiorre Delfico (1818) who, after wandering aimlessly hither and thither, fixes on the principle of expression, observing, "If it were possible to establish that expression is always an element in the beautiful, it would be a legitimate inference to regard it as the real characteristic of beauty, *i.e.* a condition without which the beautiful could not exist, and the pleasing modification which arouses the sentiment of beauty could not take place in us"; he tries to develop this principle by asserting that all other characters (order, harmony, proportion, symmetry, simplicity, unity and variety) have significance only by their subordination to the principle of expression. [13] In opposition to Malaspina's definition of beauty as "pleasure born of a representation"; and in opposition to the then fashionable threefold division of beauty into sensible, moral and intellectual, a critic of Malaspina observed that if beauty be representation, it is inconceivable that there should be intellectual beauty, which would be intelligible but not presentable. [14] Nor must Pasquale Balestrieri[1] be forgotten; he was a student of medicine who in 1847 tried to construct an Æsthetic of an exact or mathematical kind, with neither better nor worse result than many famous authors in other countries. He noticed, while turning his algebraical expressions into numerals, that such general formulæ "fulfil their object with an infinite number of systems of different ciphers[2]"; and that in art there is an element "not arbitrary, but unknown." [15] Works by German authors were frequently translated at this time, some of them, for example the writings of the two Schlegels, being reprinted several times; the *Æsthetic* of Bouterweck[3],

---

1. Pasquale Balestrieri：帕斯夸莱·巴勒斯特列里（生卒年不详），意大利美学家，著有《美学基础》。
2. ciphers：密码
3. Bouterweck：博特维克（1806—1867），德国哲学家、美学家、作家，著有《美学》《理智的宗教》等。

deriving from Kant and Schiller, [16] was read and discussed; Colecchi[1] gave an excellent statement of the æsthetic doctrines of Kant; [17] and in 1831 a certain Lichtenthal[2] adapted the *Æsthetic* of Franz Ficker[3] [18] to the use of Italian readers; later the same book was fully translated by another hand; some of Schelling's writings were translated, *e.g.* his discourses on the relation between figurative art and nature.

Rosmini and Gioberti.

It must be admitted that in Italy Æsthetic received but inadequate treatment in the revival of philosophical speculation effected by the work of Galluppi[4], Rosmini[5] and Gioberti[6]. It is treated in a merely incidental and popular manner by the first named. [19] Rosmini devotes a section of his philosophical system to the deontological sciences[7], which "treat of the perfection of being, and the method of acquiring or producing such perfection or losing it"; among these sciences is that of "beauty in the universal" under the name of Callology[8], of which a special part is Æsthetic, the science of "beauty in the sensible," establishing the "archetypes of beings." [20] In his

---

1. Colecchi：克莱基（1773—1848），意大利哲学家，著有《哲学问题》。
2. Lichtenthal：里希腾塔尔（生卒年不详），著有《美学：关于美与美术的学说》。
3. Franz Ficker：弗朗兹 · 菲克尔（1782—1849），出生于波西米亚的美学家，文史学家，著有《希腊罗马文学史》《美学》。
4. Galluppi：加卢皮（1770—1846），意大利哲学家，曾任教于那不勒斯大学，著有《人类知识批评哲学论》《哲学要素、逻辑学与形而上学教程》《超验唯心主义的哲学研究》等。
5. Rosmini：罗斯米尼（1797—1855），意大利天主教牧师、哲学家，著有《论神圣教会的五大创伤》《社会正义的体制》等。
6. Gioberti：乔贝蒂（1801—1852），意大利哲学家、政治家，著有《意大利人的道德与民权要义》《论美》《论善》等。
7. deontological sciences：义务论，又称道义论，由康德创立，强调动机的纯洁性。
8. Callology：美之学，是罗斯米尼宣扬的义务论的一个分支，研究“普遍存在的美”。

longest literary work, considered by him as his Æsthetic, [21] his essay on *The Idyl*[1], [22] Rosmini declares the aim of art to be neither imitation of nature nor direct intuition of the archetypes, but the reduction of natural things to their archetypes, which are arranged in a hierarchy of three ideals, natural, intellectual and moral. Gioberti [23] is clearly under the influence of German idealism, especially of Schelling's; for him the beautiful is "the individual union of an intelligible type with an imaginative element called into being by fancy"; the phantasm[2] gives material, while the intelligible type (concept) gives form, in the Aristotelian sense, [24] and since the ideal element predominates over the sensible or fantastic, art is a propædeutic[3] to the true and the good. Gioberti is of opinion that Hegel was wrong in detaching natural beauty from Æsthetic, for perfect beauty of nature is "the full correspondence of sensible reality with the Idea which informs and represents it," and as such "makes its appearance in the sensible universe during the second period of the primordial age described in detail by Moses in the six days of creation"; it is only through original sin that imperfection and ugliness arose in nature. [25] Art is nothing but a supplement to natural beauty, whose decadence it presupposes, and thus art is at once record and prophecy, referring to the first and last ages of the world. The Last Judgement will reintroduce perfect beauty: "organic restitution, by empowering the faculties to contemplate the intelligible in the sensible, and by refining their capabilities, will greatly intensify and purify æsthetic enjoyment. The contemplation of perfect beauty will be the beatitude of imagination[4], of which Christ gave an ineffable foretaste by appearing to his disciples

1. *The Idyl*：《论田园诗》，罗斯米尼著。
2. phantasm：幻想
3. propædeutic：预备知识
4. the beatitude of imagination：幻想的至福

visibly transfigured and shining with celestial radiance[1]." [26] Gioberti agrees with Schelling's division of art into pagan and Christian, a "heterodox beauty[2]" (Oriental and Græco-Italian art), imperfect when compared with "orthodox beauty[3]"; and between the two, a "semi-orthodox" beauty[4], [27] transitional to Christian art[5]; he also attempted a doctrine of modifications of the beautiful, wherein he held the sublime to be creator of the beautiful. Beauty is the relative intelligibility of created things apprehended by fancy: the sublime is the absolute intelligibility of time, space and infinite power as presented to itself by the faculty of imagination: "The ideal formula: the Being creates the Existing, translated into æsthetic language, gives the following formula: by means of the dynamical sublime Being creates the beautiful; and by means of the mathematical sublime contains it: this shows the ontological and psychological connexions of Æsthetic in First Science." Ugliness enters into the beautiful either as relief and counterpoise[6], or to open a way to the comic, or to depict the struggle between good and evil. The Christian ideal of artistic beauty is the figure of the God-Man, absolute union of the two forms of beauty, the sublime and the beautiful, a transfigured and divinely illuminated expression of man. [28] However carefully we sift the thoughts of Gioberti from their mythological Judaico-Christian husk[7], we find nothing of the least value to science.

Italian Romantics. Dependence of art.

On the other hand, if Italian literature of the day chose to revive

---

1. celestial radiance：天堂的灿烂光辉
2. heterodox beauty：异教美
3. orthodox beauty：正教美
4. "semi-orthodox" beauty："半正教"美
5. transitional to Christian art：作为通向基督教艺术的过渡
6. counterpoise：反衬
7. mythological Judaico-Christian husk：犹太-基督式的神话形式

and refurbish certain antiquated critical ideas, a much wider field was opened by social and political upheavals which tended to make use of literature as a practical instrument for spreading abroad the truths of history, science, religion and morality. In 1816 Giovanni Berchet[1] wrote that "poetry ... is intended to improve the habits of man and satisfy the cravings of his imagination and heart, since the tendency towards poetry, like every other desire, awakens in us moral needs"; [29] and Ermes Visconti in his *Conciliatore*[2] of 1818 says that æsthetic aims must be subordinated "to the improvement of mankind and public and private weal[3], the eminent aim of all studies." Manzoni[4], who subsequently took to philosophizing on art on the principles of Rosmini, declared in his letter on Romanticism (1823) that "poetry or literature in general should have utility as its objective, truth as its subject and interest as its means"; [30] and though noticing the vagueness of the concept of truth in poetry, he inclined always (as is seen also in his discourse on the historical novel) to its identification with historical and scientific truth. [31] Pietro Maroncelli[5] proposed as a substitute for the classic formula of art, "founded on imitation of the real and having pleasure as its object," a formula of art as "founded on inspiration, having the beautiful as means and good as end"; this doctrine he baptized "cormentalism," contrasting it with the doctrine of art for art's sake found in the writings of August Wilhelm Schlegel and Victor Hugo. [32] Tommaseo[6] defined beauty as

1. Giovanni Berchet：乔万尼·博切特（1783—1851），意大利诗人、爱国者，著有《论抒情诗人》。
2. *Conciliatore*：《调解者》，维斯孔蒂著。
3. public and private weal：公众及个人的康乐
4. Manzoni：曼佐尼（1785—1873），意大利作家、诗人和小说家，著有《约婚夫妇》《五月五日》等。
5. Pietro Maroncelli：皮埃特罗·马龙切利（1795—1846），意大利音乐家、诗人，爱国主义者，著有《狱中杂记》等。
6. Tommaseo：托马赛奥（1802—1874），意大利语言学家、记者、散文家，著有《意大利语词典》《但丁之新研究》《罗斯米尼生平》《美学词典》等。

"the union of many truths in one concept" effected by the power of feeling. [33] Giuseppe Mazzini[1], too, always conceived literature as the mediator of the universal idea or intellectual concept. [34] Attempting to restore serious content to a literature grown weak and frivolous, the Italian Romantics found themselves forced on the theoretical side, by a natural reaction, into constant and perpetual opposition to every tendency of thought likely to affirm the independence of art.

[1] Émeric-David, *Recherches sur l'art du statuaire chez les anciens*, Paris, 1805 (Ital. trans., Florence, 1857).

[2] Quatremère de Quincy, *Essai sur l'imitation dans les beaux arts*, 1823.

[3] *Recherches sur la nature et les lois de l'imagination*, 1807.

[4] *Du vrai, du beau et du bien*, 1818, many lines revised (23rd ed. Paris, 1881).

[5] *Op. cit.* lectures 6–8.

[6] *Cours d' esthétique*, ed. Damiron, Paris, 1843.

[7] *De l'art et du beau,* 1843–1846.

[8] Victor Hugo, Preface to *Cromwell*, 1827.

[9] Dugald Stewart, *Elements of the Philosophy of the Human Mind*, 1837.

[10] Gayley-Scott, *An Introd.* pp. 305–306.

[11] P. B. Shelley, *A Defence of Poetry* (in *Works*, London, 1880, vol. vii.)

[12] Parini, *Principi delle belle lettere applicati alle belle arti*, from 1773 onward; Foscolo, *Dell' origine e dell' uffizio della letteratura*, 1809, and *Saggi di critica*, composed in England.

[13] M. Delfico, *Nuove ricerche sul bello*, Naples, 1818, ch. 9.

1. Giuseppe Mazzini：朱塞佩·马奇尼（1805—1872），意大利政治家、记者、社会活动家，著有《论国家》《人的责任论集》《反人类战争》等。

[14] Malaspina, *Delle leggi del bello*, Milan, 1828, pp. 26, 233.

[15] P. Balestrieri, *Fondamenti di estetica*, Naples, 1847.

[16] Friedrich Bouterweck, *Ästhetik*, 1806, 1815 (3rd ed., Göttingen, 1824–1825).

[17] O. Colecchi, *Questions filosofiche*, vol. iii., Naples, 1843.

[18] P. Lichtenthal, *Estetica ossia dottrina del bello e delle arti belle*, Milan, 1831.

[19] *Elementi di filosofia* (5th ed., Naples, 1846), vol. ii. pp. 427–476.

[20] *Sistema filosofico*, by A. Rosmini-Serbati, Turin, 1886, § 210.

[21] Cf. *Nuovo saggio sopra l' orig. delle idee*, § v. part iv. ch. 5.

[22] *Sull' idillio e sulla nuova letteratura italiana* (*opuscoli filosofici*, vol. i.).

[23] V. Gioberti, *Del buono e del bello* (Florence ed., 1857).

[24] *Del bello*, ch. 1.

[25] *Op. cit.* ch. 7.

[26] *Op. cit.* ch. 7.

[27] *Del bello*, chs. 8–10.

[28] *Op. cit.* ch. 4.

[29] G. Berchet, *Opere*, ed. Cusani, Milan, 1863, p. 227.

[30] Words suppressed in ed. of 1870.

[31] *Epistolario*, ed. Sforza, i. pp. 285, 306, 308; *Discorso sul romanzo storico*, 1845; *Dell' invenzione*, dialogue.

[32] *Addizioni alle Miei Prigioni*, 1831 (in Pellico, *Prose*, Florence, 1858); see pp. about the *Conciliatore*.

[33] *Del bello e del sublime*, 1827; *Studî filosofici* (Venice, 1840), vol. ii. part v.

[34] Cf. De Sanctis, *Lett. Hal. nel s. XIX,* ed. Croce, Naples 1896, pp. 427–431.

XV FRANCESCO DE SANCTIS

F. de Sanctis[1]*: development of his thought.*

On the other hand, the autonomy of art found a strong supporter in Italy in the critical work of Francesco de Sanctis, who held private classes in literature at Naples from 1838 to 1848, taught at Turin and Zürich from 1852 to 1860 and in 1870 became professor in the University of Naples. He expressed his doctrines in critical essays, in monographs on Italian writers and in his classic *History of Italian Literature*. Receiving his first elements of old Italian culture in Puoti's[2] school, his natural bent towards speculation[3] led him to investigate grammatical and rhetorical doctrines with the view of reducing them to a system; but he soon began to criticize and to grow out of this phase. He pronounced Fortunio[4], Alunno[5], Accarisio[6] and Corso[7] "empirics"; he had a slightly better opinion of Bembo,

---

1. *F. de Sanctis*：弗朗西斯科·德·桑克蒂斯（1817—1883），意大利文学家、语文学家，著有《意大利文学史》《评论文集》《论皮特拉克》等。
2. Puoti：普奥迪（1782—1847），意大利文学家、词典编纂家、语法学家，著有《意大利语的基本规则》《论科学与文学之研究》《散文写作艺术的例证与理论》等。
3. natural bent towards speculation：推断的天赋
4. Fortunio：福尔图尼奥（1577—1657），意大利生理学家、哲学家，著有《语文学》《哲学》《医药哲学》等。
5. Alunno：阿鲁诺（1430—1502），意大利翁布里亚派画家。
6. Accarisio：阿卡里西奥（1497—1544），意大利政治家、哲学家、语言学家，著有《俗语的语法与正字法》《意大利语的灵魂》等。
7. Corso：科尔索（1446—1512），意大利文艺复兴时期画家。

Varchi[1], Castelvetro and Salviati[2], who introduced "method" into grammar, a process completed subsequently by Buonmattei[3], Corticelli[4] and Bartoli[5]; and he proclaimed Francisco Sanchez, author of the *Minerva*, "the Descartes of grammarians." From these his admiration spread to the French writers of the eighteenth century and the philosophical grammars of Du Marsais[6], Beauzée[7], Condillac and Gérard; following in their wake and pursuing the ideal of Leibniz, he conceived a "logical grammar"; in this effort, however, he soon began to recognize the impossibility of reducing the differences of languages to fixed logical principles. If he found the French theorists admirable in their ability to reconstitute the simple and primitive forms; from "I love" to "I am loving," something disquieted him; "Such decomposition of 'I love' into 'I am loving'" (said he) "deadens the word by depriving it of the movement proceeding from active will." [1] In the same way he read and criticized the writers of treatises on Rhetoric and Poetics from sixteenth-century men such as Castelvetro and Torquato Tasso (whom he dared to describe as an "indifferent critic," to the great scandal of Neapolitan men of letters) to Muratori and Gravina, "more acute than accurate"; and eighteenth-century Italians, Bettinelli, Algarotti[8]

---

1. Varchi：瓦尔奇（1502/1503—1565），意大利人文主义者、历史学家、诗人，著有《佛罗伦萨史》。
2. Salviati：萨尔维亚蒂（1520—1575），意大利文艺复兴后期画家。
3. Buonmattei：布翁马泰（1581—1648），意大利人文主义者、语法学家，著有《图斯卡纳语》（即意大利语）等。
4. Corticelli：科迪切利（生卒年不详），意大利语言学家。
5. Bartoli：巴尔托里（1503—1572），意大利语文学家、人文主义者，著有《论区域行政》《论雕塑》等。
6. Du Marsais：杜马塞（1676—1756），法国哲学家、语法学家，《百科全书》供稿人，著有《学习拉丁语的理性方法》《语法原理》《论修辞》等。
7. Beauzée：博泽（1717—1789），法国语言学家，著有《通用语法》《百科全书》供稿人。
8. Algarotti：阿尔贾罗蒂（1712—1764），意大利哲学家、诗人和艺术评论家，著有《与菲德烈大帝书信集》《论建筑》《论绘画》等。

and Cesarotti. Coldly rational rules found no favour with him: he urged the young to confront literary works boldly and freely absorb impressions, the only possible foundation for taste. [2]

Influence of Hegelism.

Philosophical study had not been abandoned and had not even fallen into entire decadence[1] in Southern Italy; in these days of renewed interest in philosophy the theories on Beauty from over the Alps and the new ideas of Gioberti and other Italians [3] aroused enthusiastic discussion. Vico was read again, and Bénard's[2] French translation of Hegel's Æsthetic appeared and was canvassed in Naples[3] volume by volume (the first in 1840, the second in 1843, and the rest between 1848 and 1852). In its desire for new intellectual food Italian youth set itself to learn German: De Sanctis himself had to translate the greater *Logic* of Hegel and Rosenkranz's *History of Literature* in the dungeon[4] of the Bourbon prison where he was incarcerated[5] on account of his liberal opinions. The new critical tendency was named "philosophism[6]" to distinguish it from the old grammatical criticism and from the vague, incoherent, exaggerated Romanticism. Philosophism attracted De Sanctis; to show how deeply he was imbued with[7] the Hegelian spirit a tale was told that, having devoured the first volumes of Bénard's translation, he guessed the contents of the remaining volumes and, before they could appear, was expounding

1. entire decadence: 全面衰退
2. Bénard：贝纳尔（1807—1889），法国哲学家、美学家、翻译家，译著有黑格尔《美学》。
3. canvassed in Naples：在那不勒斯风行一时
4. dungeon：地牢
5. incarcerated：关押
6. philosophism：哲学论
7. imbued with：倾注

them publicly in his classroom. [4]

His first writings show traces of metaphysical idealism and Hegelism; and they still linger here and there in the terminology of his later works. In a lecture prior to 1848 he placed the safety of criticism in the philosophic school which, in works of literature, fixed its eyes upon "that absolute part ... that uncertain idea which moves within the mind of great writers, till it appears abroad clothed in fine raiment only less beautiful than itself." [5] In a preface to Schiller's plays (1850) he wrote, "The Idea is not thought, nor is poetry reason in song, as a poet of our time is pleased to assert; the idea is at once necessity and freedom, reason and passion, and its perfect form in drama is action." [6] Elsewhere he calls attention to the death of faith and poetry, absorbed by the development of philosophy: a thesis, he remarked some years later, "imposed on our generation by Hegel with his omnipotent thought." [7] In 1856 he attempted a definition of humour as "an artistic form having for signification the destruction of limit, with consciousness of such destruction." [8] Not to dwell too long on other particulars, in the distinction to which De Sanctis always held firm throughout his critical work, that between Fancy and Imagination, the latter considered as the true and only faculty of poetry, arises undoubtedly from suggestions of Schelling and Hegel (*Einbildungskraft, Phantasie*); from the same philosophers come the phrases "prosaic content," "prosaic world," sometimes used by him.

Unconscious criticism of Hegelism.

For De Sanctis the Hegelian Æsthetic was but a lever wherewith to lift himself clear of the discussions and views of the old Italian schools. A fresh, clear spirit such as his could not escape the arbitrary shackles of grammarians and rhetoricians only to fall into those of metaphysicians, the torturers of art. He absorbed the vital part of Hegel's teaching and re-expressed the Hegelian theories in correct or somewhat attenuated interpretations; but he only maintained with hesitation, and in the end openly rebelled against, all that was artificial, formalistic and pedantic[1] in Hegel.

The following examples of such reductions and attenuations show how substantial and radical was the change he effected. "Faith has vanished and poetry is dead" (he wrote in 1856, echoing Hegel); "or it were better to say" (here is De Sanctis' own correction) "faith and poetry are immortal: what has disappeared is but one particular mode of their being. Today faith springs from conviction and poetry is the spark struck from meditation; they are not dead, they are transformed." [9] Certainly he distinguished between imagination and fancy; but for him imagination was never the mystic faculty of transcendental apperception[2], the intellectual intuition of German metaphysicians, but simply the poet's faculty of synthesis and creation, contrasting with fancy as the faculty of collecting particulars and materials in a somewhat mechanical fashion. [10] When students of Vico and Hegel understood and expounded their master's theories as emphasizing the importance of concepts in art, De Sanctis replied, "The concept does not exist in art, nature or history: the poet works unconsciously and sees no concept but only form, in which he is involved and well-nigh lost. If the philosopher, by means of abstraction, can extract the concept thence and contemplate it in all its purity, he acts in a way entirely contrary to that of art, nature and history." He warned his hearers not to misunderstand Vico, who, when he extracts concepts and exemplary types from the Homeric poems, is not writing as an art critic but as a historian of civilization: Achilles[3] is artistically

1. pedantic：学究气的
2. transcendental apperception：超验领悟
3. Achilles：阿喀琉斯。希腊神话中的英雄，海洋女神忒提斯（Thetis）和凡人英雄珀琉斯（Peleus）之子。出生后被其母亲握脚踵倒提着在冥河中浸过，因此除未浸到水的脚踵外，刀枪不入。

Achilles, not strength or any other abstraction. [11] Thus his polemic is directed in the first instance against misunderstanding what he called the true Hegelian thought, which was in fact usually a correction made upon Hegel more or less consciously by himself. He was able to boast in his latter years that even at the time when all Naples went wild over Hegel, "at the time when Hegel was master of the field," he had always "made certain reservations and refused to accept his apriorism[1], his triad or his formulæ." [12]

Criticisms of German Æsthetic.

De Sanctis also took up an independent attitude towards the other German æstheticians. The views of Wilhelm Schlegel, very advanced for the day in which they had been promulgated , seemed to him to have been already superseded. In 1856 he wrote that Schlegel strives to "transcend ordinary criticism, which leads a humdrum[2] existence among phraseology, versification and elocution, but loses its way and never comes face to face with art: whereas Schlegel throws himself headlong into the probable, the decorous and the moral; into everything save art." [13] Thrown by the hazards of life into German territory, he found himself at the Zürich Polytechnic[3], and found among his colleagues (only imagine such a thing!) Theodor Vischer. What opinion can he have formed of the ponderous Hegelian scholastic who emerged dusty and panting from the systematic labours so well known to us, and smiled disdainfully at the poetry and music of the decadent Italian race[4]? De Sanctis writes, "I went there with my opinions and my prejudices and ridiculed their ridicule. Richard

1. apriorism：先验论
2. humdrum：平凡的
3. Zürich Polytechnic：苏黎世学院，全称：苏黎世联合理工高等学院。
4. decadent Italian race：颓废的意大利民族

Wagner[1] seemed to me a corrupter of music, and nothing could be more inæsthetic than the Æsthetic of Vischer." [14] His desire to correct the distorted views of Vischer, Adolf Wagner[2], Valentin Schmidt[3] and other German critics and philosophers led him to undertake in 1858–59 a course of lectures before an international audience at Zürich upon Ariosto and Petrarch, the two Italian poets worst maltreated by these judges because hardest to reduce to philosophical allegory. He sketched a typical German critic and contrasted him with a French one, each with his own characteristic defects. "The Frenchman does not indulge in theories; he goes straight to the subject: his argument palpitates with warmth of impression and sagacity of observation[4]: he never leaves the concrete: he estimates the quality of the talent and the work, studying the man in order to understand the writer." He makes the mistake of substituting reflexion on the psychology of the author and history of his time for reflexion upon art. "Quite otherwise is your German: be a thing never so plain, he makes it his business to manipulate, distort and embroil[5]: he accumulates a mass of darkness from whose centre rays of dazzling light now and again shoot forth: truth is there at bottom, in grievous pangs of parturition[6]. Confronted with a work of art, he labours to fasten down and fix the quality which is most

1. Richard Wagner：理查德·瓦格纳（1813—1883），德国作曲家、戏剧导演、指挥家，对歌剧具有革命性影响，代表作有《漂泊的荷兰人》《尼伯龙根的指环》《特利斯坦与伊索尔德》等。
2. Adolf Wagner：阿道夫·瓦格纳（1835—1917），德国经济学家、政治家、学院派社会主义者、公共财政学者，著有《政治经济学基础》《学院派国民经济与社会主义》《财政学与国家社会主义》等。
3. Valentin Schmidt：华伦丁·施密特（1863—1927），区域史学家、外交家、档案学家，著有《论南波黑米亚反宗教改革的历史》《南波黑米亚的酿酒厂与酒业》等。
4. palpitates with warmth of impression and sagacity of observation：悸动着温暖的感受和敏锐的洞察力
5. distort and embroil：曲解和嘲讽
6. in grievous pangs of parturition：分娩的剧痛

evanescent and impalpable[1]. While nobody is more given to talk of life and the world of the living, nobody on earth takes more pains to decompose and disembody it in generalities: as consequence of this last process (last in appearance, that is to say; in reality preconceived and *a priori*), he is able to fit you the same boot on every foot and the same coat on every back." "The German school is dominated by metaphysic, the French by history." [15] About this time (1858) a Piedmontese[2] review published his exhaustive critical survey of the philosophy of Schopenhauer, [16] which was then beginning to attract disciples among his friends and companions in exile in Switzerland; the criticism provoked the philosopher himself to confess that "this Italian" had "absorbed him *in succum et sanguinem*[3]." [17] What value did De Sanctis attach to all Schopenhauer's subtleties concerning art? Having fully stated his doctrine of ideas, he contents himself with the merest reference to the third book "wherein is found an exaggerated theory of Æsthetic." [18]

Final rebellion against metaphysical Æsthetic.

This moderate resistance and opposition to the partisans of the concept and to the romantic Italian mystics and moralists (he directed criticisms equally against Manzoni, Mazzini, Tommaseo and Cantù[4] [19]) turned to open rebellion in one of his critical writings on Petrarch (1868) in which this false tendency is characterized with biting sarcasm[5]. "According to this school" (he says, meaning the

---

1. most evanescent and impalpable：转瞬即逝且难以把握的事物
2. Piedmontese：皮埃蒙特的。皮埃蒙特，位于意大利西北部，与法国和瑞士接壤，1814 年后成为意大利文艺复兴中心。
3. *in succum et sanguinem*：〈拉〉汁液和血液
4. Cantù：坎图（1804—1895），意大利历史学家，著有《论意大利文学》《意大利人的历史》等。
5. biting sarcasm：一针见血的讽刺

school of Hegel and Gioberti), "according to this school the real and living is art only in so far as it surpasses its form and reveals its concept or the pure idea. The beautiful is the manifestation of the idea. Art is the ideal, a particular idea. Under the gaze of the artist the body becomes subtilized[1] until it is nothing but the shadow of the soul, a beautiful veil. The world of poetry is peopled with phantasms; and the poet, eternal dreamer, with the eyes of one slightly intoxicated sees bodies float unsteadily around him and change their shapes. Nor do bodies merely become attenuated into forms and phantasms; these forms and phantasms themselves become free manifestations of every idea and every concept. The theory of the ideal has been driven to its last victorious limit, to the destruction of the very phantasms themselves, to concept as concept, form becoming a mere accessory." "Thus the vague, the undecided, the undulating, the vaporous, the celestial, the ærial, the veiled, the angelic, have now a high position among artistic forms: whilst criticism revels in the beautiful, the ideal, the infinite, genius, the concept, the idea, truth, the superintelligible, the supersensible, the being and the existent, and many more generalities cast into barbarous formulæ just like those of the scholastics from whose influence we had so much difficulty in escaping." All these things, instead of determining the character of art, do nothing; save illustrate the contrary of art: its feebleness and impotence, preventing it from slaying abstractions and laying hold of life. If beauty and the ideal have actually the meaning given them by these philosophers "the essence of art is neither the beautiful nor the ideal, but the living, the form; the ugly too belongs to art since ugliness lives also in nature; outside the domain of art lies nothing but the formless and the deformed. Thais in Malebolge[2] is more living

1. subtilized：变得隐约难辨
2. Thais in Malebolge：（但丁《神曲》中）地狱第八层受惩罚的泰依丝

and poetical than Beatrice[1], who is pure allegory representing abstract combinations. The Beautiful? Tell me of anything as beautiful as Iago, a form uprisen from the profundity of real life; so rich, so concrete; in every part, in each finest gradation, one of the most beautiful creations in the world of poetry." If in the course of "wrangling about[2] the idea or the concept or real, moral, or intellectual beauty, and confusing philosophical or moral truths with æsthetic" you choose to call "a great part of the poetic world ugly, granting it a permit merely that it may act as contrast, antagonist or foil to beauty, accepting Mephistopheles as a foil to Faust, or Iago as foil to Othello," you are imitating "those good folk who thought, *in illo tempore*[3], that the stars shone in the firmament in order to give light to this earth." [20]

De Sanctis' own theory

The æsthetic theory of De Sanctis himself arises entirely from the criticism of the highest manifestations of European æsthetic as known to him. Its nature is revealed by the contrast. "If you desire a statue in the vestibule of art[4]," says he, "let it be that of Form; gaze upon this, question this, begin with this. Before form is attained, that exists which existed before the creation: chaos. Chaos is no doubt a respectable thing, with a most interesting history: science has not yet uttered its last word about this pre-world of fermenting elements. Art also has its pre-world: art also has its geology, born but yesterday and as yet scarcely stretched, a science *sui generis*[5], which is neither Criticism nor Æsthetic. Æsthetic appears when form appears, in which this pre-world

1. Beatrice：贝雅特丽齐，但丁《神曲》中的圣女。
2. wrangling about：争论
3. *in illo tempore*：〈拉〉在当时
4. vestibule of art：艺术长廊
5. science *sui generis*：分类学（斜体为拉丁文）

is sunk, fused, forgotten and lost. Form is itself as the individual is himself; and no theory is so destructive to art as the continual harping upon the beautiful as manifestation, clothing, light, or veil of truth or the idea. The æsthetic world is not appearance, it is substance; to it indeed belongs everything substantial and living: its criterion, its *raison d'être*[1], lies nowhere save in this motto: I live." [21]

The concept of form.

For De Sanctis, form did not mean form "in the pedantic sense attached to it until the end of the eighteenth century," that is to say, that which first strikes a superficial observer, the words, the period, the sense, the individual image; [22] or form in the Herbartian sense, the metaphysical hypostatization[2] of the former. "Form is not *a Priori*, it is not something existing of itself and distinct from the content as though it were a kind of ornament or vesture or appearance or adjunct[3] of the content: it is generated by the content acting in the mind of the artist: such as the content is, such is the form." [23] Between form and content there is at the same time identity and diversity[4]. In a work of art the content, which had been lying in a chaotic state in the mind of the artist, appears "not as it was originally, but as it has become; the whole of it, with its own value, its own importance, its own natural beauty enriched, not weakened, by the process." Therefore content is essential for the production of concrete form; but the abstract quality of the content does not determine that of artistic form. " If the content, though beautiful and important, remain inoperative or lifeless or waste within the mind of the artist, if it have not sufficient

---

1. *raison d'être*：〈法〉存在的理由
2. hypostatization：实体化
3. adjunct：附属品
4. identity and diversity：同一性和差异性

generative power and reveal itself in the form as weak or false or vitiated, why trouble to sing its praises? In such cases the content may be important in itself, but as literature or art it is worthless. On the other hand the content may be immoral, absurd, false or frivolous: but if at certain times or in certain circumstances it has worked powerfully on in the brain of the artist, and taken form, such content is immortal. The gods of Homer are dead; the *Iliad* remains. Italy may die and, with her, every memory of Guelf and Ghibelline[1]; the *Divina Commedia* will remain. The content is subject to all the hazards of history; it is born and it dies; the form is immortal." [24] He held firmly to the independence of art, without which there can be no Æsthetic; but he objected to the exaggeration of the formula of art for art's sake in that it tended to the separation of the artist from life, to the mutilation of the content[2] and to the conversion of art into a proof of mere cleverness[3]. [25]

De Sanctis as art-critic.

For De Sanctis, the concept of form was identical with that of imagination, the faculty of expression or representation, artistic vision. So much must be said by any one anxious to express clearly the direction which his thought was taking. But De Sanctis himself never succeeded in defining his own theory with scientific exactitude[4]; and his æsthetic ideas remained the mere sketch of a system never properly interrelated and deduced. The speculative tendency shared his attention with many other lively interests, the desire to understand

1. Guelf and Ghibelline：盖尔夫派和吉贝林派，指 12 世纪至 15 世纪之间，意大利中北部城邦的教皇派和神圣罗马帝国皇帝派的政治争斗。
2. mutilation of the content：内容残缺不全
3. mere cleverness：纯粹的灵巧
4. scientific exactitude：科学的精确性

the concrete, to enjoy art and rewrite its actual history, to plunge into practical and political life; so that by turns he was professor, conspirator, journalist and statesman. "My mind inclines to the concrete," he was wont[1] to say. He philosophized just so much as was necessary to the acquisition of a point of view in problems of art, history and life; and, having procured light for his intellect, found his bearings, derived some satisfaction from the consciousness of his own activity, he plunged as quickly as possible into the particular and the determinate. To immense power of seizing the truth in the highest general principles was joined a no less intense abhorrence for the pale region of ideas in which the philosopher takes an almost ascetic delight[2]. As critic and historian of literature he is unrivalled. Those who have compared him with Lessing, Macaulay, Sainte-Beuve[3] or Taine[4] are making rhetorical comparisons.

Gustave Flaubert[5] wrote to George Sand[6]: "In your last letter you speak of criticism, and say you expect it soon to disappear. I think, on the contrary, that it is just appearing over the horizon. Criticism to-day is the exact opposite of what it was, but that is all. In the days of Laharpe[7] the critic was a grammarian; to-day he is a historian like Sainte-Beuve and Taine. When will he be an artist, a mere artist,

1. wont：倾向于
2. ascetic delight：苦行的快乐
3. Sainte-Beuve：圣佩韦（1804—1869），法国文学家、诗人、小说家，法兰西学院院士，著有《名媛肖像画》《星期一闲谈录》《圣佩韦散文选》等。
4. Taine：丹纳（1828—1893），法国评论家、美学家，著有《论批评与历史》《英国文学史》《艺术哲学》《论理智》《当代法国的起源》等。
5. Gustave Flaubert：福楼拜（1821—1880），法国小说艺术大师，对现代主义小说产生深远影响，著有《包法利夫人》《情感教育》《圣安东尼的诱惑》等。
6. George Sand：乔治·桑（1804—1876），本名奥萝尔·杜邦，法国女作家，代表作有《安蒂亚娜》《莱莉亚》《魔沼》等。
7. Laharpe：拉·阿尔普（1739—1803），法国作家、剧作家和文艺批评家，著有《美兰：不得已的誓言》《那不勒斯的珍妮》等。

but a real artist? Do you know a critic who interests himself wholeheartedly in the work itself? They analyse with the greatest delicacy the historical surroundings of the work and the causes which produced it: but the underlying poetry and its causes? the composition? the style? the author's own point of view? Never. Such a critic must have great imagination and a great goodness of heart; I mean an ever-ready faculty of enthusiasm; and then, taste; but this last is so rare, even among the best, that it is never mentioned nowadays." [26] Flaubert's ideal has been worthily reached by one critic only (that is to say, amongst critics who have given themselves to the interpretation of great writers and entire periods of literature) and that one is De Sanctis. [27] No literature of any country possesses so perfect a mirror as that possessed by Italy in the *History* and the other critical essays of Francesco de Sanctis.

De Sanctis as philosopher.

But the philosopher of art, the æsthetician in De Sanctis is less great than the critic and historian of literature. The critic is primary, the philosopher a mere accessory. The æsthetic observations scattered in aphorisms up and down his essays and monographs take various colours from various occasions, and are expressed in uncertain and often metaphorical language; this has led to his being accused of contradictions and inexactitudes which had no existence in his inmost thought and whose very appearance vanishes as soon as one takes into account the particular cases with which he was dealing. But form, forms, content, the living, the beautiful, natural beauty, ugliness, fancy, feeling, imagination, the real, the ideal, and all the other terms which he used with varying signification, demand a science both on which to rest and from which to derive. Meditation on these words stirs up doubts and problems on every side and reveals everywhere gaps and discontinuities. Compared with the few philosophical æstheticians, De

Sanctis seems wanting in analysis, in order and in system, and vague in his definitions. But these defects are outweighed by the contact he establishes between the reader and real concrete works of art, and by the feeling for truth which never leaves him. He has, too, the attraction possessed by those writers who lead one on to suspect and to divine new treasures in store beyond what they themselves reveal — living thought, which stimulates living men to pursue and prolong it.

[1] *Frammenti di scuola*, in *Nuovi saggi critici*, pp. 321–333; *La giovinezza di Fr. de S.* (autobiography), pp. 62, 101, 163–166 (works cited are those of De S. in stereotyped Naples ed. by Morano, 12 vols.).

[2] *La giovinezza di Fr. de S.* pp. 260–261, 315–316.

[3] *Saggi critici*, p. 534.

[4] De Meis, *Comm, di Fr. de S.* (in vol. *In Memoria*, Naples, 1884, p. 116).

[5] *Scritti vori*, ed. Croce, vol. ii. pp. 153–154.

[6] *Saggi critici*, p. 18.

[7] *Op. cit.* pp. 226–228; *Scritti varî,* ii. pp. 185–187; cf. vol. ii. p. 70.

[8] *Saggi critici*, ed. Imbriani, p. 91.

[9] *Saggi critici*, p. 228; cf. *Scritti varî*, vol. ii. p. 70.

[10] *Storia della letteratura*, i. pp. 66–67 *Saggi critici*, pp. 98–99; *Scritti varî*, vol. i. pp. 276–278, 384.

[11] *La giovinezza di Fr. de S.* pp. 279, 313–314, 321–324.

[12] *Scritti varî*, vol. ii. p. 83; cf. p. 274.

[13] *Op. cit.* vol. i. pp. 228–236.

[14] *Saggio sul Petrarca*, new ed. by B. Croce, p. 309 *seqq.*

[15] *Saggi critici*, pp. 361–363, 413–414; cf. as touching Klein, *Scritti varî*, vol. i. pp. 32–34.

[16] *Op. cit.*, *Schopenhauer e Leopardi*, pp. 246, 299.

[17] Schopenhauer, *Briefe*, ed. Grisebach, pp. 405–406; cf. pp. 381–

383, 403–404, 438–439.

[18] *Saggi critici*, p. 269, note.

[19] Cf. *Scritti varî*, i. pp. 39–45, and *Letterat. ital. nel sec. XIX*, lectures, ed. Croce, pp. 241–243, 427–432.

[20] *Saggio sut Petrarca*, introd. pp. 17–29.

[21] *Saggio sul Petrarca*, p. 29 *seqq*.

[22] *Scritti varî*, vol. i. pp. 276–277, 317.

[23] *Nuovi saggi critici*, pp. 239–240, note.

[24] *Nuovi saggi critici, loc. cit.*

[25] *Ibid.* and cf. *Saggio sul Petrarca*, p. 182; also *Scritti arî*, i. pp. 209–212, 226.

[26] *Lettres à George Sand*, Paris, 1884 (Letter of Feb. 2, 1869), p. 81.

[27] See above, p. 395, the judgement of De S. on French criticism.

XVI ÆSTHETIC OF THE EPIGONI[1]

Revival of Herbartian Æsthetic.

When the cry "Away with metaphysic!" was raised in Germany, and a furious reaction began against the kind of Walpurgis-night[2] to which the later Hegelians had reduced the life of science and history, the disciples of Herbart came to the front and seemed to ask, with an insinuating air: "What is all this? a rebellion against Idealism and Metaphysic? why, it is exactly what Herbart wished and undertook all by himself half a century ago! Here we stand, his legitimate descendants, and we offer you our services as allies. We shall not find it hard to agree. Our Metaphysic accords with the atomic theory, our Psychology with mechanism, and our Ethics and Æsthetic with hedonism." Herbart himself (had he not died in 1841) would most likely have spumed these disciples of his who pandered to popularity, cheapened metaphysics and gave naturalistic interpretations to his reals, his representations, his ideas, and all his highest conceptions.

With the school thus coming into fashion, the Herbartian Æsthetic too tried to put on flesh and acquire a pleasing plumpness so as not to cut too miserable a figure beside the well-nourished *corpora* of science launched upon the world by idealists. The feeding-up process was accomplished by Robert Zimmermann, professor of philosophy at

1. EPIGONI：后继者
2. Walpurgis-night：瓦普几司之夜。德国民间传说，四月三十日夜至五月一日晨，女巫和魔鬼相会狂欢。

Prague and later at Vienna, who, after years of laborious effort and an introductory sample in the shape of an ample history of Æsthetic (1858), at length produced his *General Æsthetic as Science of Form* in 1865. [1]

Robert Zimmermann.

This formalistic Æsthetic, born under bad auspices, is a curious example of servile fidelity[1] in externals combined with internal infidelity. Starting from unity, or rather from subordination of Ethics and Æsthetic to a general Æsthetic defined as "a science which treats of the modes by which any given content may acquire the right to arouse approval or disapproval" (thereby differing from Metaphysic, science of the real, and from Logic, science of right thinking), Zimmermann places such modes in form, that is to say, in the reciprocal relation of elements. A simple mathematical point in space, a simple impression of hearing or sight, a simple note, is in fact neither pleasing nor displeasing: music shows that the judgement of beauty or ugliness always depends on the relation between two notes at least. Now these relations, *i.e.* forms universally pleasing, cannot be empirically collected by induction; they must be developed by deduction. By the deductive method it can be demonstrated that the elements of an image, which in themselves are representations, may enter into relations either according to their force (quantity), or according to their nature (quality); whence we have two groups — æsthetic forms of quantity, and æsthetic forms of quality. According to the first, the strong (large) is pleasing in comparison with the weak (small), and these latter are displeasing when set beside the former; according to the other form, that pleases which is substantially

1. servile fidelity：奴性的忠诚

identical in quality (the harmonious), and that displeases which is on the whole diverse (the discordant).

But the substantial identity must not be pushed to the point of absolute identity, for in that case the harmony itself would cease to be. From harmonious form is deduced the pleasure of the characteristic or expression; for what is the characteristic but a relation of prevalent identity between the thing itself and its model? But while similarity prevailing in the distinction produces accord (*Einklang*), qualitative disharmony is as such disagreeable, and demands a resolution. (It is easy to detect the sleight of hand[1] with which Zimmermann first slips the characteristic into the relations of pure form, thereby entirely altering Herbart's original thought; and how, by a second trick, he here introduces into pure beauty the variations and modifications of the beautiful, by the help of the despised Hegelian dialectic.) If such resolution is effected by the skilful substitution of something other than the unpleasant image, we shall certainly have removed the cause of offence and established quietude[2] (not accord: *Eintracht, nicht Einklang*), but we shall have gained the mere form of correctness: it is better, then, to supersede this by means of the true image so as to reach the form of compensation (*Ausgleichung*); and, when the true image is also pleasing in itself, the final form of definitive compensation (*abschliessende Ausgleich*,) with which we exhaust the series of possible forms. And, in conclusion, what is Beauty? It is a conjunction of all these forms: a model (*Vorbild*) which has grandeur, plenitude, order, accord, correctness, definitive compensation; all this appears in a copy (*Nachbild*) in the form of the characteristic.

Putting on one side the artificial connexion Zimmermann makes between the sublime, the comic, the tragic, the ironic, the humorous and the æsthetic forms, notice must be taken (so that we may

1. sleight of hand：花招
2. quietude：平静

recognize into which of the seven heavens he is wafting us[1]) that these general æsthetic forms concern art equally with nature and morality, whose individual spheres are differentiated solely by the application of the general æsthetic forms to particular contents. These forms, applied to nature, give us natural beauty, the cosmos; applied to representation, beauty of wit (*Schöngeist*) or imagination; applied to feeling, the beautiful soul (*schöne Seele*) or taste; applied to the will, character or virtue. On one side, then, is natural beauty, on the other human beauty, in which (latter), on one hand, we have the beauty of representation, that is to say æsthetic fact in the strict sense (art); on the other, we have the beauty of will, or morality; and between the two, lastly, we have taste, common to Ethics and Æsthetic. Æsthetic in the narrow sense, as the theory of beautiful representation, determines the beauty of representations, divided into the three classes of the beauty of temporal and spatial connexion (figurative arts); the beauty of sensitive representation (music); and the beauty of thoughts (poetry). This tripartition of beauty into figurative, musical and poetical brings to a conclusion theoretical Æsthetic, the only section developed by Zimmermann.

Vischer versus Zimmermann.

Zimmermann's work was a polemic against the principal representative of Hegelian Æsthetic, Vischer, who had little difficulty in defending his own position and counter-attacking that of his assailant. He held Zimmermann up to ridicule, for example, in connexion with his view of symbolism. Zimmermann defined a symbol as the object "round which beautiful forms adhere." A painter depicts a fox simply for the sake of painting a part of animal nature. Nothing of the sort:

1. into which of the seven heavens he is wafting us：他想送我们去七重天的哪一重

this is a symbol, because the painter "makes use of lines and colours to express things other than lines and colours." "You think I'm a fox," says the animal in the picture, "but you make a great mistake: I'm a clothes-peg: I'm an appearance created by the painter with gradations of grey, white, yellow and red." Even easier was it to make game of Zimmermann's enthusiastic praises of the æsthetic quality of the sense of touch. It was a pity, the latter had written, that the pleasures of this sense were so difficult to attain; since "to touch the back of the Resting Hercules and the sinuous limbs of the Venus of Melos[1] or the Barberini Faun[2] would give to the hand a delight comparable only with that felt by the ear when listening to the majestic fugues of Bach[3] or the suave melodies of Mozart[4]." Vischer does not seem to be far wrong in declaring formalistic Æsthetic to be "a grotesque union of mysticism and mathematics." [2]

Hermann Lotze[5].

The works of Zimmermann seem to have given satisfaction to nobody save himself. Even Lotze, by no means an adversary of Herbartianism, blames him severely in his *History of Æsthetic in Germany* (1868) and other writings. Still, Lotze was unable to offer any better substitute for æsthetic formalism than of a variant of the old idealism. "Can any one persuade us," he wrote in criticism of the formalists, "that a spiritual discord expressed by a corresponding

1. Venus of Melos:《米洛斯的维纳斯》，即《断臂的维纳斯》，古希腊雕刻家阿历山德罗斯于公元前 150 年左右创作的大理石雕塑。
2. Barberini Faun：《巴尔贝里尼的农牧神》，画家阿戈斯蒂诺·卡拉奇于 1595 年创作的雕塑作品。
3. majestic fugues of Bach：巴赫强有力的赋格
4. suave melodies of Mozart：莫扎特的轻柔旋律
5. *Lotze*：洛策（1817—1881），德国哲学家、美学家、生理学家，深受费希特、谢林和黑格尔影响，著有《哲学的生物学原理的未来》《哲学体系》《德国美学史》等。

discord in external appearances may have a value equal to that of the harmonious expression of a harmonious content solely because, in both cases, the formal relation of accord is respected? Can any one persuade us that the human form is pleasing solely for its formal stereometric relations[1], irrespective of the spiritual life by which it is animated? In empirical reality the three domains of laws, facts and values invariably appear as divided; and although they are united in the Highest Good, in Goodness in itself, in the living Love of a Personal God, in the Ought which is the basis of Being, our reason is unable to attain or to know such union. Beauty alone can reveal it to us: it is in close connexion with the Good and the Holy and reproduces the rhythm of the divine ordinance and the moral government of the universe. Æsthetic fact is neither intuition nor concept; it is idea, which presents the essential of an object in the form of an end referred to the ultimate end. Art, like beauty, must include the world of values in the world of forms." [3] The war between the Æsthetic of content and that of form, having Zimmermann, Vischer and Lotze as protagonists[2], reached its culminating point between 1860 and 1870.

Efforts to reconcile Æsthetic of form and Æsthetic of content.

Several people were in favour of a reconciliation. But the reconciliations they offered were not the right one, which was at least glimpsed by a certain young Johann Schmidt, who in his thesis for doctorate observed (1875) that, with all respect for Zimmermann and Lotze, it seemed to him they were both wrong in confusing the various meanings of the word "beauty," and discussed such an absurdity as a beauty or ugliness of natural objects, that is to say, of things external to the spirit; that Lotze, following Hegel, added the second absurdity

1. stereometric relations：立体几何关系
2. protagonists：积极参与者

of an intuitive concept or conceptual intuition: lastly, that neither of them grasped the fact that the æsthetic problem does not turn upon the beauty or ugliness of the abstract content or of form understood as a system of mathematical relations, but with the beauty or ugliness of representation. Form undoubtedly must exist, but "concrete form, full of content." [4] These utterances of Schmidt met with a hostile reception: it is easy (he was told in reply) to identify beauty with artistic perfection, but the whole crux of the matter[1] lies in finding whether, beside this perfection, there exists another beauty dependent on a supreme cosmic or metaphysical principle: otherwise one is guilty of a naïve *petitio principii*[2]. [5] It was thought better, therefore, to seek other modes of reconciliation, which consisted in cooking up an appetizing dish in which a little formalism and a little contentism were mixed to taste, the latter as a rule giving the predominant flavour.

Some Herbartians were found in the ranks of the mediating or conciliatory party[3]. Hardly had Zimmermann's rigid formalism appeared, when Nahlowsky[4] jumped up to protest that it had never entered the master's head to exclude content from Æsthetic; [6] but even the ablest of the school, men such as Volkmann[5] and Lazarus, chose a middle course. [7] In the opposite camp Carrière[6], [8] and even Vischer himself (in a criticism of his own old *Æsthetic*), began to concede a larger part to the consideration of form; thus for Vischer beauty became "life appearing harmoniously," which when it appears in space is called form, and must always possess form, *i.e.* limitation

1. whole crux of the matter：关键问题
2. *petitio principii*：〈拉〉循环论证
3. mediating or conciliatory party：中间派或调和派
4. Nahlowsky：即约瑟夫·威廉·纳罗夫斯基（1812—1885），捷克裔奥地利心理学家，心理学家。
5. Volkmann：即阿尔弗雷德·威廉·福克曼（1801—1877）德国解剖学家，生理学家，哲学家，著有《动物解剖学》。
6. Carrière：即卡列尔·莫里茨（1817—1895），德国哲学家、历史学家，著有《美学》《改革时代的哲学世界观》。

(*Begrenzung*) in space and time, measure, regularity, symmetry, proportion, propriety (these characters constituting its quantitative moments) and harmony (qualitative moment), which includes variety and contrast and is therefore the most important characteristic. [9]

K. Köstlin[1].

A conciliatory Æsthetic in which formalism prevailed was attempted by Karl Köstlin, a professor at Tübingen[2] and formerly collaborator in the musical section of the works of Vischer. Köstlin [10] had been influenced by Schleiermacher, Hegel, Vischer and Herbart, but, truth to tell, does not seem to have perfectly understood the teaching of any one of his predecessors. According to him, the æsthetic object presented three requirements: richness and variety of imagery (*anregende Gestaltenfülle*), interesting content and beautiful form. Under the first we recognize, with no little difficulty, a distorted reflexion of Schleiermacher's "inspiration" (*Begeisterung*). Interesting content he defined as that which concerns man; that which he knows or does not know; that which he loves or hates (it is thus always relative to the individual and the conditions in which he exists); and he asserted that interest of content is joined to value of form, that is, he conceived content as a second value, the same of which we have heard Herbart speak. He also agreed with Herbart that form is absolute, and that its general character is determined as being easily perceptible by intuition (*anschaulich*), and by its power of giving satisfaction, pleasure and delight, in fact, as being beautiful. Its particular characteristics

---

1. *K. Köstlin*：卡尔·科斯特林（1819—1894），德国神学家、美学家、文学史学家，著有《美学绪论》《美学》等。
2. Tübingen：图宾根大学。位于德国巴登-符腾堡州的图宾根，公元 1477 年建立，是欧洲最古老的大学之一。

for Köstlin were, according to quantity, circumscription[1], simplicity (*Einheitlichkeit*), extensive and intensive size, and equilibrium[2] (*Gleichmass*); according to quality, determination (*Bestimmtheit*), unity (*Einheit*), importance (*Bedeutung*) extensive and intensive, and harmony. But when Köstlin sets himself to the empirical verification of his categories, he falls into hopeless confusion. Greatness is pleasing, but so is smallness; unity is pleasing, but so is variety; regularity is pleasing, but so, confound it, is irregularity: uncertainties and contradictions at every step; he was aware of them and made no effort to conceal them; but they should have convinced him that the abstraction of "beautiful form," whose qualities and quantities he had so laboriously collected, is a ghostly shape without body, since that alone gives æsthetic pleasure which fulfils an expressive function. But having illustrated the three demands of the æsthetic object, Köstlin wasted all his remaining breath in constructing a kingdom of intuitive imagination in the manner of Vischer, *i.e.* beauty of organic and inorganic nature; of civil life; of morality; of religion; of science; of games; of conversations; of feasts and banquets; and lastly of history, reviewing and passing æsthetic comment on its three periods, patriarchal[3], heroic and historical.

Æsthetic of content. M. Schasler[4].

Schasler, who had written as vast a history on Æsthetic as Zimmermann's own, found a starting-point for a movement toward formalism in absolute idealism, or realism-idealism, as he called it.

---

1. circumscription：界限
2. equilibrium：平衡
3. patriarchal：父系的
4. *M. Schasler*：沙斯勒（1819—1903），德国哲学家、语言学家、艺术评论家兼杂志编辑，著有《艺术体系》《讽刺的界限》等。

He began by defining Æsthetic as "the science of the beautiful and of art" (a single science ill defined as having two different objects), and proceeded to justify his unmethodical definition by saying that beauty does not exist in art alone, nor does art concern itself solely with beauty. The sphere of Æsthetic he defines as that of intuition (*Anschauung*) in which knowledge assumes a practical character and will a theoretical: the sphere of indivisible unity and absolute reconciliation of the theoretical and practical spirit, in which in a certain sense the highest human activities are developed. Beauty is the ideal, but the concrete ideal; this is why there is no ideal of a human body in abstraction from sex, no ideal of a mammal in general, but only of such and such species, as of horse or dog, and then only of determinate kind of horse or dog. Thus by descending from the more to the less abstract genus Schasler vainly attempted to reach the concrete, which inevitably escaped his grasp. In art we pass from the typical, which is natural beauty, to the characteristic, which is the typical of human feeling; hence we can frame the ideal of an old woman, a beggar or a ruffian[1]. The characteristic of art is in closer relationship to the ugly than to the beautiful in nature. On this head (passing over the remainder, which is on familiar lines) it is well to notice that Schasler has a bias towards that version of the romaunt[2] of Sir Purebeauty which ascribes the birth of the "modifications of Beauty" to the influence of the Ugly. [11] "Although," he writes, "the thought may disturb our minds, it must not be forgotten that were there no world of ugliness there could be no world of beauty; for it is only when the Ugly stirs up empty abstract Beauty, that it begins to combat the enemy and thus to produce concrete Beauty." [12] He even succeeded in converting Vischer himself, the chief supporter of the other version: "Formerly I had been accustomed to think in

1. ruffian：流氓
2. romaunt：骑士故事

the old-fashioned Hegelian style," Vischer confesses, "that unrest, fermentation and strife dwelt in the essence of Beauty; that the Idea prevails and thrusts the image forth into the infinite; so arises the Sublime; that the image, offended in its finitude, makes war on the Idea; whence arises the Comic; this finished the struggle; Beauty returned to itself from the conflict of the two moments, and was created." But now, he continues, "I must acknowledge that Schasler is right, and so are his predecessors Weisse and Ruge: the Ugly has a hand in the matter; this is the principle of movement, the ferment of differentiation: without such leaven we never reach the special forms of Beauty, for each single one presupposes' the Ugly." [13]

Eduard von Hartmann[1].

Closely allied to that of Schasler is the Æsthetic of Eduard von Hartmann (1890), preceded by a historical treatise on *German Æsthetic since Kant* [14] wherein with meticulous, critical and polemical study he upholds the definition of Beauty as "the appearance of the Idea" (*das Scheinen der Idee*). Inasmuch as he insisted on appearance (*Schein*) as the necessary characteristic of Beauty, Hartmann held himself justified in naming his Æsthetic the "Æsthetic of Concrete Idealism," and in ranging himself alongside Hegel, Trahndorff, Schleiermacher, Deutinger, Oersted, Vischer, Meising, Carrière and Schasler, against the abstract idealism of Schelling, Solger, Schopenhauer, Krause, Weisse and Lotze, all of whom, by placing beauty in the supersensible idea, overlooked the sensory element and reduced it to the rank of a mere accessory. [15] By his insistence on the idea as the other indispensable and determining element, Hartmann proclaimed

1. *Eduard von Hartmann*：爱德华·冯·哈特曼（1842—1906），德国哲学家，深受弗洛伊德及荣格心理学影响，著有《论辩证法》《不可知论的哲学》等。

himself as opposed to the Herbartian formalism. Beauty is truth; neither historical, scientific nor reflective, but metaphysical or idealistic, the very truth of Philosophy: "in proportion as Beauty is in opposition to every science and to realistic truth, so much nearer is it to Philosophy and metaphysical truth": "Beauty, with its own peculiar efficacy, remains the prophet of idealistic truth in an unbelieving age that abhors Metaphysic and recognizes no value in anything but realistic truth." Æsthetic truth, which leaps immediately from subjective appearance to ideal essence, is lacking in the control and method possessed by philosophical truth; in compensation, however, she possesses the fascinating power of conviction, the sole property of sensible intuition, and unattainable by gradual or reflected mediation. The higher Philosophy soars, the less does it need the gradual passage through the world of the senses and of science, and the slighter becomes the distance separating Philosophy and Art. The latter, for its part, will be well advised to start on its journey towards the ideal world as Bædeker's handbooks counsel the intending traveller, "with as little luggage as possible"; "not overloading herself with a weight which paralyses the wings and is made up of unnecessary and indifferent trifles," [16] Logical character, the microcosmic idea, the unconscious are immanent in beauty; by means of the unconscious, intellectual intuition operates in it, [17] and, from its being rooted in the unconscious, it is a Mystery. [18]

Hartmann and the theory of modifications.

In his employment of the exciting or reactionary influence of the Ugly, Hartmann exceeded Schasler himself. Lowest among the degrees of Beauty, indeed forming the lower limit of æsthetic fact, lies sensuous pleasure, which is unconscious formal beauty; its first true degree is formal beauty of the first order, or the mathematically pleasing (unity, variety, symmetry, proportion, the golden section, etc.); its second degree is formal beauty of the second order, the dynamically pleasing; its third is formal beauty of the third order, the passive teleological, as in the case of utensils or machinery. Indeed it may here be noted that among machines and utensils, on a level with jars, plates and cups, Hartmann placed language: it is a dead thing, said he; receiving the appearances of life (*Scheinleben*) [19] only at the very instant of utterance. Language a "dead thing," an "utensil" for the philosopher of the Unconscious, in the land of Humboldt, with a Steinthal still living! There follow, as formal beauty of the fourth order, the active teleological or living, and as formal beauty of the fifth order, conformity to species (*das Gattungsmässige*): lastly and above all, since the individual idea is superior to the specific, is beauty concrete beauty or the microcosmic individual, which is no longer formal, but beauty of content. As is to be expected, the passage from lower to Higher degrees is made by means of the Ugly: nobody has laboured like Hartmann to recount in detail the services rendered by Ugliness to Beauty. From ugliness, in the form of the destruction of the beauty of equality, arises symmetry: from ugliness in the case of the circle arises the ellipse; the beauty of a waterfall tumbling over rocks is caused by the mathematically ugly; destruction, that is to say, of a fall in a parabolic curve[1]; beauty of spiritual expression is achieved through the introduction of an ugliness relative to fleshly perfection. Beauty of a higher degree is founded on ugliness at a lower degree. When the highest degree is reached, that of individual beauty beyond which there can be nothing, even then elemental ugliness continues its work of beneficent irritation. The later phases thus produced are well known to us as the famous Modifications of the Beautiful: in this section also, nobody is so copious or detailed as Hartmann. He certainly does admit, side by side with simple or pure beauty, certain

1. parabolic curve：抛物线

modifications free from conflict, such as the sublime or graceful; but the more important modifications can arise only through conflict. There are four cases, because the resolution must be either immanent, logical, transcendent or combined: immanent in the idyllic, the melancholy, the sad, the cheerful, the moving, the elegiac; logical in the comic in all its varieties; transcendent in the tragic; combined in the humorous with the tragic-comic and its other varieties. When none of these resolutions is possible, there arises ugliness; when an ugliness of content is expressed by an ugliness of form, we have the maximum of ugliness, the real æsthetic devil.

Metaphysical Æsthetic in France: C. Levêque[1].

Hartmann is the last considerable representative of the old æsthetic school in Germany; he inspires terror by the mass of his literary production, like many others of the school, who seem to accept it as a dogma that art cannot be dealt with except in several volumes a thousand pages long. Those who are not afraid of giants and are able to attack this sort of Æsthetic, will find it a fat good-humoured Magog[2] full of vulgar prejudices, and so constituted that, despite his apparent strength, a little blow will kill him.

In other countries metaphysical Æsthetic had few followers. In France the celebrated competition of the Academy of Moral and Political Sciences in 1857 crowned with their approval and presented to the world the *Science of Beauty* by Levêque; [20] of which nobody now thinks or speaks, only remembering the author (who attitudinized as a disciple of Plato) by his eight characteristics of Beauty, derived by him from examination of a lily. The eight characteristics were as follows: —

---

1. *C. Levêque*：勒韦克（1818—1900），法国哲学家、作家，著有《美的科学》《艺术的精神主义》《希腊罗马哲学之研究》等。
2. Magog：玛各，《圣经》中的巨人族领袖。

sufficient size of form, unity, variety, harmony, proportion, normal vivacity of colour, grace and propriety; ultimately reducible to two, size and order. As supplementary proof of the truth of his theory, Levêque applied it to three beautiful things: a child playing with its mother, a symphony of Beethoven and the life of a philosopher (Socrates). Really, it is somewhat difficult (says one of his fellow-spiritualists, venturing to comment on this doctrine though speaking with the utmost deference) to imagine what may be the normal vivacity of colour in the life of a philosopher. [21] Translations and explanatory articles by Charles Bénard [22] and books by various writers belonging to French Switzerland (Töpffer[1], Pictet[2], Cherbuliez[3]) were not successful in popularizing the German systems of Æsthetic in France.

In England: J. Ruskin.

England showed even less disposition to interest herself, although John Ruskin[4] may have some claim to be considered a metaphysical æsthetician with a distinctive national stamp. But it is difficult to treat of Ruskin in a history of science, for his temperament was wholly opposed to the scientific. His disposition was that of the artist, impressionable, excitable, voluble, rich in feeling; a dogmatic tone and the appearance of theoretical form veil, in his exquisite and enthusiastic pages, a texture of dreams and fancies. The reader who recalls those pages will regard as irreverent any detailed and prosaic

1. Töpffer：特普菲尔（1799—1846），瑞士画家、小说家，著有《阿尔卑斯山漫游记》《曲折之路》等。
2. Pictet：皮克泰（1799—1875），瑞士语言学家、美学家，著有《论凯尔特语与梵语的相似性》《自然、艺术与诗歌中的美》等。
3. Cherbuliez：谢尔比列（1829—1899），瑞士作家、美学家，著有《博尔斯基历险记》《文学艺术的研究》等。
4. John Ruskin：约翰·罗斯金（1819—1900），英国维多利亚时代艺术评论家、画家、思想家和人类学家，著有《建筑的诗意》《现代画家》《绘画要素》等。

review of Ruskin's æsthetic thought, which must inevitably reveal its poverty and incoherence. Suffice it to say that, following a finalistic, mystical intuition of nature, he considered beauty as a revelation of divine intentions, the seal "God sets on his works, even upon the smallest." For him the faculty which perceives the beautiful is neither intellect nor sensibility, but a particular feeling which he names the theoretic faculty. Natural beauty, which reveals itself to a pure heart when contemplating any object untouched and unspoiled by the hand of man, asserts itself for this reason as immeasurably superior to any work of art. Ruskin was too hasty in analysis to understand the complicated psychological and æsthetic process which went on in his mind when he was moved to an artist's ecstasy by contemplating some humble natural object such as a bird's nest or a flowing rivulet. [23]

Æsthetic in Italy.

In Italy the Abate Fornasi[1] wrote a half-Hegelian, half-Catholic Æsthetic, wherein the beautiful is identified with the second person of the Trinity, the Word made man; [24] by this means he hoped to raise a bank of opposition against the liberal criticism of De Sanctis, whom he considered, from the sublime height of his own philosophy, as "a subtle grammarian." Combined Giobertian and German, especially Hegelian, influence produced several works of secondary importance; De Meis[2] developed at length the thesis of the death of Art in the historical world. [25] Somewhat later Gallo[3] also treated Æsthetic from

1. Fornasi：富尔纳里（1821—1900），意大利牧师、神学家、哲学家、作家，著有《论宇宙的和谐》《恐怖的艺术》等。
2. De Meis：德·梅伊斯（1817—1891），意大利哲学家、政治家、爱国主义者，著有《论难以治愈的医院》《论自然主义风气的学院》等。
3. Gallo：加洛（生卒年不详），意大利文学评论家，著有《唯心主义与文学》《文学的科学》等。

the Hegelian point of view, [26] and others repeated, nearly word for word, the doctrines of Schasler and Hartmann on the overcoming of the Ugly. [27]

Antonio Tari and his lectures.

The only genuine Italian teacher of metaphysical Æsthetic according to the Germans was Antonio Tari[1], who lectured on this very subject in Naples University from 1861 to 1884. He had a meticulous and superstitiously minute knowledge of everything that issued from German printing-presses, and was the author of an *Ideal Æsthetic* as well as essays on style, taste, serious work and play (*Spiel*), music and architecture, wherein he tried to keep the mean between the idealism of Hegel and the formalism of Herbart: [28] his lectures on Æsthetic attracted huge throngs and were one of the regular sights in the noisy, crowded Neapolitan university. Tari divided his treatment under three heads, Æsthesinomy, Æsthesigraphy and Æsthesipraxis[2], corresponding to the Metaphysic of the beautiful, to the doctrine of beauty in nature, and to that of beauty in art; like the German idealists, he defined the æsthetic sphere as intermediate between the theoretical and practical: he says emphatically that "in the world of spirit the temperate zone is equidistant from the glacial, peopled by the Esquimaux of thought, and from the torrid, peopled by the giants of action[3]." He pulled Beauty from her throne, substituting in

1. Antonio Tari：安东尼奥·塔里（1809—1884），意大利哲学家、作家、音乐评论家，著有《新意大利文学》《论美学与形而上学》等。
2. Æsthesinomy, Æsthesigraphy and Æsthesipraxis："美的形而上学"，"自然美说"和"艺术美说"
3. in the world of spirit the temperate zone is equidistant from the glacial, peopled by the Esquimaux of thought, and from the torrid, peopled by the giants of action 在精神世界，温带与寒带和热带是等距离的，寒带居住着抱着爱斯基摩想法的人，热带居住着行动的巨人。

her stead the Æsthetic, of which Beauty is but an initial moment, the simple "beginning of æsthetic life, eternal mortality, flower and fruit in one," whose successive moments are represented by the Sublime, the Comic, the Humorous, and the Dramatic.

Æsthesigraphy.

But the most attractive part of Tari's lectures was that devoted to Æsthesigraphy, subdivided into Cosmography, Physiography and Psychography, in the course of which he frequently quoted Vischer with great devotion; "the great Vischer" as he called him, in imitation of whom he constructed his own "æsthetic physics," brightening it with much varied erudition and enlivening it with quaint comparisons[1]. Is he speaking of beauty in inorganic nature — water, for example? He says in his fanciful manner, "When water ripples in the sunshine, in that act it has its smile; it has its frown in the breaking wave, its caprice in the fountain, its majestic fury in the foam." Is he speaking of geological configuration? "The vale, cradle perchance of the human race, is idyllic; the plain, monotonous but fat, is didactic." Of metals? "Gold is born great; iron, the apotheosis[2] of human toil, achieves greatness; the former boasts of its cradle when it does not bring it to dishonour; the latter causes it to be forgotten." He looked on vegetable life as a dream, repeating Herder's fine saying that the plant is "the new-born babe that hangs sucking upon the breast of mother nature." He divided vegetables into three types: foliaceous, ramified and umbelliferous[3]: "the foliaceous type," he says, "attains gigantic proportions in the tropics, where the queen of

1. quaint comparisons：奇特的比较
2. apotheosis：完美的典型
3. foliaceous , ramified and umbelliferous：多叶的，多枝的和伞状的

monocotyledons[1], the Palm-tree, represents despotism[2], the human scourge[3] of those desert regions. Of that solitary pinnacle, all crown, the negro may well be identified as the reptile that crawls round its base." Amongst flowers, the carnation is "symbol of betrayal, by reason of the variegation of its colours[4] and its deeply-dissected petals"; the celebrated comparison by Ariosto of a rose with a young girl is permissible only when the flower is still in bud, because "when it has unfolded its petals, disdaining the protection of thorns, displaying itself in all the pomp of its full colour, and boldly asking to be plucked by any hand, then it is woman, all woman, to call it by no harsher name, giving pleasure without feeling it, simulating love by its perfume and modesty by the crimson of its petals." He searches for and comments upon analogies between certain fruits and certain flowers; between the strawberry, for instance, and the violet; between the orange and the rose; he admired "the luxuriant spirals and the delicate architecture of a bunch of grapes": the mandarin-orange reminded him of the nobleman *qui s'est donné la peine de naître*[5]; the fig, on the contrary, was the great country bumpkin[6], "rough, rude, but profitable." In the animal kingdom, the spider symbolized primitive isolation; the bee, monasticism; the ant, republicanism. He noted, with Michelet[7], that the spider is a living paralogism[8]; it cannot feed itself without its web, and it cannot spin its web without feeding. Fish he condemns as un-æsthetic: "they are of stupid appearance with their wide-open eyes and incessant gaping, which makes them

---

1. monocotyledons：单子叶植物
2. despotism：暴政
3. scourge：祸害
4. variegation of its colours：色彩斑斓
5. *qui s'est donné la peine de naître*：〈法〉费劲地出生
6. bumpkin：土包子
7. Michelet：米什莱（1798—1874），法国历史学家，著有《罗马史》《法国史》等。
8. living paralogism：活生生的悖论

look voraciously gluttonous[1]." Not so with amphibians, for which he entertains a sympathy: the frog and the crocodile, "alpha and omega of the family[2], start from the comical, or even the scurrilous[3], and attain the sublimity of the horrid." Birds are especially æsthetic by nature, "possessing the three most genial attributes of a living being: love, song, and flight"; moreover, they present contrasts and antitheses: "opposite to the eagle, queen of the skies, stands the swan, the mild king of the marshes; the libertine vainglorious[4] cock has its contrast in the humble uxorious turtle-dove[5]; the magnificent peacock is balanced by the rude and rustic turkey." Amongst mammals, nature compensates for defects of pure beauty by dramatic value; if they cannot throw their song into the air, they have the rudiments of speech; if they have no variegated, myriad-hued plumage[6], they have dark, heavily-marked colouring, instinct with life; if they cannot fly, they have many other modes of powerful progression; and, the higher they go, the more do they attain individuality in appearance and life. "The epic of animal life is comedy in the donkey, *iniquae mentis asellus*[7]; idyl in the great wild beasts; downright tragedy in the Kaffir bull, that cloven-hoofed Codrus[8], who gives himself voluntarily to the lion in order to save the herd." As amongst birds, so amongst beasts attractive contrasts are to be made: — the lamb and the kid seem to typify Jesus and the devil; dog and cat, abnegation and egoism; hare and fox, the foolish simpleton and crafty villain. Many quaint and subtle observations does Tari let fall on human beauty and the relative

1. voraciously gluttonous：极度贪婪
2. alpha and omega of the family：这个种类的全部成员
3. scurrilous：丑陋
4. libertine vainglorious：放荡不羁自吹自擂的
5. uxorious turtle-dove：宠爱妻子的斑鸠
6. myriad-hued plumage：五光十色的羽毛
7. *iniquae mentis asellus*：〈拉〉失意的驴子
8. cloven-hoofed Codrus：偶蹄的科德鲁兹，传说中的古希腊君王，按神谕指引牺牲自我，挽救国家。

beauty of the sexes, allowing the female to have charm, not beauty: "bodily beauty is poise, and woman's body is so ill-poised that she falls easily when running; made for child-bearing, she has knock-kneed legs, adapted to support the large pelvis; her shoulders have a curve compensating the convexity of the chest." He describes the various parts of the body: "curly hair expresses physical force; straight hair, moral"; "blue, napoleonic[1] eyes have sometimes a depth like the sea; green eyes have a melancholy fascination; grey eyes are wanting in individuality; black eyes are the most intensely individual"; "a lovely mouth has been best described by Heine; two lips evenly matched; to lovers the mouth will rather seem a shell whose pearl is the kiss." [29]

How could we better take a smiling leave of metaphysical Æsthetic in the German manner than by recording this quaint vernacular version of it made by Tari, that kindly little old man, "the last jovial high-priest of an arbitrary and confused Æsthetic"? [30]

[1] *Allgemeine Ästhetik als Formwissenschaft,* Vienna, 1865; see also Meyer's *Konversations-Lexikon* (4th ed.), art. *Ästhetik*, by Zimmermann.

[2] *Kritische Gänge*, vi., Stuttgart, 1873, pp. 6, 21, 32.

[3] *Geschichte d. Ästh. i. Deutschl., passim*, esp. pp. 27, 97, 100, 125, 147, 232. 234, 265, 286, 293, 487; *Grundzüge der Ästh.* (posth., Leipzig, 1884), §§ 8–13; and two juvenile works, *Üb. d. Begriff d. Schönheit*, Göttingen, 1845, and *Üb. d. Bedingungen d. Kunstschönheit*, Göttingen, 1847.

[4] *Leibniz u. Baumgarten*, Halle, 1875, pp. 76–102.

[5] G. Neudecker, *Studien z. Gesch. d. dtschn. Ästh. s. Kant*, pp. 54–55.

[6] Polemic in *Zeitschr. f. exacte Philos.* (Herbartian organ) for 1862–

1. napoleonic：拿破仑式的

1863, ii. p. 309 *seqq.*, ii. p. 384 *seqq*, iv. pp. 26 *seqq.*, 199 *seqq.*, 300 *seqq.*

[7] Volkmann, *Lehrbuch der Psychologie*, 3rd ed., Cöthen, 1884–1885. Lazarus, *Das Leben der Seele*, 1856–1858.

[8] Moriz Carrière, *Ästhetik*, 1889 (3rd ed., Leipzig, 1885).

[9] *Kritische Gänge*, v., Stuttgart, 1866, p. 59.

[10] *Ästhetik*, Tübingen, 1869.

[11] See above, pp. 380–381.

[12] *Ästhetik*, Leipzig, 1886, i. pp. 1–16, 19–24, 70; ii. p. 52: cf. *Kritische Gesch. der Ästhetik*, pp. 795, 963, 1041–1044, 1028, 1036–1038.

[13] *Kritische Gänge*, v. pp. 112–115.

[14] *Die dtsche. Ästh. s. Kant*, 1886 (Part i. of *Ästh.*).

[15] *Philosophie des Schönen* (Part ii. of *Ästh.*), Leipzig, 1890, pp. 463–464; cf. *Deutsche Ästh. s. K.* pp. 357–362.

[16] *Phil. d. Sch.* pp. 434–437.

[17] *Op. cit.* pp. 115–116.

[18] *Op. cit.* pp. 197–198.

[19] *Op. cit.* pp. 150–152.

[20] Ch. Levêque, *La Science du beau*, Paris, 1862.

[21] E. Saisset, *L'Esthétique française* (in app. to vol. *L'Âme et la vie*, Paris, 1864), pp. 118–120.

[22] In *Revue philosophique*, vols. i. ii. x. xii. xvi.

[23] J. Ruskin, *Modern Painters* (4th ed., London, 1891); cf. De la Sizeranne, pp. 112–278.

[24] Vito Fornari, *Arte del dire*, Naples, 1866–1872; cf. vol. iv.

[25] A. C. De Meis, *Dopo la laurea*, Bologna, 1868–1869.

[26] Nic. Gallo, *L'idealismo e la letteratura*, Rome, 1880; *La scienza dell' arte*, Turin, 1887.

[27] *E.g.* F. Masci, *Psicologia del comico*, Naples, 1888.

[28] *Estetica ideale*, Naples, 1863; *Saggi di critica* (collected posthumously), Trani, 1886.

[29] A. Tari, *Lezioni di estetica generale*, collected by C. Scamaccia-Luvara, Naples, 1884; *Elementi di estetica*, compiled by G. Tommasuolo, Naples, 1885.

[30] V. Pica, *L'Arte dell' Estremo Oriente*, Turin, 1894, p. 13.

XVII ÆSTHETIC POSITIVISM AND NATURALISM

Positivism and Evolutionism.

The ground lost by idealistic metaphysic was conquered in the latter half of the nineteenth century by positivistic and evolutionary metaphysic, a confused substitution of natural for philosophical sciences, and a hotch-potch[1] of materialistic and idealistic, mechanical and theological theories, the whole crowned with scepticism and agnosticism. Characteristic of this trend of opinion was its contempt of history, especially the history of philosophy; which prevented its ever making that contact with the unbroken and age-long efforts of thinkers without which it is idle to hope for fertile work and true progress.

Æsthetic of H. Spencer[2].

Spencer (the greatest positivist of his day), whilst discussing Æsthetic, actually did not know that he was dealing with problems for all, or almost all, of which solutions had been already proposed and discussed. At the beginning of his essay on the *Philosophy of*

1. hotch-potch：混淆
2. *H. Spencer*：爱德华·斯宾塞（1831—1915），英国实证主义哲学家、思想家和历史学家。

Style[1], he remarks innocently: "I believe nobody has ever sketched a general theory of the art of writing" (in 1852); and in his *Principles of Psychology*[2] (1855), touching the æsthetic feelings he remarks that he has some recollection of observations concerning the relation of art and play made "by some German author whose name I cannot recall" (Schiller!). Had his pages on Æsthetic been written in the seventeenth century, they would have won a low position amongst the early crude attempts at æsthetic speculation; in the nineteenth century, one knows not how to judge them. In his essay on *The Useful and the Beautiful*[3](1852–1854), he shows how the useful becomes beautiful when it ceases to be useful, illustrating this by a ruined castle useless for the purposes of modern life, but a suitable scene for picnic parties and a good subject for a picture to hang on a parlour wall; which leads him to identify the principle of evolution from the useful to the beautiful as contrast. In another essay on the *Beauty of the Human Face*[4] (1852) he explains this beauty as a sign and effect of moral goodness; in that on *Grace* (1852) he considers the sentiment of the graceful as sympathy for power in conjunction with agility[5]. In the *Origin of Architectural Styles*[6](1852–1854) he discovers the beauty of architecture as consisting in uniformity and symmetry, an idea which is aroused in a man looking at the bodily equilibrium of the higher animals or, as in Gothic architecture, by analogy with the vegetable kingdom; in his essay on *Style*, he places the cause of stylistic beauty in economy of effort; in his *Origin and Function of Music* (1857) he theorizes on music as the natural language of the passions, adapted to increase sympathy between men. [1] In his *Principles of Psychology*,

---

1. 2. 3. 4. 6. *Philosophy of Style, Principles of Psychology, The Useful and the Beautiful, Beauty of the Human Face, Origin of Architectural Styles*：《风格的哲学》《心理学原理》《论实用与美》《人类面孔之美》，《建筑风格的起源》，斯宾塞著。
5. power in conjunction with agility：伴随着灵巧的力量

he maintains that the æsthetic feelings arise from the overflow of exuberant energy in the organism, and distinguishes various degrees of them, from simple sensation to that accompanied by representative elements, and so on until perception is reached, with more complex elements of representation, then emotion, and, last of all, that state of consciousness which transcends sensation and perception. The most perfect form of æsthetic feeling is attained by the coincidence of the three orders of pleasures, a coincidence produced by the full action of their respective faculties with the least possible subtraction due to the painful effect of excessive activity. But it is very rarely that we experience æsthetic excitement of this kind and strength; almost all works of art are imperfect because they contain a mixture of artistic with anti-artistic effects; now the technique is unsatisfactory, now the emotion is of a low order. These works of art which are universally admired, are found when measured by this criterion to deserve a lower place than that accorded them by popular taste. "Beginning with the Greek epic and the representations of analogous legends given by their sculptors, tending to excite egoistic or ego-altruistic[1] sentiments, and passing through the literature of the Middle Ages, equally impregnated with inferior sentiments, then through the works of the old masters, whose ideas and sentiments seldom compensate for the displeasing effect they inflict on our senses overrefined in study of appearances; and coming at last to the vaunted works of modern art, excellent for technical execution in many cases but deplorable for the emotions they arouse and express, such as Gérôme's[2] battle-pieces, alternately sensual and sanguinary[3]; — they are all far off indeed from the qualities deemed desirable, from the artistic forms corresponding to the highest forms

1. ego-altruistic：利己利他的
2. Gérôme：热罗姆（1824—1904），法国画家、雕塑家，创作主题包括希腊神话、历史绘画、东方主义，其作品风格被称为学院派。
3. alternately sensual and sanguinary：时而充满肉欲，时而充满暴力

of æsthetic feeling." [2] These last critical denunciations[1], like the theories noticed above, are mere substitutions of one word for another; "facility" for "grace"; "economy" for "beauty," and so on. Indeed, when one tries to define the exact philosophical position of Spencer, one can only possibly say that he wavers between sensationalism and moralism, and is never for a moment conscious of art as art.

Physiologists of Æsthetic: Grant Allen[2], Helmholtz[3] and others.

The same oscillation is noticeable in other English writers such as Sully[4] and Bain[5], in whom, however, we find more familiarity with works of art. [3] In his numerous essays and in *Physiological Æsthetics* (1877), Grant Allen collected a great many records of physiological experiments, all of which may be of supreme value to physiology, for aught we know to the contrary, but most assuredly are worthless from the point of view of Æsthetic. He keeps to the distinction between necessary or vital activity and the superfluous or that of play, and defines æsthetic pleasure as "the subjective concomitant[6] of the normal sum of activity, not connected directly with the vital functions, in the terminal peripheric organs[7] of the cerebrospinal nervous system[8] [4] ."

1. critical denunciations：尖酸刻薄的指责
2. *Grant Allen*：格兰特·艾伦（1848—1899），加拿大作家、小说家、心理学家、美学家，著有《心理美学》《色觉的起源与发展》《巴比伦》《魔鬼之死》等。
3. *Helmholtz*：赫姆霍兹（1821—1894），德国生理学家、物理学家、哲学家，创立能量守恒的数学公式，著有《论显微镜光学能力的局限》《论能量守恒》《论心理光学》等。
4. Sully：即詹姆斯·萨利（1842—1923），英国心理学家，主张联想主义心理学，著有《感觉与直觉》《悲观主义》《心理学大纲》等。
5. Bain：即亚历山大·贝恩（1818—1903），英国经验主义哲学家、教育家、心理学家、语言学家，著有《感觉与智力》《情绪与意志》《教育学》等。
6. subjective concomitant：主观伴随
7. peripheric organs：外围器官
8. cerebrospinal nervous system：脑髓神经系统

Physiological processes considered as causes of pleasure in art are presented under other aspects by later investigators, who assert that such pleasure arises not only "from the activity of the visual organs and the muscular systems associated with them, but also from the participation of some of the more important functions of the organism, as for instance breathing, circulation of the blood, equilibrium and internal muscular accommodation." Art, then, indubitably originated in "a prehistoric man who was habitually a deep-breather, having no call to rearrange his natural habits when scratching lines on bones or in mud and taking pains to draw them regularly spaced." [5] Physical-Æsthetic researches were pursued in Germany by Helmholtz, Brücke[1] and Stumpf[2], [6] who generally confined themselves to the narrower field of optics and acoustics, giving descriptions of the physical processes of artistic technique and the conditions to which pleasurable visual and auditive impressions must conform, without claiming to merge Æsthetic in Physics, but even pointing out the divergences between them. Degenerate Herbartians[3] hastened to disguise in physiological terms the metaphysical forms and relations of which their master had spoken, and to coquet with[4] the hedonism of the naturalists.

Method of the natural sciences in Æsthetic.

The superstitious cult of natural sciences was often accompanied (as

1. Brücke：即恩斯特·布吕克（1819—1892），德国生理学家，维也纳大学生理学与显微镜解剖学教授，著有《新德国高等诗艺中的生理学基础》《论植物生理》《人类身形之美与缺陷》等。
2. Stumpf：即卡尔·斯通弗（1848—1936），德国心理学家、哲学家、乐理学家，著有《声音心理学》《论空间认知的心理学起源》《哲学的再生》等。
3. Herbartians：赫尔巴特信徒
4. coquet with：献殷勤

is frequently the fate of superstition) by a sort of hypocrisy[1]. Chemical, physical and physiological laboratories became Sybilline grottoes[2], resounding with the questions of credulous[3] inquirers concerning the profoundest problems of the human spirit; and many of those who were really conducting their inquiries on inherently philosophic principles pretended or deluded themselves into believing that they followed the Method of Natural Science. A proof of this illusion or pretence is Hippolyte Taine's *Philosophy of Art*. [7]

H. Taine's Æsthetic.

"If by studying the art of various peoples and various epochs," says Taine, "we could define the nature and establish the conditions of the existence of each art, we should have arrived at a complete explanation of the fine arts and of art in general, *i.e.* at what is called an Æsthetic." A historical Æsthetic, not a dogmatic, which fixes characters and indicates laws "like Botany, and studies with equal attention orange and ivy, pine and birch; indeed it is a sort of botanical science applied to the works of man instead of to plants"; an Æsthetic which shall follow "the general movement which tends daily more and more to join the moral to the natural sciences and by extending to the former the principles, the safeguards and the rules of the latter, enables both to attain the same security and maintain the same progress." [8] The naturalistic prelude is followed by definitions and doctrines indistinguishable from those offered by philosophers whose infallibility[4] is not guaranteed by scientific methods, indeed, from those of the wildest of such philosophers. For, says Taine, art

1. hypocrisy：虚伪
2. Sybilline grottoes：先知的岩洞
3. credulous：轻信的
4. infallibility：可靠程度

is imitation, an imitation so carried out as to render sensible the essential character of objects; the essential character being "a quality from which all other qualities, or many others, are derived and follow unalterably from it." The essential character of a lion, for example, is to be "a great carnivore[1]"; this determines the formation of all its limbs; the essential character of Holland is to be "a country formed by alluvial soil[2]." This is why art is not restricted to objects existing in reality, but is able, as in architecture or in music, to represent essential characters without natural objects to correspond. [9]

Taine's metaphysic and moralism.

Now, in what do these essential characters, this carnivorosity and this alluviality differ, save perhaps in extravagance of example, from the "types" and "ideas" which *intellectualiste* or metaphysical Æsthetic had always considered as the proper content of art? Taine himself clears away every doubt in the matter by explicitly stating that "this character is what philosophers call the 'essence of things,' in virtue of which they affirm that the aim and end of art is to make manifest the essence of things"; he adds that, for his part, he "refuses to make use of the word 'essence' as being a technical term": [10] of the word itself, maybe; not of the concept for which it stands. There are two ways (says Taine, for all the world as though he were a Schelling) leading to the higher life of man, to contemplation: the way of science and the way of art: "the former investigates the causes and fundamental laws of reality, and expresses them in exact formulæ and abstract terms: the latter makes manifest these causes and laws, not in dry definitions inaccessible to the vulgar, and intelligible only to the select few, but in a sensible manner, appealing not merely to the reason but to the heart

1. carnivore：食肉动物
2. alluvial soil：淤积土

and senses of the most commonplace man; it has the power of being both elevated and popular, of manifesting what is most noble and elevated, and of manifesting it to every one." [11]

For Taine, as for the Hegelian æstheticians, works of art are arranged in a scale of values; so that, having begun by condemning as absurd every judgement of taste (every one to his taste [12]), he ends by asserting that "personal taste has no value whatever," and that some common measure should be abstracted and set up as a standard of progress and retrogression, ornamentation and degeneracy; a standard by which to approve and disapprove, praise and blame. [13] The scale of values set up by him is twofold or threefold, in the first instance it turns on the degree of importance of the character, *i.e.* the greater or less generality in idea, and the degree of beneficent effect (*degré de bienfaisance*), *i.e.* the greater or less moral value of the representation (two grades which are aspects of one single quality, viz. power, considered first for its own sake and then in its connexion with others): in the second instance upon the degree of convergence of effects, *i.e.* the fulness of expression, the harmony between idea and form. [14] This intellectualistic, moralistic, rhetorical doctrine is interrupted now and then by the usual naturalistic protests: "We shall, according to our custom, study this question in the manner of the natural scientist; that is to say methodically, by analysis; hoping to raise not merely a song of praise, but a code of laws," etc.; [15] as though that sufficed to alter the substance of the method adopted and the doctrine expounded. Taine finally gave himself over to dialectical treatments and solutions, and asserted that in the primitive period of Italian art, in the pictures of Giotto, we have soul without body (thesis); under the Renaissance, in Verrocchio's[1] pictures, body without soul (antithesis); in the sixteenth century, in Raphæl, there is harmony of expression and anatomy, soul

1. Verrocchio：韦罗基奥（1435—1488），意大利佛罗伦萨派雕塑家、画家，达·芬奇的老师。

and body (synthesis). [16]

G. T. Fechner[1]*: Inductive Æsthetic.*

The same protests and similar methods are to be found in the works of Gustav Theodor Fechner. In his *Introduction to Æsthetic* (1876), Fechner claims to "abandon the attempt at conceptual determination of the objective essence of beauty," since he desires to compose not a metaphysical Æsthetic from above (*von oben*), but an inductive Æsthetic from below (*von unten*) and to achieve clearness, not sublimity; metaphysical Æsthetic should bear the same relation to inductive, as the Philosophy of Nature to Physics. [17] Proceeding on inductive lines, he discovers a long series of æsthetic laws or principles: the æsthetic threshold; assistance or increment; unity in variety; absence of contradictions; clarity; association; contrast; consequence; conciliation; the correct mean; economic use; persistency; change; measure; and so on without end. This chaos of concepts he expounds with a chapter apiece, pleased and proud to show himself so highly scientific and so wholly inconclusive.

Experiments.

Next he describes the experiments he can recommend to his readers. They are of this type. Take ten rectangular pieces of white cardboard of fairly equal area (say ten square inches), but with sides variously proportioned from a ratio of 1 : 1 to one of 2 : 5, including the ratio of the golden section, 21 : 34; mix all these together on a black table and collect persons of every kind and character, but all belonging to

1. *G. T. Fechner*：古斯塔夫·提奥多·费希纳（1801—1887），德国心理学家、物理学家、自然哲学家，著有《普通有机体理论导论》《心理物理学的要素》和文中提到的《美学导论》（*Introduction to Æsthetic*）。

the educated classes, and applying the method of choice ask these people first to free their minds of all questions as to a particular use and then to pick out the pieces of cardboard which give them the highest sensation of pleasure and those which inspire them with the strongest feelings of disgust; the answers to be most carefully noted, keeping male and female subjects apart, and tabulated. Then see what follows. Fechner admits that the chosen cardboard-pickers often made reservations when questioned by himself, not knowing (very naturally) how to tell whether they liked a shape or disliked it without referring it to a definite use; sometimes they refused point-blank to make any selection at all; and they almost always seemed vague and perplexed in mind and generally, when submitted to a second test, answered in a way totally different from the first. Still, we all know that errors cancel out; and anyhow the tabulations showed that the highest sensations of delight were aroused not by the square, but by rectangular forms most nearly approaching the square, an enthusiastic rush being made for the proportion 21 : 34. [18] This method of selection received an extraordinarily felicitous definition; it was known as "an average of arbitrary judgements by an arbitrary number of persons arbitrarily selected." [19] Fechner also informs us (always in tabular form) of the result of a statistical inquiry of his own, by means of countless heaps of catalogues and gallery-guides, as to the dimensions and shapes of pictures in relation to the subjects they depict. [20]

Trivial nature of his ideas on Beauty and Art.

Nevertheless, when he tries to tell us what beauty is, he falls back on using — whether well or ill — the old speculative method, which he prefaces with the remark that for him the concept of beauty is "merely an expedient in conformity with linguistic usage for indicating briefly the link which unites the prevailing conditions of immediate pleasure." [21] He distinguishes three meanings of the word "beauty":

first, in a broad sense, the pleasing in general: secondly, in a narrow sense, a higher pleasure, but still sensuous: thirdly, in the narrowest sense, true beauty, which "not only pleases, but has the right of pleasing, possesses value in pleasing"; in it are united the concepts of beauty (the pleasing) and of goodness. [22] Beauty, in fact, is that which must please objectively and as such it corresponds with the good of action. "The Good," says Fechner, "is like a serious man, the capable organiser of his whole domestic life, sagaciously weighing[1] the present and future, setting himself to extract the greatest benefit from both. Beauty is his florid spouse, careful of the present and mindful of her husband's wishes. The Pleasing is the baby, all senses and play: the Useful is the servant who puts his hands at his master's disposal and is given bread solely in accordance with his deserts. Truth, lastly, is the preacher and teacher to the household; preacher in matters of faith, teacher in those of learning: he gives an eye to the Good and a helping hand to the Useful, and holds up a looking-glass to Beauty." [23] When speaking of art, he sums up all essential laws or rules into the following: (1) art chooses a valuable or, at any rate, an interesting, idea for representation; (2) it expresses the idea in sensible material in the manner most suitable to its contents; (3) from amongst the various means at its disposal, it selects those which in themselves are more pleasing than the others; (4) the same procedure is observed in all particulars; (5) in the event of conflict between these rules, one is made to give way to another in such a way that the greatest possible pleasure and that of highest value is attained (*das grösstmögliche und werthvollste Gefallen*). [24] But why should Fechner, who had this eudemonistic theory[2] of beauty and art (as he calls it) all ready made in advance, [25] take the trouble to enumerate principles and laws and conduct experiments

1. sagaciously weighing：敏锐地权衡利弊
2. eudemonistic theory：幸福论，主张通过由理性支配的积极生活而获得幸福。

and tabulate statistics wholly incapable of illustrating or proving it? One is tempted to believe that these pseudo-scientific operations were to him, and still are to his followers, a pastime or hobby neither more nor less important than playing Patience or collecting stamps.

Ernst Grosse[1]*: speculative Æsthetic and the Science of Art.*

Another example of the superstitious cult of the natural sciences is to be found in Professor Ernst Grosse's *Origins of Art.* [26] Contemner[2] of all philosophical research into art, which he dismisses under the title of "Speculative Æsthetic," Grosse invokes a Science of art (*Kunst wissenschaft*) whose mission is to dig out all the laws lying hidden in the mass of historical facts collected to date. It is his opinion that all ethnographic and prehistoric material should be united to historical matter proper, there being no possibility, according to him, of framing general laws when study is restricted to the art of cultured peoples "just as a theory of generation must necessarily be imperfect if founded exclusively on the form of that function predominant among mammals." [27] But immediately after his declaration of abhorrence for philosophy, and of faith in scientific methods, Grosse finds himself in the same difficulty as Taine and Fechner. Indeed, there is no escape; in order to examine the artistic productions of primitive and savage peoples, a start must be made from some sort of concept of art. All the scientific metaphors, all the verbal emollients[3] employed by Grosse cannot hide the nature of the plan he is forced to adopt, or its striking resemblance to the despised speculative

1. *Ernst Grosse*：恩斯特·格罗塞（1862—1927），德国文学家、伦理学家、艺术收藏家，著有《论文学的目的与方法》《古希腊与古罗马诗人的自然观》和文中提到的《艺术的起源》（*Origins of Art*）。
2. Contemner：蔑视者
3. verbal emollients：柔和的言辞

Æsthetic. "As a traveller who desires to explore an unknown land must provide himself with a general outline of the country and have some knowledge of the direction in which his path should lie, if he does not wish to lose his way entirely; so we, before beginning our enquiry, need a general preliminary orientation concerning the essence of the phenomena (*über das Wesen der Erscheinungen*) about to engage our attention." Most certainly "we may count upon having an exact and exhaustive answer, at earliest, when our enquiry is finished; and it is not yet begun. That characteristic which we seek to determine at the outset ... may be most radically modified by the time we reach the end:" there is no question, fie on[1] the suggestion! of imitating the old æstheticians: the only question is how "to give a definition which may serve as provisional scaffolding, to be broken away on completion of the edifice." [28] Words, words, words: the mite[2] of general ideas and artistic laws to be found in his book has been quarried by Grosse not from study of the reports brought back by travellers in savage lands, but from speculation on the forms of the spirit; and (inevitably) his interpretation of the former is reached by the light thrown on it by the latter. In his final definition, Grosse concludes by considering art as an activity which in its development or as its result, possesses immediate feeling-value (*Gefühlswerth*), and is an end to itself; practical and æsthetic activity are in direct mutual opposition between which as a middle term lies the activity of play, which like the practical activity has its end outside itself, but, like the æsthetic, finds its enjoyment not in its external end, which is more or less insignificant, but in its own activity. [29] At the end of his book he remarks that the artistic activity of primitive peoples is hardly ever unaccompanied by the practical; and that art began by being social and became individual only in civilized times. [30]

---

1. fie on：（表示不赞成）呸！
2. mite：少量的

The Æsthetics of Taine and Grosse have also been described by the epithet sociological.

Sociological Æsthetic.

But since no one knows what the science of Sociology is, we must deal with the sociological superstition as we dealt with the naturalistic; that is to say, by skipping the preface with its proposals that can never be carried out, and seeing what it is that the objective necessities of the case have forced the author to assert, and which of the possible alternative views he accepts, or between what selection of them his allegiance wavers[1]. During this examination we shall ignore the fairly common case of an author who while pretending to construct an Æsthetic simply compiles a list of facts connected with the history of art or civilization.

Proudhon[2].

Some social reformers of our day, like Proudhon, have revived the condemnations of Plato, or the mitigated moralism of antiquity and the Middle Ages. Proudhon denied the formula Art for Art's sake; he looked on art as a mere purveyor of sensuous pleasure, something which must be subordinated to legal and economical ends; poetry, sculpture, painting, music, romance, history, comedy, tragedy had for him no aim save exhortation to virtue and dissuasion from vice.[3] [31]

---

1. allegiance wavers：忠诚动摇了
2. *Proudhon*：蒲鲁东（1809—1865），法国政论家、经济学家，无政府主义创始人之一，著有《哲学的贫困》《社会问题的解决方案》《财产论》等。
3. exhortation to virtue and dissuasion from vice：劝人弃恶从善

J. M. Guyau[1].

Development of social sympathy is the whole duty of art in the estimation of J. M. Guyau, who became famous as the founder of Social Æsthetic and was, according to certain French critics, inaugurator of the third epoch in the history of Æsthetic, the first being the æsthetic of the ideal (Plato), the second that of perception (Kant), and the third that of "Social Sympathy" (Guyau). In his *Problems of Contemporary Æsthetic* (1884) Guyau combats the theory of play, and substitutes that of Life; in a posthumous publication *Art in Its Sociological Aspect* (1889) he explains more clearly that the life of which he speaks is social life. [32] If the beautiful be the intellectually pleasing, certainly it cannot be identified with the useful which is only searching for what is pleasing; but the useful (says Guyau, in the belief that he is correcting both Kant and the evolutionists) does not always exclude the beautiful, of which indeed it often forms the lowest degree. The study of art is embraced partly, [33] not wholly, by Sociology: for art fulfils two ends, firstly and primarily that of provoking pleasant sensations (of colour, sound, etc.) and in this sense finds itself in the presence of practically incontestable scientific laws which connect Æsthetic with the physics (optics, acoustics, etc.), mathematics, physiology and psychophysics. Sculpture, in fact, rests especially on anatomy and physiology: painting on anatomy, physiology and optics: architecture on optics (golden section, etc.): music on physiology and acoustics: poetry on metrics, whose most general laws are acoustical and physiological. The second function of art is to produce the phenomena of "psychological induction," which bring to a head ideas and sentiments of most complex nature (sympathy with personages

1. *J. M. Guyau*：让·居约（1854—1888），法国唯心主义哲学家、诗人，社会美学创立者，著有《论文学的伦理》《英国当代伦理学》《当代美学的问题》等。

represented, interest, pity, indignation, etc.), in short all the social feelings, which constitute it "the expression of life." Whence are derived the two tendencies recognised in art; one inclining towards harmony, consonance, and everything delightful to ear and eye: the other towards the transfusion of life into the domain of art. Genius, true genius is destined to preserve the balance of the two tendencies: decadents and degenerates deprive art of its social sympathetic aim by setting æsthetic sympathy at war against human sympathy. [34] Translating all this into familiar terms, we may say that Guyau asserts one purely hedonistic art, above which he superimposes another art, also hedonistic, but serviceable to the cause of morality.

M. Nordau[1].

The same polemic against decadents, degenerates and individualists is carried on by another writer, Max Nordau, who gives art the task of re-establishing the wholeness of life amongst the fragmentary specialisation characteristic of industrial society[2]; he asserts that art for art's sake, art as the simple expression of internal states or the objectification of the artist's feelings, no doubt exists, but is merely "the art of Quaternary man, the art of the cave-dweller[3]." [35]

Naturalism. C. Lombroso[4].

---

1. *M. Nordau*：麦克斯·诺尔道（1849—1923），原名西蒙·马克西米利安·修德费尔德，犹太复国主义领袖、生理学家、作家、社会评论家，著有《堕落》《悖论》《世纪的病态》等。
2. fragmentary specialisation characteristic of industrial society：工业社会的碎片式专业化特征。
3. the art of Quaternary man, the art of the cave-dweller：第四纪人类艺术与穴居人艺术的区别。第四纪是人类祖先进化诞生的时代。
4. *C. Lombroso*：隆布罗索（1835—1909），意大利犯罪学家、精神病学家，实证主义犯罪学学派的创始人和主要代表，著有《天才与疯子》《无政府主义者》等。

Naturalistic is the best term with which to qualify the Æsthetic derived from that identification of genius with degeneracy which made the fortune of Lombroso and his school. This identification derives its chief strength from the following piece of reasoning. Great mental efforts, total absorption in one dominating thought, often bring about physiological disorders in the bodily organism and weakness or atrophy of various vital functions[1]. But such derangements[2] come under the head of the pathological concept of illness, degeneration, madness. Therefore genius is identical with illness, degeneration and madness. A syllogism from particular to general, in which case, according to traditional Logic, *non est consequentia*[3]. But with sociologists such as Nordau, Lombroso and company, we almost overstep the line separating respectable error from that grosser form which we call a blunder[4].

A mere confusion between scientific analysis and historical inquiry or description is visible in the works of certain sociologists and anthropologists. Thus one of them, Carl Bücher[5], in studying the life of primitive peoples, asserts that poetry, music and work were originally fused in one single act; that poetry and music were used to regulate the rhythms of labour. [36] This may be historically true or false, important or no: it has nothing whatever to do with æsthetic science. In the same way Andrew Lang[6] maintains that the doctrine concerning the origin of art as disinterested expression of the mimetic faculty finds no

1. atrophy of various vital functions：各种重要功能的退化
2. derangements：错乱
3. *non est consequentia*：〈拉〉无法推导出结论
4. blunder：大错
5. Carl Bücher：卡尔·比歇尔（1847—1930），德国经济学家，创立了非市场经济学，德国新闻学的倡导者与实践者，著有《国民经济的兴起》《工作与节律》等。
6. Andrew Lang：安德鲁·朗格（1844—1912），英国诗人、小说家、人类学家，著有《习俗与神话》《神话、礼仪与宗教》《法国古代歌谣抒情诗集》等。

confirmation from what we know of primitive art, which is decorative rather than expressive: [37] as though primitive art, which is a mere fact awaiting interpretation, could ever be converted into a criterion for the interpretation of art in general.

Decline of Linguistic.

The same vague naturalism exercised a baneful influence[1] on Linguistic, which of late years has been wholly lacking in such profound research as that inaugurated by Humboldt and followed up by Steinthal. But Steinthal never succeeded in founding a school. Max Müller[2], popular and inaccurate, maintained the indivisibility of speech and thought, confounding, or at least not distinguishing, æsthetic and logical thought; although at one time he had noted that the formation of names had a closer connexion with wit, in the sense of Locke, than with judgement. He maintained, moreover, that the science of language is not a historical but a natural science, because language is not the invention of man: the dilemma of "historical" and "natural" was canvassed and resolved over and over again with little result. [38] Another philologist, Whitney[3], attacked the "miraculous" theory of Müller and denied that thought is indivisible from speech: "The deaf-mute does not speak, but he can think," he observes; "thought is not function of the acoustic nerve." By this means Whitney relapsed into the ancient doctrine that speech is a symbol or means of expression, of human thought, subject to the will, the result of a

---

1. baneful influence：恶劣影响
2. Max Müller：麦克斯·缪勒（1823—1900），英国语言学家，西方宗教学的创始人。著有《梵语文学史》《语言科学演讲集》《以印度宗教为例论宗教的起源与发展》等。
3. Whitney：惠特尼（1827—1894），美国语言学家、语文学家、词典编纂家，以编著梵语语法闻名，著有《语言学原理十二讲》《论语言的材料与形式》《达尔文主义与语言》等。

synthesis of faculties and of a capacity for intelligent adaptation of means to end. [39]

Signs of revival. H. Paul[1].

Philosophical spirit reappeared in Paul's *Principles of the History of Language* (1880), [40] though the author's efforts to defend himself from the terrifying accusation of being a philosopher led him to hunt out a fresh title to replace the scandalous "Philosophy of Language." But if Paul is vague about the relation of Logic to Grammar, he must be given every credit for identifying, as Humboldt had already done, the question of the origin of language with that of its nature; and reasserting that language is created afresh whenever we speak. He must also be given credit for having conclusively criticized the Ethnopsychology (*Völkerpsychologie*) of Steinthal and Lazarus, showing that there is no such thing as collective psyche and that there can be no language other than of the individual.

The linguistic of Wundt[2].

Wundt [41] on the other hand attached the study of language, mythology and customs to this non-existent science of Ethnopsychology; in his latest work, on this very subject of language, [42] he foolishly echoes Whitney's gibes[3] and denounces as a "miracle theory" (*Wundertheorie*) that glorious doctrine inaugurated by Herder and

---

1. *H. Paul*：赫尔曼·保罗（1846—1921），德国语言学家、词典编纂家，曾任德国弗莱堡大学德国语言文学教授，著有《元音演变》《德国语文学纲要》《语言史原理》等。
2. *Wundt*：即威廉·冯特（1832—1920），德国生理学家、心理学家、哲学家，被公认为是实验心理学之父。著有《生理心理学原理》《民族心理学》《关于人类和动物心灵的讲演录》。
3. gibes：嘲讽

Humboldt, whom he accuses of "mystical obscurity" (*mystiche Dunkel*): he observes that this view may have had some justification before the principle of evolution had reached its triumphant application to organic nature in general and to man in particular. He has not the faintest notion of the function of imagination, or of the true relation between thought and expression; he finds no substantial difference between expression in the naturalistic, and expression in the spiritual and linguistic sense; he considers language as a special highly developed form of the vital psychophysical manifestations and of the expressive movements of animals. Out of these facts language is developed by imperceptible gradations; so that, beyond the general concept of expressive movement (*Ausdrucksbewegung*) "there is no specific mark by which language can be distinguished in any but an arbitrary manner." [43] The philosophy of Wundt betrays its own weakness by showing its inability to master the problem of language and art. In his *Ethics* æsthetic facts are presented as a complex of logical and ethical elements; the existence of æsthetic as a special normative science is denied, not for the good and sufficient reason that there are no such things as "normative sciences," but because this special science is said by him to be absorbed by the two sciences of Logic and Ethics, [44] which amounts to denying the existence of Æsthetic and the originality of art.

[1] *Essays, Scientific, Political, and Speculative, 1858–1862.*

[2] *Principles of Psychology*, 1855; 2nd ed. 1870, part viii. ch. 9, §§ 533–540.

[3] J. Sully, *Outlines of Psychology*, London, 1884; *Sensation and Intuition, Studies in Psychology and Æsthetics,* London, 1874; cf. *Encycl. Britannica*, ed. 9, art. "Æsthetics"; Alex. Bain, *The Emotions and the Will,* London, 1859, ch. 14.

[4] *Physiological Æsthetics*, London, 1877; various arts, in *Mind,* vols.

iii. iv. v. (o. s.).

[5] Vernon Lee and C. Anstruther-Thomson, "Beauty and Ugliness," in *Contemp. Review*, October-November, 1897: (abstract in Arréat, *Dix années de philosophie*, pp. 80–85); same author's *Le Rôle de l'élément moteur dans la perception esthétique visuelle, Mémoire et questionnaire soumis au Congrès de Psychologie*, reprinted Imola, 1901.

[6] H. Helmholtz, *Die Lehre von der Tonempfindungen als physiologische Grundlage für die Théorie der Musik*, 1863, 4th ed., 1877; Brücke-Helmholtz, *Principes scientifiques des beaux arts*, Fr. ed., Paris, 1881; C. Stumpf, *Tonpsychologie*, Leipzig, 1883.

[7] *Philosophie de l'art*, 1866–1869 (4th ed. Paris, 1885).

[8] *Op. cit.* i. pp. 13–15.

[9] *Philosophie de l'art*, i. pp. 17–54.

[10] *Op. cit.* i. p. 37.

[11] *Op. cit.* i. p. 54.

[12] *Op. cit.* i. p. 15.

[13] *Op. cit.* ii. p. 277.

[14] *Philos. de l'art*, ii. pp. 257–400.

[15] *Op. cit.* ii. pp. 257–258.

[16] *Op. cit.* ii. p. 393.

[17] *Vorschule der Ästhetik*, 1876 (2nd ed. Leipzig, 1897–1898).

[18] *Vorschule der Ästhetik*, i. ch. 19.

[19] Schasler, *Krit. Geschichte d. Ästh.* p. 1117.

[20] *Vorschule der Ästh.* ii. pp 273–314.

[21] *Op. cit.* pref. p. iv.

[22] *Op. cit.* i. pp. 15–30.

[23] *Op. cit.* i. p. 32.

[24] *Vorschule der Ästh.* ii. pp. 12–13.

[25] *Op. cit.* i. p. 38.

[26] *Die Anfänge der Kunst*, Freiburg i. B. 1894.

[27] *Op. cit.* p. 19.

[28] *Die Anfänge der Kunst*, pp. 45–46.

[29] *Op. cit.* pp. 46–48.

[30] *Op. cit.* pp. 293–301.

[31] *Du principe de l'art et de sa destination sociale*, Paris, 1875.

[32] M. Guyau, *L'Art au point de vue sociologique*, 1889 (3rd ed. Paris, 1895); *Les Problèmes de l'esthétique contemporaine*, Paris, 1884; cf. Fouillée, pref. to the former work, pp. xli–xliii.

[33] *L'Art au point de vue sociologique,* pref. p. xlvii.

[34] *Op. cit., passim,* esp. ch. 4; cf. pp. 64, 85, 380.

[35] Max Nordau, *Social Function of Art*, 2nd ed., Turin, 1897.

[36] Karl Bücher, *Arbeit u. Rhythmus*, 2nd ed., Leipzig, 1899.

[37] *Custom and Myth*, p. 276; quoted by Knight, *The Philosophy of the Beautiful*, vol. i. pp. 9–10.

[38] *Lectures on the Science of Language*, 1861 and 1864 (Fr. tr., Paris, 1867).

[39] William Dwight Whitney, *The Life and Growth of Language*, London, 1875 (It. tr., Milan, 1876).

[40] Hermann Paul, *Principien der Sprachgeschichte*, 1880 (2nd ed., Halle, 1886).

[41] Wilh. Wundt, *Über Wege u. Ziele d. Völkerpsychologie*, Leipzig, 1886.

[42] *Die Sprache*, Leipzig, 1900, 2 vols, (part i. Of *Völkerpsychologie, eine Untersuchung der Entwicklungsgesetze von Sprache, Mythus und Sitte*).

[43] *Die Sprache, passim;* cf. i. p. 31 *seqq.*, ii. pp. 599, 603–609.

[44] *Ethik*, ed. 2, Stuttgart, 1892, p. 6.

XVIII ÆSTHETIC PSYCHOLOGISM AND OTHER RECENT TENDENCIES

Neo-criticism and empiricism.

The neo-critical or neo-Kantian movement was powerless to make headway against hedonistic, psychological and moralistic views of the æsthetic fact, although it made every effort to save the concept of spirit from the invading rush of naturalism and materialism. [1] Kant bequeathed to neo-criticism his own failure to understand creative imagination, and the neo-Kantians do not seem to have had the faintest notion of any form of cognition other than the intellectual.

Kirchmann[1].

Amongst German philosophers of any renown who clung to æsthetic sensationalism and psychologism was Kirchmann, promoter of a so-called realism, and author of *Æsthetic on a Realistic Basis* (1868). [2] In his doctrine the æsthetic fact is an image (*Bild*) of a real; an animated (*seelenvolles*) image, purified and strengthened, that is, idealized, and divided into the image of pleasure, which is the beautiful, and that of pain, which is the ugly. Beauty admits of a threefold series of varieties or modifications, being determined according to the content as sublime,

1. *Kirchmann*：基尔希曼（1802—1884），德国法学家、哲学家、政治家，曾担任柏林哲学协会主席，著有《论不朽》《作为科学的法学的无价值性》《基于现实主义的美学》等。

comic, tragic, etc.; according to the image, as beauty of nature or of art; and according to the idealization as idealistic or naturalistic, formal or spiritual, symbolical or classical. Not having grasped the nature of æsthetic objectification, Kirchmann takes the trouble to draw up a new psychological category of ideal or apparent feelings, arising from artistic images and being attenuations of the feelings of real life. [3]

Metaphysic translated into Psychology. Vischer.

To the evolution or involution of the Herbartians into physiologists of æsthetic pleasure corresponds a similar evolution or involution of the idealists into adherents of psychologism. The first place must be given to the veteran Theodor Vischer, who in a criticism of his own work pronounced Æsthetic to be "the union of mimics and harmonics" (*vereinte Mimik und Harmonik*), and Beauty the "harmony of the universe," never actually realized because realized only at infinity, so that when we think to seize it in the Beautiful, we are under an illusion: a transcendent illusion, which is the very essence of the æsthetic fact. [4] His son Robert Vischer[1] coined the word *Einfühlung*[2] to express the life with which man endows natural objects by means of the æsthetic process. [5] Volkelt[3], when treating of the *Symbol* [6] and joining symbolism to pantheism, opposed associationism and favoured a natural teleology immanent in Beauty.

Siebeck[4].

1. Robert Vischer：罗伯特·费肖尔（1847—1933），弗里德里希·提奥多·费肖尔之子，德国哲学家，提出移情说。
2. *Einfühlung*：〈德〉移情
3. Volkelt：伏尔盖特（1848—1930），德国哲学家、美学家，曾任莱比锡大学哲学与教育学教授，"移情说"美学的主要代表人物。著有《美学体系》《审美意识论》等。
4. *Siebeck*：西贝克（1842—1920），德国哲学家，著有《柏拉图〈菲力布篇〉中的理念教条》《作为思想家的歌德》等。

The Herbartian Siebeck (1875) abandoned the formalistic theory and tried to explain the fact of beauty by the concept of the appearance of personality. [7] He distinguishes between objects which please by their content alone (sensuous pleasures), those which please by form alone (moral facts), and those which please by the connexion of content with form (organic and æsthetic facts). In organic facts the form is not outside the content, but is the expression of the reciprocal action and conjunction of the constitutive elements: whereas in æsthetic facts the form is outside the content, and as it were its mere surface; not a means to the end, but an end in itself. Æsthetic intuition is a relation between the sensible and the spiritual, matter and spirit, and is thus form regarded as the appearance of personality. Æsthetic pleasure arises from the spirit's consciousness of discovering itself in the sensible. Siebeck borrows the theory of modifications of the beautiful from the metaphysical idealists, who held that only in such modifications can beauty be found in the concrete, just as humanity can only exist as a man of determinate race and nationality. The sublime is that species of beauty wherein the formal moment of circumscription is lost, and is therefore the unlimited, which is a kind of extensive or intensive infinity; the tragic arises when the harmony is not given but is the result of conflict and development; the comic is a relation of the small to the great; and so on. These traces of idealism, together with his firm hold on the Kantian and Herbartian absoluteness of the judgement of taste, make it impossible to regard Siebeck's Æsthetic as purely psychological and empirical and wholly devoid of philosophical elements.

M. Diez[1].

---

1. *M. Diez*：麦克斯·迪兹（生卒年不详），德国美学家，德语文学教授，著有《作为美学基础的感觉理论》《德语入门》等。

It is the same with Diez, who, in his *Theory of Feeling as Foundation of Æsthetic* (1892), [8] tries to explain the artistic activity as a return to the ideal of feeling (*Ideal des fühlenden Geistes*), parallel with science (ideal of thought), morality (ideal of will) and religion (ideal of personality). But whatever is this so-called feeling? is it the empirical feeling of the psychologists, irreducible to an ideal, or the mystic faculty of communication and conjunction with the Infinite and the Absolute? the absurd "pleasure-value" of Fechner, or the "judgement" of Kant? One is inclined to say that these writers, and others like them, still under the influence of metaphysical views, lack the courage of their opinions: they feel themselves to be in an atmosphere of hostility and speak under reservations or compromises. The psychologist Jodi asserts the existence of elementary æsthetic feelings, as discovered by Herbart, and defines them as "immediate excitations not resting upon associative or reproductive activity or on the fancy," although "in ultimate analysis they must be reduced to the same principles." [9]

Psychological tendency. Teodor Lipps[1].

The purely psychological and associationistic tendency becomes clearly defined in Professor Teodor Lipps and his school. Lipps criticizes and rejects a whole series of æsthetic theories: *a* of play; *b* of pleasure; *c* of art as recognition of real life, even if displeasing; *d* of emotion and passional excitation; *e* syncretism[2], attributing to art beside the primary purpose of play and pleasure the further ends of recognition of life, in its reality, revelation of individuality, commotion, freedom from a weight, or free play of the imagination. His theory differs little at bottom from that of Jouffroy, for in his thesis he assumes artistic beauty to be the sympathetic. "The object of sympathy

---

1. *Teodor Lipps*：西奥多·利普斯（1851—1914），德国哲学家、美学家。
2. syncretism：混合论

is our objectified ego, transposed into others and therefore discovered in them. We feel ourselves in others and we feel others in ourselves. In others, or by means of them, we feel ourselves happy, free, enlarged, elevated, or the contrary of all these. The æsthetic feeling of sympathy is not a mere mode of æsthetic enjoyment, it is that enjoyment itself. All æsthetic enjoyment is founded, in the last analysis, singly and wholly upon sympathy; even that caused by geometrical, architectonic, tectonic, ceramic, etc., lines and forms." "Whenever in a work of art we find a personality (not a defect of the man, but something positively human) which harmonizes with and awakes an echo in the possibilities and tendencies of our own life and vital activities: whenever we find positive, objective humanity, pure and free from all real interests lying outside the work of art, as art only can reproduce it and æsthetic contemplation alone can demand; the harmony, the resonance, fills us with joy. The value of personality is ethical value: outside it there is no possibility or determination of ethical character. All artistic and in general æsthetic enjoyment is, therefore, the enjoyment of something which has ethical value (*eines ethische Werthvollen*); not as element of a complex, but as object of æsthetic intuition." [10]

The æsthetic fact is thus deprived of all its own value and allowed merely a reflexion from the value of morality.

Without lingering over Lipps's pupils (such as Stern[1] and others [11]) and writers of similar tendency (such as Biese[2], with his theory of anthropomorphism[3] and universal metaphor; [12] or Konrad Lange[4],

1. Stern：即保罗·斯特恩（生卒年不详），德国美学家，著有《新美学中的移情与联想》。
2. Biese：即阿尔弗雷德·比塞（1856—1930），德国文学史学家、美学家，著有《美学中的联想原理与拟人论》《德国文学史》等。
3. anthropomorphism：拟人论，以人的能力、行为或经验的术语来解释动物或非生物的有关特性。
4. Konrad Lange：孔拉德·朗格（1855—1921），德国艺术史学家，著有《艺术的本体》《德国青年的艺术教育》《雅典的国王大殿》《自我幻觉是艺术享受的核心》等。

who propounds a thesis that art is conscious self-deception), [13] we will call attention to Professor Karl Groos[1] (1892), who comes within measurable distance of the concept of æsthetic activity as a theoretic value. [14] Between the two poles of consciousness, sensibility and intellect, are several intermediate grades, amongst which lies intuition or fancy, whose product, the image or appearance (*Schein*), is midway between sensation and concept. The image is full like sensation, but regulated like the concept; it has neither the inexhaustible richness of the former, or the barren nudity[2] of the latter. Of the nature of image or appearance is the æsthetic fact; which is distinguished from the simple, ordinary image not by its quality, but by its intensity alone: the æsthetic image is merely a simple image occupying the summit of consciousness. Representations pass through consciousness like a crowd of people hurrying over a bridge, each bent on his own business; but when a passer-by halts on the bridge and looks at the scene, then is it holiday, then arises the æsthetic fact. This is therefore not passivity but activity; according to the formula adopted by Groos it is internal imitation (*innere Nachahnung*). [15] It may be objected against the theory that every image, so far as it is an image at all, must occupy the summit of consciousness if only for an instant; and that the mere image is either the product of an activity just as is the æsthetic image, or it is not a real image at all. It may also be objected that the definition of the image as something sharing in the nature of sensation and concept may lead back to intellectual intuition and the other mysterious faculties of the metaphysical school, for which Groos professes abhorrence. His division of the æsthetic fact into form and content is even less happy. He recognizes four classes of content: associative (in the strict sense), symbolic, typical, individual: [16] and

1. Karl Groos：卡尔·格鲁斯（1861—1946），德国哲学家、心理学家、美学家，著有《美学导论》《动物的游戏》《人类的游戏》等。
2. barren nudity：贫瘠荒芜

into his inquiries he introduces, quite unnecessarily, the concepts of infusion of personality and of play. In connexion with the latter he remarks that "internal imitation is the noblest game of man," [17] and adds that "the concept of play applies fully to contemplation, but not to æsthetic production, save in the case of primitive peoples." [18]

The modifications of the Beautiful in Groos and Lipps.

Groos does however free himself from the "modifications of Beauty," because, æsthetic activity having been identified with internal imitation, it is clear that whatever is not internal imitation is excluded from that activity as something different. "All Beauty (beauty understood in the sense of 'sympathetic') belongs to the æsthetic activity, but not every æsthetic fact is beautiful." Beauty, then, is the representation of the sensuously pleasant; ugliness, the representation of the unpleasant; the sublime, that of a mighty thing (*Gewaltiges*) in a simple form; the comic, that of an inferiority which arouses in us a pleasing sense of our own superiority. And so forth. [19] With great good sense Groos holds up to derision the office assigned to the ugly by Schasler and Hartmann with their superficial dialectic. To say that an ellipse contains an element of ugliness in comparison with the circle because it is symmetrical about its two axes only and not about infinite diameters is like saying "wine has a relatively unpleasant taste because in it is lacking (*ist aufgehoben*) the pleasant taste of beer." [20] Lipps too, in his writings upon Æsthetic, recognizes that the comic (of which he gives an accurate psychological analysis) [21] has in itself no æsthetic value; but his moralistic views lead him to outline a theory of it not unlike that of the overcoming of the ugly; he explains it as a process leading to a higher æsthetic value (*i.e.* sympathy). [22]

E. Véron and the double form of Æsthetic.

Work such as that of Groos and, occasionally, of Lipps is of some value towards the elimination of errors, as well as confining æsthetic research to the field of internal analysis. Merit of the same kind belongs to the work of a Frenchman, Véron[1], [23] who controverts the Absolute Beauty of academical Æsthetic and, after accusing Taine of confounding Art with Science and Æsthetic with Logic, remarks that if it be the duty of art to make manifest the essence of things, their one dominating quality, then "the greatest artists would be those who have best succeeded in exhibiting this essence ... and the greatest works would resemble each other more closely than any others and would clearly demonstrate their common identity, whereas the exact opposite happens." [24] But one looks in vain for scientific method in Véron; a precursor of Guyau, [25] he asserts that art is at bottom two different things; there are two arts: one decorative, whose end is beauty, that is to say the pleasure of eye and ear resulting from determinate dispositions of lines, forms, colours, sounds, rhythms, movements, light and shade, without necessary interventions of ideas and feelings, and capable of being studied by Optics and Acoustics: the other, expressive, which gives "the agitated expression[2] of human personality." He considers that decorative art prevails in the ancient world, and expressive art in the modern. [26]

We cannot here examine in detail the æsthetic theories of artists and men of letters; the scientific and historicist prejudices, the theory of experiment and human document, which underlie the realism of Zola[3], or the moralism which underlies the problem-art of Ibsen[4] and

---

1. Véron：即尤金·维龙（1825—1889），法国哲学家、美学家、评论家、修辞学家，著有《美学》《人性的理智发展》《现代艺术相比于古代艺术的优势之处》等。
2. agitated expression：狂躁不安的表达
3. Zola：左拉（1840—1902），法国现实主义作家，提倡自然主义，代表作有《萌芽》《崩溃》等。
4. Ibsen：易卜生（1828—1906），挪威剧作家，现实主义戏剧创始人，代表作有《玩偶之家》《人民公敌》等。

the Scandinavian school. Gustave Flaubert wrote of art profoundly, better perhaps than any other Frenchman has ever written, not in special treatises but throughout his letters, which were published after his death. [27]

L. Tolstoy[1].

Under the influence of Véron and his hatred for the concept of beauty, Leo Tolstoy wrote his book on art, [28] which, according to the great Russian artist, communicates feelings in the same way in which words communicate thoughts. The meaning of this theory is made clear by the parallel he drew between Art and Science, and his conclusion that "the mission of art is to render sensible and capable of assimilation that which could not be assimilated under the form of argumentation"; and that "true science examines truths considered as important for a certain society at a given epoch and fixes them in the consciousness of man, whereas art transports them from the domain of knowledge to that of feeling." [29] There is therefore no such thing as art for art's sake, any more than science for science' sake. Every human function should be directed to increase morality and to suppress violence. This amounts to saying that nearly all art, from the beginning of the world, is false. Aeschylus, Sophocles, Euripides, Aristophanes, Dante, Tasso, Milton, Shakespeare, Raphæl, Michæl Angelo, Bach, Beethoven are (according to Tolstoy) "artificial reputations created by critics." [30]

F. Nietzsche[2].

---

1. *L. Tolstoy*：列夫·托尔斯泰（1828—1910），俄罗斯批判现实主义作家、思想家，代表作有《战争与和平》《安娜·卡列尼娜》《复活》等。
2. *F. Nietzsche*：弗里德里希·尼采（1844—1900），德国哲学家、诗人，唯意志论代表人物，著有《悲剧的诞生》《权力意志》《查拉图斯特拉如是说》等。

Amongst artists rather than amongst philosophers must be reckoned Friedrich Nietzsche, whom we should wrong (as we said of Ruskin) by trying to expound his æsthetic doctrines in scientific language and then holding them up to the facile criticism which, so translated, they would draw upon themselves. In none of his books, not even in his first, *The Birth of Tragedy*, [31] in spite of the title, does he offer us a real theory of art; what appears to be theory is the mere expression of the author's feelings and tendencies. He shows a kind of anxiety concerning the value and aim of art and the problem of its inferiority or superiority to science and philosophy, a state of mind characteristic of the Romantic period of which Nietzsche was, in many respects, a belated but magnificent representative. To Romanticism, as well as to Schopenhauer, belong the elements of thought which issued in the distinction between Apollinesque art[1] (that of serene contemplation, to which belong the epic and sculpture) and Dionysiac art[2] (the art of agitation and tumult, such as music and the drama). The thought is vague and does not bear criticism; but it is supported by a flight of inspiration which lifts the mind to a spiritual region seldom if ever reached again in the second half of the nineteenth century.

An æsthetician of music: E. Hanslick[3].

The most notable æsthetic students of that time were perhaps a group of persons engaged in constructing theories of particular arts. And since — as we have seen [32] — philosophical laws or theories of individual arts are inconceivable, it was inevitable that the ideas

1. Apollinesque art：阿波罗型艺术，即日神型艺术，讲究规则与秩序。
2. Dionysiac art：狄奥尼索斯型艺术，即酒神型艺术，以激情和癫狂为主要内容。
3. *E. Hanslick*：爱德华·汉斯里克（1825—1904），德国音乐家、音乐评论家，著有《论音乐美》《维也纳音乐会历史》《现代歌剧》等。

presented by such thinkers should be (as indeed they are) nothing more than general æsthetic conclusions. First may be mentioned the acute Bohemian critic Eduard Hanslick, who published his work *On Musical Beauty* in 1854; it was often reprinted and was translated into various languages. [33] Hanslick waged war against Richard Wagner and in general against the pretension of finding concepts, feelings and other definite contents in music. "In the most insignificant musical works, where the most powerful microscope can discover nothing, we are now asked to recognize a *Night Before the Battle*, a *Summer Night in Norway*, a *Longing for the Sea*, or some such absurdity, should the cover have the audacity[1] to affirm that this is the subject of the piece." [34] With equal vivacity he protests against the sentimental hearers who, instead of enjoying the work of art, set themselves to extract pathological effects[2] of passionate excitement and practical activity. If it be true that Greek music produced effects of this kind, "if it needed but a few Phrygian strains[3] to animate troops with courage in the face of the enemy, or a melody in the Dorian mode[4] to ensure the fidelity of a wife whose husband was far away, then the loss of Greek music is a melancholy thing for generals and husbands; but æstheticians and composers need not regret it." [35] "If every senseless *Requiem*[5], every noisy funeral march, every wailing *Adagio*[6] had the power of depressing us, who could put up with existence under such conditions? But let a real musical work confront us, clear-eyed and glowing with beauty, and we feel ourselves enslaved by its invincible fascination even if its material is all the sorrows of the age." [36]

---

1. audacity：无畏
2. pathological effects：病理效果
3. Phrygian strains：弗里吉亚式曲调（古希腊慷慨激扬的音乐）
4. Dorian mode：多利亚调式，是古希腊简朴、庄重的乐调。
5. *Requiem*：追思弥撒
6. wailing *Adagio*：哀怨的慢板

Hanslick's concept of form.

Hanslick maintained that the sole aim of music is form, musical beauty. This affirmation won him the goodwill of the Herbartians, who hastened to welcome such a vigorous and unexpected ally; by way of returning the compliment, Hanslick felt obliged in later editions of his work to mention Herbart himself and his faithful disciple Robert Zimmermann who had given (so he said) "full development to the great æsthetic principle of Form." [37] The praises of the Herbartians and the courteous declarations of Hanslick both arose from a misunderstanding: for the words "beauty" and "form" have one meaning for the former and quite another for the latter. Hanslick never thought that symmetry, purely acoustical relations and pleasures of the ear constituted musical beauty; [38] mathematics, he held, are utterly useless to musical Æsthetic. [39] Musical beauty is spiritual and significative[1]: it has thoughts, undoubtedly; but those thoughts are musical. "Sonorous forms[2] are not empty, but perfectly filled; they cannot be compared with simple lines delimiting a space; they are the spirit assuming body and extracting from itself the stuff of its own incarnation[3]. Rather than an arabesque, music is a picture; but a picture whose subject can neither be expressed in words nor enclosed in precise concept. There are in music both meaning and connexion, but these are of a specifically musical nature; music is a language we understand and speak, but which it is not possible to translate." [40] Hanslick asserts that though music does not portray the quality of feelings, it does portray their dynamic aspect or tone: if not the substantives, then the adjectives: it depicts not "murmuring tenderness" or "impetuous courage," but the "murmuring" and the

1. significative：意味深长的
2. sonorous forms：声音形式
3. incarnation: 化身

"impetuous." [41] The backbone of the book is the denial that form and content can ever be separated in music. "In music there can be no content in opposition to the form, since there can be no form outside the content." "Take a motive, the first that comes into your head; what is its content, what its form? where does this begin, and that end? ... What do you wish to call content? The sounds? Very well: but they have already received a form. What will you call form? Also the sounds? but they are form already filled; form supplied with content." [42] Such observations denote acute penetration of the nature of art, though not scientifically formulated or framed into a system. Hanslick thought he was dealing with peculiarities of music, [43] instead of with the universal and constitutive character of every form of art, and this prevented him from taking larger views.

Æstheticians of the figurative arts. C. Fiedler[1].

Another specialist æsthetician is Conrad Fiedler, author of many essays on the figurative arts, the most important being his *Origin of Artistic Activity* (1887). [44] No one, perhaps, has better or more eloquently emphasized the activistic character of art, which he compares with language. "Art begins exactly where intuition (perception) ends. The artist is not differentiated from other people by any special perceptive attitude enabling him to perceive more or with greater intensity, or endowing his eye with any special power of selecting, collecting, transforming, ennobling or illuminating; but rather by his peculiar gift of being able to pass immediately from perception to intuitive expression; his relation with nature is not perceptive, but expressive." "A man standing passively at gaze

1. *C. Fiedler*：孔拉德·费德勒（1841—1895），德国艺术史学家、造型美学家、作家，19 世纪德国重要艺术理论家之一，著有《艺术史学家词典》《艺术性活动的起源》等。

may well imagine himself in possession of the visible world as an immense, rich, varied whole: the entire absence of fatigue with which he traverses the infinite mass of visual impressions, the rapidity with which representations dart across his consciousness, convince him that he stands in the midst of an immense visible world, although he may quite well be unable at any one instant to represent it to himself as a whole. But this world, so great, so rich, so immeasurable, disappears the moment art seeks to become its master. The very first effort to emerge from this twilight and arrive at clear vision restricts the circle of things to be seen. Artistic activity may be conceived as continuation of that concentration by which consciousness makes the first step towards clear vision, which it reaches only by self-limitation." Spiritual process and bodily process are here an indivisible whole, which is expression.

Intuition and Expression.

"This activity, simply because it is spiritual, must consist of forms wholly determinate, tangible, sensibly demonstrative." Art is not in a state of subjection to science. Like the man of science, the artist desires to escape from the natural perceptive state and to make the world his own; but there are regions to which we can penetrate not by the forms of thought and science but only through art. Art is, strictly speaking, not imitation of nature; for what is nature save this confused mass of perceptions and representations, whose real poverty has been demonstrated already? In another sense, however, art may be called imitation of nature inasmuch as its aim is not to expound concepts or to arouse emotions, that is to create values of intellect and feeling. Art does create both these values, if you like to say so; but only in one quite peculiar quality, which consists in complete visibility (*Sichtbarkeit*). Here we have the same sane conception, the same lively comprehension of the true nature of art which we found in Hanslick, only expressed in a more rigorous and philosophical manner. With Fiedler is connected his friend Adolf Hildebrand[1], who brought into high relief the activistic, or architectonic[2] as opposed to imitative, character of art, illustrating his theoretical discussions especially from sculpture, the art which he himself followed. [45]

Narrow limits of these theories.

What we chiefly miss in Fiedler and others of the same tendency is the conception of the æsthetic fact not as something exceptional, produced by exceptionally gifted men, but as a ceaseless activity of man as such; for man possesses the world, so far as he does possess it, only in the form of representation-expressions, and only knows in so far as he creates. [46] Nor are these writers justified in treating language as parallel with art, or art with language; for comparisons are drawn between things at least partially different, whereas art and language are identical.

H. Bergson[3].

The same criticism can be made in the case of the French philosopher Bergson, who in his book on *Laughter* [47] states a theory of art very similar to that of Fiedler and makes the same mistake of conceiving the artistic faculty as something distinct and exceptional in comparison with the language of everyday use. In ordinary life, says Bergson, the individuality of things escapes us; we see only as

1. Adolf Hildebrand：阿道夫 · 希尔德布兰德（1847—1921），德国雕塑家。
2. architectonic：自成体系的
3. *H. Bergson*：亨利 · 柏格森（1859—1941），法国哲学家、心理和生物学家，1927 年诺贝尔文学奖获得者。提出创造进化论，著有《创造进化论》《时间与自由意志》《宗教与道德的两个源泉》等。

much of them as our practical needs demand. Language helps this simplification; since all names, proper names excepted, are names of kinds or classes. Now and then, however, nature, as if in a fit of absence of mind, creates souls of a more divisible and detached kind (artists), who discover and reveal the riches hidden under the colourless signs and labels of everyday life, and help others (non-artists) to catch a glimpse of what they themselves see, employing for this purpose colours, forms, rhythmic connexions of words, and those rhythms of life and breath even more intimate to man, the sounds and notes of music.

Attempts to return to Baumgarten. C. Hermann.

A healthy return to Baumgarten, a revival and correction of the old philosopher's theories in the light of later discoveries, might perhaps have given Æsthetic some assistance, after the collapse of the old idealistic metaphysic, towards thinking the concept of art in its universality and discovering its identity with pure and true intuitive knowledge. But Conrad Hermann, who preached the return to Baumgarten [48] in 1876, did bad service to what might have been a good cause. According to him Æsthetic and Logic are normative sciences; but Logic does not contain, as does Æsthetic, "a definite category of external objects exclusively and specifically adequate to the faculty of thought"; and on the other hand "the products and results of scientific thought are not so external and sensibly intuitive as those of artistic invention." Logic and Æsthetic alike refer not to the empirical thinking and feeling of the soul, but to pure and absolute sensation and thought. Art constructs a representation standing midway between the individual and the universal. Beauty expresses specific perfection, the essential or, so to speak, the rightful (*seinsollend*) character of things. Form is "the external sensible limit, or mode of appearance of a thing, in opposition to the kernel of the thing itself and to its essential and substantial content." Content and form are both æsthetic, and the æsthetic interest concerns the entirety of the beautiful object. The artistic activity has no special organ such as thought possesses in speech. The æsthetician, like the lexicographer[1], has the task of compiling a dictionary of tones and colours and of the different meanings which may possibly be attached to them. [49] We can see that Hermann accepted side by side the most inconsistent propositions. He welcomes even the æsthetic law of the golden section, and applies it to tragedy; the longer segment of the Une is the tragic hero; the punishment which overtakes him (the entire line) exceeds his crime in the same proportion in which he oversteps the common measure (the shorter segment of the line). [50] It reads almost like a joke.

Without direct reference to Baumgarten, a proposal that Æsthetic be reformed and treated as the "science of intuitive knowledge" was made in a miserable little work by one Willy Nef[2] (1898), [51] who makes the dumb animals share his "intuitive knowledge," in which he distinguishes a formal side (intuition) and a material side or content (knowledge), and considers the everyday relations between men, their games and their art, as belonging to intuitive knowledge.

Eclecticism. B. Bosanquet[3].

The English historian of Æsthetic, Bosanquet (1892) tried to find a reconciliation between content and form in unity of expression. "Beauty," says Bosanquet in the Introduction to his *History*, "is that

1. lexicographer：词典编纂者
2. Willy Nef：威利·内夫（1876—?）德国哲学家，美学家，著有《威廉·冯特对康德认知理论的立场》。
3. *B. Bosanquet*：伯纳德·鲍桑葵（1848—1923），英国哲学家、美学家，表现主义美学代表人物之一。著有《美学史》《国家的哲学理论》《美学三讲》等。

which has characteristic and individual expressiveness for sensuous perception or imagination, subject to the conditions of general or abstract expressiveness by the same means." In another passage he observes: "The difficulty of real Æsthetic is to show how the combination of decorative forms in characteristic representations, by intensifying the essential character immanent in them from the beginning, subordinates them to a central signification which stands to their complex combination as their abstract signification stands to each one of them taken singly." [52] But the problem, as propounded in a way suggested by the antithesis between the two schools (contentism and formalism) of German Æsthetic, is in our opinion insoluble.

Æsthetic of expression: present state.

De Sanctis founded no school of æsthetic science in Italy. His thought was quickly misunderstood and mutilated by those who presumed to correct it, and, in fact, only returned to the outworn rhetorical conception of art as consisting of a little content and a little form. Only within the last ten years has there been a renewal of philosophical studies, arising out of discussions concerning the nature of history [53] and the relation in which it stands to art and science, and nourished by the controversy excited by the publication of De Sanctis' posthumous works. [54] The same problem of the relation between history and science, and their difference or antithesis, reappeared also in Germany, but without being put in its true connexion with the problem of Æsthetic. [55] These inquiries and discussions, and the revival of a Linguistic impregnated by philosophy in the work of Paul and some others, appear to us to offer much more favourable ground for the scientific development of Æsthetic than can be found on the summits of mysticism or the low plains of positivism and sensationalism.

[1] A. F. Lange, *Geschichte des Materialismus, u. Kritik seiner Bedeutung i. d. Gegenwart*, 1866.

[2] J. F. v. Kirchmann, *Ästhetik auf realistischer Grundlage*, Berlin, 1868.

[3] *Ästh. auf real. Grund.* vol. i. pp. 54–57; see above, pp. 80–81.

[4] *Kritische Gänge*, vol. v. pp. 25–26, 131.

[5] R. Vischer, *Über das optische Formgefühl*, Leipzig, 1873.

[6] *Der Symbol-Begriff in der neuesten Ästh.*, Jena, 1876.

[7] *Das Wesen d. ästh. Anschauung, Psychologische Untersuchungen z. Theorie d. Schönen u. d. Kunst*, Berlin, 1875.

[8] Max Diez, *Theorie des Gefühls z. Begründung d. Ästhetik*, Stuttgart, 1892.

[9] Friedr. Jodi, *Lehrb. der Psychologie*, Stuttgart, 1896, § 53, pp. 404–414.

[10] *Komik und Humor, eine psychol. ästhet. Untersuch.*, Hamburg-Leipzig, pp. 223–227.

[11] Paul Stern, *Einfühling u. Association i. d. neueren Ästh.*, 1898, in *Beiträge z. Ästh.*, ed. Lipps and R. M. Werner (Hamburg-Leipzig).

[12] Alfr. Biese, *Das Associationsprincip u. d. Anthropomorphismus i. d. Ästh.*, 1890; *Die Philosophie des Metaphorischen*, Hamburg-Leipzig, 1893.

[13] Konrad Lange, *Die bewusste Selbsttäuschung als Kern des künstlerischen Genusses*, Leipzig, 1895.

[14] Karl Groos, *Einleitung i. d. Ästhetik*, Giessen, 1892.

[15] *Op. cit.* pp. 6–46, 83–100.

[16] *Einleitung i. d. Ästh.* pp. 100–147.

[17] *Op. cit.* pp. 168–170.

[18] *Op. cit.* pp. 175–176.

[19] *Op. cit.* pp. 46–50, and all part iii.

[20] *Einleitung i. d. Ästh.* p. 292, note.

[21] See above, pp. 123–124.

[22] *Komik und Humor*, p. 199 *seqq.*

[23] Eug. Véron, *L'Esthétique*, 2nd ed. Paris, 1883.

[24] *Op. cit.* p. 89.

[25] See above, pp. 431–432.

[26] *Esthétique*, pp. 38, 109, 123 *seqq.*

[27] *Correspondance*, 1830–1880, 4 vols., new ed., Paris, 1902–1904.

[28] *What is Art?* Eng. tr.

[29] *Op. cit.* pp. 171–172, 308.

[30] *Op. cit.* pp. 201–202.

[31] *Die Geburt der Tragödie oder Griechenthum und Pessimismus*, 1872 (Ital. trans., Bari, 1907).

[32] See above, p. 114.

[33] *Vom Musikalisch-Schönen*, Leipzig, 1854; 7th ed. 1885 (French trans., *Du beau dans la musique*, Paris, 1877).

[34] *Vom Musikalisch-Schönen*, p. 20.

[35] *Op. cit.* p. 98.

[36] *Op. cit.* p. 101.

[37] *Op. cit.* p. 119, note.

[38] *Op. cit.* p. 50.

[39] *Op. cit.* p. 65.

[40] *Vom Musikalisch-Schönen*, pp. 50–51.

[41] *Op. cit.* pp. 25–39.

[42] *Op. cit.* p. 122.

[43] *Op. cit.* pp. 52, 67, 113, etc.

[44] Conrad Fiedler, *Der Ursprung der künstlerischen Thätigkeit*, Leipzig, 1887. Collected with others of same author in *Schriften tiber die Kunst*, ed. H. Marbach, Leipzig, 1896.

[45] *Das Problem der Form in der bildenden Kunst*, 2nd ed. 1898 (4th ed., Strassburg, 1903).

[46] See above, pp. 44–45.

[47] H. Bergson, *Le Rire, essai sur la signification du comique,* Paris, 1900, pp. 153–161 (Eng. tr., London).

[48] Conrad Hermann, *Die Ästhetik in ihrer Geschichte und ah wissenschaftliches System*, Leipzig, 1876.

[49] *Die Ästhetik*, etc., *passim*.

[50] *Die Ästhetik*, § 56.

[51] Willy Nef, *Die Ästhetik als Wissenschaft der anschaulichen Erkenntniss*, Leipzig, 1898.

[52] *A History of Æsthetics*, pp. 4–6, 372, 391, 447, 458, 466.

[53] B. Croce, *La storia ridotta sotto il concetto generale dell'arte*, 1893 (2nd ed. entitled *Il concetto della storia nelle sue relazioni col concetto dell' arte*, Rome, 1896); P. R. Trojano, *La storia come scienza sociale*, vol. i., Naples, 1897; G. Gentile, *Il concetto della storia* (in Crivellucci's *Studî storici*, 1889); see also F. de Sarlo, *Il problema estetico*, in *Saggi di filosofia*, vol. ii., Turin, 1897; and by same author, *I dati dell' esperienza psichica*, Florence, 1903, concluding chapter.

[54] *La letteratura italiana nel secolo XIX*, edited by B. Croce, Naples, 1896; also *Scritti varî*, ed. Croce, Naples, 1898, 2 vols.

[55] H. Rickert, *Die Grenzen der naturwissenschaftlichen Begriffsbildung*, Freiburg i. B., 1896–1902.

XIX HISTORICAL SKETCHES OF SOME PARTICULAR DOCTRINES

Result of the history of Æsthetic.

We have reached the end of our history. Having passed in review the travail and doubt through which the discovery of the æsthetic concept was achieved, the vicissitudes first of neglect, then of revival and rediscovery to which it was exposed, the various oscillations and failures in its exact determination, the resurrection, triumphant and overwhelming, of ancient errors supposed to be dead and buried; we may now conclude, without appearing to assert anything unproven, that of Æsthetic in the proper sense of the word we have seen very little, even including the last two centuries' active research. Exceptional intellects have hit the mark and have supported their views with energy, with logic, and with consciousness of what they were doing. It would no doubt be possible to extract many true affirmations leading to the same point of view from the works of non-philosophical writers, art-critics and artists, from commonly received opinions and proverbial sayings; such a collection would show that this handful of philosophers does not stand alone, but is surrounded by a throng of supporters and is in perfect agreement with the general mind and universal common sense. But if Schiller was right in saying that the rhythm of philosophy is to diverge from common opinion in order to return with redoubled vigour, it is evident that such divergence is necessary, and constitutes the growth of science, which is science itself. During this tedious process Æsthetic made mistakes which were at once deviations from the truth and attempts to reach it: such were the hedonism of the sophists and rhetoricians of antiquity and of the sensationalists of the eighteenth and second half of the nineteenth century; the moralistic hedonism of Aristophanes, of the Stoics, of the Roman eclectics, of the mediæval and Renaissance writers; the ascetic and logical hedonism of Plato and the Fathers of the Church, of some mediæval and even some quite modern rigorists; and finally, the æsthetic mysticism which first appeared in Plotinus and reappeared again and again until its last and great triumph in the classical period of German philosophy. In the midst of these variously erroneous tendencies, ploughing the field of thought in every direction, a tenuous golden rivulet[1] seems to flow, formed by the acute empiricism of Aristotle, the forceful penetration of Vico, the analytical work of Schleiermacher, Humboldt, De Sanctis and others who echoed them with weaker voice. This series of thinkers suffices to remind us that æsthetic science no longer remains to be discovered; but at the same time the fact that they are so few and so often despised, ignored or controverted, proves that it is in its infancy.

History of science and history of the scientific criticism of particular errors.

The birth of a science is like that of a living being: its later development consists, like every life, in fighting the difficulties and errors, general and particular, which lurk in its path on every side. The forms of error are numerous in the extreme and mingle with each other and with the truth in complications equally numerous: root out one, another appears in its stead; the uprooted ones also reappear, though never in the same shape. Hence the necessity for perpetual scientific criticism and the impossibility of repose or finality in a science and of an end to further discussion. The errors which

1. tenuous golden rivulet：若隐若现的金色小溪

may be described as general, negations of the concept of art itself, have been touched on from time to time in the course of this History; whence it may be gathered a simple affirmation of the truth has not always been accompanied by any considerable recapture of enemy territory. As to what we have called particular errors, it is clear that when freed from confusing admixture of other forms and divested of fanciful expression, they reduce themselves to three heads, under which they have already been criticized in the first or theoretical part of this work. That is to say, errors may be directed (*a*) against the characteristic quality of the æsthetic fact; (*b*) against the specific; (*c*) against the generic: they may involve denial of the character of intuition, of theoretic contemplation, or of spiritual activity, which together constitute the æsthetic fact. Among the errors which fall into these three categories we are now to sketch in outline the history of those which have had, or have to-day, the greatest importance. Rather than a history it will be a historical essay, sufficient to show that, even in the criticism of individual errors, æsthetic science is in its infancy. If among these errors some appear to be decadent and nearly forgotten, they are not dead; they have not accomplished a legal demise[1] at the hands of scientific criticism. Oblivion or instinctive rejection is not the same thing as scientific denial.

---

1. demise：死亡

I RHETORIC: OR THE THEORY OF ORNATE FORM

Rhetoric in the ancient sense.

Proceeding according to rank in importance, we inevitably head the list of theories for examination with the theory of Rhetoric, or Ornate Form.

It will not be superfluous to observe that the meaning given in modern times to the word Rhetoric, namely, the doctrine of ornate form, differs from that which it had for the ancients. Rhetoric in the modern sense is above all a theory of elocution, while elocution (λέξις, φράσις, ἑρμηνεία, elocutio) was but one portion, and not the principal one, of ancient Rhetoric. Taken as a whole, it consisted strictly of a manual or *vade-mecum*[1] for advocates and politicians; it concerned itself with the two or the three "styles" (judicial, deliberative, demonstrative), and gave advice or furnished models to those striving to produce certain effects by means of speech. No definition of the art is more accurate than that given by its inventors the earliest Sicilian rhetoricians, scholars of Empedocles[2] (Corax[3], Tisias[4], Gorgias): Rhetoric is the creator of persuasion (πειθος δημιουργός). It devoted

1. *vade-mecum*：〈法〉手册
2. Empedocles：恩培多克勒（公元前 490—前 430），希腊哲学家，主张元素说，认为所有物质都是由水、火、土、气构成的，所有变化都是运动引起的。
3. Corax：克拉克斯，希腊修辞学创始人之一。
4. Tisias：提西阿斯，希腊修辞学创始人之一。

itself to showing the method of using language so as to create a certain belief, a certain state of mind, in the hearer; hence the phrase "making the weaker case stronger" (τὸ τὸν ἥττω λόgον κρείττω ποιεῖν); the "increase or diminution according to circumstances" (*eloquentia in augendo minuendoque consistit*); the advice of Gorgias to "turn a thing to a jest if the adversary takes it seriously, or to a serious matter if he takes it as a jest[1]," [1] and many similar well-known maxims.

Criticism from moral point of view.

He who acts in this manner is not only æsthetically accomplished, as saying beautifully that which he wishes to say; he is also and especially a practical man with a practical end in view. As a practical man, however, he cannot evade moral responsibility for his actions; this point was fastened upon by Plato's polemic against Rhetoric, that is to say against fluent political charlatans and unscrupulous lawyers and journalists. Plato was quite right to condemn Rhetoric (when dissociated from a good purpose) as blameworthy and discreditable, directed to arouse the passions, a diet ruinous to health, a paint disastrous to beauty. Even had Rhetoric allied herself to Ethics, becoming a true guide of the soul (ψυχαγωγία τις διὰ τῶν λόγον); had Plato's criticism been directed solely against her abusers (everything being liable to abuse save virtue itself, says Aristotle); had Rhetoric been purified, producing such an orator as Cicero desired, *non ex rhetorum officinis sed ex academiae spatiis*[2] [2] and imposing on him, with Quintilian, the duty of being *vir bonus dicendi peritus*[3]; [3] yet the unalterable fact remains that Rhetoric can never be considered a regular science, being formed of a congeries of widely

---

1. jest：笑话
2. *non ex rhetorum officinis sed ex academiae spatiis*：〈拉〉（演说家）不应从修辞工厂里制造出来，而应从学院里培养出来。
3. *vir bonus dicendi peritus*：〈拉〉擅长演讲的善良人

dissimilar cognitions.

Accumulation without system.

It included descriptions of passions and affections, comparisons of political and judicial institutions, theories of the abbreviated syllogism or enthymeme[1] and of proof leading to a probable conclusion, pedagogic and popular exposition, literary elocution, declamation and mimicry, mnemonic[2], and so forth.

Its fortunes in the Middle Ages and Renaissance.

The rich and heterogeneous content of this ancient Rhetoric (which reached its highest development in the hands of Hermagoras of Temnos[3] in the second century B.C.) gradually diminished in volume with the decadence of the ancient world and the change in political conditions. This is not the place to dwell on its fortunes in the Middle Ages or its partial replacement by formularies and *Artes dictandi*[4] (and later by treatises upon the art of preaching), or to quote the reasons given by such writers as Patrizzi and Tassoni for its disappearance from the world of their day; [4] such history would be well worth writing, but would be out of place here. We will merely state that whilst conditions were at work on every side corroding this complex of cognitions, Louis Vives[5],

1. abbreviated syllogism or enthymeme：简洁的三段论或三段论省略式
2. declamation and mimicry, mnemonic：激辩和模仿，记忆术
3. Hermagoras of Temnos：泰姆诺斯的赫玛戈拉斯，公元前 2 世纪的希腊修辞学家，提出“争点论”。
4. *Artes dictandi*：〈拉〉《说话的艺术》
5. Louis Vives：路易斯·维威斯（1493—1540），法国人文主义哲学家、逻辑学家和修辞学家，对情绪、记忆和习得有独到见解，被誉为“现代心理学之父”，著有《论灵魂与生命》《智慧入门》《反对伪辩证法》等。

Peter Ramus[1] and Patrizzi himself were busy criticizing it from the point of view of systematic science.

Criticisms by Vives, Ramus and Patrizzi.

Vives emphasized the confused methods of the ancient treatise-writers, who embraced *omnia*[2], united eloquence with morality, and insisted that the orator must be *vir bonus*[3]. He rejected four-fifths of ancient Rhetoric as extraneous: namely, memory, which is necessary in all arts; invention, which is the matter of each individual art; recitation, which is external; and disposition, which belongs to invention. He retained elocution only, not that which treats of *quid dicendum*[4], but of *quem ad modum*[5], extending it beyond the three styles or kinds to include history, apologue, epistles[6], novels and poetry. [5] Antiquity furnishes us with few and faint attempts at such extension; now and then a Rhetorician ventures to suggest that the γένος ἱστορικόν and ἐπιστολικόν be included in Rhetoric, and even (in spite of opposition) "infinite" questions, that is to say merely theoretical questions with no practical application, which amounts to a scientific or philosophical genus; [6] others agreed with Cicero [7] that when one had mastered the most difficult of all arts, forensic eloquence, all else seemed child's-play (*ludus est homini non hebeti* ...). Ramus and his pupil Omer Talon[7] reproached Aristotle, Cicero and Quintilian with having confused

1. Peter Ramus：彼得·拉米斯（1515—1572），法国人文主义者、逻辑学家、数学家、教育改革家，开创了教材的范式，著有《辩证法》《算术》等。
2. *omnia*：〈拉〉所有的
3. *vir bonus*：〈拉〉善者
4. *quid dicendum*：〈拉〉说了什么
5. *quem ad modum*：〈拉〉怎样说
6. apologue, epistles：寓言、书信
7. Omer Talon：奥美尔·塔龙（1510—1562），法国人文主义者、修辞学家，拉姆斯弟子，著有《演讲学院》。

Dialectic and Rhetoric; and they assigned invention and disposition to the former, agreeing with Vives that "elocution" alone should be allowed to Rhetoric. [8] Patrizzi, on the other hand, refused the name of science to either, recognizing them as simple faculties, containing no individual matter (not even the three genera), and differentiating them only by attaching the term Dialectic to the dialogue form and proof of the necessary, and Rhetoric to connected discourse directed to persuasion in matters of opinion. Patrizzi observes that "conjoined speech" is used by historians, poets and philosophers, no less than by orators; and thus approaches the view of Vives. [9]

Survival into modern times.

In spite of these opinions the body of rhetorical doctrine continued to flourish in the schools. Patrizzi was forgotten; if Ramus and Vives had some followers (such as Francisco Sanchez and Keckermann), they were generally held up to odium[1] by the traditionalists. In the end, Rhetoric found a supporter in philosophy when Campanella made the following declaration in his *Rational Philosophy*: "*quodammodo Magiae portiuncula, quae affectus animi moderator et per ipsos voluntatem ciet ad quaecumque vult sequenda vel fugienda.*[2]" [10] Baumgarten owed to it his tripartition of Æsthetic into heuristic, methodology and semeiotic (invention, disposition and elocution), adopted later by Meier. Among Meier's numerous works is a little book entitled *Theoretic Doctrine of Emotional Disturbances in General*[3], [11] considered by him to be a psychological

1. were generally held up to odium：普遍遭到诋毁
2. *quodammodo Magiae portiuncula, quae affectus animi moderator et per ipsos voluntatem ciet ad quaecumque vult sequenda vel fugienda.*：〈拉〉如同某种形式的"幻术"的一部分，它改变人的情感，逼迫意志去追寻或躲避人想要的东西。
3. *Theoretic Doctrine of Emotional Disturbances in General*：《通常情况下的情绪干扰理论依据》，梅尔著。

introduction to æsthetic doctrine. On the other hand, Immanuel Kant in his *Critique of Judgment* observes that eloquence, in the sense of *ars oratoria* or art of persuasion by means of beautiful appearance and dialectical form, must be distinguished from beautiful speaking (*Wohlredenheit*); and that the art of oratory, playing upon the weakness of men to gain its own ends, "is worthy of no esteem" (*gar keiner Achtungwürdig*) [12] . But in the schools it flourished in many celebrated compilations, including one by the French Jesuit Father Dominique de Colonne, which was in use until some few decades ago. Even to-day, in so-called Literary Institutions, we come across survivals of ancient Rhetoric, notably in chapters devoted to the art of oratory; and fresh manuals on judicial or sacred eloquence (Ortloff[1], Whately[2], etc. [13]) are actually appearing, though rarely, to-day. Still, Rhetoric in the ancient sense may be said to have disappeared from the system of the sciences; to-day no philosopher would dream of following Campanella in dedicating a special section of rational philosophy to Rhetoric.

Modern signification of Rhetoric. Theory of literary form.

In compensation for this process, the theory of elocution and beautiful speech has been in modern times progressively emphasized and thrown into scientific form. But the idea of such a science is ancient, as we have seen; and equally ancient is the style of exposition, consisting in the doctrine of a double form and the concept of ornate form.

Concept of ornament.

The concept of "ornament" must have occurred spontaneously to

1. Ortloff：奥特洛夫（1797—1868），德国法学家，著有《罗马法宝典》《德国民法体系的基本特征》等。
2. Whately：维特利（1787—1863），英国神学家、修辞和逻辑学家，著有《论基督教的一些独特性》《修辞学要素》《逻辑学要素》等。

the mind as soon as attention was directed to the values of speech by listening to poets reciting [14] or to oratorical contests in public gatherings. It must very early have been thought that the difference between good speaking and bad, or between that which gave more pleasure and that which gave less, between grave or solemn, and commonplace or colloquial, consisted in something additional superimposed upon the canvas of ordinary speech like an embroidery by a skilful orator. These considerations led the Græco-Roman rhetoricians to adopt the practice, like the Indians, who arrived at the distinction independently, to distinguish the bare (ψιλή) or purely grammatical form from another form containing an addition which they called ornament, κόσμος: *ornatum est* (Quintilian will serve, as typical of all the rest) *quod perspicuo ac probabili plus est*[1]. [15]

The notion of ornament as something added on from outside forms the basis of the theory which Aristotle, the philosopher of Rhetoric, gave of the queen of ornaments, Metaphor. According to him the high pleasure aroused by metaphor arises from the collocation of different terms and the discovery of relations between species and genera, producing "learning and knowledge by means of the genus" (μαθησιν και γνῶσιν διὰ τοῦ γένους), and that easy learning which is the greatest of human pleasures, [16] which amounts to saying that metaphor adds to the concept under consideration a group of minor incidental cognitions, as a kind of diversion and relief and pleasant instruction for the mind.

Classes of ornament.

Ornaments were divided and subdivided in a number of different ways. Aristotle (and previously Socrates, rather differently) classified the ornaments which diversify bare or nude form, under the heads of

1. *ornatum est ... quod perspicuo ac probabili plus est*：〈拉〉修辞是让人更清楚、更可能相信的形式。

dialect forms, substitutions and epithets, prolongations, truncations[1] and abbreviations of words, and other departures from common usage, and, finally, rhythm and harmony. Substitutions were of four classes: species for genus; genus for species; species for species; and proportionate.[17] After Aristotle, elocution was especially studied by Theophrastus[2] and Demetrius Phalereus[3]; these rhetoricians and their followers further solidified the classification of ornament by distinguishing tropes from figures (σχήματα) and dividing figures into figures of speech (scheimata τῆς λέχεως) and of thought (τῆς διανοίας), figures of speech into grammatical and rhetorical, and figures of thought into pathetic and ethic. Substitutions were divided into fourteen principal forms: metaphor, synecdoche, metonymy, antonomasia, onomatopeia, catachresis, metalepsis, epithet, allegory, enigma, irony, periphrase, hyperbaton and hyperbole[4]; each divided into subspecies and contrasted with its relative vice. Figures of speech amounted to a score or so (repetition, anaphora, antistrophe, climax, asyndeton, assonance, etc.[5]); figures of thought to about the same number (interrogation, prosopopœia, ætiopœia, hypotyposis, commotion, simulation, exclamation, apostrophe, aposiopesis, etc.[6]). If these divisions have any value as aids to memory in relation to articular literary forms,

1. truncations：缩短
2. Theophrastus：泰奥弗拉图斯（约公元前 371—287），希腊哲学家，逍遥学派领袖，著有《三段论分析》《论诡辩》《论肯定与否定》等。
3. Demetrius Phalereus：德米特里乌斯（约公元前 350—280），希腊辩论家，泰奥弗拉图斯弟子，逍遥学派人物之一。
4. Substitutions were divided into fourteen principal forms: metaphor, synecdoche, metonymy, antonomasia, onomatopeia, catachresis, metalepsis, epithet, allegory, enigma, irony, periphrase, hyperbaton and hyperbole：替代分成十四种重要形式：隐喻、提喻、换喻、反义、拟声、借喻、转喻、形容喻、讽喻、隐晦语、反语、迂回、倒装、夸张。
5. repetition, anaphora, antistrophe, climax, asyndeton, assonance, etc.：重复、照应、同语颠倒复用、递进、连词省略、类韵等。
6. interrogation, prosopopœia, ætiopœia, hypotyposis, commotion, simulation, exclamation, apostrophe, aposiopesis, etc.：疑问法、拟人法、归因法、直叙法、激动法、拟态法、感叹法、顿呼法、顿绝法等。

considered rationally they are simply capricious, as is evidenced by the fact that many classes of the ornate appear now under the heading of tropes[1], now of figures; sometimes under figures of speech, then as those of thought, no reason for the alteration is given except the arbitrary caprice of an individual rhetorician which so decrees and disposes. And since one function which may be fulfilled by the rhetorical categories is to point out the divergence between two ways of expressing the same thing, one of which is arbitrarily selected as "proper," [18] it is easy to see why the ancients defined metaphor as "*verbi vel sermonis a propria significatione in aliam cum virtute mutatio*[2]," and figure as "*conformatio quaedam orationis remota a communi et primum se offerenti ratione*[3]." [19]

The concept of the Fitting.

So far as we know, antiquity raised no revolt against the theory of ornament or of double form. We do sometimes hear Cicero, Quintilian, Seneca and others saying, *Ipsae res verba rapiunt, Pectus est quod disertos facit et vis mentis, Rem tene, verba sequentur, Curam verborum rerum volo esse sollicitudinem*[4], or *Nulla est verborum nisi rei cohaerentium virtus.*[5] But these maxims did not bear the weighty meanings which we moderns might attach to them; they were perhaps in contradiction with the theory of ornament, but as the contradiction was unheeded, it was ineffective: they were the protests of common sense, powerless to combat the fallacies of school doctrine.

1. tropes：转义
2. *verbi vel sermonis a propria significatione in aliam cum virtute mutatio*：〈拉〉言语片段的原意向另一含义的转变
3. *conformatio quaedam orationis remota a communi et primum se offerenti ratione*：〈拉〉非同一般或者更明确的造句方式
4. *Ipsae res verba rapiunt, Pectus est quod disertos facit et vis mentis, Rem tene, verba sequentur, Curam verborum rerum volo esse sollicitudinem*:〈拉〉引发言语的是事实本身，赋予言语说服力，给予言语力量的是心灵。
5. *Nulla est verborum nisi rei cohaerentium virtus.*：〈拉〉如与事实不符，言词便毫无意义。

Moreover, the latter was fitted with a safety-valve, a sage contrivance[1] to disguise its inherent absurdity. If the ornate consisted of a *plus*[2], in what degree should it be used? if it gave pleasure, must we not conclude that the more it were used, the greater the pleasure derived? would its extravagant use be attended by extravagant pleasure? Herein was peril: instinctively the rhetoricians hastened to the defence, snatching up the first weapon that came to hand, namely, the fitting (πρέπον). Ornament must be used carefully; neither too much too little; *in medio virtus*[3]; as much as is fitting (ἀλλά πρέπον). Aristotle recommends a style seasoned with "a certain dose" (δεῖ ἄρα κεκρᾶσθαί πως τούτοις.) for ornament should be a condiment[4], not a food (ἥδυσμα, οὐκ ἔδεσμα). [20] The fitting was a concept quite inconsistent with that of ornament; it was a rival, and enemy, destined to destroy it. Fitting to what? to expression of course; but that which is fitting to expression cannot be called an ornament, an external addition; it coincides with expression itself. But the rhetoricians contented themselves with maintaining peaceful relations between the ornate and the fitting, without troubling to mediate them through a third concept. The pseudo-Longinus alone in answer to an observation of his predecessor Cæcilius that more than two or three metaphors must not be used in the same place, remarked that a larger number ought to be used where passion (τὰ πάθη) rushes headlong like a torrent, carrying with it as necessaries (ὡς ἀναγκαῖον) a multitude of such substitutions. [21]

The theory of ornament in the Middle Ages and Renaissance.

Preserved in the compilations of later antiquity (such as the works

---

1. sage contrivance：精巧装置
2. a *plus*：有利因素
3. *in medio virtus*：〈拉〉适量即美德
4. condiment：调味品

of Donatus[1] and Priscian[2] and the celebrated allegorical tract of Marcianus Capella[3]), and in the compendia of Bede[4], Rhabanus Maurus[5] and others, the theory of ornament passed to the Middle Ages. Throughout this period Rhetoric, Grammar and Logic continued to form the *trivium*[6] of the schools. The theory was to some extent favoured in mediæval times by the fact that writers and scholars made use of a dead language; this helped to reinforce the idea that beautiful form was not a spontaneous thing but consisted in an addition or embroidery. Under the Renaissance the theory continued to flourish and was revived by study of the best classical sources; to the works of Cicero were added the *Institutiones* of Quintilian and the *Rhetoric* of Aristotle, with the host of minor Latin and Greek rhetoricians, amongst whom was Hermogenes[7] with his celebrated *Ideas*, brought into fashion by Giulio Camillo[8]. [22]

Even those writers who dared to criticize the organism of ancient Rhetoric left the theory of ornament unassailed. Vives lamented over the "exaggerated subtlety of the Greeks" which had multiplied distinctions to infinity in this matter without diffusing light, [23] but he never took up a definite stand against the theory of ornament. Patrizzi

---

1. Donatus：多纳图斯，生活在 4 世纪中期，古罗马语法学家、修辞学家著有《语法艺术》《主要艺术》《次要艺术》等。
2. Priscian：普里西安，生活于公元 500 年前后，古罗马著名拉丁文语法学家，为思辨语法提供大量素材，著有《语法学院》等。
3. Marcianus Capella：马尔西亚诺·卡佩拉（生卒年不详），七大人文学科体系开创者之一，著有《语文学与墨丘利的联姻》《爱的寓言》等。
4. compendia of Bede：比得的手册。比得（672—735），英国圣彼得修道院修士，著有《英国人民教会史》《论理性时代》等。
5. Rhabanus Maurus：拉巴努斯·马乌鲁斯（780—856），德国本笃会修士、神学家，著有《物性论》。
6. *trivium*：〈拉〉三大学科
7. Hermogenes：赫尔莫根尼斯，生活于公元前 5 世纪到 4 世纪之间，古希腊哲学家，希腊军事统帅希波尼库斯之子，著有《论观念》。
8. Giulio Camillo：朱利奥·卡米洛（1480—1544），意大利哲学家，著有《记忆的戏剧》《戏剧的理念》等。

was dissatisfied with the insufficient definition of ornament given by the ancients; but he asserted the existence of ornaments and metaphors as well as seven different modes of "conjoined speech," — narrative, proof, amplification, diminution, ornament with its contrary, elevation and depression. [24] The school of Ramus continued to entrust Rhetoric with the "embellishment" of thought. Owing to the vast extension and intensification of life and literature in the sixteenth century, it would be easy to quote phrases, as we have done from ancient authors, asserting the strict dependence of speech upon the things it wishes to express, and lively attacks on pedants and pedantic forms and rules for beautiful speech. But what would be the use? The theory of ornament was always in the background, tacitly admitted as indisputable by all. Juan de Valdés[1], for instance, makes the following confession of stylistic faith: "*Escribo como hablo; solamente tengo cuidado de usar de vocablos que sinifiquen bien lo que quiero decir, y dígolo cuanto más llanamente me es posible, porqué, á mi parecer, en ninguna lengua está bien la afectación.*[2]" But Valdés also says that beautiful language consists "*en que digais lo que quereis con las menos palabras que pudiéredes, de tal manera que ... no se pueda quitar ninguna sin ofender á la sentencia, ó l encarescimiento, ó á la elegancia.*[3]" [25] Here it seems that amplification and elegance are conceived as extraneous to the meaning or content. — A gleam of truth is visible in Montaigne[4], who, confronted by the laboured

1. Juan de Valdés：胡安·德·瓦尔德斯（1490—1541），西班牙宗教作家、新教改革家，著有《语言谈话录》《〈马太福音〉评注》等。
2. *Escribo como hablo; solamente tengo cuidado de usar de vocablos que sinifiquen bien lo que quiero decir, y dígolo cuanto más llanamente me es posible, porqué, á mi parecer, en ninguna lengua está bien la afectació*:〈西〉我很注意一些词的使用；这些词能明白无误地表达我的想法，而且尽可能把话说得通顺直白。因为我认为，任何语言中都不存在矫揉造作。
3. *en que digais lo que quereis con las menos palabras que pudiéredes, de tal manera que ... no se pueda quitar ninguna sin ofender á la sentencia, ó l encarescimiento, ó á la elegancia.*：〈西〉用尽量少的语言将你的想法表达出来。如此……，不损害句子本身，不改变句子长度，更不影响句子的优雅。
4. Montaigne：蒙田（1533—1592），法国文艺复兴后期作家、人文主义思想家、怀疑论者，著有《蒙田随笔》。

categories into which rhetoricians divide ornament, observes: "*Oyez dire Métonymie, Métaphore, Allégorie et aultres tells noms de la Grammaire; semble il pas qu'on signifie quelque forme de langage rare et pellegrin? Ce sont tiltres qui touchent le babil de vostre chambrière.*[1]"[26] That is to say, they are anything but language remote from the *primum se offerens ratio*[2].

Reductio ad absurdum[3] *in the seventeenth century.*

The impossibility of upholding the theory of ornament was first noticed during the decadence of Italian literature in the seventeenth century, when literary production became but a play of empty forms, and the convenient, long violated in practice, was abandoned and forgotten even in theory, and came to be looked on as a limit arbitrarily imposed on the fundamental principle of ornamentation. The opponents of that style loaded with conceits which is known as "secentismo[4]" from its prevalence in the seventeenth century (Matteo Pellegrini, Orsi and others) felt the viciousness of the literary production of their day; they were aware that decadence was due to the fact that literature was no longer the serious expression of a content; but they were embarrassed by the reasoning of the champions of bad taste, who were able to demonstrate that the whole business conformed in every particular with the literary theory of ornament, the common ground of both parties. In vain did the former appeal to the "convenient," the "moderate," the "avoidance of affectation," to

1. *Oyez dire Métonymie, Métaphore, Allégorie et aultres tells noms de la Grammaire; semble il pas qu'on signifie quelque forme de langage rare et pellegrin? Ce sont tiltres qui touchent le babil de vostre chambrière.*：〈法〉提到换喻、隐喻、寓喻和其他语法术语的时候，你不会觉得这是些稀奇古怪的说法吧？它们就是你家女佣们嚼舌头的时候用的玩意儿。
2. *primum se offerens ratio*：〈拉〉主流（语言）环境
3. *Reductio ad absurdum*：〈拉〉归谬法
4. secentismo：〈意〉非常讲究技巧的文体

ornament as "condiment, not food," and all the other weapons which had sufficed in times when healthy literary production and sound æsthetic taste had automatically corrected faulty theory: the other party replied, there was no reason to be sparing in use of ornament when it lay in masses ready to hand, or to avoid an ostentatious display of wit[1] when one had an inexhaustible supply. [27]

Polemic concerning the theory of ornament.

The same reaction against the abuse of ornament, against "Spanish and Italian conceits" (whose supporters had been Gracian in Spain and Tesauro in Italy), took place in France. " ... *Laissez à l'Italie De tous ces faux brillants l'éclatante folie*[2]"; "*Ce que l'on conçoit bien s'énonce clairement. Et les mots, pour le dire, arrivent aisément.*[3]" [28] Among the sharpest critics of conceits was the Jesuit Bouhours, already quoted, author of the *Manière de bien penser dans les œuvres d'esprit.*[4] The rhetorical forms were the subject of warm controversy. Orsi, on national grounds the opponent of Bouhours (1703), asserted that all the ornamental devices of wit rested on a middle term and could be reduced to a rhetorical syllogism, and that wit consists of a truth which appears false or a falsehood which appears true. [29] If this controversy produced no great scientific result at the time, at least it prepared the mind for greater liberty; and, as we have remarked elsewhere, [30] it may have influenced Vico, who, in framing his new concept of poetical

1. ostentatious display of wit：毫无节制的智巧炫耀
2. *Laissez à l'Italie De tous ces faux brillants l'éclatante folie*：〈法〉让意大利带着他的光辉和优雅文风发疯去吧！
3. *Ce que l'on conçoit bien s'énonce clairement. Et les mots, pour le dire, arrivent aisément.*：〈法〉心中有底，就能表达清楚。找到合适的词语来表达简直易如反掌。
4. *Manière de bien penser dans les œuvres d'espri.*：〈法〉《精神作品中的良好思维方式》，鲍赫斯著。

imagination, recognized that it necessitated a wholesale reconstruction of the theory of rhetoric and the conclusion that its figures and tropes are not "caprices of pleasure" but "necessities of the human mind." [31]

Du Marsais and metaphor.

We find the theory of rhetorical ornament jealously kept intact by Baumgarten and Meier, while in France it was as vigorously assailed by César Chesneau du Marsais, who published in 1730 a treatise on *Tropes* (the seventh part of his *General Grammar*) [32] wherein he develops, on the subject of metaphor, the observation already made by Montaigne: indeed he was perhaps inspired by Montaigne, although he does not mention his name. Du Marsais remarks that it is said that figures are modes of speech and turns of expression removed from the ordinary and common; which is an empty phrase, as good as saying "the figured differs from the non-figured and figures are figures and not non-figures." On the other hand it is wholly untrue that figures are removed from ordinary speech, for "nothing is more natural, ordinary and common than figures: more figures of speech are used in the town square on a market-day than in many days of academical discussion"; and no speech, however short, can be composed entirely of non-figurative expressions. And Du Marsais gives instances of quite obvious and spontaneous expressions in which Rhetoric cannot refuse to recognize the figures of apostrophe, congeries, interrogation, ellipsis, prosopopœia: "The apostles were persecuted and suffered their persecutions with patience. What can be more natural than the description given by St. Paul? *Maledicimur et benedicimus; persecutionem patimur et sustinemus; blasphemamur et obsecramus.*[1] Yet the apostle makes

---

1. *Maledicimur et benedicimus; persecutionem patimur et sustinemus; blasphemamur et obsecramus.*：〈拉〉我们被诅咒，却在祝福；遭了迫害，却在忍受；受了欺负，却在祈祷。

use of a fine figure of antithesis; cursing is the opposite to blessing; persecution to endurance; blasphemy to prayer." But further, the very language of the figure is figured, since it is a metaphor. — But after such acute observations, Du Marsais ends by himself becoming confused and defines figures as "manners of speech differing from others in a particular modification by which it is possible to reduce each one to a species apart, and give a more lively, noble or pleasing effect than can be gained by a manner of speech expressing the same content of thought without such particular modification." [33]

Psychological interpretation.

But the psychological interpretation of figures of speech, the first stage towards their æsthetic criticism, was not allowed to drop here. In his *Elements of Criticism*, Home says that he had long questioned whether that part of Rhetoric concerning figures might not be reduced to rational principles, and had finally discovered that figures consist in the passional element; [34] he set himself therefore to analyse prosopopœia, apostrophe and hyperbole[1] in the light of the passional faculty. From Du Marsais and Home is derived everything of value in the *Lectures on Rhetoric and belles lettres* of Hugh Blair[2], professor at Edinburgh University from 1759 onwards; [35] published in book form, these lectures had an immense vogue in all the schools of Europe including those of Italy, and replaced advantageously, by their "reason and good sense," works of a much cruder type. Blair defined figures in general as "language suggested by imagination or passion." [36] Similar ideas were promulgated

1. hyperbole：夸张法
2. Hugh Blair：修・布莱尔（1718—1800），苏格兰布道家、修辞学家，著有《论自然法的基础与义务》《修辞学与美的文学教程》等。

in France by Marmontel[1] in his *Elements of Literature.* [37] In Italy Cesarotti was contrasting the logical element or "cypher-terms" of language with the rhetorical element or "figure-terms," and rational eloquence with imaginative eloquence. [38] Beccaria[2], though a shrewd psychological analyst, held to the view of literary style as "accessory ideas or feelings added to the principal in any discourse"; that is, he failed to free himself from the distinction between the intellectual form intended for the expression of the principal ideas, and the literary form, modifying the first by the addition of accessory ideas. [39] In Germany an effort was made by Herder to interpret tropes and metaphors as Vico had done, that is to say as essential to primitive language and poetry.

Romanticism and Rhetoric: present day.

Romanticism was the ruin of the theory of ornament, and caused it practically to be thrown on the scrap-heap[3], but it cannot be said to have gone under for good or to have been superseded by a new and accurately stated theory. The chief philosophers of Æsthetic (not only Kant, who as we know remained in bondage to the mechanical and ornamental theory; not only Herder, whose knowledge of art seems to have been confined to a little music and a great deal of rhetoric; but such romantic philosophers as Schelling, Solger and Hegel) still retained the sections devoted to metaphor, trope and allegory for tradition's sake, without severe scrutiny. Italian Romanticism with Manzoni at its head destroyed the belief in beautiful and elegant

---

1. Marmontel：马蒙泰尔（1723—1799），法国历史学家、作家、文学家，著有《法国诗学》《论语言使用的权威性》《论文学要素》等。
2. Beccaria：贝卡利亚（1738—1794），意大利经济学家、法理学家和刑罚改革者，西方近代刑事古典学派的创始人，著有《论犯罪与惩罚》等。
3. scrap-heap：废物堆

words, and dealt a blow at Rhetoric: but was it killed by the stroke? Apparently not, judging by the concessions unconsciously made by the scholastic treatise-writer Ruggero Bonghi[1], whose *Critical Letters* assert the existence of two styles or forms, which at bottom are nothing else than the plain and the ornate. [40] German schools of philology have pretty generally accepted the stylistic theory of Gröber[2], who divides style into logical (objective) and affective (subjective): [41] an ancient error masked by terminology borrowed from the psychological philosophy in fashion at modern universities. In the same spirit a recent writer rechristens[3] the rhetorical doctrine of tropes and figures by the title "Doctrine of the Forms of Æsthetic Apperception," and divides them into the four categories (the ancient wealth of categories reduced to a paltry four!) of personification, metaphor, antithesis, and symbol. [42] Biese[4] has devoted an entire book to metaphor; but one searches it in vain for a serious æsthetic analysis of this category. [43]

The best scientific criticism of the theory of ornament is found scattered throughout the writings of De Sanctis, who when lecturing on rhetoric preached what he called anti-rhetoric. [44] But even here the criticism is not conducted from a strictly systematic point of view. It seems to us that the true criticism should be deduced negatively from the very nature of æsthetic activity, which does not lend itself to partition; there is no such thing as activity type *a* or type *b*, nor can the same concept be expressed now in one way, now in another. Such is the only way of abolishing the double monster of bare form which is, no one knows how, deprived of imagination, and ornate

1. Ruggero Bonghi：鲁杰罗·邦吉（1826—1895），意大利学者、政治家、作家，著有《持之以恒》《新本体论》等。
2. Gröber：戈洛博（1844—1911），德国语文学家，著有《法国古代传奇小说与田园诗》《中世纪法国文学纲要》等。
3. rechristens：重新命名
4. Biese：比耶塞（1816—1881），德语教师、语文学家。

form which contains, no one knows how, an addition on the side of imagination. [45]

[1] For Gorgias' saying see Aristotle, *Rhet.* iii. ch. 18.

[2] Cicero, *Orat. ad Brut.*, introd.

[3] Quintilian, *Inst. orat.* xii. c. i.

[4] Fran. Patrizzi, *Della rhetorica*, ten dialogues, Venice, 1582, dial. 7; Tassoni, *Pensieri diversi*, bk. x. ch. 15.

[5] *De causis corruptarum artium*, 1531, bk. iv.; *De ratione dicendi*, 1533.

[6] Cicero, *De or at:* i. chs. 10–11; Quintil. *Inst. oral.* iii. ch.5.

[7] *De orat.* ii. chs. 16–17.

[8] P. Ramus, *Instil, dialecticæ,* 1543; *Scholæ in artes liberales*, 1555 etc.; Talæus, *Instit. orator.*, 1545.

[9] *Della rhetorica*, dial. 10, and *passim.*

[10] *Ration. Philos.*, part iii. *Rhetoricorum liber unus juxta propria dogmata* (Paris, 1636), ch. 3.

[11] *Theoretische Lehre von den Gemüthsbewegungen überhaupt*, Halle, 1744.

[12] *Kritik d. Urtheils kraft*, § 53 and *n.*

[13] H. F. Ortloff, *Die gerichtliche Redekunst*, Neuwied, 1887; R. Whately, *Rhetoric*, 1828 (for *Encyd. Brit.*); Ital. trans., Pistoia, 1889.

[14] Aristotle, *Rhet.* iii. ch. 1.

[15] Quintil. *Inst. orat.* viii. ch. 3.

[16] *Rhet.* iii. ch. 10.

[17] *Poet.* chs. 19–22; cf. *Rhet.* iii. cc. 2, 10.

[18] See above, pp. 100–101.

[19] Quintilian, *Inst. orat.* viii. ch. 6; ix. ch. 1.

[20] Aristotle, *Rhet.* iii. ch. 2; *Poet.* ch. 22.

[21] *De sublimitate* (in *Rhet. græci*, ed. Spengel, vol. 1. § 32.)

[22] Giulio Camillo Delminio, *Discorso sopra le Idee di Ermogene* (in

Opere, Venice, 1560); and trans. of Hermogenes (Udine, 1594).

[23] *De causis corruptarum artium, loc. cit.*

[24] *Della rhetorica*, dial. 6.

[25] *Diálogo de las lenguas* (ed. Mayans y Siscar, *Origines de la lengua espanola*, Madrid, 1873), pp. 115, 119.

[26] *Essais*, i. ch. 52 (ed. Garnier, i. 285); ci. *ibid.* chs. 10, 25, 39; 10.

[27] Croce, *I trattatisti italiani del concettismo*, pp. 8–22.

[28] Boileau, *Art poétique*, i. 11. 43–44, 153–154.

[29] G. G. Orsi, *Considerazioni sopra la maniera di ben pensare*, etc., 1703 (reprinted Modena, 1735, with all polemics relating thereto).

[30] See above, pp. 262–263.

[31] See above, pp. 257–258.

[32] *Des tropes ou des différens sens dans lesquels on peut prendre un même mot dans une même langue*. Paris, 1730 (*Œuvres de Du Marsais*, Paris, 1797, vol. i.).

[33] *Des tropes ou des différens sens dans lesquels on peut prendre un même mot dans une même langue*, part i. art. 1; cf. art. 4.

[34] *Elem. of Criticism*, iii. ch. 20.

[35] Hugh Blair, *Lectures on Rhetoric and belles lettres* (London, 1823).

[36] *Lect. on Rhet. and belles lettres*, lecture 14.

[37] Marmontel, *Éléments de littéral*, (in *Œuvres*, Paris, 1819), iv. p. 559.

[38] Cesarotti, *Saggio sulla filos. del linguaggio*, part ii.

[39] *Ricerche intorno alla natura dello stile* (Turin, 1853), ch. 1.

[40] R. Bonghi, *Lettere critiche*, 1856 (4th ed., Naples, 1884), pp. 37, 65–67, 90, 103.

[41] Gustav Gröber, *Grundriss d. romanischen Philologie*, vol. 1. pp. 200–250, K. Vossler, *B. Cellinis Stil in seiner Vita, Versuch einer psychol. Stilbetrachtung*, Halle a. S., 1899; cf. the self-criticism of Vossler, *Positivismus u. Idealismus in der Sprachwissenschaft*, Heidelberg, 1904 (It. trans., Bari, Laterza, 1908).

[42] Ernst Elsteb, *Principien d. Literaturwissenschaft*, Halle a. S., 1097. vol. i. pp. 359–413.

[43] Biese, *Philos, des Metaphorischen*, Hamburg-Leipzig, 1893.

[44] *La Giovinezza di Fr. de S.* chs. 23, 25; *Scritti varî*, ii. pp. 272–274.

[45] See above, pp. 67–73.

II HISTORY OF THE ARTISTIC AND LITERARY KINDS

The kinds in antiquity. Aristotle.

The theory of artistic and literary kinds and of the laws or rules proper to each separate kind has almost always followed the fortunes of the rhetorical theory.

Traces of the threefold division into epic, lyric and dramatic are found in Plato; and Aristophanes gives an example of criticism according to the canon of the kinds, particularly that of tragedy. [1] But the most conspicuous theoretical treatment of the kinds bequeathed us by antiquity is precisely the doctrine of Tragedy which forms a large part of the Aristotelian fragment known as the Poetics. Aristotle defines such a composition as an imitation of a serious and complete action, having size, in language adorned in accordance with the requirements of the different parts, its exposition to be by action and not by narration, and using pity or terror as means to free or purify us from these same passions; [2] he gives minute details as to the six parts of which it is composed, especially the plot and the tragic character. It has been often said, ever since the days of Vincenzo Maggio in the sixteenth century, that Aristotle treated of the nature of poetry, or particular forms of poetry, without claiming to give precepts. But Piccolomini answered that "all these things and other similar ones are shown or asserted with no other purpose but that we may see in what way their precepts and laws must be obeyed and carried out," just as, to make a hammer or saw, one begins by describing the parts of which they are composed. [3] The error of which we take Aristotle as representative lies in transmuting[1] abstractions and empirical partitions into rational concepts: this was almost inevitable at the beginnings of æsthetic reflexion, and the Sanskrit[2] theory of poetry employed the same method independently when, for example, it defines and legislates for ten principal and eighteen secondary styles of drama; forty-eight varieties of hero; and we know not how many kinds of heroines. [4]

In the Middle Ages and Renaissance.

After Aristotle, the theory of poetic kinds does not seem to have been completely or elaborately developed in antiquity. The Middle Ages may be said to have expressed the doctrine in treatises of the kind known as "rhythmic arts" or "methods of composition." When the Aristotelian fragment was first noticed, it is curious to see the way in which the paraphrase of Averroes distorted the theory of kinds. Averroes conceives tragedy as the art of praise, comedy as that of blame, which amounts to identifying the former with panegyric, the latter with satire; and he believes the *peripeteia*[3] to be the same thing as antithesis, or the artifice of beginning the description of a thing by describing its opposite. [5] This distortion demonstrates afresh the merely historical character of these kinds and their unintelligibility by the methods of pure logic to a thinker living in times and under customs different from those of the Hellenic world. The Renaissance seized upon Aristotle's text, partly expounded it, partly distorted it and partly thought it out afresh, and thus succeeded in establishing a long list of kinds and sub-kinds rigidly defined and

---

1. transmuting：转变
2. Sanskrit：梵语的
3. *peripeteia*：（戏剧、小说中）命运突变

subjected to inexorable laws. Controversy now began over the correct understanding of the unities of epic or dramatic poetry; over the moral quality and social standing proper to the characters in this kind of poem and in that; over the nature of the plot, and whether it includes passions and thoughts, and whether lyrics should or should not be received as true poetry; whether the material of tragedy should be historical; whether the dialogue of comedy may be in prose; whether a happy ending may be allowed in tragedy; whether the tragic character may be a perfect gentleman; what kind and number of episodes is admissible in the poem, and how they should be incorporated in the main plot; and so on. Great anguish was caused by the mysterious rule of catharsis found in black and white in Aristotle's text, and Segni naïvely predicted that tragic poetry would be revived in its perfect spectacular entirety for the sake of experiencing the effect spoken of by Aristotle, that "purgation[1]" which causes "the birth of tranquillity in the soul and of freedom of all perturbation[2]." [6]

The doctrine of the three unities.

Amongst the many undertakings brought to a glorious end by the critics and treatise-writers of the sixteenth century, the best known is the establishment of the three unities of time, place and action. One cannot indeed see why they are called unities, for in strictness they could at most be spoken of as shortness of time, straitness of space and limitation of tragic subjects to a certain class of action. It is well known that Aristotle prescribed unity of action only, and reminded his hearers that theatrical custom alone imposed on the action a time-limit of one day. On this last point the critics of the sixteenth century accorded six, eight, or twelve hours according to individual taste or

1. purgation：净化
2. perturbation：心绪不宁

humour: some of them (amongst them Segni) allowed twenty-four hours, including the night as particularly propitious[1] to assassinations and the other acts of violence which usually form the plot of tragedies; others extended the limit to thirty-six or forty-eight hours. The last, and most curious, unity, that of place, was slowly developed by Castelvetro, Riccoboni and Scaliger until the Frenchman Jean de la Taille[2] joined it as a third to the existing two in 1572, and in 1598 Angelo Ingegneri[3] finally formulated it more explicitly.

Poetics of the kinds and rules. Scaliger.

The Italian treatises were widely read and regarded as authoritative all over Europe, and awakened the first effort towards a learned theory of poetry in France, Spain. England and Germany. A good representative of his class is Julius Cæsar Scaliger, who has been considered, with some exaggeration, as the true founder of French pseudo-classicism or neo-classicism; as one who (it has been said) "laid the first stone of the classical Bastille." But if he was neither the first nor the only one, he certainly helped greatly to reduce "to a system of doctrines the principal consequences of the sovranty of Reason[4] in works of literature," with his minute distinctions and classifications of kinds, the insurmountable barriers he erected between them, and his distrust of free inspiration and imagination. [7] Scaliger numbers among his descendants (beside Daniel Heinsius[5]) d'Aubignac[6],

1. propitious：有利的
2. Jean de la Taille：让·德·拉·塔耶（1540—1607），法国作家、诗人、戏剧家。
3. Angelo Ingegneri：安吉罗·因杰涅里（1550—1613），意大利作曲家。
4. sovranty of Reason：理性至上
5. Daniel Heinsius：丹尼尔·海因修斯（1580—1655），荷兰文艺复兴时期著名学者、诗人、剧作家，著有《虱子的赞美诗》《受伤的自由》《无辜者的大屠杀》等。
6. d'Aubignac：奥比纳克神父（弗朗索瓦·黑德林），法国作家，著有《论戏剧诗》《圣凯特琳的烈士》等。

Rapin[1], Dacier and ther tyrants of French literature and drama: Boileau turned the rules of neo-classicism into neat verses.

Lessing.

It has been noticed that Lessing entered the same field; his opposition to the French rules (which was an opposition of rule to rule, in which he had been forestalled by Italian writers, for example by Calepio in 1732) is anything but radical. Lessing maintained that Corneille[2] and other authors had misinterpreted Aristotle, to whose laws even the Shakespearian drama could be shown to conform; [8] but on the other hand he strongly opposed the abolition of all rules and those who shouted "genius, genius," placing genius above the law and saying that genius makes the law. For the very reason that genius is law, replied Lessing, laws have their value and can be determined: negation of them would entail the confinement of genius to its first trial flights, making example or practice useless. [9]

Compromises and extensions.

But the "kinds" and their "limits" could be maintained for centuries solely by means of infinitely subtle interpretations, analogical extensions and more or less concealed compromises. The Italian Renaissance critics, while working at their Poetics in the style of Aristotle, found themselves confronted with chivalric poetry, and had to make the best of it; this they did by assigning it to a kind of poem

---

1. Rapin：拉潘（1535—1608），法国文艺复兴时期朝臣、翻译家、诗人和讽刺作家，著有《哲学的爱》《离别挽歌》等。
2. Corneille：高乃依（1606—1684），法国剧作家，与莫里哀、拉辛并称17 世纪法国三大剧作家，代表作有《熙德》《贺拉斯》《西拿》等。

not foreseen by antiquity (Giraldi Cintio[1]). [10] Here and there indeed a rigorist was heard protesting that romances were in no way different from heroic poetry, and were only "badly written heroics" (Salviati). And since it was impossible to deny a place in Italian literature to Dante's poem, Iacopo Mazzoni, in his *Defence of Dante*, overhauled once more the categories of Poetics in order to find a niche for the sacred poem. [11] Farces[2] made their appearance at this time, and Cecchi[3] (1585) declares "Farce is a third novelty, occupying a place between tragedy and comedy ... " [12] The *Pastor fido* of Guarini[4] was published, neither tragedy nor comedy, but tragicomedy; and discovering no heading among the kinds deduced from moral or civil philosophy suitable for the intruder, Jason de Nores[5] proceeded to rule it out of existence; Guarini made a valiant defence and claimed special protection for his beloved *Pastor*[6] under a third, or mixed, style, representative of real life. [13] Another rigorist, Fioretti[7] (Udeno Nisieli) proclaimed the poem "a poetic monster, so huge and deformed that centaurs, hippogriffs and chimæras[8] are comparatively graceful and charming ..., fit to bring a blush to the cheek of the muse, a disgrace to poetry, a mixture of ingredients in themselves discordant, inimical

---

1. Giraldi Cintio：吉拉尔迪·辛迪奥（1504—1573），意大利小说家、诗人、剧作家，曾任自然哲学教授，著有史诗《埃尔克莱》、悲剧《奥博切》等。
2. farces：滑稽剧
3. Cecchi：切基（1518—1587），意大利诗人、剧作家，著有《医生》《魔咒》等。
4. Guarini：瓜里尼（1538—1612），意大利诗人、剧作家、文艺理论家。首创"悲喜混杂剧体诗"与"悲喜混杂"之说。著有悲喜混合剧《忠实的牧羊人》《悲喜混杂剧体诗的纲要》等。
5. Jason de Nores：亚松·德·诺莱斯（1530—1590）意大利学者，作家，帕多瓦大学哲学教授。
6. *Pastor*：田园剧，指瓜里尼 1590 年出版的田园诗作《忠实的牧羊人》。
7. Fioretti：贝内德托·菲奥莱蒂（1579—1642），又名乌德诺·尼西利，意大利神学家、哲学家、诗人、文学评论家。
8. centaurs, hippogriffs and chimæras：半人半马、半鹰半马和狮头羊

and incompatible"; [14] but will this bluster drive the delicious *Pastor fido* from the hands of lovers of poetry? The same thing occurred in the case of Marino's *Adone*, described by Chapelain[1] as "a poem of peace" for want of a better definition, though other supporters called it "a new form of epic poem"; [15] and the same thing happened again in the case of the comedy of art and musical drama. Corneille, who had called down a furious tempest from Scudéry and the Academicians on the head of his *Cid*, remarked in his discourse on Tragedy, though basing his position on that of Aristotle, that there was necessity for *"quelque modération, quelque favorable interprétation, ... pour n'être pas obligés de condamner beaucoup de poèmes, que nous avons vu réussir sur nos théâtres." "Il est aisé de nous accommoder avec Aristote ...* "[2] [16] he says in another place: a piece of literary hypocrisy which startles by its verbal resemblance to "*les accommodements avec le Ciel*[3]" of the Tartuffian[4] ethics. The following century saw the accepted kinds augmented by "bourgeois tragedy[5]" and pathetic comedy, nicknamed "lachrymose[6]" by its enemies; de Chassiron[7] [17] attacked, and Diderot[8], Gellert[9] and

---

1. Chapelain：夏普兰（1595—1674），法国诗人、评论家，因参与组建法兰西学院而闻名，著有《圣女》（史诗）《古代小说阅读心得》等。
2. "*quelque modération, quelque favorable interprétation, ... pour n'être pas obligés de condamner beaucoup de poèmes, que nous avons vu réussir sur nos théâtres.*" "*Il est aisé de nous accommoder avec Aristote ...* "：〈法〉"为了不再责难我们在舞台上取得巨大成就的诗作，做一些修改和有益的解释。""与亚里士多德保持一致很容易……"
3. *les accommodements avec le Ciel*：〈法〉与上天一致
4. Tartuffian：伪君子的
5. bourgeois tragedy：资产阶级悲剧
6. lachrymose：催泪剧
7. de Chassiron：即查尔斯·德·沙西龙男爵（1818—1871），法国外交官，著有《中国—日本—印度游记》。
8. Diderot：狄德罗（1713—1784），法国启蒙思想家、哲学家、戏剧家，百科全书派代表人物，主张艺术应反映现实生活。著有《哲学沉思》《拉摩的侄儿》《修女》等。
9. Gellert：热莱尔（1715—1769），德国启蒙运动作家、诗人，著有《精神颂歌》《祈祷的姐妹》《生病的女人》等。

Lessing [18] defended the new arrival. In this way the schematism of the kinds continued to suffer violence and to cut a very poor figure; nevertheless, in spite of adversity, it made every effort to retain power even at the sacrifice of dignity: just as an absolute king turns constitutional by force of circumstance, and chooses the lesser evil of squaring his divine right with the will of the nation.

Rebellion against rules in general.

This retention of power would have been more difficult had any success attended the attempts at rebellion against all laws, against law in general, which broke out in varying degrees at the end of the sixteenth century. Pietro Aretino[1] made mock of the most sacred precepts: in a prologue to one of his comedies he remarks derisively, "If you see more than five characters on the stage at once, do not laugh; for chains which would fasten water-mills to the river could not hold the fools of to-day." [19]

G. Bruno. Guarini.

A philosopher, Giordano Bruno[2], entered the lists against the "regulators of poetry": rules, said he, are derived from poetry: "there are as many genera and species of true rules as there are genera and species of true poets"; such an individualization of kinds dealt them a deathblow. "How then" (asks the interlocutory opponent) "shall veritable poets be recognized?" "By their singing of verse" (answers

---

1. Pietro Aretino：皮耶特罗·阿雷帝诺（1492—1556），意大利剧作家、诗人，著有《奢华十四行诗》《爱的疑问》等。
2. Giordano Bruno：乔达诺·布鲁诺（1548—1600），意大利文艺复兴时期思想家、自然科学家、哲学家和文学家。他反对地心说，宣传日心说和宇宙观、并为捍卫真理而死，著有《论无限宇宙和世界》《诺亚方舟》等。

Bruno); "of that which, being sung, either delights or instructs, or delights and instructs at the same time." [20] In much the same way Guarini defended his *Pastor fido* in 1588, declaring "the world is the judge of poets; against its sentence there is no appeal." [21]

Spanish critics.

Amongst European countries, Spain was perhaps the sturdiest in her resistance to the pedantic theories of the writers of treatises; Spain was the land of freedom in criticism from Vives to Feijóo, from the sixteenth to the middle of the eighteenth century when decadence of the old Spanish spirit allowed Luzán[1], with others, to introduce neo-classical poetry of Italian and French origin. [22] That rules must change with the times and with actual conditions; that modern literature demands modern poetics; that work carried out contrary to established rule does not signify that it is contrary to all rule or unwilling to submit itself to a higher law; that nature should give, not receive, laws; that the laws of the three unities are as ridiculous as it would be to forbid a painter to paint a large landscape in a small picture; that the pleasure, taste, approbation of readers and spectators are the deciding element in the long run; that notwithstanding the laws of counterpoint, the ear is the true judge of music; these affirmations and many like them are frequent in Spanish criticism of the period. One critic, Francisco de la Barreda (1622), went so far as to compassionate the strong wits of Italy bound by fear and cowardice (*temerosos y acobardados*) to rules that hampered them on every side; [23] he may have been thinking of Tasso, a memorable case of such degradation. Lope de Vega wavered between neglect of rules in practice, and obsequious acceptance of them in theory, alleging in

1. Luzán：鲁赞（1702—1754），西班牙评论家、诗人，著有《诗学：诗歌在通常意义和主要分类的 6 个规则》。

excuse for his conduct that he was forced to yield to the demands of the public who paid money to see his plays; he said, "when I write my comedies, I lock and double-lock the door against the precept-mongers, that they may not rise up and bear witness against me"; "Art (that is, Poetics) speaks truth which is contradicted by the vulgar ignorant"; "may the rules forgive us when we are induced to violate them." [24] But a contemporary admirer of Lope's work writes of him that "*en muchas partes de sus escritos dice que el no guardar el arte antiguo lo hace por conformarse con el gusto de la plebe ... dicelo por su natural modestia, y porqué no atribuya la malicia ignorante à arrogancia lo que es politica perfeccion.*[1]" [25]

G. B. Marino.

Giambattista Marino also protested "I assert that I have a more thorough knowledge of the rules than have all the pedants in the world; but the only true rule is to know how to break the rules at the right place and time, and to conform with the custom and taste of the day." [26] The drama of Spain, the comedy of art, and other literary novelties of the seventeenth century caused Minturno[2], Castelvetro and other rigid treatise-writers of the preceding century to be looked at with contemptuous pity as "antiquaries"; this may be seen in Andrea Perucci[3] (1699), the theorist of improvised comedy. [27] Pallavicino criticized the writers on "the disciplines of beautiful speech" on

---

1. *en muchas partes de sus escritos dice que el no guardar el arte antiguo lo hace por conformarse con el gusto de la plebe ... dicelo por su natural modestia, y porqué no atribuya la malicia ignorante à arrogancia lo que es politica perfeccio*：〈西〉在很多作品中，维加都说没必要为了满足平民的口味保留古代艺术。他是出于天生谦逊才这么说的，没有把无知的恶意当成高雅，这是精明的完美。
2. Minturno：明图尔诺（1500—1574），意大利文艺复兴时期的批评家。原名安东尼奥·塞巴斯蒂亚诺。著有《论诗人》《诗的艺术》等。
3. Andrea Perucci：安德烈亚·贝鲁奇（1651—1704），意大利剧作家、诗人、耶稣会士，著有诗集《缪斯的想法》、剧本《石头上的客人》《赎罪的人》等。

the ground that they "generally base their precepts on observing by experience what things in writers give pleasure, rather than pointing out what would naturally conform to the particular affections and instincts implanted by the Creator in the souls of men." [28]

G. V. Gravina.

A note of distrust towards the fixed kinds may be heard in the *Discorso sull' Endimione*[1] (1691), wherein Gravina severely blames the "ambitious and miserly precepts" of rhetoricians, and makes the penetrating comment: "No work can see the fight without finding itself confronted by a tribunal of critics specially convened to examine it, and questioned firstly as to its name and nature. Next begins the action which lawyers call prejudicial, and controversy arises as to its status, whether it is a poem, a romance, a tragedy, a comedy, or another of the prescribed kinds. And if the said work have ignored the slightest precept ... they decree forthwith its exile and perpetual banishment. And yet, however they recast and expand their aphorisms, they will never be able to include all the different kinds that can be freshly created by the varied and ceaseless motion of human wit. For this reason I cannot see why we should not free ourselves from this insolent curb on the soaring grandeur of our imaginations, and allow them to follow an open road amongst those immeasurable spaces they are fitted to explore." He remarks on the work of Guidi[2] which forms the subject of his discourse, "I know not whether it be tragedy, comedy, tragicomedy, or anything else invented by rhetoricians. It is a representation of the loves of Endymion and Diana. If those terms have sufficient breadth of extension, they will comprehend this

1. *Discorso sull' Endimione*：《论恩底弥翁》，格拉维纳著。
2. Guidi：圭迪（1650—1712），意大利诗人、剧作家，罗马学院创始人之一，代表作有《致命运》《达芙妮与恩底弥翁》等。

work; if they have not, let another be framed (a power which may be granted to any one in so unimportant a matter); if no such term can be invented, let us not, for want of a word, deprive ourselves of a thing so beautiful." [29] These remarks have quite a modern ring, but Gravina can hardly have thought out their implications very deeply, for later on he wrote a special treatise on the rules of the tragic kind. [30] Antonio Conti too declared at times his antagonism towards the rules, but he referred to the Aristotelian rules only. [31]

Fr. Montani.

More courage was displayed by Count Francesco Montani of Pesaro[1] in the polemic roused by Orsi's book against Bouhours; in 1705 he wrote: "I know that there are immutable and eternal rules, founded on such sound good sense and solid reason as will remain unshaken as long as mankind lives. But these rules, whose incorruptibility gives them authority to guide our spirits to the end of time, are rare enough to be counted with the nose, and it seems to me somewhat arbitrary to claim to test and regulate our new works by old laws now wholly abrogated and annulled[2]." [32]

Critics of the eighteenth century.

In France the rigorism of Boileau was followed by the rebellion of Du Bos, who unhesitatingly declared that "men will always prefer poetry which moves them to that composed according to rule," [33] and the like heresies. In 1730, De la Motte made war against the unities of time and place, asserting as the most general, and even superior to that of

---

1. Count Francesco Montani of Pesaro：佩萨罗伯爵弗兰切斯科·蒙塔尼（1680—1755），意大利 18 世纪早期文学评论家。
2. wholly abrogated and annulled：完全废止

action, the unity of interest. [34] Batteux tended to make free with the rules; and Voltaire, though he opposed De la Motte and declared the three unities to be the "three great laws of good sense," uttered some bold sentiments in his *Essay on Epic Poetry*, and it was he who remarked that "*tous les genres sont bons hors le genre ennuyeux*,[1]" and that the best kind is "*celui qui est le mieux traité*.[2]" Diderot was in certain respects a forerunner of Romanticism, and with him must be mentioned Friedrich Melchior Grimm[3], who was influenced by him. A breath of liberty was wafted into Italy by Metastasio[4], Bettinelli, Baretti[5] and Cesarotti: in 1766 Buonafede[6] notes in his *Epistola della libertà poetica* that when erudite persons[7] "define epic poetry, or comedy, or odes, they ought to frame as many definitions as there are compositions and authors." [35] In Germany the first to rise in rebellion against the rules (opposing Gottsched and his disciples) were the representatives of the Swiss school. [36] In England, after examining the definitions by which critics endeavoured to distinguish epic poetry from other compositions, Home wrote, "It affords no little diversion to watch so many profound critics hunting after that which does not exist. They presuppose — without shadow of proof — that there exists a precise

---

1. *tous les genres sont bons hors le genre ennuyeux*：〈法〉除了令人厌烦的那一类之外，其他种类都是好的
2. *celui qui est le mieux traité.*：〈法〉研究得最好的那种
3. Friedrich Melchior Grimm：弗里德里希·梅尔奇奥·格林（1723—1807），德裔法国记者、外交家、艺术评论家，《百科全书》撰稿人之一，著有《抒情诗》《文学、哲学、批评书信集》等。
4. Metastasio：梅塔斯塔西奥（1698—1782），本名 Pietro Antonio Domenico Trapassi，意大利诗人、剧作家，著有《德梅特里奥》《叙利亚的阿德里亚诺》等。
5. Baretti：巴雷蒂（1719—1789），意大利文学评论家、诗人、词典编纂家，著有《意大利人的图书馆》《文学鞭策》等。
6. Buonafede：博纳菲迪（1716—1793），意大利哲学家，著有《以人的天赋权利对著名征服进行的研究》以及文中提到的《论诗歌自由的书信集》（*Epistola della libertà poetica*）。
7. erudite persons：饱学之士

criterion by which to distinguish epic poetry from all other kinds of composition. But literary compositions melt one into another like colours: and if in their stronger shades it is easy to recognize them, they are susceptible of such variety and of so many different forms that it is impossible to say where one ends and another begins." [37]

Romanticism and the "strict kinds": Berchet, V. Hugo.

Literary thought between the late eighteenth and the first decades of the nineteenth century, that is to say from " the period of genius" to that of romanticism properly so called, rose in rebellion against separate individual rules and against all rules as such. But to describe the battles fought, and their more important episodes; to recount the names of captains victorious or discomfited, or to deplore the excesses committed by the conquerors, is no part of our present task. Upon the ruins of the strict kinds, the "*genres tranché*[1]" beloved by Napoleon [38] (a Romanticist in the art of war, but a Classicist in poetry), flourished the drama, the romance and every other mixed kind: upon the ruins of the three unities, flourished the unity of *ensemble*. Italy made her protest against rules of style in Berchet's[2] famous *Lettera semiseria di Grisostomo* (1816); and France made hers somewhat later in Victor Hugo's preface to *Cromwell* (1827). Henceforth men discussed not the kinds, but Art. What is the unity of *ensemble* but the demand of art itself, which is always an *ensemble*, a synthesis? What else is the principle, introduced by August Wilhelm Schlegel and adopted by Manzoni and other Italian romanticists, to the effect that form of component parts must be "organic not mechanical, resulting from the nature of the subject and its interior development ... not from the

1. *genres tranché*: 〈法〉 纯种
2. Berchet：贝尔切特（1783—1851），意大利诗人、爱国主义者，著有《格里索托姆的半严肃书信集》《帕尔嘎难民》等。

impress of an external and extraneous stamp"? [39]

Their persistence in philosophical theories.

But it would be quite wrong to suppose that this victory over the rhetoric of kinds was either the cause or the consequence of a final victory over its philosophical presuppositions. In pure theory, none of the critics above named wholly abandoned the kinds and the rules. Berchet admitted four elementary forms, that is four fundamental kinds, in poetry; lyrical, didactic, epic and dramatic, claiming for the poet only the right of "uniting and fusing together the elementary forms in a thousand fashions." [40] Manzoni's only real quarrel was with those rules "founded on special facts instead of on general principles; on the authority of rhetoricians instead of reason." [41] Even De Sanctis was satisfied with a concept somewhat vague, though true enough at bottom: "the most important rules are not those capable of being applied to every content, but those which draw their force *ex visceribus caussæ*[1], from the very heart of the content itself." [42] Even more diverting than the spectacle which had delighted Home, is the sight of German philosophy according the honour of a dialectical deduction to the empirical classification of kinds. We shall give two examples, each representing one extreme end of the chain:

Fr. Schelling.

Schelling at the beginning of the century (1803), and Hartmann at the end (1890). One section of Schelling's *Philosophy of Art* is devoted to "the construction of individual poetic kinds"; in it he remarks that were he to follow the historical order, Epic would come first; whereas

1. *ex visceribus caussæ*：〈拉〉从内容深处

in the scientific order the Lyric occupies the first place: indeed, if poetry is the representation of the infinite in the finite, the Lyric, in which difference prevails (the finite, the subject), is its first moment, corresponding with the first power of the ideal series, reflexion, knowledge, consciousness, whereas Epic corresponds with the second power, action. [43] From Epic, which is *par excellence* the objective kind (as being the identity of subjective and objective), derive the Elegy and the Idyl[1] if subjectivity be placed in the object and objectivity in the poet: if objectivity be placed in the object and subjectivity in the poet, didactic poetry[2] results. [44] To these differentiations of the Epic, Schelling adds the romantic or modern Epic, the poem of chivalry; the novel; and the experiments in an epic of ordinary life such as the *Luisa* of Voss[3] and the *Hermann and Dorothea* of Goethe; and, co-ordinate with all the foregoing, the *Comedia* of Dante, "an epic kind in itself" (*eine epische Gattung für sich*). Finally, from the union on a higher plane of Lyric with Epic, liberty with necessity, arises the third form, the Drama, the reconciliation of antitheses in a totality, "supreme incarnation of the essence and the in-itself of all art." [45]

E. von Hartmann.

In Hartmann's *Philosophy of the Beautiful*, poetry is divided into spoken poetry and read poetry. The former is subdivided into Epic, Lyric and Dramatic, with further subdivisions of Epic into plastic Epic, or strictly epic Epic, and pictorial or lyrical Epic; of Lyric into epical Lyric, lyrical Lyric and dramatic Lyric; of Dramatic into lyrical Drama, epic Drama and dramatic Drama. Read poetry (*Lese poesie*)

1. Elegy and the Idyl：挽歌和田园诗
2. didactic poetry：教诲诗
3. Voss：福斯（1768—1832），德国作家，被誉为“第一位德国科幻小说家”，著有《21 世纪的罗马人》《剧院中的戴帽子女士》等。

is again subdivided into predominantly epical, lyrical or dramatic form with tertiary partitions of the affecting, the comic, the tragic and humorous; and into poems "to be read at a sitting" (like the short story) or to be taken up again and again (like the novel). [46]

The kinds in the schools.

Without these highly philosophical trivialities the divisions of kinds still wander through the books called *Institutions of Literature*, written by philologists and men of letters, and the ordinary school-books of Italy, France and Germany; and psychologists and philosophers still persist in writing about the Æsthetic of the tragic, of the comic and of the humorous. [47] The objectivity of literary kinds is frankly maintained by Ferdinand Brunetière[1], who looks on literary history as "the evolution of kinds," [48] and gives sharply defined form to a superstition which, seldom confessed so truthfully or applied so rigorously, survives to contaminate modern literary history. [49]

[1] *Republic*, iii. 394; see also E. Müller, *Gesch. i. Th. d. Kunst*, i. pp. 134–206; ii. pp. 238–239, note.
[2] *Poet.* ch. 6
[3] *Annotazioni*, introd.
[4] Cf. for Sanskrit poetry S. Levi, *Le Théâtre indien*, pp. 11–152.
[5] Cf. Menendez y Pelayo, *op. cit.* I., i. pp. 126–154, 2nd ed.
[6] Introd. to his tr. of the *Poetics.*
[7] Lintilhac, *Un Coup d'état*, etc., p. 543.
[8] *Hamburg. Dramat.* Nos. 81, 101–104.

1. Ferdinand Brunetière：费迪南·布吕内提埃（1849—1906），法国作家、批评家，著有《法国文学史的批评研究》《历史与文学》《批评的问题》等。

[9] *Op. cit.* Nos. 96, 101–104.

[10] G. B. Giraldi Cintio, *De' romanzi, delle comedie e delle tragedie,* 1554 (ed. Dælli, 1864).

[11] Iacopo Mazzoni, *Difesa della commedia di Dante,* Cesena, 1587.

[12] G. M. Cecchi, prologue to *Romanesca,* 1585.

[13] Cf. besides the two *Veratti,* the *Compendio della poesia tragicomica,* Venice, 1601.

[14] *Proginn. poet.,* Florence, 1627, iii. p. 130.

[15] Cf. A. Belloni, *Il seicento,* Milan, 1898, pp. 162–164.

[16] *Examens,* and *Discours du poème dramatique, de la tragédie, des trois unités,* etc.

[17] *Réflexions sur le comique larmoyant,* 1749 (trans. by Lessing, *Werke, vol. cit.*).

[18] Gellert, *De comœdia commovente,* 1751; Lessing, *Abhandlungen von den weinerlichen oder rührenden Lustspiele,* 1754 (in *Werke,* vol. vii.).

[19] Prologue to the *Cortigiana,* 1534.

[20] *Degli eroici furori* in *Opere italiane,* ed. Gentile, ii. pp. 310–311.

[21] *Il Veratto* (against Jason de Nores), Ferrara, 1588.

[22] Menendez y Pelayo, *op. cit.* iii. pp. 174–175 (1st ed.), i.

[23] Menendez y Pelayo, *op. cit.* iii. p. 468 (2nd ed.).

[24] *Arte nuevo de hacer comedias* (1609), ed. Morel Fatio, 11. 40–41, 138–140, 157–158.

[25] Menendez y Pelayo, *op. cit.* iii. p. 459.

[26] Marino, letter to G. Preti, in *Lettere,* Venice, 1627, p. 127.

[27] *Dell' arte rappresentiva meditata e all' improvviso,* Naples, 1699; cf. pp. 47, 48, 65.

[28] *Trattato dello stile e del dialogo,* 1646, preface.

[29] *Discorso su l' Endimione* (in *Opere italiane, ed. cit.*), ii. pp. 15–16.

[30] *Della tragedia,* 1715 (*ibid.* vol. i.).

[31] *Prose e poesie, cit.,* pref. and *passim.*

[32] In Orsi, *Considerazioni, ed. cit.* ii. pp. 8, 9.

[33] *Réflexions, cit.* sect. 34.

[34] *Discours sur la tragédie*, 1730.

[35] *Opuscoli* of Agatopisto Cromaziano, Venice, 1797.

[36] Danzel, *Gottsched*, p. 206 *seqq*.

[37] *Elements of Criticism*, iii. pp. 144–145, note.

[38] See conversation of Napoleon with Goethe, in Lewes, *The Life and Works of Goethe*, ii. p. 441.

[39] Manzoni, *Epistol.* i. pp. 355–356; cf. *Lettera sul romanticismo, ibid.* pp. 293–299.

[40] *Lettera di Grisostomo, opere*, ed. Cusani, p. 227.

[41] *Lettera sul romanticismo, ibid.* p. 280.

[42] *La giovinezza di F. de S.* chs. 26–28.

[43] *Philos, d. Kunst*, pp. 639–645.

[44] *Op. cit.* pp. 657–659.

[45] *Op. cit.* p. 687.

[46] *Philosophie d. Schönen*, ch. 2, § 2.

[47] See, *e.g.*, Volkelt, *Ästh. d. Tragischen*, Munich, 1897; Lipps, *Der Streit über Tragödie*, etc.

[48] See his other works, *L'évolution des genres dans l'histoire de la littérature*, Paris, 1890 *seqq*., and *Manuel de l'hist. de la littér. française, ibid.*, 1898.

[49] Croce, *Per la storia della critica e storiografia letter*, pp. 23–25.

III THE THEORY OF THE LIMITS OF THE ARTS

To Lessing must be ascribed the merit and the sole glory of having discovered that every art has its special character and inviolable limits. But his merit lies not in his own theory, which, in itself, is scarcely tenable[1], [1] but in having, though by an error, aroused discussion of a highly important æsthetical point till then wholly overlooked. After some slight notice from Du Bos and Batteux, some preparation of the field by Diderot [2] and Mendelssohn, [3] and long disquisitions by Meier and other Wolffians upon natural and conventional symbols, [4] Lessing was the first to raise clearly the question of the value attaching to the distinction between the various arts. Antiquity, the Middle Ages and the Renaissance had enumerated the arts according to denominations of current phraseology, and had composed numbers of technical hand-books distinguishing major and minor arts; but in Aristoxenus[2] or Vitruvius, Marchetto da Padova[3] or Cennino Cennini[4], Leonardo da Vinci or Leon Battista Alberti, Palladio[5] or Scamozzi[6],

---

1. scarcely tenable：很难成立的
2. Aristoxenus：亚里士多塞努斯（前 375—3 前 35），逍遥派哲学家，亚里士多德弟子，大多数著作已遗失，仅存残本《和谐的要素》。
3. Marchetto da Padova：马尔凯特・达・帕多瓦（? — 约 1319），意大利音乐理论家、作曲家，著有《音乐艺术变革的边界》等。
4. Cennino Cennini：切尼诺・切尼尼（1360—1427），意大利画家、艺术评论家，著有《艺术之书》《论绘画》等。
5. Palladio：帕拉迪奥（1508—1580），意大利建筑家，活跃于威尼斯，代表作有奥利匹克大剧院。
6. Scamozzi：斯卡莫齐（1548—1616），意大利建筑家、作家，在帕拉迪奥去世后接任承建奥利匹克大剧院。

it would be vain to look for the problem proposed by Lessing, for the spirit of these technical treatise-writers is entirely different. Some rudiments of the question may be detected in the comparisons made, and the questions of precedence raised, between poetry and painting or painting and sculpture, to be found now and then in stray paragraphs of their books (Leonardo da Vinci pressed the claims of painting, Michæl Angelo those of sculpture): the theme eventually became a favourite one for academic discussion, and was not despised by Galileo himself. [5]

The limits of the arts in Lessing. Arts of space and arts of time.

Lessing was induced to raise the question in the attempt to controvert the strange views of Spence concerning the close union between painting and poetry among the ancients, and of Count Caylus[1], who held that the excellence of a poem must be judged by the number of subjects it offers to the brush of the painter. He was further instigated[2] by the comparisons between poetry and painting upon which were commonly founded the most ridiculous rules for tragedy: the maxim *Ut picture poësis*, whose original motive was to emphasize the representative or imaginative character of poetry, and the community of nature among the arts, had been converted by superficial interpretation into a defence of the most vicious intellectualistic and realistic prejudices. Lessing argued in this wise: "If painting in its imitations employs precisely a medium or symbol different from that of poetry (the former employing spatial forms and colours, the latter temporal articulated sounds), since the symbol must certainly be in

1. Count Caylus：开鲁斯伯爵（1692—1765），法国文物研究家、考古学家，著有《绘画与雕塑的新主题》《〈伊利亚特〉〈奥德修斯〉〈埃涅伊德〉中的绘画》等。
2. was instigated：由……来推动

close relation with that which is signified, coexistent symbols can only express coexistent objects or parts of objects, and consecutive symbols can only express consecutive objects or parts of objects. Objects mutually coexistent, or having mutually coexistent parts, are called bodies. Bodies, then, through their quality of visibility, are the true objects of painting. Objects successively consecutive amongst themselves, or whose parts are consecutive, are called in general actions. Actions, then, are the suitable objects of poetry." Painting, undoubtedly, may represent action, but only by means of bodies which indicate it; and poetry may represent bodies, but only by indicating them by means of actions. When a poet using language, *i.e.* arbitrary symbols, sets himself to describe bodies, he is no longer a poet but a prose-writer, since a true poet only describes bodies by the effect they produce on the soul. [6] Retouching and developing this distinction, Lessing described action or movement in a picture as an addition made by the imagination of the beholder; so true is this, says he, that animals perceive nothing save immobility in a picture. He further studied the various unions of arbitrary with natural symbols, such as that of poetry with music (in which the former is subordinate to the latter), of music with dancing, of poetry with dancing, and of music and poetry with dancing (union of arbitrary consecutive audible symbols with natural visible symbols): of the pantomime of antiquity[1] (union of arbitrary consecutive visible symbols with natural consecutive visible symbols): of the language of the dumb (the only art that employs arbitrary consecutive visible symbols): and, lastly, of imperfect unions, such as that of painting with poetry. If not every use to which language is put is poetic, Lessing holds that not every use of natural coexistent signs is pictorial: painting, like language, has its prose. Prosaic painters are those who represent consecutive objects

1. the pantomime of antiquity：古代哑剧

notwithstanding the character of coexistence in their signs, allegorical painters those who make arbitrary use of natural signs, and those who pretend to represent the invisible or the audible by means of the visible. Desirous of preserving the naturalness of symbolism, Lessing ended by condemning the custom of painting objects on a diminished scale, and concludes: "I think that the aim of an art should be that only to which it is specially adapted, not that which can be performed equally well by other arts. I find in Plutarch a comparison which illustrates this admirably: he who would split wood with a key and open the door with an axe not only spoils both utensils but deprives himself of the unity of each alike." [7]

Limits and classifications of the arts in later philosophy.

The principle of limitations or of the specific character of individual arts, as laid down by Lessing, occupied the attention of philosophers in later days, who, without discussing the principle itself, employed it in classifying the arts and arranging them in series.

Herder and Kant.

Herder here and there continued Lessing's examination in his fragment on *Plastic*[1] (1769); [8] Heydenreich wrote a treatise (1790) on the limits of the six arts (music, dance, figurative arts gardening, poetry and representative art), and criticized the *clavecin oculaire* of Father Castel[2], a contrivance for the combination of colours which should act in the same way as the series of musical notes in harmony and melody, [9] Kant appealed to the analogy of a speaking man, and

1. *Plastic*：《论造型艺术》
2. the *clavecin oculaire* of Father Castel：卡斯特尔教父发明的羽管键琴上用色谱校正音律的装置

classified the arts according to speech, gesture and tone as arts of speech, figurative arts, and arts producing a mere play of sensations (mimicry and colouring). [10]

Schelling.

Schelling differentiated the artistic identity according as it consisted in the infusion of the infinite into the finite, or of the finite into the infinite (ideal art or real art): into poetry and art proper. Under the heading of real arts he included the figurative arts, music, painting, plastic (which comprehended architecture, bas-relief[1] and sculpture): in the ideal series were the three corresponding forms of poetry, lyrical, epical and dramatic. [11]

Solger.

With a similar method, Solger placed poetry, the universal art, side by side with art strictly so called, which is either symbolical (sculpture) or allegorical (painting), and, in either case, is a union of concepts and bodies: if you take corporality without concept, you have architecture; if concept without matter, music. [12] Hegel makes poetry the bond of union between the two extremes of figurative art and of music. [13]

Schopenhauer.

We have already seen how Schopenhauer destroyed the accepted limitations of art and built them up again, following the order of the ideas which they represent. [14] Herbart clung to Lessing's two groups, simultaneous arts and successive arts, and defined the former as

---

1. bas-relief：浅浮雕

"permitting themselves to be inspected from every side," the latter as "rejecting complete investigation and remaining in semi-darkness": in the first group he placed architecture, plastic, church music and classical poetry; in the second ornamental gardening, painting, secular music and romantic poetry. [15]

Herbart.

Herbart was implacable against those who look in one art for the perfections of another; who "look on music as a sort of painting, painting as poetry, poetry as an elevated plastic and plastic as a species of æsthetic philosophy," [16] while admitting that a concrete work of art, such as a picture, may contain elements of the picturesque, the poetic and other kinds, held together by the skill of the artist. [17]

Weisse. Zeising.

Weisse divided the arts into three triads, intended to recall the nine Muses. [18] Zeising invented a cross-division into figurative arts (architecture, sculpture, painting), musical arts (instrumental music, song, poetry), and arts of mimicry (dance, musical mimicry, representative art), and into macrocosmic arts[1] (architecture, instrumental music, dance), microcosmic arts[2] (sculpture, song, musical mimicry) and historical arts (painting, poetry and representative art). [19]

Vischer.

Vischer classified them according to the three forms of imagination

---

1. macrocosmic arts：宏观艺术
2. microcosmic arts：微观艺术

(figurative, sensuous and poetic), into objective arts (architecture, plastic and painting), a subjective art (music) and an objective-subjective art [20] (poetry). Gerber[1] proposed to recognize a special "art of language" (*Sprachkunst*), distinguishable alike from prose and poetry and consisting in the expression of simple movements of the soul. Such an art would correspond with plastic in the following scheme: arts of the eye — (*a*) architecture, (*b*) plastic, (*c*) painting; arts of the ear — (*a*) prose, (*b*) the art of language, (*c*) poetry. [21]

M. Schasler.

The two most recent systems of classification are furnished by Schasler and Hartmann, who have also submitted the schemes of their predecessors to searching criticism. Schasler [22] arranges the arts in two groups, adopting the criterion of simultaneity and succession: the arts of simultaneity are architecture, plastic and painting; of succession, music, mimicry and poetry. He says that by following the series in the order indicated, it will be seen that simultaneity, originally predominant, yields place to succession, which predominates in the second group and subordinates without wholly displacing the other. Parallel with this, another division is evolved, deduced from the relation between the ideal and material elements in each separate art, between movement and repose; which begins with architecture "materially the heaviest, spiritually the lightest of all the arts," and ends with poetry, in which the opposite relation is observed. Curious analogies are established by this method between the first and second group of arts: between architecture and music; between plastic and mimicry; between painting in its three forms of landscape, *genre* and historical, and poetry in its three forms of lyric (declamatory), epic

1. Gerber：即古斯塔夫·盖尔贝尔（1820—1901），德国教育家、哲学家，著有《作为艺术的语言》等。

(rhapsodic) and drama (representative).

E. v. Hartmann.

Hartmann [23] divides the arts into arts of perception and arts of imagination: the former tripartite into spatial or visual (plastic and painting), temporal or auditory (instrumental music, linguistic mimicry, expressive song) and temporal-spatial or mimic (pantomime, mimic dances, art of the actor, art of the opera-singer); the second contains but one single species, which is poetry. Architecture, decoration, gardening, cosmetic and prosewriting are excluded from this system of classification and lumped together[1] as non-free arts.

The supreme art. Richard Wagner.

Parallel with this search for a classification of the arts, the same philosophers were led into the quest of the supreme art. Some favoured poetry, others music or sculpture; others again claimed the supremacy for combined arts, especially for Opera, according to the theory of it already advanced in the eighteenth century [24] and maintained and developed in our day by Richard Wagner. [25] One of the latest philosophers to raise the question "whether single arts, or arts in combination, had the greater value," concluded that single arts as such possess their own perfection, yet the perfection of united arts is still greater, notwithstanding the compromises and mutual concessions enforced upon them by their union; that single arts, from another point of view, have the greater value; and lastly, that both single and combined arts are necessary to the realisation of the concept of art. [26]

---

1. lumped together：归为一类

Lotze's attack on classifications.

The capriciousness, emptiness and childishness of such problems and their solutions must have excited feelings of impatience and disgust, but we rarely find a doubt thrown on their validity. One such dissentient[1] is Lotze when he writes: "It is difficult to see the use of such attempts. Knowledge of the nature and laws of individual arts is but little increased by indication of the systematic place allotted to each." He further observed that in real life the arts are variously conjoined, forming themselves into no systematic series, while in the world of thought an immense variety of orders can be created; he therefore selected one of these possible orders, not because it was the sole legitimate one, but because it was convenient (*bequem*). His series begins with music, "the art of free beauty, determined only by the laws of its matter, not by conditions imposed by a given task of purpose or of imitation"; followed by architecture, "which no longer plays freely with forms, but subjects them to the service of an end"; and then by sculpture, painting and poetry, excluding minor arts which cannot be co-ordinated with the others, since they are incapable of expressing with any approach to completeness the totality of the spiritual life. [27] A recent French critic, Basch[2], opens his treatise with the following excellent remarks: "Is it necessary to show there is no such thing as an absolute art, differentiating itself later by means of one knows not what immanent laws[3]? What exists is the particular forms of art, or rather artists who have striven to translate, as best they can, according to the material means at their command, the song of the ideal in their

1. dissentient：持不同意见者
2. Basch：巴施（1863—1944），法国哲学家、美学家，联合创办"人权联盟"并担任主席，著有《康德的美学》《席勒的诗学》《现代个人主义的起源》等。
3. immanent laws：内在规律

souls." But later on he thinks it possible to effect a division of the arts by starting "from the artist, instead of the art in itself," by proceeding "according to the three great types of fancy, visual, motor and auditory"; and as for the debated point of the supreme art, he thinks it must be settled in favour of music. [28]

Schasler is not altogether wrong in his spirited counterattack on Lotze's criticism; he protests against the principle of indifference and convenience, and remarks that "the classification of the arts must be regarded as the real touchstone, the real differential test of the scientific value of an æsthetic system; for on this point all theoretical questions are concentrated and crowd together to find a concrete solution." [29]

Contradictions in Lotze.

The principle of convenience may be excellent as applied to the approximative grouping of botanical or zoological classifications, but it has no place in philosophy; and as Lotze, in common with Schasler and other æstheticians, conformed to Lessing's principle of the constancy, limits and peculiar nature of each art, and therefore held that the concepts of the individual arts were speculative and not empirical concepts, he could not evade the duty of fixing the mutual relations of these concepts, arranging them in series, subordinating and co-ordinating them, and arriving at each of them either deductively or dialectically. He ought, in order to get definitely rid of these barren attempts at classification and at discovering the supreme art, to have criticized and dissolved Lessing's principle itself: to keep the principle and deny the need for a classification, as Lotze did, was obviously inconsistent. But not a single æsthetician has ever re-examined or investigated the scientific foundation of the distinctions enunciated by Lessing in his fluent and elegant prose; no one has probed to the bottom the truth which was illumined by Aristotle in a single lightning-flash, when he refused to allow an extrinsic difference, that of metre, as the real distinction between prose and poetry: [30] no one, that is to say, save perhaps Schleiermacher, who at least called attention to the difficulties of the current doctrine.

Doubts in Schleiermacher.

He proposed to start from the general concept of art and prove by deduction the necessity of all its forms; and after finding two sides to artistic activity, the objective consciousness (*gegenständliche*) and the immediate consciousness (*unmittelbare*), and observing that art stands wholly neither in the one nor in the other and that the immediate consciousness or representation (*Vorstellung*) gives rise to mimicry and music, while the objective consciousness or image (*Bild*) gives rise to the figurative arts, he then, proceeding to analyse a painting, found the two forms of consciousness to be in this case inseparable, and remarks: "Here we arrive at the precise opposite: searching for distinction, we find unity." Nor did the traditional division of the arts into simultaneous and successive seem to him very solid, for "when looked at attentively, it evaporates entirely[1]"; in architecture or gardening, contemplation is successive, while in the arts labelled as successive, such as poetry, the chief thing is coexistence and grouping: "from whichever side we look at it, the difference is but secondary and the antithesis between the two orders of art[2] merely means that every contemplation, like every act of production, is always successive, but, in thinking out the relation of the two sides in a work of art, both seem indispensable: coexistence (*Zugleichsein*) and successive existence (*das Successivsein*)." In another passage he observes: "The reality of art as external appearance is conditioned by the mode, depending on our

1. evaporates entirely：彻底消失
2. antithesis between the two orders of art：两个艺术种类的冲突

physical and corporeal organism, in which the internal is externalised: movements, forms, words ... That which is common to all arts is not the external, which is rather the element of diversification." When these observations are compared with the sharp distinction he himself drew between art and technique, it would be easy to deduce that he held the partitions of the arts and the concepts of the particular arts to be devoid of æsthetic value. But Schleiermacher does not draw this logical inference, he wavers and hesitates: he recognizes the inseparability of the subjective and objective, musical and figurative, elements in poetry, yet he struggles to discover the definitions and limits of the individual arts; sometimes he dreams of a union of the various arts from which a complete art would spring; and when composing the syllabus of his lectures[1] on Æsthetic, he arranged the arts into arts of accompaniment (mimicry and music), figurative arts (architecture, gardening, painting, sculpture) and poetry. [31] Nebulous, vague, contradictory as this may be, Schleiermacher had the acumen[2] to distrust the soundness of Lessing's theory and to inquire by what right particular arts are singled out from art in general.

[1] See above, pp. 145–147.

[2] D. Diderot, *Lettre sur les aveugles*, 1749; *Lettre sur les sourds et muets*, 1751; *Essai sur la peinture*, 1765.

[3] M. Mendelssohn, *Briefe über Empfind.*, 1755; *Betrachtungen, cit.*, 1757.

[4] J. Chr. Wolff, *Psychol. empirica*, §§ 272–312; Meier, *Anfangsgründe*, §§ 513–528, 708–735; *Betrachtungen*, § 126.

[5] Letter to Lodovico Cardi da Cigoli, June 26, 1612.

[6] *Laokoon*, §§ 16–20.

---

1. compose the syllabus of his lectures：整理讲课提纲
2. acumen：敏锐

[7] *Laokoon*, appendix, § 43.
[8] *Laokoon*, appendix, § 43.
[9] *Plastik einige Wahrnehmungen über Form und Gestalt aus Pygmalions bildenden Träume*, 1778 (Select Works of Herder in the collection *Deutsche Nationlitteratur*, vol. 76, part iii. § 2).
[10] *System der Ästhetik*, pp. 154–236.
[11] *Kritik d. Urtheilskr.* § 51. 5 *Phil. d. Kunst*, pp. 370–371.
[12] *Vorles. üb. Ästh.* pp. 257–262.
[13] *Op. cit.* ii. p. 222.
[14] See above, pp. 337–338.
[15] *Einleitung*, § 115, pp. 170–171.
[16] *Schriften z. prakt. Phil*, in *Werke*, viii. p. 2.
[17] *Einleitung*, § 110, pp. 164–165.
[18] Cf. Hartmann, *Dtsche. Ästh. s. Kant*, pp. 539–540.
[19] *Ästh. Forsch.* pp. 547–549.
[20] *Ästh.* §§ 404, 535, 537, 838, etc.
[21] Gustav Gerber, *Die Sprache als Kunst*, Bromberg, 1871–1874.
[22] *Das System der Künste*, 2nd ed., Leipzig-Berlin, 1881.
[23] *Phil. d. Sch.* chs. 9, 10.
[24] *E.g.* by Sulzer, *Allg. Theorie*, on word *Oper.*
[25] Rich. Wagner, *Oper und Drama*, 1851.
[26] Gustav Engel, *Ästh. der Tonkunst,* 1884, abstracted in Hartmann, *Dtsche. Ästh. s. Kant,* pp. 579–580.
[27] Lotze, *Geschichte d. Ästh.* pp. 458–460; cf. p. 445.
[28] *Essai critique sur l'Esth. de Kant*, pp. 89–496.
[29] *Das System der Künste*, p. 47.
[30] *Poet.* ch. i.
[31] *Vorles. üb. Ästh.* pp. 11, 122–129, 137, 143, 151, 167, 172, 284–286, 487–488, 508, 635.

IV OTHER PARTICULAR DOCTRINES

The æsthetic theory of Natural Beauty.

I. Schleiermacher also rejected the concept of Natural Beauty, giving Hegel greater praise than he deserved in the matter, because Hegel's denial of this concept was, as we have seen, more verbal than real. At all events, Schleiermacher's radical denial of the existence of a natural beauty external to and independent of the human mind marked a victory over a serious error, and appears to us imperfect and one-sided only so far as it seems to exclude those æsthetic facts of imagination which are attached to objects given in nature. [1] Important contributions towards the correction of this imperfect and one-sided element were supplied by the historical and psychological study of the "feeling for nature," promoted successfully by Alexander Humboldt in his dissertation to be found in the second volume of *Cosmos*, [2] and continued by Laprade[1], Biese, and others in our own time. [3] In his criticism of his own *Ästhetik*, Vischer completes the passage from the metaphysical construction of beauty in nature to the psychological interpretation of it, and recognizes the necessity of suppressing the section devoted to Natural Beauty in his first æsthetic system, and incorporating it with the doctrine of imagination: he says that such treatments do not belong to æsthetic science, being a medley

1. Laprade：即维克多·德·拉普拉多（1812—1883），法国诗人、评论家，写有诗集《玛德琳的体香》《耶稣的愤怒》和散文集《道德与艺术的问题》《律师的理智习惯》等。

of zoology, sentiment, fantasy and humour[1], worthy of development in monographs[2] in the style of the poet G. G. Fischer's[3] on the life of birds, or Bratranek's on the æsthetic of the vegetable world. [4] Hartmann, as heir of the old metaphysics, reproaches Vischer for this exclusion, and maintains that, in addition to the beauty of imagination introduced by man into natural things (*hineingelegte Schönheit*), there exist a formal and a substantial beauty in nature, coinciding with realisation of the immanent ends or ideas of nature. [5] But the way chosen ultimately by Vischer is the only one by which Schleiermacher's thesis can be successfully developed so as to show the precise meaning which may be given to the assertion of (æsthetic) beauty in nature.

The theory of æsthetic senses.

II. That æsthetic senses or superior senses exist and that beauty attaches to certain senses only, not to all, is a very old opinion. We have seen already [6] that Socrates, in the *Hippias maior*, mentions the doctrine of beauty as "that which pleases hearing and sight" (τὸ καλὸν εστὶ τὸ δι' ἀκοῆs τε καὶ ὄψεως ἡδύ): and he adds, it seems impossible to deny that we take pleasure in looking at handsome men and fine ornaments, pictures and statues with our eyes, and hearing beautiful songs or beautiful voices, music, speeches and conversations with our ears. Nevertheless Socrates himself in the same dialogue confutes this theory by perfectly valid arguments, amongst which is that, besides the difficulty arising from the fact that beautiful things may be found outside the range of the sensible impressions of eye and ear, there is

1. medley of zoology, sentiment, fantasy and humour：动物学，情感，幻想和幽默感的综合体
2. monographs：专论
3. G. G. Fischer：乔治・费歇尔（1816—1897），德国诗人、剧作家，代表作有《索尔》《弗里德里希二世》《佛罗瑞安・盖尔》等。

no reason for creating a special class for the pleasure arising from impressions on these two senses, to the exclusion of others. He also states the more subtle and philosophical objection that that which is pleasing to the sight is not so to the hearing, and *vice versa*; whence it follows that the ground of beauty must not be sought in visibility or audibility, but in something differing from either and common to both. [7]

The problem was never again, perhaps, attacked with such acumen and seriousness as in this ancient dialogue. In the eighteenth century Home remarked that beauty depended on sight, and that impressions received by the other senses might be agreeable but were not beautiful, and distinguished sight and hearing as superior to those of touch, taste and smell, the latter being merely bodily in nature and without the spiritual refinement of the other two. He held these to produce pleasures superior to organic pleasures though inferior to intellectual; decorous pleasures, that is to say; elevated, sweet, moderately exhilarating; as far removed from the turbulence of the passions as from the languor of indolence, and intended to refresh and soothe the spirit. [8] Following suggestions of Diderot, Rousseau and Berkeley, Herder drew attention to the importance of the sense of touch (*Gefühl*) in plastic art: of this "third sense, which perhaps deserves to be investigated first of all, and is unjustly relegated to a place amongst the grosser senses." Certainly "touch knows nothing of surface or colour," but "sight, for its part, knows nothing of forms and configurations." Thus "touch cannot be so gross a sense as it is reputed, if it is the very organ by which we sensate all other bodies, and rules over a vast kingdom of subtle and complex concepts. As the surface stands to the body, so does sight stand in respect of touch, and it is merely a colloquial abbreviation[1] to speak of seeing bodies as surfaces and to

1. colloquial abbreviation：口头简称

suppose that we see with our eyes that which we have gradually learnt in infancy simply by the sense of touch." Every beauty of form or corporeity[1] is a concept not visible, but palpable[2]. [9] From the triad of æsthetic senses thus established by Herder (sight for painting; hearing for music; touch for sculpture), Hegel returned to the customary dyad[3], saying that "the sensory part of art has reference only to the two theoretic senses of sight and hearing"; that smell, taste and touch must be excluded from artistic pleasures, since they are connected with matter as such and the immediate sensible quality it may possess (smell with material volatilization[4]; taste with material solution of objects; and touch with hot, cold, smooth and so forth); and that hence they can claim no concern with the objects of art, which are obliged to keep themselves in real independence, rejecting all relation with the merely sensory. That which pleases these senses is not the beautiful of art. [10]

It was Schleiermacher once more who recognized the impossibility of disposing of the matter in this summary fashion. He refused to admit the distinction between confused senses and clear senses, and asserted that the superiority of sight and hearing over the other senses lay in the fact that the others "are not capable of any free activity, and indeed represent the maximum of passivity, whereas sight and hearing are capable of an activity proceeding from within, and are able to produce forms and notes without having received impressions from outside"; were eye and ear merely means of perception, there would be no visual or auditory arts, but they also operate as a function of voluntary movements which supply a content to the dominion of the senses. From another standpoint, however, Schleiermacher thinks that

1. corporeity：形体
2. palpable：可触碰的
3. customary dyad：传统的二分法，即将感觉分为视觉和听觉。
4. volatilization：挥发性

"the difference seems to be one rather of degree or quantity, and a minimum of independence must be recognized as existing in the other senses as well." [11] Vischer remains faithful to the traditional "two æsthetic senses," "free organs and no less spiritual than sensuous," which "have no reference to the material composition of the object," but allow this "to subsist as a whole and work upon them." [12] Köstlin was of opinion that the inferior senses offer "nothing intuitible separate from themselves, and are only modifications of ourselves, but taste, smell and touch are not devoid of all æsthetic importance, since they assist the superior senses; without touch an image could not be recognized by the eye as being hard, resistant or rough; without smell certain images could not be represented as sweet or scented." [13]

We cannot go into a detailed account of all doctrines connected with sensationalistic principles, [14] for all the senses are naturally accepted as æsthetic by the sensationalists, who use "æsthetic" interchangeably with "hedonistic": it will suffice if we recall the "learned" Kralik[1], who was ridiculed by Tolstoy for his theory of the five arts of taste, smell, touch, hearing and sight. [15] The few quotations already given show the embarrassing difficulty caused by the use of the word "æsthetic" as a qualification of "sense," compelling writers to invent absurd distinctions between various groups of senses, or to recognize all senses as being æsthetic, thus giving æsthetic value to every sensory impression, as such. No way out of this labyrinth can be found save by asserting the impossibility of effecting a union between such wholly disparate orders of ideas as the concept of the representative form of the spirit and that of particular physiological organs or a particular matter of sense-impressions. [16]

The theory of kinds of style.

---

1. Kralik：克拉利克（1852—1934），德国作家、哲学家，著有《理想的智慧》《古希腊音乐》《世界之美——普通美学之探究》等。

III. A variety of the error of literary kinds is to be found in the theory of modes, forms or kinds of style (χαρακτῆρες τῆς φράσεως), considered by the ancients as consisting of three forms, the sublime, the medium and the tenuous, a tripartition due, it would seem, to Antisthenes[1], [17] modified later into *subtile, robustum* and *floridum*[2], or amplified into a fourfold division, or designated by adjectives of historic origin as in the Attic, Asiatic or Rhodian styles[3]. The Middle Ages preserved the tradition of a tripartite division, sometimes giving it a curious interpretation, to the effect that the sublime style treats of kings, princes and barons (*e.g.* the *Aeneid*); the mediocre, of middle-class people (*e.g. Georgies*); the humble, of the lowest class (*e.g. Bucolics;*) and the three styles were for this reason also called tragic, elegiac and comic. [18] It is a well-known fact that kinds in style have never ceased to afford matter for discussion in rhetorical text-books down to modern times; for instance, we find Blair distinguishing styles by such epithets as the diffuse, the concise, the nervous, the daring, the soft, the elegant, the flowery, etc. In 1818 the Italian Melchiorre Delfico, in his book on *The Beautiful*, energetically criticized the "endless division of styles," or the superstition "that there could be so many kinds of style"; saying that "style is either good or bad," and adding that it is not possible "it should exist as a preconceived idea in the artist's mind," but that "it should be the consequence of the principal idea, *i.e.* that conception which determines the invention and the composition." [19]

The theory of grammatical forms or parts of speech.

IV. The same error reappears in the philosophy of language, as the

---

1. Antisthenes：安提西尼（约公元前 445—前 365），古希腊哲学家，苏格拉底弟子，悲观主义哲学创立人，著有《物理》等。
2. *subtile, robustum* and *floridum*：〈拉〉细腻的、粗犷的和华丽的
3. Attic, Asiatic or Rhodian styles：雅典式、亚细亚式或罗德岛式

theory of grammatical forms or parts of speech, [20] first created by the sophists (Protagoras[1] is credited with having first distinguished the gender of nouns), adopted by the philosophers, notably by Aristotle and the Stoics (the former was acquainted with two or three parts of speech, the latter with four or five), developed and elaborated by the Alexandrian grammarians in the famous and endless controversy between the analogists and the anomalists[2]. The analogists (Aristarchus[3]) aimed at introducing logical order and regularity into linguistic facts, and described as deviations all such as seemed to them irreducible to logical form. These they called pleonasm, ellipsis, enallage, parallage, and metalepsis[4]. The violence thus wrought by the analogists upon spoken and written language was such that (as Quintilian tells us) some one wittily (*non invenuste*) remarked that it appeared to be one thing to talk Latin and quite another to talk grammar (*aliud esse latine, aliud grammatice loqui*). [21] The anomalists must be credited with restoring to language its free imaginative movement: the Stoic Chrysippus[5] composed a treatise to prove that one thing (one same concept) may be expressed by different sounds, and one and the same sound may express different concepts (*similes res dissimilibus verbis et similibus dissimiles esse vocabulis notatas.*) Another anomalist was the celebrated grammarian Apollonius Dyscolus[6], who rejected the metalepsis, the schemes, and the other artifices by which

---

1. Protagoras：普罗泰戈拉（约公元前 490—前 420），古希腊哲学家，智者派的著有代表人物，著有《论神》等。
2. the analogists and the anomalists：类比推理论者和变规推理论者
3. Aristarchus：阿里斯塔克（约公元前 310—前 230），古希腊杰出的天文学家、数学家、文献学家。
4. pleonasm, ellipsis, enallage, parallage, and metalepsis：赘语、省略、换用、对句和进一步转喻法
5. Chrysippus：克吕西普（约公元前 279—前 206），斯多葛学派的代表人物，对斯多葛派哲学体系的形成和传播做出很大贡献。
6. Apollonius Dyscolus：阿波罗尼乌斯·迪斯克鲁斯，生活于约公元前 2 世纪，希腊语法学家。

the analogists tried to explain facts which did not fit their categories, and pointed out that the use of one word for another, or one part of speech for another, is not a grammatical figure, but a blunder, a thing hardly to be attributed to a poet such as Homer. The upshot of the dispute[1] between anomalists and analogists was the science of Grammar (τεχνη γραμματική), as handed down by the ancients to the modern world, which is justly considered as a sort of compromise between the two opposed parties because, if the schemes of inflection (κανόνες) satisfy the demands of the analogists, their variety satisfies those of the anomalists; hence the original definition of Grammar as theory of analogy was changed subsequently to "theory of analogy and anomaly" (ὁμοίον τε καὶ ἀνομοίου θεωρία). The concept of correct usage, with which Varro[2] hoped to settle the controversy, fell into the trap (common to compromises), merely stating the contradiction in set terms, like the "convenient ornament" of Rhetoric or the kinds accorded a "certain licence" in the literature of precept. If language follows usage (that is to say, the imagination), it does not follow reason (or logic); if it follows reason, it does not follow usage. When the analogists upheld logic as supreme at least inside the individual kinds and sub-kinds, the anomalists hastened to show that even this was not the case. Varro himself was forced to confess that "this part of the subject really is very difficult" (*hic locus maxime lubricus est*). [22]

In the Middle Ages grammar was cultivated to the point of superstition. Divine inspiration was found lurking in the eight parts of speech because "*octavus numerus frequenter in divinis scripturis sacratis invenitur,*[3]" and in the three persons of verbal conjugation, created

---

1. upshot of the dispute：论战的产物
2. Varro：瓦罗（公元前 116—前 27），古罗马学者、作家，传说著有 74 部作品，仅有一部《农业三书》完整流传在世。
3. *octavus numerus frequenter in divinis scripturis sacratis invenitur*：〈拉〉数字 8 在《圣经》中时常出现

simply "*ut quod in Trinitatis fide credimus, in eloquiis inesse videatur.*[1]" [23] Grammarians of the Renaissance and later recommenced the study of linguistic problems and worked to death ellipsis, pleonasm, licence, anomaly and exception; only in comparatively recent times has Linguistic begun to question the very validity of the concept of parts of speech (Pott, Paul and others). [24] If they still survive, the reason may lie in the facts that empirical, practical grammar cannot do without them; that their venerable antiquity disguises their illegitimate and shady origin; and that energetic opposition has been worn down by the fatigue of an endless war.

Theory of æsthetic criticism.

V. The relativity of taste is a sensationalistic theory which denies a spiritual value to art. But it is rarely maintained by writers in the ingenuous categorical garb of the old adage: *De gustibus non est disputandum*[2] (concerning which it would be useful to enquire when the saying was born, and what it fust meant: whether, too, the word *gustibus*[3] referred solely to impressions of the palate, and was only later extended to include æsthetic impressions); as though sensationalists, as if dimly conscious of the higher nature of art, have never been able to resign themselves to the complete relativity of taste. Their torments in the matter really move one to pity. "Is there," Batteux asks, "such a thing as good taste, and is it the only good taste? In what does it consist? Upon what depend? Does it depend upon the object itself or the genius at work upon it? Are there, or are there not, rules? Is wit alone, or heart alone, the organ

1. *ut quod in Trinitatis fide credimus, in eloquiis inesse videatur.*〈拉〉语言中存在着我们信仰的三位一体
2. *De gustibus non est disputandum*：〈拉〉鉴赏力毋庸置疑
3. *gustibus*：〈拉〉鉴赏力

of taste, or both together? How many questions have been raised on this familiar often-treated subject, how many obscure and involved answers have been given!" [25] This perplexity is shared by Home. Tastes, he says, must not be disputed; neither those of the palate nor those of other senses. A remark which seems highly reasonable from one point of view; but, from another, somewhat exaggerated. But yet how can one dispute it? how can one maintain that what actually pleases a man ought not to please him? The proposition then must be true. But now no man of taste will assent to it. We speak of good taste and bad taste; are all criticisms which turn upon this distinction to be considered absurd? have these everyday expressions no meaning? Home ends by asserting a common standard of taste, deduced from the necessity of a common life for mankind or, as he says, from a "final cause"; for without uniformity of taste, who would trouble to produce works of art, build elegant and costly edifices, or lay out beautiful gardens and so forth? He does not fail to draw attention to a second final cause; that of the advisability of attracting citizens to public shows and uniting those whom class-differences and diversity of occupation tend to keep apart. But how shall a standard of taste be established? This is a new perplexity, which one cannot think to be escaped by observing that, as in framing moral rules we seek the counsel of the most honourable of educated men, not of savages; so to determine the standard of taste we should have recourse to the few who are not worn out by degrading bodily labour, not corrupted in taste, and not rendered effeminate by pleasure, who have received the gift of good taste from nature, and have brought it to perfection by the education and practice of a lifetime: if, notwithstanding, controversies arise, then reference must be made to the principles of Criticism as set forth by Home himself in his own book. [26] Similar contradictions and vicious circles reappear in David Hume's *Essay on Taste*, where Hume tries in vain to define the distinctive characteristics of the man of taste whose judgement must be law, and, while asserting the uniformity of the general principles of taste as founded in human nature, and warning the reader against giving undue weight to individual perversions and ignorances, at the same time asserts that divergences in taste may be irreconcilable, insuperable, and yet blameless. [27]

But a criticism of æsthetic relativism cannot be based upon the opposite doctrine which, by its affirmation of absoluteness, resolves taste into concepts and logical inferences. The eighteenth century offers examples of this mistake in Muratori, one of the first to maintain the existence of a rule of taste and a universal beauty whose rules are furnished by Poetics; [28] in André, who said that "the beauty in a work of art is not that which pleases at the first glance of fancy through certain individual dispositions of the mental faculties or bodily organs, but that which has a right to please the reason and reflexion by its own inherent excellence or rightness and, if the expression be allowed, by its intrinsic agreeableness"; [29] in Voltaire, who recognized a "universal taste" which was "intellectual"; [30] and in very many others. This intellectualistic error, no less than the sensationalistic, was attacked by Kant; but even Kant, by making beauty consist in a symbolism of morality, failed to grasp the concept of an imaginative absoluteness of taste. [31] Succeeding generations of philosophers met the difficulty by passing it over in silence.

Nevertheless, this criterion of an imaginative absoluteness, the idea that in order to judge works of art one must place oneself at the artist's point of view at the moment of production, and that to judge is to reproduce, gathered weight little by little from the beginning of the eighteenth century, when its first appearance is seen in the work of the Italian Francesco Montani already quoted (1705), and by the English poet Alexander Pope in his *Essay on Criticism.* ("A perfect judge will read each work of wit With the same spirit that its author writ." [32]) A few years later Antonio Conti recognized part of the truth in the *règle du premier aspect*[1] advised by Terrasson[2] as a test for judging poetry, while noting it to be more applicable to modern than to ancient works: "*quand on n'a pas l'esprit prévenu, et que d'ailleurs on l'a assez pénétrant, on peut voir tout d'un coup si un poète a bien imité son objet; car, comme on connaît l'original, c'est-à-dire les hommes et les mœurs de son siècle, on peut aisément lui confronter la copie, c'est-à-dire la poésie qui les imite.*[3]" In judging ancient writers something more is necessary: "*cette règle du premier aspect n'est presque d'aucun usage dans l'examen de l'ancienne poésie, dont on ne peut pas juger qu'après avoir longtemps réfléchi sur la religion des anciens, sur leurs lois, leur mœurs, sur leurs manières de combattre et d'haranguer, etc. Les beautés d'un poème, indépendantes de toutes ces circonstances individuelles, sont très rares, et les grands peintres les ont toujours évitées avec soin, car ils voulaient peindre la nature et non pas leurs idées;*[4]" [33] the necessary criterion, therefore, is to be found in history. The end of the same century saw the concept of congenial reproduction sufficiently defined by Heydenreich: "A philosophical critic of art must himself be possessed of genius for art;

---

1. *règle du premier aspect*：〈法〉《第一阶段规则》
2. Terrasson：即让·特拉松（1670—1750），法国牧师、作家，法兰西学院院士，著有《荷马〈伊利亚特〉批评论》《萨索斯的一生》等。
3. *quand on n'a pas l'esprit prévenu, et que d'ailleurs on l'a assez pénétrant, on peut voir tout d'un coup si un poète a bien imité son objet; car, comme on connaît l'original, c'est-à-dire les hommes et les mœurs de son siècle, on peut aisément lui confronter la copie, c'est-à-dire la poésie qui les imite.*：〈法〉当我们的评论不带成见，并且有深入内心的认识，那立刻就能看出诗人对描述对象的摹仿逼真与否；因为只要我们对原貌有一定认识，摹本与原貌的对比就轻而易举了。
4. *cette règle du premier aspect n'est presque d'aucun usage dans l'examen de l'ancienne poésie, dont on ne peut pas juger qu'après avoir longtemps réfléchi sur la religion des anciens, sur leurs lois, leur mœurs, sur leurs manières de combattre et d'haranguer, etc. Les beautés d'un poème, indépendantes de toutes ces circonstances individuelles, sont très rares, et les grands peintres les ont toujours évitées avec soi car ils voulaient peindre la nature et non pas leurs idées*:〈法〉上述的第一阶段规则在评价古诗时几乎毫无用武之地。要评价古代诗歌，首先必须对古人的宗教信仰、法律、习俗、斗争辩论的方式进行长时间的思考。脱离这些背景，诗歌之美便无从谈起。伟大的画家必须时刻避免此类状况，因为他想描绘的是大自然，而非他的思想。

reason exacts this qualification and grants no dispensation, just as she will refuse to appoint a blind man as judge of colours. The critic must not pretend to be able to feel the attraction of beauty by means of syllogisms (*Vernunftschlüsse*); beauty must manifest itself to feeling with irresistible self-evidence and, attracted by its fascination, reason must find no time to linger over the why and wherefore; the effect, with its delightful and unexpected possession and domination of the whole being, should suffocate at birth any inquiry into origins or causes. But this state of fanatical admiration cannot last long; reason must inevitably recover consciousness of itself and direct its attention upon the state in which it was during the enjoyment of beauty and upon its present memories of that state ..." [34] This was the wholesomely impressionistic theory which prevailed among the Romanticists and was accepted even by De Sanctis. [35] Still there was even then no definite theory of criticism, which demanded as its condition of existence a precise concept of art and of the relations of the work of art with its historical antecedents. [36] The very possibility of æsthetic criticism was questioned in the second half of the nineteenth century, when taste was relegated to a place amongst the facts of individual caprice, and a so-called historical criticism as proclaimed the sole scientific criticism and expounded in works of irrelevant learning or buried beneath the preconceptions of positivists and materialists. Those who reacted against such extremalism and materialism generally made the mistake of supporting themselves by a kind of intellectualistic dogmatism [37] or an empty æstheticism. [38]

Distinction between taste and genius.

VI. We have seen that in the seventeenth century, when the words "taste" and "genius" or "wit" were in fashion, the facts they designated were sometimes interchanged amongst themselves and came to be considered as one single fact, while sometimes each was conceived as distinct in itself, genius being the faculty of production, and taste the faculty of judgement, taste being further subdivided into the sterile and the fertile: a terminology adopted by Muratori [39] in Italy and Ulrich König [40] in Germany. Batteux said, "*le goût juge des productions du génie*[1]" [41] ; and Kant speaks of defective works having genius without taste or taste without genius, and of others in which taste alone suffices; [42] now we find him distinguishing the two concepts as the judging and producing faculties, now he speaks of them as a single faculty existing in various degrees. An inherent difference between taste and genius was accepted by later writers on Æsthetic and assumed its most rigid form in the hands of Herbart and his followers.

Concept of artistic and literary history.

VII. The evolutionary theory of art made its appearance towards the end of the eighteenth century. This was the time when the distinction between classical and romantic art was first made; a classification later augmented by an introductory section on Oriental art, owing to the increase of knowledge concerning the pre-Hellenic world. Towards the end of his life Goethe told his friend Eckermann that the concepts of classical and romantic had been formed by himself and Schiller, for he himself had upheld the objective method in poetry, whilst Schiller, in order to champion the subjective form to which he inclined, had written the essay *On Naïve and Sentimental Poetry*, in which the word naïve (*naiv*) expresses the style later called classical and the word sentimental (*sentimentalisch*) that later called romantic. "The Schlegels," continues Goethe, "seized upon these ideas and disseminated them, so that to-day everyone uses them and speaks of classical and romantic,

1. *le goût juge des productions du génie*：〈法〉天才的作品是由鉴赏力来判断的

things perfectly unknown fifty years ago" [43] (Goethe was speaking in 1831). Schiller's essay bears the imprint of Rousseau's influence and is dated 1795–6. [44] It contains such statements as this: "Poets are above all things the preservers of nature; and when they cannot be so entirely, and have tried upon themselves the destructive force of arbitrary and artificial forms or have fought against such forms, they stand up to bear witness on her behalf. Poets, therefore, either are nature or, having lost her, seek her. Hence arise two wholly distinct kinds of poetic composition, exhausting between them the whole field of poetry; all poets who are worthy of the name must belong, according to the times and conditions in which they flourish, either to the category of naïve or to that of sentimental poets." Schiller recognized three kinds of sentimental poetry: satirical, elegiac and idyllic[1]; he defined a satirical poet as one "who takes as his object the desertion of nature and the contrast of the real with the ideal." The weak point of this division is the concept of two distinct kinds of poetry, the reduction of the infinite forms in which poetry appears to individuals, to two kinds. If one of these two kinds be taken the perfect and the other as the imperfect kind, the mistake is made of converting imperfection into a kind or species, the negative into a positive. Wilhelm von Humboldt pointed out to his friend that if form is the essence of art, there cannot be a kind of poetry, such as the sentimental or romantic is supposed to be, in which matter preponderates over form, for that would constitute a pseudo-art, not a separate kind of art. [45] Schiller attached no historical meaning to his classification, in fact he declared explicitly that in using the words "ancient" and "modern" as equivalent to "ingenuous" and "sentimental" he did not mean to deny that some "ancient" poets, in his sense of the word, could be found among contemporary writers; the two characters might even be united in the same poet or the same

1. sentimental poetry: satirical, elegiac and idyllic：抒情诗分为：讽刺诗、挽歌与田园诗。

poetical work, as (to give Schiller's own example) in *Werther.* [46] The first to assign a historical meaning to the division were Friedrich and Wilhelm von Schlegel; the former in an early work of 1795, the latter in his celebrated lectures on literary history given at Berlin in 1801–4. But the two senses, systematic and historical, were variously alternated and mixed by literary men and critics, and other distinctions were added; "classical" was sometimes used to describe poetry of a rigid and imitative style, while "romantic" poetry was the inspired; in some countries the word "romantic" came to mean a political reactionary, in Italy it stood for "liberal"; and so forth. In 1815, when Friedrich Schlegel spoke of ancient Persian romantic poems, or when in our times attention is called to the romanticism of the Greek, Latin or French classics, the historical signification is lost in the theoretical, the sense originally intended by Schiller.

But the historical sense was prevalent in German idealism, which inclined towards the construction of a universal history, including that of literature and art, upon a scheme of ideal evolution. Schelling made a sharp division between pagan and Christian art; the second being held an advance upon the former which was the lowest step. [47] Hegel accepted this division and introduced a final regress by dividing the history of art into three periods: symbolic (Oriental) art, classical (Hellenic) and romantic (modern). Just as he conceived Roman art (with its introduction of satire and other kinds indicative of a failure to maintain harmony between form and content) as the dissolution of classical art, a thought suggested by Schiller, so he found in the subjective humour of Cervantes and Ariosto [48] the dissolution of romantic art; and he regarded this series as completing the possibilities of art, though some interpreters think that by a self-contradiction he admitted the possibility of a fourth period, an art of the modern or future world. Indeed amongst his disciples we find Weiss rejecting the Oriental period in order to save the triadic division, and placing as third the modern period, synthesis of the ancient and the mediæval: [49]

Vischer too inclines to recognize a modern or progressive period. [50]

These arbitrary constructions reappear in the works of positivist metaphysicians[1] in the shape of an evolutionary or progressive history of art. Spencer dreamed of writing some sort of treatise on the subject, and in the published programme of his system (1860) we read that the third volume of his *Principles of Sociology* was to contain amongst other things a chapter on æsthetic progress "with the gradual differentiation of fine arts from primitive institutions and from each other, with their increasing variety in development, their progress in reality of expression and superiority of end." No grief need be felt that the chapter was left unwritten when we remember the samples of it preserved in the *Principles of Psychology* and already reviewed in these pages. [51]

The strong historical sense of our own day is leading us further and further away from the evolutionary or abstractly progressive theories which falsify the free and original movement of art. Fiedler remarked not without justice that unity and progress cannot be introduced into a history of art, and that the works of artists must be judged discretely as so many fragments of the life of the universe. [52] In recent times a remarkable student of the history of figurative art, Venturi, has tried to bring evolutionism into fashion, and has illustrated it in a *History of the Madonna*, in which the presentment of the Virgin is conceived as an organism which is born, grows, attains perfection, grows old and dies! Others have claimed for artistic history its true character, intolerant of outward curb and rule, drawing her ever-varied productions from the well-head of the infinite Spirit. [53]

Conclusion.

These hurried notes may suffice to show in how narrow a circle

1. positivist metaphysicians：实证主义形而上学者

has hitherto moved the scientific criticism of the errors we have called "particular." Æsthetic needs to be surrounded and nourished by a watchful and vigorous critical literature drawing its life from her and forming in turn her safeguard and strength.

[1] See above, pp. 130–131.

[2] *Das Naturgefühl nach Verschiedenheit der Zeiten und Volksstämme*, in *Cosmos*, ii.

[3] V. Laprade, *Le Sentiment de la nature avant le christianisme*, 1866; also *chez les modernes*, 1867; Alfred Biese, *Die Entwicklung des Naturgefühls den Griechen und Römern*, Kiel, 1882–1884; *Die Entwicklung des Naturgefühls im Mittelalter und in der Neuzeit*, 2nd ed., Leipzig, 1892.

[4] *Kritische Gänge*, v. pp. 5–23.

[5] *Dtsche. Ästh. s. Kant*, pp. 217–218; cf. *Philos, d. Schönen*, bk. ii. ch. 7.

[6] See above, pp. 196–197.

[7] *Hippias maior, passim.*

[8] *Elements of Criticism*, introd., and cf. ch. 3.

[9] Herder, *Kritische Wälder* (in *Werke, ed. cit.* iv.), pp. 47–53; cf. *Kaligone* (*ibid.* vol. xxii.), *passim*; and fragment on *Plastic.*

[10] *Vorles. üb. Ästh.* i. pp. 50–51.

[11] *Op. cit.* p. 92 *seqq.*

[12] *Ästh.* i. p. 181.

[13] *Ästh.* pp. 80–83.

[14] *E.g.* Grant Allen, *Physiological Æsthetics*, chs. 4 and 5.

[15] Tolstoy, *What is Art?* pp. 19–22. Kralik is the author of *Weltschönheit, Versuch einer allgemeinen Ästhetik,* Vienna, 1894.

[16] See above, pp. 50–52.

[17] Cf. Volkmann, *Rhet. d. G. u. Röm.* pp. 532–544.

[18] Comparetti, *Virgilio net M. E.* i. p. 172.

[19] *Nuove ricerche sul hello*, ch. 10.

[20] See above, pp. 177–178.

[21] *Inst. Oral.* i. ch. 6.

[22] For all this cf. the works of Lersch and of Steinthal, which contain the more important texts.

[23] Comparetti, *Virgilio nel M. E.*, i. pp. 169–170.

[24] Pott, introd. to Humboldt, *cit.* Paul, *Principien d. Sprachgeschichte*, ch. 20.

[25] Batteux, *Les Beaux Arts,* part ii. p. 54.

[26] *Elem. of Criticism*, iii. ch. 25.

[27] *Essays, Moral, Political and Literary* (London ed., 1862), ch. 23: *On the Standard of Taste.*

[28] *Perfetta poesia*, bk. v. ch. 5.

[29] *Essai sur le beau*, disc. 3.

[30] *Essai sur le goût, cil.*

[31] See above, pp. 312–314.

[32] *Essay on Criticism*, 1711, part ii. 11. 233–234.

[33] Letter to Maffei, in *Prose e poesie*, ii. pp. cxx–cxxi.

[34] *System d. Ästhetik,* pref. pp. xxi–xxv.

[35] Amongst other places *Saggi critici*, pp. 355–358.

[36] See above, pp. 355–359.

[37] *E.g.* A. Ricardou, *La Critique littéraire*, Paris, 1896.

[38] *E.g.* A. Conti, *Sul fiume del tempo*, Naples, 1907.

[39] *Perf. poesia*, bk. v. ch. 5.

[40] *Untersuchung v. d. guten Geschmack*, 1727.

[41] *Les Beaux Arts*, part ii. ch. 1.

[42] *Krit. d. Urtheilskr.* § 48.

[43] Eckermann, *Gespräche mit Goethe*, under date March 21, 1831.

[44] *Über naive und sentimentalische Dichtung*, 1795–1796 (in *Werke*, ed. Goedecke, vol. xii.).

[45] Quoted in Danzel, *Ges. Aufs.* pp. 21–22.

[46] *Üb. naive u. sentim. Dicht., ed. cit.*, p. 155, note.

[47] See above, p. 323.

[48] *Vorles. üb. Ästh.*, vols. ii. and iii.

[49] Cf. von Hartmann, *Dtsche. Ästh. s. Kant*, pp. 99–101.

[50] *Ästh.* part iii.

[51] See above, pp. 420–430.

[52] C. Fiedler, *Ursprung d. künstl. Thätigkeit*, p. 136 *seqq.*

[53] Ad. Venturi, *La Madonna*, Milan, 1899. Cf. B. Labanca, in *Rivista polit, e lett.* (Rome), Oct. 1899, and in *Rivisla di filos. e pedag.* (Bologna), 1900; and B. Croce, in *Nap. nobiliss., Rivista di lopografia e storia dell' arte*, viii. pp. 161–163, ix. pp. 13–14 (reprinted in *Probl. di estetica*, pp. 265–272). On the theory of method in artistic and literary history cf. above, pp. 160–171.

BIBLIOGRAPHICAL APPENDIX

The first attempt at a history of Æsthetic is the work of J. Roller mentioned by Zimmermann (*Gesch. d. Ästh.* pref., p. v) as being so exceedingly rare that he had never been able to see a copy of the book. We ourselves have had the good fortune to find the book in the Royal Library of Munich in Bavaria, by the help of our friend Dr. Arturo Farinelli of Innsbruck University, and to obtain the loan of it. It bears the title *Entwurf* | *zur* | *Geschichte und Literatur* | *der Æsthetik* | *von Baumgarten auf die* | *neueste Zeit.* | *Herausgegeben* | *von* | *J. Koller.* | *Regensburg* | in der Montag und Weissischen Buchhandlung | 1799 (pp. viii–107, small 8vo); in the preface the author declares his intention of supplying young men attending Lectures on the Criticism of Taste and the Theory of the Fine Arts in the German Universities with a "lucid summary of the origin and later progress of these studies," premising that he will treat of general theories only and that his judgements are frequently derived from reviews in literary periodicals. The introduction treats of æsthetic theories from antiquity down to the beginning of the eighteenth century; Koller observes that "the names and form of a general Theory of Fine Art and Criticism of Taste were unknown to the ancients, whose imperfect ethical theory prevented their producing anything in this field." He dedicates § 5 to the Italians, "who have produced little in theory"; indeed the only Italian books mentioned are the *Entusiasmo* of Bettinelli and the small work of Jagemann, *Saggio di buon gusto nelle belle arti ove si spiegano gli elementi dell' estetica*, di Fr. Gaud. Jagemann, Regente agostiniano, In Firenze, MDCCLXXI, Presso Luigi Bastianelli e compagni; 60 pp. (concerning this, see B. Croce, *Problemi di estetica*, pp. 387–390).

The section on the History and Literature of Æsthetic begins with the oft-quoted passage from Bülffinger ("*Vellem existerent*, etc.") and passes at once to Baumgarten: "the theoretical epoch owes its existence undeniably to Baumgarten; to him belongs the inalienable merit of having first conceived an Æsthetic founded on principles of reason and wholly developed, and of having tried to put it into practice by the means offered him by his own philosophy." Immediately after this, Meier is mentioned, followed by the titles, accompanied by brief extracts and remarks — a sort of *catalogue raisonné* — of many German books on Æsthetic from those of K. W. Müller (1759) to one by Ramier (1799), mixed with various French and English writings under the dates of their German translations. Special emphasis is laid on Kant (pp. 64–74), with the remark that, prior to the appearance of the *Critique of Judgment*, æstheticians were divided into sceptics, dogmatics and empiricists: the most powerful intellects of the nation inclined towards empiricism, so much so that had Kant himself "been asked by what literature he had been most strongly influenced in the development of his own thought, he would certainly have named the acute empirical writers of England, France and Germany"; but "by no pre-Kantian method had it been possible to establish an agreement (*eine Einhelligkeit*) between men upon matters of taste." The last pages call attention to the revival of interest in æsthetic studies, which nobody would now dare call a waste of time as in former days. "May Jacobi, Schiller and Mehmel soon enrich literature by publication of their theories!"

The rarity of Koller's book has led us to notice it at some length. Apart from this the first general history of Æsthetic worthy the name is that written by Robert Zimmermann, *Geschichte der Ästhetik als philosophischer Wissenschaft*, Vienna, 1858. It is divided into four books: "the first of these contains the history of philosophical concepts concerning the beautiful and art from the Greeks down to the constitution of Æsthetic as a philosophical science through the labours of Baumgarten"; the second runs from Baumgarten down to the reform of Æsthetic brought about by the *Critique of Judgment*; the third, from Kant to the Æsthetic of idealism; the fourth, from the beginnings of idealistic Æsthetic down to the author's own day (1798–1858). The work is on Herbartian lines, and is remarkable for solid research and lucid exposition, although the erroneous point of view and neglect of all æsthetic movement other than Græco-Roman or German are grave defects; besides, it is now sixty years out of date.

Less solid and more compilatory in nature, whilst retaining all the defects of the foregoing, is the history by Max Schasler, *Kritische Geschichte der Ästhetik*, Berlin, 1872, divided into three books treating of ancient Æsthetic and that of the eighteenth and nineteenth centuries. The author belongs to the Hegelian school and conceives his history as a propædeutic to theory, "in order, that is, to attain a supreme principle for the construction of a new system"; he schematizes the material of facts for each period into three grades of Æsthetic of sensation (*Empfindungsurtheil*,) of intellect (*Verstandsurtheil*) and of reason (*Vernunfturtheil*.)

English literature has Bernard Bosanquet's *History of Æsthetics*, London, 1892; a sober and well-arranged work, written from an eclectic point of view between the Æsthetic of content and the Æsthetic of form. The author, however, is wrong in believing he has passed over "no writer of the first rank"; he has passed over not only writers but some important movements of ideas, and in general he shows insufficient knowledge of the literature of the Latin races. Another general history of Æsthetic in English is the first volume of *The Philosophy of the Beautiful, being Outlines of the History of Æsthetics*, by William Knight, London, Murray, 1895: it consists mainly of a rich collection of extracts and abridgements of ancient and modern books treating of Æsthetic. In this respect the most noteworthy chapters are those on Holland, Great Britain and America (10–13); the second volume, published in 1898, has in an appendix, pp. 251–281, notices upon Æsthetic in Russia and Denmark. Another recent publication is George Saintsbury's *A History of Criticism and Literary Taste in Europe from the Earliest Times to the Present Day*; vol. i., Edinburgh and London, 1900, concerning classical and mediæval criticism; vol. ii., 1902, criticism from the Renaissance to end of the eighteenth century: vol. iii., 1904, modern criticism. The writer of this History, equally skilled in literature and innocent of philosophy, has thought it possible to exclude æsthetic science in the strict sense, "the more transcendental Æsthetic, those ambitious theories of Beauty and artistic pleasure in general which seem so noble and fascinating until we discover them to be but cloud-appearances of Juno," and to limit his treatise to "lofty Rhetoric and Poetic, to the theory and practice of Criticism and literary taste" (book i. ch. I). Thus is produced a book instructive in many ways but wholly deficient in method and definite object. What is lofty Rhetoric and Poetic, the theory of Criticism and literary taste, if not Æsthetic pure and simple? how can the history of these be composed without due notice of metaphysical Æsthetic and other manifestations whose interaction and development are the fabric of history itself? Perhaps Saintsbury hoped to be able to write a History of Criticism as distinct from that of Æsthetic; if that be the case, he has been unsuccessful in writing either one or the other. Cf. *La Critica*, ii. (1904), pp. 59–63.

The generosity of the Hungarian Academy of Science has enabled us to handle the History of Æsthetic (*Az Æsthetika története*) of Bela Janosi, Budapesth, 1899–1901, in three volumes; the first volume treats the Æsthetic of Greece; the second, of Æsthetic from the Middle Ages to Baumgarten; the third, from Baumgarten to the present day. For us it is a book sealed with seven seals, save for reviews which have appeared in the *Deutsche Litteraturzeitung* of Berlin, August 25, 1900, July 12, 1902, and May 2, 1903.

Amongst Latin countries, France has no special history of Æsthetic, for this title cannot be given to the portion of the second volume (pp. 311–570) of the work by Ch. Levêque, *La Science du beau* (Paris, 1862), under the heading *Examen des principaux systems d'esthétique anciens et modernes*, where eight chapters are devoted to an exposition of the theories of Plato, Aristotle, Plotinus and St. Augustine, Hutcheson, André and Baumgarten, Reid, Kant, Schelling and Hegel. Spain, on the other hand, possesses the work of Marcelino Menendez y Pelayo, *Historia de las idéas estéticas en España*, 2nd ed., Madrid, 1890–1901 (5 vols., variously distributed amongst the 1st ed., 1883–1891, and the 2nd), which is not restricted, as the title suggests, to Spain alone or to Æsthetic alone but, as the author observes in his preface (i. pp. xx–xxi), includes the metaphysical disquisitions on the beautiful, the speculations of mystics on the beauty of God and on love; the theories of art scattered through the pages of philosophers; the æsthetic considerations found in treatises upon individual arts (Poetics and Rhetoric, works on painting, architecture, etc.); and, finally, ideas enunciated by artists concerning their own particular arts. This work is of capital importance on everything to do with Spanish authors, and also in its general part contains good treatments of matters generally passed over by historians. Menendez y Pelayo inclines to metaphysical idealism, yet seems not disinclined to welcome elements from other systems, even empirical theories: in our opinion this vagueness has an unfortunate effect on the work as a whole. Some years ago Professor V. Spinazzola announced the forthcoming publication of a course of lectures given by Francesco de Sanctis in Naples in 1845 on *Storia della critica da Aristotele ad Hegel*. For the history of Æsthetic in Italy cf. Alfredo Rolla, *Storia delle idee estetiche in Italia*, Turin, 1904; on which see Croce, *Problemi di estetica*, pp. 401–415.

We need take no notice of the historical remarks or chapters that generally stand at the beginning of treatises on Æsthetic; the most important occur in the volumes of Solger, Hegel and Schleiermacher. A general history of Æsthetic, from the rigorous point of view of the principle of Expression, has not been attempted before the present work.

For the bibliography down to the end of the eighteenth century, Sulzer's *Allgemeine Theorie der schönen Künste*, 2nd ed., with additions by von Blankenburg, Leipzig, 1792, in four volumes, is practically complete and is an inexhaustible mine of information. For the nineteenth century much material is collected by C. Mills Gayley and Fred Newton Scott in *An Introduction to the Methods and Materials of Literary Criticism. The Bases in Æsthetics and Poetics*, Boston, 1899. Besides Sulzer, we may mention æsthetic dictionaries by Gruber, *Wörterbuch z. Behuf d. Ästh. d. schönen Künste*, Weimar, 1810: Jeithles, *Ästhetisches Lexikon*, vol. i. A–K, Vienna, 1835: Hebenstreit, *Encyklopädie d. Ästhetik*, 2nd ed., Vienna, 1848.

The following notes contain for the convenience of the student several books which the author has not been able to see.

I. Concerning ancient Æsthetic no better or more comprehensive work can be found than the *Geschichte der Theorie der Kunst bei den Alten*, by Ed. Müller, Breslau, 1831–1837, 2 vols. For inquiries concerning the Beautiful special reference should be made to Julius Walter, *Die Geschichte der Ästhetik im Alterthum ihren begrifflichen Entwicklung nach*, Leipzig, 1893. See also Em. Egger, *Essai sur l'histoire de la critique chez les Grecs*, 2nd ed., Paris, 1886: Zimmermann, Bk. I.: Bosanquet, ch. ii.–v. and Saintsbury, vol. i.

Of the innumerable special monographs: for Plato's Æsthetic see Arn. Ruge, *Die platonische Ästhetik*, Halle, 1832: for Aristotle's, Döring, *Die Kunstlehre des Aristoteles*, Jena, 1876: C. Bénard, *L'Esthétique d'Aristote et de ses successeurs*, Paris, 1890: S. H. Butcher, *Aristotle's Theory of Poetry and Fine Art*, 3rd ed., London, 1902. For Plotinus, E. Vacherot, *Histoire critique de l'école d'Alexandrie*, Paris, 1846: E. Brenning, *Die Lehre vom Schönen bei Plotin im Zusammenhang seines Systems dargestellt*, Göttingen, 1864. On the *Ars Poetica* of Horace, A. Viola, *L'arte poetica di Orazio nella critica italiana e straniera*, 2 vols. Naples, 1901–1907.

For the history of ancient Psychology see H. Siebeck, *Geschichte der*

Psychologie, 1880; A. E. Chaignet, *Histoire de la psychologie des Grecs*, Paris, 1887; L. Ambrosi, *La psicologia dell' immaginazione nella storia della filosofia*, Rome, 1898. For the history of the philosophy of language see H. Steinthal, *Geschichte der Sprachwissenschaft bei den Griechen und Römern mit besonderer Rücksicht auf die Logik*, 2nd ed. Berlin, 1890–1891, 2 vols.

II. For the æsthetic ideas of St. Augustine and early Christian authors see Menendez y Pelayo, *op. cit.* pp. 193–266. For Thomas Aquinas, L.Taparelli, *Delle ragioni del bello seconde la dottrina di san Tommaso d'Aquino* (in *Civiltà cattolica* for 1859–1860): P. Vallet, *L'Idée du beau dans la philosophie de St. Thomas d'Aquin*, 1883: M. de Wulf, *Études historiques sur l'esthétique de St. Thomas*, Louvain, 1896.

For the literary doctrines of the Middle Ages see D. Comparetti, *Virgilio nel medio evo*, 2nd ed. Florence, 1893, vol. i., and G. Saintsbury, *op. cit.*, vol. i. pp. 369–486. For the early Renaissance see K. Vossler, *Poetische Theorien in d. italien. Frührenaissance*, Berlin, 1900. For the Poetics of the high Renaissance see J. E. Spingarn, *History of Literary Criticism in the Renaissance, with special reference to the influence of Italy*, New York, 1899 (Italian trans. with corrections and additions, Bari, 1905). See also F. de Sanctis, *Storia della letteratura italiana*, Naples, 1870, *passim.*

For the traditions of Platonic and neo-Platonic ideas in the Middle Ages and Renaissance, for best and fullest information see Menendez y Pelayo, *op. cit.*, vol. i. part ii. and vol. ii. For Italian treatises on beauty and love see Michele Rosi, *Saggi sui trattati d' amore del cinquecento*, Recanati, 1899, and F. Flamini, *Il cinquecento*, Milan, Vallardi, N.D., ch. iv. pp. 378–381. For Tasso see Alfredo Giannini, *Il "Minturno" di T. Tasso*, Ariano, 1899: see also E. Proton *Rass. crit. lett. ital.* vi. (Naples, 1901) pp. 127–145. For Leone Ebreo see Edm. Solmi, *Benedetto Spinoza e L. E., studio su una fonte italiana dimenticata dello spinozismo*, Modena, 1903: cf. G. Gentile in *Critica*, ii. pp. 313–319.

On J. C. Scaliger see Eug. Lintilhac, *Un Coup d'État dans la république des lettres: Jules César Scaliger, fondateur du classicism cent ans avant*

Boileau (in the *Nouv. Revue*, 1890, vol. lxiv. pp. 333–346, 528–547). On Fracastoro, Giuseppe Rossi, *Girolamo Fracastoro in relazione all' aristotelismo e alla scienza nel Rinascimento*, Pisa, 1893. On Castelvetro, Ant. Fusco, *La poetica di Ludovico Castelvetro*, Naples, 1904. On Patrizzi, Oddone Zenatti, *Fr. Patrizzi, Orazio Ariosto, e Torquato Tasso*, etc. (Verona, per le nozze Morpurgo-Franchetti, N.D.).

III. For this period of ferment see H. von Stein, *Die Entstehung der neueren Ästhetik*, Stuttgart, 1886: K. Borinski, *Die Poetik der Renaissance und die Anfänge der litterarischen Kritik in Deutschland*, Berlin, 1886 (esp. the last chapter): also same author's *Baltasar Gracian und die Hofliteratur in Deutschland*, Halle a. S., 1894, B. Croce, *I trattatisti italiani del Concettismo e B. Gracian*, Naples, 1899 (in *Atti dell' Acc. Pont.* vol. xxix., reprinted in *Problemi di estetica*, pp. 309–345), *Elizabethan Critical Essays*, edited with an introduction by G. Gregory Smith, Oxford, 1904, 2 vols.: *Critical Essays of the Seventeenth Century*, edited by J. E. Spingam, Oxford, 1908, 2 vols.: Leone Donati, *J. J. Bodmer und die italienische Litteratur* (in the vol. *J. J. Bodmer, Denkschrift z. C. C. Geburtstag*, Zürich, 1900, pp. 241–312): see also *Probl. di estetica*, pp. 371–380.

On Bacon see K. Fischer, *Franz Baco von Verulam*, Leipzig, 1856 (2nd ed. 1875), cf. P. Jacquinet, *Fr. Baconis in re litteraria iudicia*, Paris, 1863. On Gravina, Em. Reich, *G. V. Gravina als Ästhetiker* (in the Trans, of the Viennese Academy, vol. cxx. 1890): B. Croce, *Di alcuni giudizi sul Gravina considerate come estetico*, Florence, 1901 (in *Miscellanea d' Ancona*, pp. 456–464), reprinted in *Probl. di est.* pp. 360–370. On Du Bos, Morel, *Étude sur l'abbé du Bos*, Paris, 1849: P. Petent, *J. B. Dubos*, Tramelan, 1902. On Bouhours, Doncieux, *Un jésuite homme de lettres au XVIIe siècle*, Paris, 1886. On the Bouhours-Orsi controversy, F. Fottano, *Una polemica nel settecento*, in *Ricerche letterarie*, Leghorn, 1897, pp. 313–332: A. Boeri, *Una contesa letteraria franco-italiana nel secolo XVIII*, Palermo, 1900 (cf. *Giorn. stor. lett. ital.* xxxvi. pp. 255–256): B. Croce, *Varietà di storia dell' estetica*, § § 1–2, in *Rass. crit. lett. ital.* cit., vi. 1901, pp. 115–126, reprinted in *Probl. di est.* pp. 346–359.

IV. On Cartesianism in literature see É. Krantz, *L'Esthétique de Descartes étudiée dans les rapports de la doctrine cartésienne avec la littérature classique française au XVIIIe siècle*, Paris, 1882; see also the chapter on André, pp. 311–341, and the introduction by V. Cousin to the *œuvres philosophiques du p. André*, Paris, 1843: on Boileau, Borinski, *Poetik d. Renaissance*, c. 6, pp. 314–329; J. Brunetière, *L'Esthétique de B.* in *Revue des Deux Mondes*, June 1, 1899.

On the English intellectualist æstheticians see Zimmermann, *op. cit.* pp. 273–301; also von Stein, *op. cit.* pp. 185–216. On Shaftesbury and Hutcheson see esp. Gid. Spicker, *Die Philosophie d. Grafen v. Shaftesbury*, Freiburg i. B., 1872, part iv. on art and literature, pp. 196–233: T. Fowler, S. *and Hutcheson*, London, 1882: William Robert Scott, *Francis Hutcheson, his life, teaching and position in the history of philosophy*, Cambridge, 1900.

On Leibniz, Baumgarten and contemporary German writers see Th. W. Danzel, *Gottsched und seine Zeit*, 2nd ed., Leipzig, 1855: H. G. Meyer, *Leibnitz und Baumgarten als Begründer der deutschen Ästhetik*, Inaugural Dissertation, Halle, 1874: Joh. Schmidt, *L. und B.*, Halle, 1875: Ém. Grucker, *Histoire des doctrines littéraires et esthétiques en Allemagne* (from Opitz to the Swiss writers), Paris, 1883: Fr. Braitmaier, *Geschichte der poetischen Theorie und Kritik von den Diskursen der Maler his auf Lessing*, Frauenfeld, 1888–1889. In the last-named book the first part treats of the beginning of Poetics and criticism in Germany, considered in their relation to the doctrines of classical, French and English writers: the second part treats of an attempt to found an æsthetic philosophy and theory of poetry upon a basis of Leibnitian-Wolffian psychology: which includes a long discussion of Baumgarten and quotations from two dissertations, Raabe's *A. G. Baumgarten, æstheticæ in disciplinæ formam parens et auctor*, and Prieger's *Anregung u. metaphysische Grundlage d. Ästh. von A. G. Baumgarten*, 1875 (cf. vol. ii. p. 2).

V. On Vico as æsthetician see B. Zumbini, *Sopra alcuni principia di critica letteraria di G. B. V.* (reprinted in *Studî di letter. italiana*, Florence,

1894, pp. 257–268): B. Croce, *G. B. V. primo scopritore della scienza estetica*, Naples, 1901 (reprinted from *Flegrea*. April 1901), incorporated in the present volume as has been mentioned already: see also G. Gentile in *Rass. crit. della lett. ital.*, cit., vi. pp. 254–265: E. Bertana, in *Giorn. stor. lett. ital.* xxxviii. pp. 449–451: A. Martinazzoli, *Intorno alle dottrine vichiane di ragion poetica*, in *Riv. di filos. e sc. aff.* of Bologna, July 1902: also the reply of B. Croce, *ibid.*, August 1902: Giovanni Rossi, *Il pensiero di G. B. V. intorno alla natura della lingua e all'ufficio dette lettere*, Salerno, 1901. The important position occupied by Vico in respect to Æsthetic had been remarked earlier by C. Marini, *G. B. V. al cospetto del secolo XIX*, Naples, 1852, c. 7, § 10. For the influence exercised by Vico, B. Croce, *Per la storia della critica e storiografia letteraria*, Naples, 1903 (in *Atti d. Acc. Pont.*, vol. xxxiii.), pp. 7–8, 26–28 (reprinted in *Probl. di est.* pp. 423–425), and G. A. Borgese, *Storia della critica romantica in Italia*, Naples, 1905, *passim.*

On Vico's thought in general, as well as on his Æsthetic, see B. Croce, *La filosofia di Giambattista Vico*, Bari, 1911: English translation by R. G. Collingwood, 1913. The copious literature concerning Vico is given by B. Croce in *Bibliografia vichiana*, Naples, 1904 (reprinted from *Atti dell' Acad. Pont.* vol. xxxiv.), and *Supplemento, ibid.* 1907, and *Secondo Supplemento*, 1910 (*Atti* cit., vols, xxxvii. And xli.).

VI. On the literary doctrines of Conti see G. Brognoligo, *L'opera letteraria di A. Conti*, in *Arch. veneto*, 1894, vol. i. pp. 152–209: on Cesarotti, Vitt. Alemanni, *Un filosofo delle lettere*, vol. i. Turin, 1894: on Pagano, B. Croce, *Varietà di storia dell' estetica*, § 3; *Di alcuni estetici italiani della seconda metà del secolo XVIII*, in *Rass. crit.* cit. vii. 1902, pp. 1–17 (reprinted in *Probl. di est.* pp. 381–450).

On the German æstheticians, in addition to the various general histories already quoted, see R. Sommer, *Grundzüge einer Geschichte der deutschen Psychologie u. Ästhetik von Wolff-Baumgarten his Kant-Schiller*, Würzburg, 1892. Greatly inferior is M. Dessoir, *Geschichte d. neueren deutschen Psychologie*, 2nd ed., Berlin, 1897 (the first half only is published, down to Kant exclusive).

On Sulzer, Braitmaier, *op. cit.* ii. pp. 55–71: on Mendelssohn, *ibid.* pp. 72–279: for Elias Schlegel, *op. cit.* i. p. 249 *seqq.*; on Mendelssohn see also Th. Wilh. Danzel, *Gesammelte Aufsätze*, Leipzig, Jahn, 1855, pp. 85–98: Kannegiesser, *Stellung Mendelssohns in d. Gesch. d. Ästh.*, 1868. On Riedel, K. F. Wize, *F. J. Riedel u. seine Ästhetik*, Diss., Berlin, 1907. On Herder, Ch. Joiet, *H. et la renaissance littéraire en Allemagne au XVIIIe siècle*, Paris, 1875: R. Haym, *H. nach seinem Leben u. seinen Werken*, 2 vols., Berlin, 1880: G. Jacobi, *H.'s und Kant's Ästh.*, Leipzig, 1907. For the ideas of Hamann and Herder concerning the origins of poetry see Croce in *Critica*, ix. (1911), pp. 469–472. On the history of Linguistic, see Th. Benfey, *Geschichte d. Sprachwissenschaft in Deutschland*, Munich, 1869, introd.: H. Steinthal, *Der Ursprung der Sprache im Zusammenhange mit d. letzen Fragen alles Wissens, eine Darstellung, Kritik und Fortentwicklung der vorzüglichsten Ansichten*, 4th ed., Berlin, 1888.

VII. On Batteux see E. v. Danckelmann, *Charles Batteux, sein Leben u. sein ästhetisches Lehrgebäude*, Rostock, 1902. On Hogarth, Burke and Home, Zimmermann, *op. cit.* pp. 223–273; Bosanquet, *op. cit.* pp. 202–210. On Home esp. J. Wohlgemüth, *H. Home's Ästhetik*, Rostock, 1894: W. Neumann, *Die Bedeutung Homes für d. Ästhetik, u. Sein Einflüss auf d. deutschen Ästhetik*, Halle, 1894. On Hemsterhuis, Ém. Grucker, *François H., sa vie et ses oeuvres*, Paris, 1866.

On Winckelmann, Goethe, *W. u. sein Jahrhundert*, 1805 (in *Werke*, ed. Goedecke, vol. xxxi.): C. Justi, *W. u. seine Zeitgenossen*, 2nd ed., Leipzig, 1898. A criticism of Winckelmann's theory, by H. Hettner, appeared in the *Revue Moderne*, 1866. On Mengs, Zimmermann, *op. cit.* pp. 338–355. On Lessing, Th. Wilh. Danzel, *G. E. Lessing, sein Leben und seine Werke*, Leipzig, 1849–1853: Kuno Fischer, *L. als Reformater d. deutschen Litteratur*, Stuttgart, 1881: Ém. Grucker, *Lessing*, Paris, 1891: Erich Schmidt, *Lessing*, 2nd ed., Berlin, 1899: K. Borinski, *Lessing*, Berlin, 1900.

On Spalletti see B. Croce, *Var.*, cit., § 3 (*Probl. d. est.* pp. 392–

398). On Meier, Hirth and Goethe, Danzel, *Goethe und die Weimarsche Kunstfreunde in ihrem Verhältniss z. Winckelmann*, in *Gesamm. Aufs.* pp. 118–145. On Goethe's Æsthetic esp. see Wilh. Bode, *Goethes Ästhetik*, Berlin, 1901.

VIII. Critical expositions of Kant's Æsthetic are very numerous even in Italy: for example, O. Colecchi, *Questioni filosofiche*, Naples, 1843, vol. iii.; C. Cantoni, *E. Kant*, Milan, 1884, vol. iii. In German, esp. H. Cohen, *Kants Begründung der Ästhetik*, Berlin, 1889; also an important chapter in Sommer, *op. cit.* pp. 337–352; a sufficient representative of a host of others is the elaborate work of Victor Basch, *Essai critique sur l'esthétique de Kant*, Paris, 1896. See also, on an Italian trans. of the *Kr. d. Urth.*, B. Croce in *Critica*, v. (1907), pp. 160–164.

For Kant's lectures and the historical antecedents of his *Critique of Judgment* (besides the dissertations of H. Falkenheim, *Die Entstehung der kantischen Ästhetik*, Heidelberg, 1890, and Rich. Grundmann, *Die Entwickel d. Ästh. Kants*, Leipzig, 1893) see the exhaustive work of Otto Schlapp, *Kant's Lehre vom Genie und die Entstehung d. Kritik d. Urtheilskraft*, Göttingen, 1901.

IX. For the whole of this period, beside the general histories already quoted which treat of it in great detail, see Th. Wilh. Danzel, *Über den gegenwärtigen Zustand d. Philosophie d. Kunst u. ihre nächste Aufgabe* (in the *Zeitschr. f. Phil*, of Fichte, 1844–1845, and reprinted in *Gesammelte Aufsätze*, pp. 1–84): this treats of Kant, Schiller, Fichte, Schelling, Hegel and, more particularly, of Solger, pp. 51–84: Herm. Lotze, *Geschichte der Ästhetik in Deutschland*, Munich, 1868 (in the coll. "History of the Sciences in Germany," published by the Royal Academy of Sciences of Munich in Bavaria): first book, history of general points of view from Baumgarten to the Herbartian school: second book, history of individual fundamental æsthetic concepts: third book, contributions to the history of the theory of the arts: Ed. v. Hartmann, *Die deutsche Ästhetik s. Kant* (first part, historico-critical), Berlin, 1886, divided into two books. The first book discusses the doctrine of the chief æstheticians and, after an introduction on the foundation of philosophical æsthetic by Kant, treats of the Æsthetic of the content, divided into that of abstract idealism (Schelling, Schopenhauer, Solger, Krause, Weisse, Lotze); of concrete idealism (Hegel, Trahndorff, Schleiermacher, Deutinger, Oersted, Vischer, Zeising, Carrière, Schasler); of the Æsthetic of feeling (Kirchmann, Wiener, Horwicz); the Æsthetic of form, subdivided into abstract formalism (Herbart, Zimmermann), and concrete formalism (Köstlin, Siebeck). The second book is concerned with the more important special problems.

On the Æsthetic of Schiller specially see, amongst numerous monographs, Danzel, *Schillers Briefwechsel mit Körner*, in *Ges. Aufs.* pp. 227–244: G. Zimmermann, *Versuch einer schillerschen Ästhetik*, Leipzig, 1889: F. Montargis, *L'Esthétique de Schiller*, Paris, 1890: the chapter in Sommer, *op. cit.* pp. 365–432: V. Basch, *La Poétique de Schiller*, Paris, 1901.

On the Æsthetic of Romanticism, R. Haym, *Die romantische Schule: ein Beitrag z. Geschichte d. deutschen Geistes*, Berlin, 1870 (cf. on Tieck, book i.; on Novalis, book iii.: for criticism of the two Schlegels, bk. ii. and bk. iii. ch. 5): N. M. Pichtos, *Die Ästhetik Aug. W. v. Schlegel in ihrer geschichtlichen Entwickelung*, Berlin, 1893. On the Æsthetic of Fichte, G. Tempel, *Fichtes Stellung z. Kunst*, Metz, 1901.

On the Æsthetic of Hegel, Danzel, *Über d. Ästhetik der hegelschen Philosophie*, Hamburg, 1844: R. Haym, *Hegel u. seine Zeit*, Berlin, 1857, pp. 433–443: J. S. Kedney, *Hegel's Æsthetics: a critical exposition*, Chicago, 1885: Kuno Fischer, *Hegels Leben u. Werke*, Heidelberg, 1898–1901, chs. 38–42, pp. 811–947: J. Kohn, *Hegels Ästhetik* in *Zeitschrift für Philosophie*, 1902, vol. 120, fasc. ii.: see also B. Croce, *Cio che è vivo e cio che è morto della filosofia di Hegel*, Bari, 1907, ch. 6; Engl. tr. by D. Ainslie, 1915.

X. For the Æsthetic of Schopenhauer, Fr. Sommerlad, *Darstellung u. Kritik d. ästh. Grundanschauungen Schopenhauers*, Diss., Giessen, 1895: Ed. v. Mayer, *Schopenhauers Ästhetik u. ihr Verhältniss z. d. ästh. Lehren Kants u. Schellings*, Halle, 1897: Ett. Zoccoli, *L'estetica di A. Sch.: propedeutica all' estetica Wagneriana*, Milan, 1901: G. Chialvo, *L'estetica di A. Sch., saggio esplicativo-critico*, Rome, 1905.

For the Æsthetic of Herbart, beside Zimmermann, *op. cit.* pp. 754–804, see O. Hostinsky, *Herbarts Ästhetik in ihrer grundlegenden Theilen quellenmässig dargestellt u. erläutert*, Hamburg-Leipzig, 1891.

XI. Of the Æsthetic of Schleiermacher, the fullest treatment is given by Zimmermann, pp. 609–634, and von Hartmann, pp. 156–169.

XII. For the history of the theory of Language, beside Benfey, *op. cit.* introd., see Max. Leop. Loewe, *Historiæ criticæ grammatices universalis seu philosophicæ lineamenta*, Dresden, 1839: A. F. Pott, *W. v. Humboldt und die Sprachwissenschaft*, introd. to the reprint of Humboldt's *Verschiedenheit d. menschl. Sprachbaues* (2nd ed., Berlin, 1880, vol. i.).

On Humboldt see esp. Steinthal, *Der Ursprung der Sprache*, pp. 59–81, and Pott's introd. cit., *Wilh. v. Humboldt u. die Sprachwissenschaft.*

XIII. For this period, treated with unnecessary fulness, see von Hartmann, *op. cit.* bk. i.: more concisely by Menendez y Pelayo, vol. iv. (1st ed.), part i. chs. 6–8.

For the doctrine of the modifications of beauty see Zimmermann, *op. cit.* pp. 715–744: Schasler, *op. cit.* §§ 517–546: Bosanquet, *op. cit.* ch. 14, pp. 393–440: in greater detail, v. Hartmann, bk. ii. Part i. pp. 363–461.

For the history of the Sublime see also F. Unruh, *Der Begriff des Erhabenen seit Kant*, Königsberg, 1898. For Humour see B. Croce, *Dei varî significanti della parola umorismo e del suo uso nella critica letteraria*, in the *Journal of Comparative Literature* of New York, 1903, fasc. iii. (reprinted in *Probl. di est.* pp. 275–286) F. Baldensperger, *Les Définitions de l'humour*, in *Études; d'hist. litt.* Paris, 1907. For the history of the concept of the Graceful, F. Torraca, *La grazia secondo il Castiglione e secondo lo Spencer* (in Morandi, *Antol. della critica lett. ital.* 2nd ed., Città di Castello, 1885, pp. 440–444): F. Braitmaier, *op. cit.* ii. pp. 166–167.

XIV. For the history of Æsthetic in France during the nineteenth century there is nothing so good as Menendez y Pelayo, vol. iii. part

ii. chs. 3–9; *ibid.* chs. 1–2 give full information concerning Æsthetic in England.

For Æsthetic in Italy in the first half of the nineteenth century, Karl Werner, *Idealistische Theorien des Schönen in d. italienischen Philosophie des neunzehnten Jahrhunderts*, Vienna, 1884 (from Trans, of the Imperial and Royal Viennese Academy). On Rosmini see esp. P. Bellezza, *Antonio Rosmini e la grande questione letteraria del secolo XIX* (in the collection *Per Antonio Rosmini nel primo centenario*, Milan, 1897, vol. i. pp. 364–385). On Gioberti, Ad. Faggi, *Vinc. Gioberti esteta e letterato*, Palermo, 1901 (from the *Atti della R. Accad. di Palermo*, s. iii. vol. vi.). On Delfico, G. Gentile, *Dal Gcnovesi al Galluppi*, Naples, 1903, ch. ii. On Leopardi, E. Bertana in *Giorn. stor. lett. ital.* xli. pp. 193–283; R. Giani, *L'estetica nei pensieri di G. Leopardi*, Turin, 1904 (cf. G. Gentile in *Critica*, ii. pp. 144–147). See also a book quoted by A. Rolla and B. Croce, *loc. cit.*, containing a catalogue of Italian books on Æsthetic of the nineteenth century (*Probl. di est.* pp. 401–415).

On the theories of the Italian Romanticists, F. De Sanctis, *La poetica del Manzoni*, in *Scritti varî*, ed. Croce, i. pp. 23–45; and the same author's *La letteratura italiana nel secolo XIX*, ed. Croce, Naples, 1897, on Tommaseo, pp. 233–243: on Cantù, pp. 244–273: on Berchet, pp. 479–493: on Mazzini, pp. 424–441. On Mazzini esp. F. Ricitari, *Concetto dell' arte e della critica letteraria nella mente di G. Mazzini*, Catania, 1896. For all these see G. A. Borgese, *Storia della critica romantica in Italia*, cit.

XV. For the life of De Sanctis and the bibliography of his works see *Scritti varî*, ed. Croce, ii. pp. 267–308, also the volume *In memoria di Fr. de S.* edited by M. Mandalari, Naples, 1884.

On De Sanctis as literary critic, P. Villari, *Commemorazione*: A. C. de Meis, *Commem.*, in the above-mentioned vol. *In memoria*: Marc Monnier in *Revue des Deux Mondes*, April I, 1884: Pio Ferrieri, *Fr. de S. e la critica letteraria*, Milan, 1888: B. Croce, *La critica letteraria*, Rome, 1896, ch. 5; *Fr. de S. e i suoi critici recenti* (in *Atti dell' Accad. Pontan.* vol. xxviii. reprinted in *Scritti varî*, append, ii. 309–352), and prefs. to vols, already quoted, *La lett. ital. nel sec. XIX*, and *Scritti varî; De Sanctis e Schopenhauer*, in *Atti della Pontaniana*, xxxii. 1902: Enr. Cocchia, *Il pensiero critico di Fr. de S. nell' arte e nella politica*, Naples, 1899: G. A. Borgese, *op. cit.* last chapter and *passim*.

XVI. On the last phase of metaphysical Æsthetic, G. Neudecker, *Studien z. Geschichte d. deutschen Ästhetik s. Kant*, Würzburg, 1878, which discusses and criticises more particularly Vischer (self-criticism), Zimmermann, Lotze, Köstlin, Siebeck, Fechner and Deutinger. On Zimmermann, von Hartmann, *op. cit.* pp. 267–304: Bonatelli, in *Nuova Antologia*, October 1867. On Lotze, Fritz Kogel, *Lotzes Ästhetik*, Göttingen, 1886: A. Matragrin, *Essai sur l'esthétique de Lotze*, Paris, 1901. On Köstlin, von Hartmann, pp. 304–317. On Schasler, see the same, pp. 248–252, also Bosanquet, pp. 414–424. On Hartmann, Ad. Faggi, *Ed. H. e l' estetica tedesca*, Florence, 1895. On Vischer see M. Diez, *Fried. Vischer u. d. ästh. Formalismus*, Stuttgart, 1889. For French and English æstheticians, besides Menendez y Pelayo, *op.cit.*, on Ruskin, see J. Milsand, *L'Esthétique anglaise, étude sur J. Ruskin*, Paris, 1864: R. de la Sizeranne, *Ruskin et la religion de la beauté*, 3rd ed., Paris, 1898; cf. part iii. On Fornari, V. Imbriani, *Vito Fornari estetico* (reprinted in *Studî letterarî e bizzarri e satiriche*, ed. Croce, Bari, 1907). On Tari see Nic. Gallo, *Antonio Tari, studio critico,* Palermo, 1884: Croce, in *Critica*, v. (1907), pp. 357–361; also in pref. to vol.: *A. Tari, saggi di estetica e metafisica,* Bari, 1910.

XVII. For positivist Æsthetic see Menendez y Pelayo, *op. cit.* iv. (1st ed.) vol. ii. pp. 120–136, 326–369: N. Gallo, *La scienza dell'arte*, Turin, 1887, chs. 6–8, pp. 162–216.

XVIII. On Kirchmann, von Hartmann, pp. 253–265. For various recent German æstheticians, Hugo Spitzer, *Kritische Studien z. Ästhet. Der Gegenwart*, Leipzig, 1897. On Nietzsche, Ettore G. Zoccoli, *Fred. Nietzsche*, Modena, 1898, pp. 268–344: Jul. Zeitler, *Nietzsches Ästhetik*, Leipzig, 1900. On Flaubert, A. Fusco, *La teoria dell' arte in G. F.*, Naples, 1907: cf. *Critica*, vi. (1908), pp. 125–134. For books on Æsthetic published during the last decade of the nineteenth century see Luc. Arréat, *Dix années de philosophie*, 1891–1900, Paris, 1901, pp. 74–116. A few remarks on contemporary Æsthetic are made by K. Groos in *Die Philosophie im Beginn. des XXen Jahrh.*, ed. by W. Windelband, Heidelberg, 1904–1905. For latest books on Æsthetic see *Critica*, ed. B. Croce (Naples), from 1903 onward, which publishes reviews of them. There is also a review, started in 1906, published at Stuttgart (ed. F. Enke), *Zeitschrift für Ästhetik und allgemeine Kunstwissenschaft*, edited by Max Dessoir.

XIX. The history of particular problems is usually omitted, or, at best, erroneously treated in histories of Æsthetic: for example, see the difficulty experienced by Ed. Müller, *Gesch.*, cit., ii. pref. pp. vi–vii, in connecting his treatment of the history of Rhetoric with that of Poetics. Some writers attach Rhetoric to the individual arts or to artistic technique; others treat the doctrines of the modification of beauty and of natural beauty (in the metaphysical sense) as special problems; others, again, discuss the kinds or classifications in art in an incidental manner, without seeking to incorporate them in the principal æsthetic problem.

§ 1. On the history of Rhetoric in the ancient sense see Rich. Volkmann, *Die Rhetorik der Griechen und Römer in systematischer Übersicht dargestellt*, 2nd ed., Leipzig, 1885, of capital importance: A. Ed. Chaignet, *La Rhétorique et son histoire*, Paris, 1888; rich in material, but ill-arranged and with the preconception that Rhetoric is still a defensible body of science. For special treatment see Ch. Benoist, *Essai historique sur les premiers manuels d'invention oratoire, jusqu'à Aristote*, Paris, 1846: Georg Thiele, *Hermagoras, ein Beitrag z. Geschichte d. Rhetorik*, Strasburg, 1893. There is no history of rhetoric in modern times. For criticism of Vives and other Spaniards see Menendez y Pelayo, *op. cit.* iii. pp. 211–300 (2nd ed.). For Patrizzi see B. Croce, *F. Patrizzi e la critica della rettorica antica*, in the vol. of *Studî* in honour of A. Graf, Bergamo, 1903 (*Probl. d. est.* pp. 297–308).

For Rhetoric as theory of literary form in antiquity see Volkmann, *op. cit.* pp. 393–566: Chaignet, *op. cit.* pp. 413–539: also Egger, *passim*, and Saintsbury, bks. i. ii. For purposes of comparison see Paul Reynaud, *La Rhétorique sanskrite exposée dans son développement historique et ses rapports avec la rhétorique classique*, Paris, 1884. For the Middle Ages, Comparetti, *Virgilio nel medio evo*, vol. i., and Saintsbury, bk. iii. There is need for a work on modern Rhetoric in this sense also. For the form it assumed ultimately according to the theory of Gröber see B. Croce, *Di alcuni principî di sintassi e stilistica psicologiche del Gröber*, in *Atti dell' Accad. Pontan.* vol. xxix. 1899: K. Vossler, *Literaturblatt für germ. u. roman. Philologie*, 1900, N.I.: B. Croce, *Le categorie rettoriche e il prof.* Gröber, in *Flegrea*, April 1900: K. Vossler, *Positivismo e idealismo nella scienza del linguaggio*, Ital. trans. Bari, 1908, pp. 48–61 (cf. *Probl. d. est.* pp. 143–171). Very incomplete observations on the history of the concept of metaphor are made by A. Biese, *Philosophie d. Metaphorischen*, Hamburg-Leipzig, 1893, pp. 1–16; but this book has the merit of calling attention to the importance of the views and influence of Vico.

§ 2. For the history of the literary kinds in antiquity see the works above quoted by Müller, Egger, Saintsbury, and the vast literature on Aristotle's *Poetics*. For comparison with Sanskrit poetics, Sylvain Levi, *Le Théâtre indien*, Paris, 1890, esp. pp. 11, 152. For mediæval poetry see esp. Gio. Mari, *I trattati medievali di ritmica latina*, Milan, 1899; and his recent edition of *Poetica magistri Iohannis anglici*, 1901.

For the history of the kinds under the Renaissance see principally Spingarn, *op. cit.* i. chs. 3–4; ii. ch. 2; iii. ch. 3. Also Menendez y Pelayo, Borinski, Saintsbury, *passim*.

Special works: on Pietro Aretino, De Sanctis, *Storia della letteratura italiana*, ii. pp. 122–144: A. Graf, *Attraverso il cinquecento*, Turin, 1888, pp. 87–167: K. Vossler, *P. A.'s künstlerisches Bekenntniss*, Heidelberg, 1901. On Guarini, V. Rossi, *G. B. Guarini e il Pastor Fido*, Turin, 1886, pp. 238–250. On Scaliger, Lintilhac, *Un Coup d'État*, cit. For the three unities, L. Morandi, *Baretti* contro Voltaire, 2nd ed., Città di Castello,

1884: Breitinger, *Les Unités d'Aristote avant le Cid de Corneille*, 2nd ed., Geneva-Basle, 1895: J. Ebner, *Beitrag z. einer Geschichte d. dramatischen Einheiten in Italien*, Munich, 1898. On the Spanish polemic concerning comedy see A. Morel Fatio on the defenders of comedy and of the *Arte nuevo*, in the *Bulletin Hispanique* of Bordeaux, vols. iii. and iv.: on the dramatic theories see Arnaud, *Les Théories dramatiques au XVIIe siècle, étude sur la vie et les œuvres de l'abbé D'Aubignac*, Paris, 1888: Paul Dupont, *Un Poète philosophe au commencement du XVIII e siècle, Houdar de la Motte*, Paris, 1898: Alfredo Galletti, *Le teorie drammatiche e la tragedia in Italia nel secolo XVIII*, part i. 1700–1750, Cremona, 1901. On the history of French Poetics, F. Brunetière, *L'Évolution des genres dans l'histoire de a littérature*, Paris, 1890, vol. i. introd.: "*L'évolution de la critique depuis la Renaissance jusqu'à nos jours.*" On that of English Poetics, Paul Hamelius, *Die Kritik in d. engl. Literatur des XVII en u. XVIII ⟨sup⟩ en ⟨/sup⟩ Jahrh.*, Leipzig, 1897: also the well-filled chapter in Gayley-Scott, *op. cit.* pp. 382–422, the sketch of a book on the subject. For the romantic period see Alfred Michiels, *Histoire des idées littéraires en France au XIXe siècle, et de leurs origins dans les siècles antérieures*, 4th ed., Paris, 1863. For Italy see G. A. Borgese, *op. cit.* § 3. For the early history of the distinction and classification of the arts see the literature quoted above in relation to Lessing, and his *Laokoon*, with notes by Blümner. For subsequent history, H. Lotze, *Geschichte*, cit., bk. iii.: Max Schasler, *Das System der Künste auf einem neuen, im Wesen der Kunst begründeten Gliederungsprincip*, 2nd ed., Leipzig-Berlin, 1881, introd.: Ed. v. Hartmann, *Deutsche Ästh. s. Kant*, bk. ii. part ii. especially pp. 524–580: V. Basch, *Essai sur l'esth. de Kant*, pp. 483–496.

§ 4. For the doctrine of styles in antiquity see Volkmann, *op. cit.* pp. 532–566. The history of grammar and parts of speech is treated fully so far as Græco-Roman antiquity is concerned in Laur. Lersch, *Die Sprachphilosophie der Alten*, Bonn, 1838–1841: better still by Steinthal, *Geschichte*, cit. vol. ii. For Apollonius Dyscolus see Egger, *Apollon Dyscole*, Paris, 1854. For the history of grammar in the Middle

Ages see Ch. Thurot, *Extraits de divers manuscrits latins pour servir à l'histoire des doctrines grammaticales au moyen âge*, Paris, 1869. For modern times, C. Trabalza, *Storia della grammatical italiana*, Milan, 1908. For the history of Criticism several books mentioned under § 2 may be consulted: in addition to these, B. Croce, *Per la storia della critica e storiografia letteraria*, containing Italian examples (*Probl. d. est.* pp. 419–448): for the theories of recent French criticism see Ém. Hennequin, *La Critique scientifique*, Paris, 1888, and Ernest Tissot, *Les évolutions de la critique française*, Paris, 1890. On the concept of "romanticism" see G. Muoni, *Note per una poetica storica del romanticismo*, Milan, 1906: cf. B. Croce, *Le definizioni del romanticismo*, in *Critica*, iv. pp. 241–245 (reprinted in *Probl. di estetica*, pp. 285–294).

术语汇编与简释

《美学理论》

æsthetic reproduction：审美再造

审美判断是把艺术品在自己心中再造，艺术鉴赏是一种审美再造。审美再造的过程和审美创造的过程是一致的。鉴赏如果不能“再造”出作者所“创造”的东西，就会成为被动的接受，就体会不到艺术作品的真正的妙处。

association：联想

联想是心灵的活动，是综合，是心灵有意识地把各种元素综合、拼接到一起。

expression：表现

表现指心灵赋予物质以形式、使之对象化并产生具体形象的过程，表现是主动的。克罗齐认为直觉即表现，表现即艺术。

externalization：外射

外射就是把内在的审美的直觉品，借助物理媒介，如语言、颜色、声音、线条等，放射到外面去，留下可以让人欣赏的痕迹。艺术家的创作就是外射活动。

identity of linguistic and æsthetic：语言学与美学的同一

克罗齐认为艺术的科学与语言的科学都是表现。语言是声音组合起来传递意义，达成表现，语言在本质上是一种表现。因此，语言学与美学是同一的。

intuition：直觉

直觉是心灵的一种能力，是见到一个事物时，不假思索，不进行推理和分析而领会其意象的能力。

intuitive knowledge：直觉知识

知识有两种形式：直觉知识和逻辑。直觉知识来自于想象，是有关个体的，直觉知识产生意象。

logic: 逻辑

逻辑是知识的两种形式之一，逻辑知识是理性的，是关于整体的，关于事物之间的关系。逻辑产生概念。

mystical æsthetic：神秘主义美学

神秘主义美学从神秘主义哲学出发来思考美和艺术的问题，把直觉理解为一种带有神秘主义色彩的特殊认知能力，认为审美经验具有“不可言说性”。

representation：表象

表象是事物投射在心灵的形象，那个形象便代表着事物的本身。没有在表现中对象化了的东西就不是直觉或表象。

system of the spirit：精神系统

精神活动可分认识活动和实践活动，认识活动又可分为直觉和概念，实践活动又可分为经济和道德两方面。认识活动先于实践活动，实践活动基于认识活动，它们彼此又循环相生，构成一个既循环往复又不断发展的运动。

utilitarianism：功利主义

功利主义是道德哲学范畴中的理论，提倡追求效用和目的，趋利避害，对事物取功利性的态度。

《美学史》

æsthetic hedonism：**审美快感主义**

审美快感主义认为快感是艺术的一种性质，艺术活动（或审美活动）和快感是同一的，快感对于审美活动和对精神活动的任何其他形式都是共有的。

æsthetic positivism：**实证主义美学**

实证主义美学强调感觉经验，突出审美的社会性因素，主张用社会学的观点来观察艺术，从时代和环境的背景来分析艺术，强调审美快感是植根于人的生理结构的自然需要。

Cartesianism：**笛卡尔主义**

笛卡尔提出身心二元论，确立了主体的独立地位，认为所有主体都具有相同的心智框架，我们所感知的世界并不一定是真实的物质世界，而是精神感知的反馈。笛卡尔主义通过理性的推演，在形而上学的思辨中寻求美的本质和来源，认为美是对象与主体（判断）之间的一种关系，当审美对象的刺激与审美主体感官的反应处于协调和谐状态时，才能让人感到愉快，才能产生美。

Hegelian æsthetic：**黑格尔美学**

黑格尔的美学思想是其唯心主义哲学体系的一个组成部分。黑格尔提出美是理念的感性显现，强调美是理性和感性的统一、内容和形式的统一、主观和客观的统一这三个基本原则。他认为艺术的根本特点是理念通过感性的形象来显现自己、认识自己。

Herbartian æsthetic：**赫尔巴特美学**

赫尔巴特是形式主义美学的主要代表人物，他主张美只能通过形式来检验，而形式则产生于作品各组成要素的关联中，包括音调、颜色、线条、思想和意志的关系。美与快乐之间也有着巨大的不同，快乐不需要表现，而美需要用关系表现出来。

neo-criticism：新批判主义

新批判主义，又称新康德主义。该学说继承和发展了康德的先验论和不可知论，采取心理学或生理学的观点，把先天的认识形式归结为先天的生理结构，用认识主体的心理或生理的结构说明知识的问题。

transcendental æsthetic：先验形式美学

康德认为事物本身是实在的、客观的，但我们认识事物的“形式”却是先验的、主观的。认识与知识之所以可能，就在于人类先天地具备一套将感觉材料加工成知觉的能力，即“先验形式”。康德把人的能力分为三种：认识能力、感受愉快和不快的情感能力、欲求能力，这三种能力投射到认知上，对应“知性”“判断力”和“理性”。这三种能力面对外部对象的时候，都受制于先验观念。“美”不在客观事物本身，而在审美主体头脑中的先验形式，审美能力是先验的。

Benedetto Croce

Theory of Aesthetic / History of Aesthetic

图书在版编目（CIP）数据

美学理论；美学史 = Theory of Aesthetic / History of Aesthetic：导读注释版：英文 /（意）克罗齐著；赵美娟注释．—上海：上海译文出版社，2022.9

（世界学术经典系列）

ISBN 978-7-5327-9011-1

Ⅰ.①美… Ⅱ.①克… ②赵… Ⅲ.①美学理论—英文 ②美学史—西方国家—英文 Ⅳ.①B83 ②B83-095

中国版本图书馆CIP数据核字（2022）第142962号

美学理论 / 美学史

［意］克罗齐　著　赵美娟　导读　注释

责任编辑 / 谭春惠　装帧设计 / 张志全工作室

上海译文出版社有限公司出版、发行

网址：www.yiwen.com.cn

201101　上海市闵行区号景路159弄B座

江阴市机关印刷服务有限公司印刷

开本 890×1240　1/32　印张 21.5　插页 6　字数 614,000

2022年10月第1版　2022年10月第1次印刷

印数：0,001—1,500册

ISBN 978-7-5327-9011-1/B·523

定价：168.00元

World Academic Classics
世界学术经典·英文版

Marxism
马克思主义

The Communist Manifesto (Marx & Engels)
共产党宣言（马克思、恩格斯）

Chinese Classics
中国经典

论语（汉英对照）
The Analects of Confucius

道德经（汉英对照）
Dao De Jing

孟子（汉英对照）
Mencius

庄子（汉英对照）
Zhuangzi

孙子兵法（汉英对照）
Sun Tzu on the Art of War

大学・中庸（汉英对照）
The Great Learning & The Doctrine of the Mean

Philosophy
哲　学

Republic (Plato)
理想国（柏拉图）

Metaphysics (Aristotle)
形而上学（亚里士多德）

Meditations on First Philosophy (René Descartes)
第一哲学沉思录（笛卡尔）

A Discourse on Method (René Descartes)
方法论（笛卡尔）

Pascal's Pensées (Blaise Pascal)
思想录（帕斯卡尔）

A Treatise of Human Nature (David Hume)
人性论（休谟）

Critique of Pure Reason (Kant)
纯粹理性批判（康德）

Critique of Judgment (Kant)
判断力批判（康德）

The Phenomenology of Mind (Hegel)
精神现象学（黑格尔）

The Logic of Hegel (Hegel)
小逻辑（黑格尔）

The World as Will and Representation (Schopenhauer)
作为意志与表象的世界（叔本华）

Thus Spake Zarathustra (Nietzsche)
查拉图斯特拉如是说（尼采）

Either / Or (Kierkegaard)
非此即彼（克尔凯郭尔）

Ideas: General Introduction to Pure Phenomenology (Husserl)
纯粹现象学导论（胡塞尔）

On First Philosophy (Husserl)
论第一哲学（胡塞尔）

Tractatus Logico-Philosophicus (Wittgenstein)
逻辑哲学论（维特根斯坦）

Philosophical Investigations (Wittgenstein)
哲学研究（维特根斯坦）

Pragmatism (William James)
实用主义（威廉・詹姆斯）

Reconstruction in Philosophy (John Dewey)
哲学的改造（杜威）

Time and Free Will (Henri Bergson)
时间与自由意志（柏格森）

A Discourse on the Positive Spirit (Auguste Comte)
论实证精神（孔德）

Novum Organum (Francis Bacon)
新工具（培根）

Ethics
伦理学

The Nicomachean Ethics of Aristotle (Aristotle)
尼各马可伦理学（亚里士多德）

The Theory of Moral Sentiments (Adam Smith)
道德情操论（亚当·斯密）

A Discourse on Inequality (Rousseau)
论人类不平等的起源与基础（卢梭）

Critique of Practical Reason (Kant)
实践理性批判（康德）

Utilitarianism (Mill)
功利主义（密尔）

Principia Ethica (Moore)
伦理学原理（摩尔）

Religion
宗教学

Confessions (St. Augustine)
忏悔录（圣·奥古斯丁）

Basic Writings of Saint Thomas Aquinas (Thomas Aquinas)
圣托马斯基本著作（阿奎那）

Basic Theological Writings (Martin Luther)
路德基本著作选（马丁·路德）

The Courage to Be (Paul Tillich)
存在的勇气（蒂利希）

The Essence of Christianity (Ludwig Feuerbach)
基督教的本质（费尔巴哈）

Politics
政治学

The Politics of Aristotle (Aristotle)
政治学（亚里士多德）

The Prince (Machiavelli)
君主论（马基雅维利）

The Social Contract (Rousseau)
社会契约论（卢梭）

Leviathan (Thomas Hobbes)
利维坦（霍布斯）

Two Treatises of Government (John Locke)
政府论（洛克）

The Spirit of the Laws (Montesquieu)
论法的精神（孟德斯鸠）

Democracy in America (Tocqueville)
论美国民主（托克维尔）

Considerations on Representative Government (Mill)
代议制政府（密尔）

The Federalist Papers (Hamilton)
联邦党人文集（汉密尔顿）

The Ancien Régime and the Revolution (Tocqueville)
旧制度与大革命（托克维尔）

Utopia (Thomas More)
乌托邦（莫尔）

On Liberty (Mill)
论自由（密尔）

Reflections on the Revolution in France (Edmund Burke)
评法国革命（伯克）

The Will to Power (Nietzsche)
权力意志（尼采）

On War (Carl von Clausewitz)
战争论（克劳塞维茨）

Mutual Aid: A Factor of Evolution (Kropotkin)
互助论（克鲁泡特金）

Economics
经济学

The Economics of Welfare (Pigou)
福利经济学（庇古）

Principles of Economics (Alfred Marshall)
经济学原理（马歇尔）

The General Theory of Employment, Interest and Money (Keynes)
就业、利息与货币通论（凯恩斯）

The Theory of Economic Development (Schumpeter)
经济发展理论（熊彼特）

An Inquiry into the Nature and Causes of the Wealth of Nations (Adam Smith)
国民财富的性质和原因的研究（亚当・斯密）

Capitalism, Socialism and Democracy (Schumpeter)
资本主义、社会主义与民主（熊彼特）

An Essay on Population (Thomas Robert Malthus)
人口原理（马尔萨斯）

The Principles of Scientific Management (Frederick Winslow Taylor)
科学管理原理（泰勒）

Sociology
社会学

Suicide: A Study in Sociology (Durkheim)
自杀：社会学研究（迪尔凯姆）

The Protestant Ethic and the Spirit of Capitalism (Max Weber)
新教伦理与资本主义精神（韦伯）

The Philosophy of Money (Georg Simmel)
货币哲学（西美尔）

The Mind and Society: A Treatise on General Sociology (Vilfredo Pareto)
思想与社会：社会学总论（帕累托）

Ideology and Utopia (Mannheim)
意识形态与乌托邦（曼海姆）

First Principles (Herbert Spencer)
基本原理（斯宾塞）

The Theory of the Leisure Class (Thorstein B. Veblen)
论闲逸阶级（凡勃伦）

Anthropology
人类学

The Golden Bough (James George Frazer)
金枝（詹姆斯・乔治・弗雷泽）

Argonauts of the Western Pacific (Malinowski)
西太平洋的航海者（马林诺夫斯基）

Structure and Function in Primitive Society (Radcliffe-Brown)
原始社会的结构和功能（拉德克利夫–布朗）

Race, Language and Culture (Franz Boas)
种族、语言与文化（弗朗茨・博厄斯）

The Savage Mind (Claude Lévi-Strauss)
野性思维（克劳德・列维–斯特劳斯）

Psychology
心理学

The Principles of Psychology (William James)
心理学原理（威廉・詹姆斯）

Principles of Physiological Psychology (Wundt)
生理心理学原理（冯特）

The Interpretation of Dreams (Sigmund Freud)
释梦（弗洛伊德）

The Archetypes and the Collective Unconscious (Jung)
原型与集体无意识（荣格）

Law
法　学

Ancient Law (Maine)
古代法（梅因）

English Law and Political Philosophy (Maitland)
英国法律与政治哲学（梅特兰）

Lectures on Jurisprudence (Austin)
法理学演讲录（奥斯丁）

Fundamental Principles of the Sociology of Law (Ehrlich)
法律社会学基本原理（埃利希）

History
历史学

The Histories (Herodotus)
历史（希罗多德）

The Peloponnesian War (Thucydides)
伯罗奔尼撒战争史（修昔底德）

The Annals of Imperial Rome (Tacitus)
编年史（塔西佗）

The City of God (St. Augustine)
上帝之城（圣·奥古斯丁）

On Heroes, Hero Worship, and the Heroic in History (Thomas Carlyle)
论历史上的英雄、英雄崇拜和英雄业绩（卡莱尔）

The Idea of History (Collingwood)
历史的观念（科林伍德）

History of Civilization in Europe (Francois Guizot)
欧洲文明史（基佐）

The Influence of Sea Power (Alfred T. Mahan)
海权对历史的影响（马汉）

Theory & History of Historiography (Benedetto Croce)
史学理论与史学史（克罗齐）

A Short Account of the Destruction of the Indies (Bartolomé de Las Cases)
西印度毁灭述略（卡萨斯）

Literary Theory and Aesthetics
文学理论与美学

On the Art of Poetry (Aristotle)
诗学（亚里士多德）

On the Art of Poetry (Horace)
诗艺（贺拉斯）

On the Sublime (Longinus)
论崇高（朗吉努斯）

Theory of Aesthetic / History of Aesthetic (Benedetto Croce)
美学理论 / 美学史（克罗齐）

Hegel's Aesthetics Lectures on Fine Art (Hegel)
黑格尔美学讲演录（黑格尔）

The Birth of Tragedy (Friedrich Nietzsche)
悲剧的诞生（尼采）

Linguistics
语言学

Course in General Linguistics (Saussure)
普通语言学教程（索绪尔）

On Language (Wilhelm von Humboldt)
论语言（威廉·冯·洪堡特）

Scientific Philosophy
自然科学哲学

On the Origin of Species (Charles Darwin)
物种起源（查尔斯·达尔文）

Mathematical Principles of Natural Philosophy and Its System of the World (Issac Newton)
自然哲学的数学原理（牛顿）

Natural History (Pliny the Elder)
自然史（老普林尼）